Transcription Factors

The HUMAN MOLECULAR GENETICS series

Series Advisors

D.N. Cooper, *Institute of Medical Genetics, University of Wales College of Medicine, Cardiff, UK*

S.E. Humphries, *Division of Cardiovascular Genetics, University College London Medical School, London, UK*

A. Wolffe, *Sangamo Biosciences Inc, Point Richmond Tech Center, Richmond, CA, USA*

Human Gene Mutation
From Genotype to Phenotype
Functional Analysis of the Human Genome
Molecular Genetics of Cancer
Environmental Mutagenesis
HLA and MHC: Genes, Molecules and Function
Human Genome Evolution
Gene Therapy
Molecular Endocrinology
Venous Thrombosis: from Genes to Clinical Medicine
Protein Dysfunction in Human Genetic Disease
Molecular Genetics of Early Human Development
Neurofibromatosis Type 1: from Genotype to Phenotype
Analysis of Triplet Repeat Disorders
Molecular Genetics of Hypertension
Human Gene Evolution
Analysis of Multifactorial Disease
Transcription Factors

Forthcoming title
B Cells

Transcription Factors

J. Locker
Albert Einstein College of Medicine, Yeshiva University, New York, USA

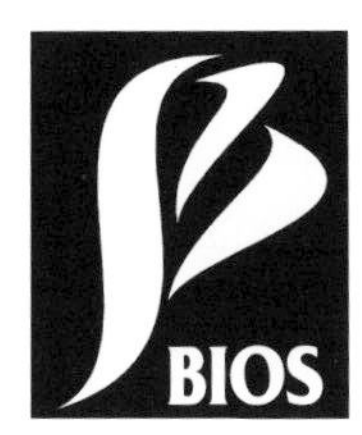

First published in 2001

A CIP catalogue record for this book is available from the British Library.

ISBN 1 859961 82 7

BIOS Scientific Publishers Ltd
9 Newtec Place, Magdalen Road, Oxford OX4 1RE, UK
Tel. +44 (0)1865 726286. Fax +44 (0)1865 246823
World Wide Web home page: http://www.bios.co.uk/

Published in the United States, its dependent territories and Canada by Academic Press, Inc., A Harcourt Science and Technology Company, 525 B Street, San Diego, CA 92101–4495. www.academicpress.com

Production Editor: Paul Barlass
Typeset by Saxon Graphics Ltd, Derby, UK.
Printed by Biddles Ltd, Guildford, UK, www.biddles.co.uk

Contents

Contributors

Benecke, A. Institut de Génétique et de Biologie Moléculaire et Cellulaire (IGBMC)/CNRS/INSERM/ULP, BP 163, 67404 Illkirch CEDEX, CU de Strasbourg, France

Bowser, R. Department of Pathology, University of Pittsburgh School of Medicine, 3500 Terrace Street, Pittsburgh, PA 15261, USA

Courey, A.J. Department of Chemistry and Biochemistry, University of California, Los Angeles, 405 Hilgard Avenue, Los Angeles, CA 90095-1569, USA

Dynlacht, B. Harvard University, Department of Molecular and Cellular Biology, 16 Divinity Avenue, Cambridge, MA 02138, USA

Fairall, L. MRC Laboratory of Molecular Biology, Hills Road, Cambridge, CB2 2QH, UK

Gaudon, C. Institut de Génétique et de Biologie Moléculaire et Cellulaire (IGBMC)/CNRS/INSERM/ULP, BP 163, 67404 Illkirch CEDEX, CU de Strasbourg, France

Gronemeyer, H. Institut de Génétique et de Biologie Moléculaire et Cellulaire (IGBMC)/CNRS/INSERM/ULP, BP 163, 67404 Illkirch CEDEX, CU de Strasbourg, France

Hagman, J. Department of Immunobiology, National Jewish Medical and Research Center, 1400 Jackson Street, K516B, Denver, CO 80206, USA

Hunger, S.P. Section of Pediatric Hematology/Oncology, Department of Pediatrics, University of Colorado Health Sciences Center, Campus Box C229, Denver, CO 80262, USA

Locker, J. Albert Einstein College of Medicine, 1300 Morris Park Avenue, Bronx, New York 10461, USA

Lufkin, T. Brookdale Center for Developmental and Molecular Biology, Mount Sinai School of Medicine, One Gustave Levy Place, New York, NY 10029, USA

Monaghan-Nichols, A.P. Department of Neurobiology and Psychiatry, University of Pittsburgh, Pittsburgh, PA 15261, USA

Pugh, B.F. Department of Biochemistry and Molecular Biology, 452 N. Frear, The Pennsylvania State University, University Park, PA 16802, USA

Schwabe, J.W.R. MRC Laboratory of Molecular Biology, Hills Road, Cambridge, CB2 2QH, UK

Sharrocks, D. School of Biological Sciences, Stopford Building, University of Manchester, Oxford Road, Manchester, M13 9PT, UK

Shore, P. School of Biological Sciences, Stopford Building, University of Manchester, Oxford Road, Manchester, M13 9PT, UK

Siu, G. Department of Microbiology, Columbia University College Physicians and Surgeons, 701 W. 168th Street, New York, NY 10032, USA

Wolffe, A.P. Laboratory of Molecular Embryology, National Institute of Child Health and Human Development, NIH, Bethesda, MD 20892-5431, USA

Abbreviations

Aβ	amyloid β
AD	Alzheimer's disease
AHR	aryl hydrocarbon receptor
AHRNT	AHR nuclear transporter
ALLs	acute lymphoblastic leukemias
AP	anteroposterior
APC	antigen presenting cells
ATF-2	activating transcription factor 2
BAMs	branchial arch muscles
bHLH	basic helix-loop-helix (domain)
BMP	bone morphogenetic protein
Brg1	brahma-related gene product
BSAP	B cell-specific activaor protein
bZip	basic-region-leucine-zipper domain
CAD	carbamyl-phosphate synthase/aspartate carbamyltransferase/dihydoorotase
CAK	CDK-activating kinase
CaMKIV	calmodulin-dependent kinase IV
cAMP	cyclic adenosine monophosphate
CAP	catabolite activator protein
CAR	constitutive androstane receptor
CBF	core-binding factor
CBP	CREB binding protein
CDK	cyclin-dependent kinases
C/EBP	CCAAT/enhancer-binding protein
CHOP	C/EBP homologous protein
CKI	cyclin-dependent kinase inhibitor
CKII	casein kinase II
CNS	central nervous system
COUP-TF	chick ovalbumin upstream promotor transcription factor
CREB	cAMP response element-binding protein
CRM	chromosomal region maintenance
CTD	carboxyl-terminal domain of RNA polymerase II
DBP	D-site binding protein
DN	double negative
DNA-PK	DNA-dependent protein kinase
DP	double positive
DPE	downstream promotor element
DRIP	vitamin D receptor-interacting protein
Dsh	dishevelled (protein)
DV	dorsoventral

EBF	early B-cell factor
EGF	epidermal growth factor
ERB	estrogen receptor beta
Exd	extradentical
Ext	extension (proteins)
FGF	fibroblast growth factor
FTF	fetoprotein transcription factor
FXR	farnesyl X receptor
gadd	growth arrest and DNA damage (inducible gene)
GE	ganglionic eminence
GRIP	glucocorticoid receptor interacting protein
GSK	glycogen synthase kinase
GTPase	guanosine triphosphatase
HAT	histone acetyltransferase
hBrm	human brahma
HCM	Hox cooperating motif
HLF	hepatic leukemia factor
HMG	high mobility group (proteins)
HNF	hepatocyte nuclear factor
HOM-C	*Drosophila* homeotic gene complex
HSF	heat shock factor
HTH	helix-turn-helix
IHF	integration host factor
IKK	IkB kinase
IL	interleukin
IRF	interferon regulatory factor
IZ	intermediate zone
JAK	Janus kinase
KID	kinase-inducible domain
LBD	ligand-binding domain
LCR	locus control region
LEF-1	lymphoid enhancer factor-1
LGE	lateral ganglionic eminence
LIP	liver-enriched inhibitory protein
LRH	liver receptor homolog
LXR	liver X receptor
MAP	mitogen activated protein
MAPK	MAP kinase
MARs	matrix attachment regions
MAZ	Myc-associated zinc (finger)
MDM	murine double mutant
MEF	myocyte enhancer biding factor
MGE	medial ganglionic eminence
MHC	major histocompatibility complex
ML	mediolateral
MOZ	monocytic zinc-finger protein
MZ	marginal zone

NCOR	nuclear corepressor
NE	neural ectoderm
NFAT	nuclear factor of activated T cells
NFIL6	nuclear factor IL-6
NFκB	nuclear factor κB
NGF	nerve growth factor
NLS	nuclear localization signal
NMR	nuclear magnetic resonance
PAH	paired amphipathic helix
PAR	proline and acidic amino acid-rich
Pbx	pre B-cell transcription factor
PCAF	p300/CBP-associated factor
PEPCK	phosphoenolpyruvate carboxykinase
PIC	pre-initiation complex
PKA	protein kinase A
PKC	protein kinase B
PLZF	promyelocytic leukemia zinc-finger protein
pol I–III	RNA polymerase I–III
POU	DNA binding domain named for Pit1, Oct1 and UNC86 transcription factors
POZ	poxvirus and zinc finger (domain)
PPAR	peroxisome proliferator activated receptor
pRB	retinoblastoma protein
Prep1	PBX-regulating protein 1
PTF	pancreas transcription factor
PXR	pregnant X receptor
RAP	RNA polymerase-associated protein
RARE	retinoic acid response element
RB	retinoblastoma
RHD	runt homology domain
RTK	receptor tyrosine kinase
RXR	retinoid X receptor
SAF	silencer associated factors
SAGA	Spt-Ada-Gen5-acetyltransferase (complex)
SARs	scaffold attachment regions
SCG	superior cervical ganglion
SF1	steroidogenic factor 1
SHH	sonic hedgehog
SIN	SWI/SNF independent
SIV	simian immunodeficiency virus
SMCC	SRB/Mediator-protein-containing complex
SMRT	silencing mediator for retinoid and thyroid hormone receptors
SNAP	small nuclear activator protein
SNF	sucrose non-fermenting (gene)
SP	single positive
SRB	suppressor of RNA polymerase B
SRE	serum response element

SRF	serum response factor
STAT	signal transducers and activators of transcription
SVZ	subventricular zone
SWI	switch (gene)
TAF	TBP-associated factors
TALE	three-amino acids-loops
TBP	TATA-binding protein
TCF	ternary complex factor
TCPOBOP	1,4-bis[2-(3,5-dichloropyridyloxy)]benzene
TEF	transcription enhancer factor
TFIIA–I,s	basal transcription factors associated with RNA pol II
TGF	transforming growth factor
TGIF	TG interacting factor
THAD	TEF/HLF activation domain
TIF	transcription intermediary factors
TR	thyroid hormone receptor
TRAP	thyroid hormone receptor-interacting protein
TTF1	thyroid transcription factor 1
UAS	upstream activating sequence
UBF	upstream binding factor
USF	upstream stimulatory factor
VBP	vitellogenin gene binding protein
VE	visceral endoderm
VZ	ventricular zone
YY1	yin yang 1
ZLI	zona limitans intrathalamica

Preface

This book focuses on the DNA-binding transcription factors and the proteins with which they directly interact. Much more than a topic of limited specialized interest, these factors are the critical regulators that make gene expression specific.

In principal, all human biological processes can be explained by the control systems that regulate gene transcription. For each gene, the program of regulation is encoded in a set of binding sites, which may be at the transcription start site, near the start site, or far away from it. Those at the start interact with a complex but limited set of basal factors, while in each gene the other sites bind a few among thousands of specific transcription factors. The specific factors may be generally expressed or limited to specific cell type, and their expression and function may be constitutive or regulated. Both the basal and specific factors also interact with cofactors that do not directly bind DNA. The number of cofactors is large, but not as large as the number of DNA-binding factors, and relatively few cofactors are cell specific.

The basal machinery is common to all eukaryotic organisms, while the specific human factors have counterparts in all metazoans. Thus, even though the number of transcription factors increases with the complexity of the organism, simple organisms such as yeast provide direct information about the mechanisms of human transcription, while more complex organisms such as *Drosophila* provide direct counterparts to specific human factors and regulation. Research on these other organisms provides a major resource for deciphering human gene regulation and, from this perspective, has been included in this book on human factors.

Research to date has analyzed the controls of hundreds of genes and a large number of transcription factors. This book is designed to provide enough information to make sense out of current knowledge for both the newcomer and the experienced researcher. Nevertheless, the best is yet to come. Despite the wealth of current information, the transcriptional regulatory elements have probably not been completely determined for even a single gene. There are hundreds to thousands of transcription factors yet to be discovered and 100 000 new genes to be analyzed. Many processes, for example how distant control elements regulate genes, are poorly understood. However, powerful new resources are available to approach these challenges. The complete human genome sequence contains all gene regulatory elements (if they can be identified) and should enable identification of all transcription factors (at least those with recognizable DNA-binding domains). New hybridization array studies will define comprehensive patterns of gene expression that will include both transcription factors and their regulatory targets.

In a book of this length, it is impossible to cover the entire field of transcription factor research comprehensively. Instead, we have attempted to cover the important paradigms through discussion of selected topics. While introducing the field of transcription factor research, the presentation highlights work in progress and important unanswered questions.

Most of the general introduction to gene regulation is presented in the first four chapters. Chapter 1 introduces the core promoter and basic machinery of RNA polymerase II transcription. Chapter 2 presents the general components and principals of transcription control beyond the basal promoter, although the rapidly growing knowledge of transcriptional coactivators is incorporated into Chapter 8. Chapter 4 elaborates the concepts of specific DNA binding and introduces the general families of transcription factors as grouped by their characteristic DNA binding domains. Recent work has focused on the interaction between the transcription machinery and chromatin, which is covered in the more specialized presentation of Chapter 3.

The next section, consisting of Chapters 5 and 6, contains detailed presentations of more specialized transcriptional regulation that coordinates the cell cycle and signal transduction. These processes are present in all cells of higher organisms.

The final chapters present detailed discussions of more selected topics of cell-specific regulation, but all deal with related aspects of developmental regulation. These chapters also provide specific examples of the general mechanisms introduced in the earlier chapters. Thus, Chapter 7 discusses cell-specific regulation in the immune system, but focuses on transcriptional regulation of cell lineage progression, and also presents a transcriptional silencer. Chapter 8 amplifies the discussions of all of the general chapters through its comprehensive presentation of the nuclear receptor family of transcription factors.

Chapter 9 introduces complex regulation by Hox factors, transcriptional regulators of body plan development that activate genes through multimerization with other homeobox factors. At another level, the Hox genes are present in unusual multigene arrays that regulate their expression through coordinated developmental transcription controls. Chapter 10 discusses regulation of the differentiated phenotype, focusing on the liver, the tissue in which the largest number of specific genes have been studied. Chapter 11 analyzes the central problem in transcriptional regulation of the nervous system, through the sequential pattern of expression of numerous developmental factors. Finally, Chapter 12 highlights all of the general mechanisms of transcriptional regulation by presenting a comprehensive review of the abnormal transcription factors that lead to neoplastic transformation.

Each chapter represents the distinctive views of its author(s), provides a comprehensive view of an important topic, and can be read independently of the others. Nevertheless, the coverage is also complementary. We hope this book provides a manageable and contemporary overview of the rapidly developing field of transcriptional regulation, useful as both a general introduction and a specific reference.

Joseph Locker

1

RNA polymerase II transcription machinery

B. Franklin Pugh

1. Introduction: promoter function and strategies in gene control

The set of instructions or genes that define an organism constitutes its genome. The genomes of free-living, single-celled bacteria and eukaryotes contain 4000–6000 genes (Goffeau *et al.*, 1996). In higher eukaryotes, including man, that number can rise to about 70 000. Each gene encodes the information to synthesize a particular protein (via a messenger RNA intermediate) or a particular functional RNA such as transfer RNAs or ribosomal RNAs. Not all genes are turned 'on' to the same level or at the same time. While many genes are required to make the common basic constituents of a cell, and thus are constitutively 'on', others are induced only when needed to make specific cell types or to respond to a changing environment. Selective gene expression is the end result of a variety of signal transduction pathways that originate from signaling molecules. Both constitutive and inducible gene expression are controlled by the transcriptional regulators, activators and repressors described in the ensuing 11 chapters. DNA recognition occurs through precise molecular complementarity between amino acids in the regulator protein and nucleotide bases in the DNA. A promoter-bound activator then facilitates the recruitment of the transcription machinery to initiate the decoding of the gene. Binding sites for transcriptional regulators, in general, are not confined to precise locations near the transcriptional start site, but may be located as far as 10 000 base-pairs upstream or downstream of the transcription start site. However, due to the compaction of DNA in chromatin, these sites may be physically quite close to the transcription start site.

The core of the promoter is composed of one or more DNA elements that are recognized by components of the general transcription machinery, and thus is the assembly site of the transcription machinery. In some cases, multiple core promoters at various locations along a single gene may be employed to provide flexibility in transcribing parts of the gene and generating functionally related but distinct protein products. This chapter will focus on the assembly of the transcription machinery at core promoters. It is important to keep in mind that *in vivo* transcriptional regulators orchestrate this assembly process.

Transcription Factors, edited by J. Locker.

2. Architecture of the core promoter

A critical feature of many but not all core promoters is the TATA box, which is recognized by the TATA-binding protein (TBP) subunit of transcription factor, TFIID (Hernandez, 1993). A typical strong TATA box sequence is TATAAAA. RNA transcripts that require a particular TATA box begin 25–30 nucleotides (~60 in yeast) downstream of this element. Genes that are highly transcribed typically possess strong TATA boxes. A weak TATA box might have one or more nucleotide substitutions in the DNA element. A large subset of promoters are 'TATA-less' in that they lack any TATA-like consensus sequence in their –30 region, and are generally unaffected by mutations in this region. TATA-less promoters are often, but not always, associated with genes that are transcribed constitutively at low levels.

TATA-less promoters rely largely on an 'initiator' element. The initiator was originally characterized as having a pyrimidine (Py) rich consensus: PyPyANA/TPyPy, with the 'A' nucleotide being the transcriptional start site (Smale and Baltimore, 1989). This element may be recognized in part by RNA polymerase II (pol II) (Carcamo *et al.*, 1991). More complex initiators have been described, and include additional downstream recognition sequences for gene-specific regulators such as YY1 (yin yang I) (Usheva and Shenk, 1994), TFII-I (Roy *et al.*, 1993), E2F (Means *et al.*, 1992) and upstream stimulatory factor (USF) (Du *et al.*, 1993), as well as components of the core transcription machinery such as the TBP-associated factor (TAF) $TAF_{II}150$ subunit of TFIID (Verrijzer *et al.*, 1994). Both the TATA box and the initiator may work independently or cooperatively (Emami *et al.*, 1997). Many TATA-less promoters also contain a conserved downstream promotor element (DPE) promoter element located approximately 30 nucleotides downstream of the transcriptional start site (consensus: (AG)G(AT)CGTG), which is recognized by the histone-like $TAF_{II}60$ and $TAF_{II}40$ subunits of TFIID (Burke and Kadonaga, 1997; 1996).

In vitro, core promoters can drive the efficient expression of an attached gene. However, *in vivo*, core promoters alone are transcriptionally silent (and do not naturally exist). The difference can be attributed to repression of both the core promoter and components of the transcription machinery *in vivo*, which are lost upon biochemical fractionation *in vitro*. In addition to the core promoter, gene expression *in vivo* is critically dependent upon gene-specific activator sequences or enhancers (Chapter 2). Activators that bind to these elements facilitate the removal of inhibitors as well as directly assisting in the loading of the transcription machinery at the core promoter.

3. Rate-limiting steps in gene expression

The assembly of a pol II transcription complex can be recapitulated *in vitro* at a core promoter lacking activators. Such biochemical fractionation and reconstitution of promoter-specific transcription has been instrumental in identifying the general transcription factors, and has provided insight into the assembly pathway and the interacting partners. While biochemical experiments have defined potential mechanistic steps in transcription complex assembly and their regulation, ascertaining whether such steps are rate-limiting for gene expression *in vivo* is more difficult.

Within the natural state of a cell, some genes are 'off', others are transcribed at low levels, while others are fully induced. Any particular gene can move between these states in response to cellular and environmental signals. Consequently, the step that is rate-limiting for the expression of a gene will vary depending upon the incoming signals. In general, genes that are 'off' do not have transcription complexes assembled at their promoters, and the rate-limiting step may be the removal of bound repressors by transcriptional activators. Such repressors include histone proteins that wrap DNA into nucleosomes. Some activators function in part by recruiting nucleosome remodeling complexes, which by some unknown mechanism, increases the accessibility of the nucleosomal DNA to the general transcription machinery (Grant *et al.*, 1997).

While accessibility of core promoter DNA is essential for transcription complex assembly, it does not appear to be sufficient. Thus, in general, a transcription complex will not assemble at a core promoter *in vivo*, even under conditions where the promoter might be accessible. Despite the ability of TBP to recognize the TATA box, it appears to be unable to do so *in vivo* unless directed by a promoter-bound transcriptional activator (Jackson-Fisher *et al.*, 1999; Kuras and Struhl, 1999; Li *et al.*, 1999; Selleck and Majors, 1987). A number of genetic screens have identified mutant proteins that allow high levels of activator-independent transcription, suggesting that TBP and other components of the general transcription machinery are intrinsically repressed (Blair and Cullen, 1997; Prelich and Winston, 1993). Thus a second potential function of activators is to alleviate repressors associated with the general transcription machinery. A third potential function of transcriptional activators is to direct the general transcription machinery to the proper promoter.

4. Assembly and regulation of the TATA-binding protein and associated factors

TBP is one of the first components of the general transcription machinery to be recruited to the promoter (*Figure 1*). It is a highly conserved protein found in two of the three domains of life: archaea and eukarya. TBP is unique in that it is required for the expression of all genes, even those transcribed by pol I and pol III. Interestingly, not all genes contain a TATA box, but evidence suggests that TBP nevertheless binds promoter DNA in approximately the same location where the TATA box normally resides (Pugh and Tjian, 1991; Zenzie-Gregory *et al.*, 1993). Apparently, other sequence-specific DNA-binding proteins that directly or indirectly interact with TBP properly position TBP at such TATA-less promoters.

While it is unclear what essential role TBP plays at a promoter, two interesting properties stand out. First, TBP bends DNA by nearly 90°, so it might function to bring transcription factors located distally on either side of TBP together (Kim *et al.*, 1993a; 1993b). Second, TBP binds to a number of the components of the transcription machinery, so it might orchestrate the proper assembly of the transcription machinery.

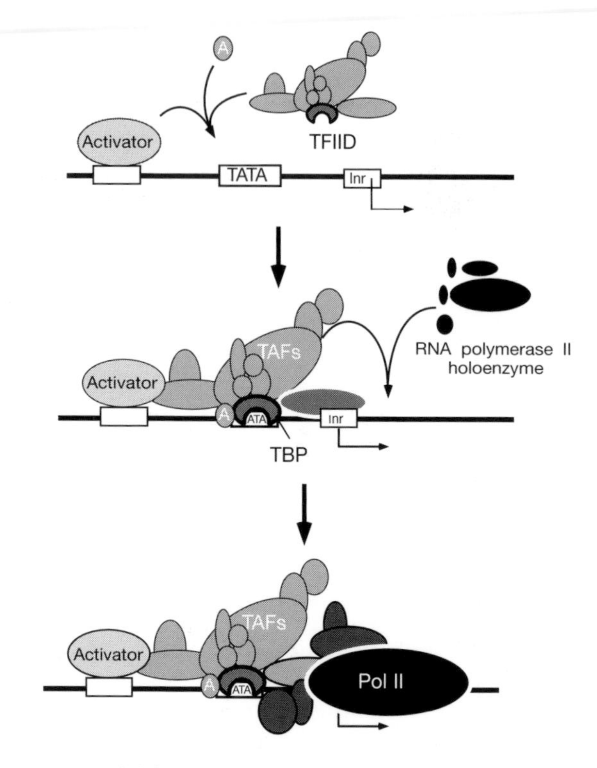

Figure 1. A two-stage model for the assembly of an RNA polymerase II (pol II) transcription complex. A transcriptional activator binds to its cognate site on a promoter and recruits transcription factors TFIID and TFIIA approximately 30 nucleotides upstream from the transcriptional start site. The combined action of this complex recruits the RNA polymerase holoenzyme. Reprinted from *Gene* 255, Pugh, Control of gene expression through regulation of the TATA-binding protein, 1–14, 2000, with permission from Elsevier Science.

TBP is a component of a number of functionally distinct multi-subunit complexes that are illustrated in *Figure 2* (Hernandez, 1993). Some of these complexes appear to function only with certain RNA polymerases: SL1, TFIID, and TFIIIB are specific for pols I, II, and III, respectively. SNAP appears to function with both pol II and III, but only at a subset of genes, which includes the *snRNA* genes. Another TBP-containing complex, SAGA (Spt-Ada-Gcn5-acetyltransferase), has important roles in chromatin remodeling. Other TBP complexes, containing factors such as NC2 and Mot1, appear to represent a repressed state of TBP.

Perhaps the best characterized of the TBP complexes is TFIID, which is recruited to pol II transcribed promoters by transcriptional activators (*Figure 1*). Apart from TBP, TFIID is composed of about a dozen distinct subunits that are referred to as TAF_{II}s. TAF_{II}s fall into a variety of categories. First, some TAF_{II}s form structures akin to histones, and like histones may assemble into a nucleosome-like structure containing two copies each of the four different histone-like TAF_{II}s (Xie *et al.*, 1996). Second, a number of the TAF_{II}s, including the histone-like TAF_{II}s, appear to reside in the SAGA complex, and thus both SAGA and

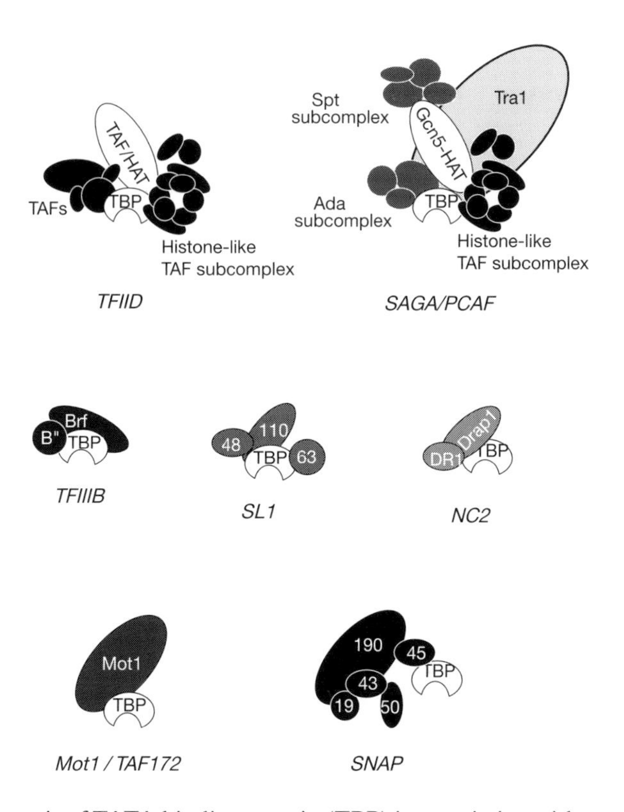

Figure 2. Schematic of TATA-binding protein (TBP) in association with a variety of functionally distinct complexes. Note that TFIID and SAGA/PCAF share some of the same subunits. SAGA, Spt-Ada-Gcn5-acetyltransferase; PCAF, p300/CBP-associated factor; TAF, TBP-associated factor; HAT, histone acetyltransferase. Reprinted from *Gene* 255, Pugh, Control of gene expression through regulation of the TATA-binding protein, 1–14, 2000, with permission from Elsevier Science.

TFIID are likely to share some common functions (Grant *et al.*, 1998; Imhof *et al.*, 1997; Martinez *et al.*, 1998). Third, some TFIID TAF_{II}s appear to have either structural or functional homologs in SAGA. For example, the $TAF_{II}28/TAF_{II}18$ subunits have a structure homolog in SAGA called Spt3 (Birck *et al.*, 1998). The $TAF_{II}250$ subunit of TFIID has histone acetyltransferase activity, and may be functionally equivalent to the Gcn5 subunit of SAGA (Mizzen *et al.*, 1996). Fourth, some TAF_{II}s appear to be the direct target of transcriptional activators, and thus have been termed coactivators. For example, gene specific activators such as p53 or the viral activator VP16 interact with $TAF_{II}40$ to enhance the recruitment of TFIID to a promoter (Farmer *et al.*, 1996; Goodrich *et al.*, 1993).

Once TBP has assembled at a promoter, the remainder of the transcription machinery assembles quite rapidly. Therefore, the recruitment of TBP to TATA is an important rate-limiting step in gene expression and is subjected to an extraordinary degree of regulation. Since TBP can bind both TATA DNA and nearly any DNA sequence with fairly high affinity, mechanisms must exist to keep TBP off the DNA unless it is directed to bind by transcriptional activators. Indeed, at least four distinct mechanisms of TBP repression can be envisioned, which are shown in *Figure 3*:

(i) DNA targets of TBP may be sequestered into nucleosomes or other repressors, thereby preventing TBP from assembling onto DNA. Depletion of histones *in vivo* leads to elevated expression of certain genes (Han and Grunstein, 1988; Wyrick *et al.*, 1999).

(ii) TBP autorepresses itself through dimerization. The DNA-binding surface of TBP interacts with the same DNA-binding surface on a second TBP molecule, thus preventing both from binding DNA. TBP mutations that interfere with dimerization (and to a lesser extent DNA binding) lead to increased levels of activator-independent gene expression *in vivo* (Jackson-Fisher *et al.*, 1999).

(iii) Once TBP binds to DNA, if activator-catalyzed transcription complex assembly does not immediately follow, then factors such as $TAF_{II}250$ and Mot1 might induce TBP to dissociate from DNA (Auble and Hahn, 1993; Bai *et al.*, 1997; Chicca *et al.*, 1998; Kokubo *et al.*, 1998). Interestingly, a portion of $TAF_{II}250$ forms a DNA-like structure, which interacts with the DNA-binding surface of TBP thereby possibly leading to the disruption of the TBP/DNA complex (Liu *et al.*, 1998). In contrast to $TAF_{II}250$, Mot1 uses the energy of ATP hydrolysis to dissociate TBP from DNA (Auble and Hahn, 1993; Chicca *et al.*, 1998). Mutations in Mot1 lead to elevated levels of gene expression (Auble *et al.*, 1994; Davis *et al.*, 1992).

(iv) Another mechanism that involves NC2, does not directly lead to the dissociation of TBP from DNA. Instead, it appears that the TBP/DNA complex is sequestered into an inactive form that prevents it from forming a functional transcription complex. Another complex termed NOT, may also sequester TBP/DNA complexes. Mutations in these repressor subunits lead to elevated levels of transcription *in vivo* (Collart and Struhl, 1994; Prelich, 1997).

Where any repression mechanism exists, there must also be anti-repressors. TFIIA is one such anti-repressor that appears to counteract many of the repressors just described. It induces TBP dimers to dissociate, thereby accelerating TBP/TATA binding (Coleman *et al.*, 1999). It competes with Mot1 and $TAF_{II}250$, thereby preventing the dissociation of TBP from TATA DNA (Kokubo *et al.*, 1998; Ozer *et al.*, 1998). TFIIA also counteracts repression by NC2, and induces a conformational change in TFIID that may facilitate transcription complex assembly (Chi and Carey, 1996; Inostroza *et al.*, 1992). *In vitro*, biochemical removal of these repressors alleviates the need for TFIIA, which further attests to its role as an anti-repressor.

By itself, TFIIA may be fairly inefficient as an anti-repressor, and it appears to work best in the context of transcriptional activators. Thus, TFIIA is also considered a transcriptional co-activator. It is possible that transcriptional activators enhance the interactions between TFIIA and TBP (or TFIID), allowing TFIIA to function more efficiently.

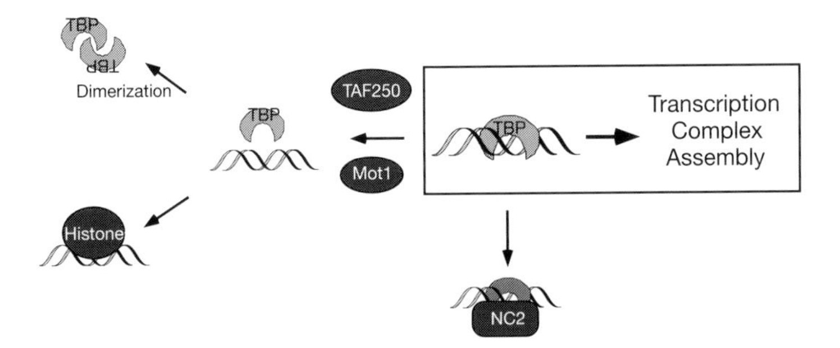

Figure 3. Mechanisms of TATA-binding protein (TBP) repression that counteract transcription complex assembly. TBP dimerization and histones bound over the TATA box each prevent TBP from loading onto DNA. $TAF_{II}250$ and Mot1 each function to dissociate TBP from DNA. NC2 prevents the RNA polymerase II holoenzyme from loading onto the promoter. TAF, TBP-associated factors. Reprinted from *Gene* 255, Pugh, Control of gene expression through regulation of the TATA-binding protein, 1–14, 2000, with permission from Elsevier Science.

5. Recruitment of the RNA polymerase II holoenzyme

It is not clear whether TBP arrives at promoters as part of TFIID, or free of TAFs. Similarly it is not understood whether other complexes, such as SAGA in yeast and p300/CBP-associated factor (PCAF) in humans, help deliver TBP to the promoter, in addition to, or instead of, TFIID. Once TBP has docked at the promoter, the activator–TFIIA–TBP complex is then in position to recruit the pol II holoenzyme. While the pol II holoenzyme has been purified in a number of forms, either lacking or containing certain subunits, it is clear that the holo-enzyme is massive, weighing in at about the size of a ribosome (Koleske and Young, 1994). For simplicity, we will consider the holoenzyme as an amalgamation of a variety of complexes (*Figure 4*), including the core catalytic activity of pol II (~12 subunits), the SRB/Mediator complex involved in transcriptional activation (~20 subunits),

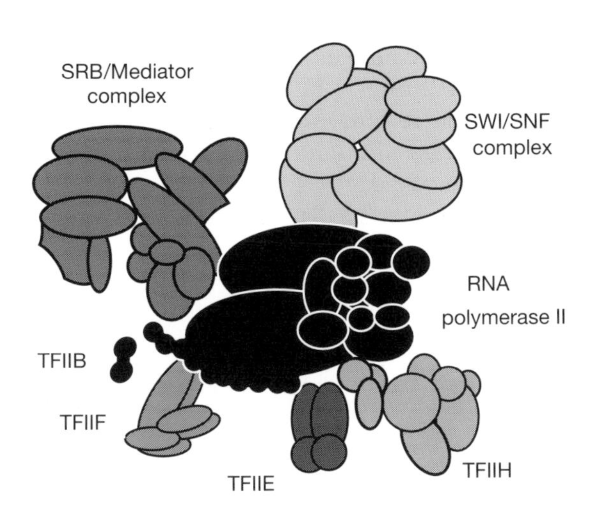

Figure 4. A schematic of the RNA polymerase II holoenzyme, including a variety of distinct complexes that comprise the general transcription factors. SWI, Switch; SNF, sucrose non-fermenting; TFII, transcription factor of RNA polymerase II. Reprinted from *Gene* 255, Pugh, Control of gene expression through regulation of the TATA-binding protein, 1–14, 2000, with permission from Elsevier Science.

the chromatin remodeling SWI/SNF (Switch–Sucrose non-fermenting) complex (~9 subunits), and each of the general initiation factors TFIIB, TFIIE, TFIIF, and TFIIH (~18 subunits total). Some of these complexes may be composed of a number of loosely associated subcomplexes. For example, Srb8 to Srb11 appear to form a dissociable subcomplex within the SRB/Mediator complex (Myers *et al.*, 1998). The ability of subunits to readily dissociate may reflect the existence of multiple interchangeable and regulated forms of the subunit.

Current hypotheses posit that promoter-bound activators and coactivator, along with TFIID subunits, recruit components of the SRB/Mediator complex and the general initiation factors (most notably TFIIB) through direct physical association (Koh *et al.*, 1998; Orphanides *et al.*, 1996; Pugh, 1996).

5.1 Transcription factor TFIIB

TFIIB is one of the few components of the transcription machinery that can be isolated as a single subunit (Ha *et al.*, 1991). It is highly conserved from archaea to eukaryotes. In organisms that utilize pol III, a pol III-specific paralog of TFIIB, termed Brf1, is used (Colbert and Hahn, 1992; Qureshi *et al.*, 1995). The conserved core of TFIIB is structurally well defined, having two repeated globular domains that straddle the carboxyl terminal stirrup of TBP (Nikolov *et al.*, 1995). TFIIB contacts DNA both upstream and downstream of TBP. Upstream DNA interactions include binding to a BRE (consensus: (GC)(GC)(GA)CGCC) (Lagrange *et al.*, 1998). Not only does TFIIB dock onto TBP, but it also makes contact with $TAF_{II}40$ (Goodrich *et al.*, 1993). TBP–TFIID dimerization is likely to inhibit the association of TFIIB with TBP/TFIID prior to DNA binding. Although a TBP/TATA complex is prepared to accept TFIIB, an amino-terminal region of TFIIB interferes with the docking process (Roberts and Green, 1994). Acidic activators, however, alleviate this inhibitory interaction and facilitate the loading of TFIIB and presumably the entire pol II holoenzyme (Roberts and Green, 1994).

5.2 Transcription factor TFIIF

TFIIF in higher eukaryotes is composed of two subunits each of the RNA polymerase-associated proteins, Rap74 and Rap30 (Burton *et al.*, 1988; Flores *et al.*, 1990). Yeast TFIIF appears to have an extra subunit, which is also present in TFIID and the SWI/SNF chromatin-remodeling complex (Cairns *et al.*, 1996). TFIIF is more tightly associated with the core pol II enzyme than the other general transcription factors, and in a number of respects may be the eukaryotic homolog to eubacterial sigma factors for a number of reasons:

(i) portions of TFIIF share limited sequence homology to *E. coli* σ^{70} (Sopta *et al.*, 1989);
(ii) like σ^{70}, TFIIF minimizes the binding of pol II to non-specific DNA, thereby freeing up polymerase for promoter specific interactions (Killeen and Greenblatt, 1992);
(iii) TFIIF can bind *E. coli* RNA polymerase and be displaced by σ^{70} (McCracken and Greenblatt, 1991).

Since Rap30 binds both TFIIB and pol II, and is sufficient to deliver the latter to the promoter, Rap30 might function as the next link in the chain between TBP and pol II. The known interaction between Rap74 and $TAF_{II}250$ might provide an additional link, as well as source of regulation, inasmuch as $TAF_{II}250$ phosphorylates TFIIF (Dikstein *et al.*, 1996; Ruppert and Tjian, 1995).

5.3 The CTD and the SRB/Mediator complex

The largest subunit of pol II contains a repeating (~30 amino acids in yeast and ~50 amino acids in humans) heptad motif at its carboxyl terminal end, termed the CTD (Corden, 1990). The CTD may help tether pol II to the promoter via interactions with TBP, although this possibility has not been demonstrated *in vivo* (Usheva *et al.*, 1992). The CTD might be the home of the SRB/Mediator complex. The SRB/Mediator complex initially was an overlooked component of the general transcription machinery, largely because the stringent purification protocols for pol II stripped away the complex and also because the complex was unnecessary for *in vitro* transcription with purified factors. Two distinct approaches – genetics and biochemistry – converged to discover this important regulatory component. SRB gene mutations were identified in suppressor screens of CTD truncations that generate a slow growth phenotype (Nonet and Young, 1989). If the CTD plays a positive role in transcription, then many of the SRBs are likely to encode transcriptional repressors. If SRB proteins diminish the functionality of pol II, and CTD truncations further diminish the functionality to the point of producing a growth defect, then mutations in SRBs that relieve this inhibitory activity will relieve the growth defect. Out of this genetic screen came Srb2, and Srb4 to Srb11 (Hengartner *et al.*, 1995). Temperature-sensitive mutations in Srb4 result in a severe growth arrest at the restrictive temperature, as well as a rapid loss of total mRNA, indicating that at least some components of the SRB/Mediator complex are likely to be important for all genes (Thompson and Young, 1995).

At about the same time that the SRB genes were being isolated, Mediator was identified as a biochemical activity that overcame the squelching effect of a potent viral activator of transcription, and was later found to stimulate (or mediate) the transcriptional activation process (Kelleher *et al.*, 1990). Purification of the Mediator complex resulted in the identification of several previously described proteins – Gal11, Sin4, Rox3, Rgr1 – as well as many novel subunits (Med1 through to Med8), and many Srb proteins (Gustafsson *et al.*, 1997; Li *et al.*, 1995; Myers *et al.*, 1998). Srb8 to Srb11 were absent, indicating that these proteins might form a separate and dissociable subcomplex (Myers *et al.*, 1998). Srb4 appears to interact with a number of proteins within the SRB/Mediator complex, but also interacts with the activation domain of the sequence-specific transcriptional activator Gal4 (Koh *et al.*, 1998). Thus, recruitment of the RNA polymerase II holoenzyme may occur via the SRB/Mediator components, as well as through a TBP-TAF-TFIIB-TFIIF connection.

5.4 Transcription factors TFIIE and TFIIH

TFIIE is a heterotetramer of two distinct proteins (Maxon and Tjian, 1994; Ohkuma *et al.*, 1990; Serizawa *et al.*, 1994). Recent evidence suggests that TFIIE may play a role in loading TBP and TFIIA onto a promoter (Yokomori *et al.*, 1998). Much of TFIIE's activity, however, has been associated with DNA strand separation (or promoter melting) once the entire pre-initiation complex has assembled. Whether TFIIE catalyzes strand opening or simply stabilizes an open configuration induced by pol II is not clear. DNA supercoiling appears to abrogate the need for TFIIE (Holstege *et al.*, 1995; Parvin and Sharp, 1993).

TFIIH is a large multi-subunit complex that is involved in both transcription and DNA repair (Drapkin *et al.*, 1994; Feaver *et al.*, 1993). TFIIH possesses an ATP-dependent helicase activity that may catalyze promoter melting (Drapkin and Reinberg, 1994; Holstege *et al.*, 1997). At many promoters that are not supercoiled, TFIIE, TFIIH and ATP are required for promoter melting and thus transcription initiation (Parvin and Sharp, 1993; Timmers, 1994). The respective roles of TFIIE, TFIIH and pol II in strand separation have not been fully elucidated. It is plausible that negative supercoiling facilitates strand opening in such a way that the pol II active site can engage the nucleotide bases without the need for TFIIE, TFIIH and ATP. Under conditions where the DNA helix is more stable, pol II might still be able to melt the helix, but only transiently. In this case, TFIIE which binds pol II near the separated strands, might lock it into an open configuration. Finally, under conditions where the helix is very stable, possibly caused by the DNA sequence and/or the physical arrangement of the pre-initiation complex, the energy of ATP hydrolysis catalyzed by the Ercc3 subunit of TFIIH might be coupled to the separation of the DNA strands. TFIIE might assist in this process by stabilizing the open structure once ATP hydrolysis is complete. Once the open complex has formed, TFIIH may provide further helicase assistance by propagating the open strands as pol II starts to clear the promoter and moves into the elongation phase.

TFIIH may play an additional role in transcription initiation (see Chapter 5), where its cyclin-dependent kinase subunit, Cyclin H/Cdk7 (yeast MO15), phosphorylates the CTD (Feaver *et al.*, 1994; Lu *et al.*, 1992; Roy *et al.*, 1994; Serizawa *et al.*, 1995). The unphosphorylated form of the CTD is associated with the pre-initiation complex, while the phosphorylated form is associated with the elongating polymerase (Laybourn and Dahmus, 1989; Lu *et al.*, 1991). Interestingly, Cyclin C/Cdk8 (yeast Srb10 and Srb11) constitute a second Cyclin/Cdk pair that is responsible for phosphorylation of the CTD (Liao *et al.*, 1995). Yeast strains lacking Srb10–Srb11 kinase function are deficient in galactose-induced transcription and possess an under-phosphorylated CTD (Liao *et al.*, 1995). It is possible that one or both kinases induce transcription initiation and promoter clearance by phosphorylating the CTD and causing the CTD to disengage its targets.

6. Events subsequent to transcription initiation

Once pol II has moved out of the promoter and into the elongation phase it will face numerous obstacles set up by the DNA and nascent mRNA, such as pause sites and arrest sites (Reines *et al.*, 1996). A variety of elongation factors present in the pre-initiation complex or arriving in response to the phosphorylated state of the polymerase will assist in moving it through the pause sites (Moreland *et al.*, 1998; Shilatifard *et al.*, 1997). Cleavage factors, such as TFIIS, assist arrested RNA polymerases by inducing the polymerase to back up, cleave the nascent RNA in an RNA/DNA hybrid, and restart RNA synthesis off the 3′ end of the main chain of the cleaved RNA (Izban and Luse, 1993; Reines *et al.*, 1993). Once pol II has transcribed its gene and has transcribed over a cleavage and polyadenylation site (core consensus: AAUAAA), cleavage and polyadenylation factors that entered the preinitiation complex through TFIID cleave the RNA (Dantonel *et al.*, 1997). Loss of the transcript results in a rapid termination and dissociation of pol II.

When pol II holoenzyme initiates transcription and moves down the template, TFIID is left behind at the promoter to recruit another holoenzyme. High levels of gene expression are due, in large part, to the rapid and continuous reloading of RNA polymerase.

The ability of the RNA polymerase to rapidly reload at the promoter under conditions of activated transcription might preclude the entry of certain transcriptional inhibitors, such as those that target TBP. However, once the gene-activating signal is removed, and the activators no longer recruit RNA polymerase, then TBP repressors, such as Mot1, might reverse the assembly process leading to the dissociation of TFIIA and TBP (TFIID), ultimately shutting down expression of the gene.

7. Conclusions and prospects

The past 10 years have seen an explosive growth in the number of identified general transcriptional regulatory proteins that are distinct from sequence-specific activators. What role these factors play in orchestrating gene expression is largely unresolved. It is possible that some general transcriptional regulatory proteins are recruited to specific promoters via sequence-specific activators and repressors. Other general regulators might be operating at most, if not all, promoters.

It is difficult to fully ascertain whether or not a given transcriptional regulatory protein is involved in the expression of a particular gene. If one disables a regulator via genetic manipulation, and changes in the expression of a specific gene follows, then one can conclude that the regulator is somehow involved in the expression of that gene. However, if disabling the regulator has no effect on a particular gene, it does not necessarily follow that the regulator is not involved in the expression of that gene. A lack of an effect might be due to several factors:

(i) another regulatory protein that is functionally redundant to the first may be present. One would need to disable both proteins in order to see an effect;
(ii) for regulatory proteins that are essential for cell viability, one must often use conditional mutants of that regulator (e.g. temperature-sensitive mutants), and look for gene-specific effects under the restrictive conditions. However, it is possible that the mutation affects only part of a multifunctional regulator. Therefore, some genes might be affected (hence the conditional phenotype) while others are unaffected but nevertheless utilize the regulator;
(iii) a transcriptional regulator might be involved in the expression of a gene, but its activity may not be rate-limiting. Thus, decreasing its activity by as much as 90% might have little noticeable impact on the expression of certain genes. Only upon 100% inactivation would an effect be detected.

With the advent of total genome sequencing and microarray technology, it is now feasible to determine how a particular factor affects the expression of all genes. As it turns out, some TAFs and pol II holoenzyme components affect the expression of most genes in yeast, while others seem to affect only a small fraction of genes (Holstege *et al.*, 1998). The latter observation is subject to the limitation described earlier. Nevertheless, by monitoring total genome expression patterns it should eventually be possible to define the entire regulatory network of a cell. Understanding how all these factors dance in harmony to achieve gene-specific and cell-type-specific regulation will continue to be a challenge for at least another 10 years.

References

Auble, D.T. and Hahn, S. (1993) An ATP-dependent inhibitor of TBP binding to DNA. *Genes Dev.* **7**: 844–856.

Auble, D.T., Hansen, K.E., Mueller, C.G., Lane, W.S., Thorner, J. and Hahn, S. (1994) Mot1, a global repressor of RNA polymerase II transcription, inhibits TBP binding to DNA by an ATP-dependent mechanism. *Genes Dev.* **8**: 1920–1934.

Bai, Y., Perez, G.M., Beechem, J.M. and Weil, P.A. (1997) Structure–function analysis of TAF130: identification and characterization of a high-affinity TATA-binding protein interaction domain in the *N* terminus of yeast TAF(II)130. *Mol. Cell. Biol.* **17**: 3081–3093.

Birck, C., Poch, O., Romier, C., *et al.* (1998) Human TAF(II)28 and TAF(II)18 interact through a histone fold encoded by atypical evolutionary conserved motifs also found in the SPT3 family. *Cell* **94**: 239–249.

Blair, W.S. and Cullen, B.R. (1997) A yeast TATA-binding protein mutant that selectively enhances gene expression from weak RNA polymerase II promoters. *Mol. Cell. Biol.* **17**: 2888–2896.

Burke, T.W. and Kadonaga, J.T. (1997) The downstream core promoter element, DPE, is conserved from *Drosophila* to humans and is recognized by TAFII60 of *Drosophila*. *Genes Dev.* **11**: 3020–3031.

Burke, T.W. and Kadonaga, J.T. (1996) *Drosophila* TFIID binds to a conserved downstream basal promoter element that is present in many TATA-box-deficient promoters. *Genes Dev.* **10**: 711–724.

Burton, Z.F., Killeen, M., Sopta, M., Ortolan, L.G. and Greenblatt, J. (1988) RAP30/74: a general initiation factor that binds to RNA polymerase II. *Mol. Cell. Biol.* **8**: 1602–1613.

Cairns, B.R., Henry, N.L. and Kornberg, R.D. (1996) TFG3/TAF30/ANC1, a component of the yeast SWI/SNF complex that is similar to the leukemogenic proteins ENL and AF-9. *Mol. Cell. Biol.* **16**: 3308–3316.

Carcamo, J., Buckbinder, L. and Reinberg, D. (1991) The initiator directs the assembly of a transcription factor IID-dependent transcription complex. *Proc. Natl Acad. Sci. USA* **88**: 8052–8056.

Chi, T.H. and Carey, M. (1996) Assembly of the isomerized TFIIA-TFIID-TATA ternary complex is necessary and sufficient for gene activation. *Genes Dev.* **10**: 2540–2550.

Chicca, J.J., 2nd, Auble, D.T. and Pugh, B.F. (1998) Cloning and biochemical characterization of TAF-172, a human homolog of yeast Mot1. *Mol. Cell. Biol.* **18**: 1701–1710.

Colbert, T. and Hahn, S. (1992) A yeast TFIIB-related factor involved in RNA polymerase III transcription. *Genes Dev.* **6**: 1940–1949.

Coleman, R.A., Taggart, A.K.P., Burma, S., Chicca II, J.J. and Pugh, B.F. (1999) TFIIA regulates TBP and TFIID dimers. *Mol. Cell* **4**: 451–457.

Collart, M.A. and Struhl, K. (1994) NOT1(CDC39), NOT2(CDC36), NOT3, and NOT4 encode a global-negative regulator of transcription that differentially affects TATA-element utilization. *Genes Dev.* **8**: 525–537.

Corden, J.L. (1990) Tails of RNA polymerase II. *Trends Biochem. Sci.* **15**: 383–387.

Dantonel, J.C., Murthy, K.G., Manley, J.L. and Tora, L. (1997) Transcription factor TFIID recruits factor CPSF for formation of 3′ end of mRNA. *Nature* **389**: 399–402.

Davis, J.L., Kunisawa, R. and Thorner, J. (1992) A presumptive helicase (MOT1 gene product) affects gene expression and is required for viability in the yeast *Saccharomyces cerevisiae*. *Mol. Cell. Biol.* **12**: 1879–1892.

Dikstein, R., Ruppert, S. and Tjian, R. (1996) TAFII250 is a bipartite protein kinase that phosphorylates the base transcription factor RAP74. *Cell* **84**: 781–790.

Drapkin, R., Reardon, J.T., Ansari, A., *et al.* (1994) Dual role of TFIIH in DNA excision repair and in transcription by RNA polymerase II. *Nature* **368**: 769–772.

Drapkin, R. and Reinberg, D. (1994) The multifunctional TFIIH complex and transcriptional control. *Trends Biochem. Sci.* **19**: 504–508.

Du, H., Roy, A.L. and Roeder, R.G. (1993) Human transcription factor USF stimulates transcription through the initiator elements of the HIV-1 and the Ad-ML promoters. *EMBO J.* **12**: 501–511.

Emami, K.H., Jain, A. and Smale, S.T. (1997) Mechanism of synergy between TATA and initiator: synergistic binding of TFIID following a putative TFIIA-induced isomerization. *Genes Dev.* **11**: 3007–3019.

Farmer, G., Colgan, J., Nakatani, Y., Manley, J.L. and Prives, C. (1996) Functional interaction between p53, the TATA-binding protein (TBP), and TBP-associated factors *in vivo*. *Mol. Cell. Biol.* **16**: 4295–4304.

Feaver, W.J., Svejstrup, J.Q., Henry, N.L. and Kornberg, R.D. (1994) Relationship of CDK-activating kinase and RNA polymerase II CTD kinase TFIIH/TFIIK. *Cell* **79**: 1103–1109.

Feaver, W.J., Svejstrup, J.Q., Bardwell, L., *et al.* (1993) Dual roles of a multiprotein complex from *S. cerevisiae* in transcription and DNA repair. *Cell* **75**: 1379–1387.

Flores, O., Ha, I. and Reinberg, D. (1990) Factors involved in specific transcription by mammalian RNA polymerase II. Purification and subunit composition of transcription factor IIF. *J. Biol. Chem.* **265**: 5629–5634.

Goffeau, A., Barrell, B.G., Bussey, H., *et al.* (1996) Life with 6000 genes. *Science* **274**: 546, 563–547.

Goodrich, J.A., Hoey, T., Thut, C.J., Admon, A. and Tjian, R. (1993) Drosophila $TAF_{II}40$ interacts with both a VP16 activation domain and the basal transcription factor TFIIB. *Cell* **75**: 519–530.

Grant, P.A., Duggan, L., Cote, J., *et al.* (1997) Yeast Gcn5 functions in two multisubunit complexes to acetylate nucleosomal histones: characterization of an Ada complex and the SAGA (Spt/Ada) complex. *Genes Dev.* **11**: 1640–1650.

Grant, P.A., Schieltz, D., Pray-Grant, M.G., *et al.* (1998) A subset of TAF(II)s are integral components of the SAGA complex required for nucleosome acetylation and transcriptional stimulation. *Cell* **94**: 45–53.

Gustafsson, C.M., Myers, L.C., Li, Y., *et al.* (1997) Identification of Rox3 as a component of mediator and RNA polymerase II holoenzyme. *J. Biol. Chem.* **272**: 48–50.

Ha, I., Lane, W.S. and Reinberg, D. (1991) Cloning and structure of a human gene encoding the general transcription initiation factor IIB. *Nature* **352**: 689–694.

Han, M. and Grunstein, M. (1988) Nucleosome loss activates yeast downstream promoters *in vivo*. *Cell* 55: 1137–1145.

Hengartner, C.J., Thompson, C.M., Zhang, J., *et al.* (1995) Association of an activator with an RNA polymerase II holoenzyme. *Genes Dev.* **9**: 897–910.

Hernandez, N. (1993) TBP, a universal eukaryotic transcription factor? *Genes Dev.* **7**: 1291–1308.

Holstege, F.C.P., Jennings, E.G., Wyrick, J.J., *et al.* (1998) Dissecting the regulatory circuitry of a eukaryotic genome. *Cell* **95**: 717–728.

Holstege, F.C., Fiedler, U. and Timmers, H.T. (1997) Three transitions in the RNA polymerase II transcription complex during initiation. *EMBO J.* **16**: 7468–7480.

Holstege, F.C., Tantin, D., Carey, M., van der Vliet, P.C. and Timmers, H.T. (1995) The requirement for the basal transcription factor IIE is determined by the helical stability of promoter DNA. *EMBO J.* **14**: 810–819.

Imhof, A., Yang, X.J., Ogryzko, V.V., Nakatani, Y., Wolffe, A.P. and Ge, H. (1997) Acetylation of general transcription factors by histone acetyltransferases. *Curr. Biol.* **7**: 689–692.

Inostroza, J.A., Mermelstein, F.H., Ha, I., Lane, W.S. and Reinberg, D. (1992) Dr1, a TATA-binding protein-associated phosphoprotein and inhibitor of class II gene transcription. *Cell* **70**: 477–489.

Izban, M.G. and Luse, D.S. (1993) SII-facilitated transcript cleavage in RNA polymerase II complexes stalled early after initiation occurs in primarily dinucleotide increments. *J. Biol. Chem.* **268**: 12864–12873.

Jackson-Fisher, A.J., Chitikila, C., Mitra, M. and Pugh, B.F. (1999) A role for TBP dimerization in preventing unregulated gene expression. *Mol. Cell* **3**: 717–727.

Kelleher, R.D., Flanagan, P.M. and Kornberg, R.D. (1990) A novel mediator between activator proteins and the RNA polymerase II transcription apparatus. *Cell* **61**: 1209–1215.

Killeen, M.T. and Greenblatt, J.F. (1992) The general transcription factor RAP30 binds to RNA polymerase II and prevents it from binding nonspecifically to DNA. *Mol. Cell. Biol.* **12**: 30–37.

Kim, J.L., Nikolov, D.B. and Burley, S.K. (1993a) Co-crystal structure of TBP recognizing the minor groove of a TATA element. *Nature* **365**: 520–527.

Kim, Y., Geiger, J.H., Hahn, S. and Sigler, P.B. (1993b) Crystal structure of a yeast TBP/TATA-box complex. *Nature* **365**: 512–520.

Koh, S.S., Ansari, A.Z., Ptashne, M. and Young, R.A. (1998) An activator target in the RNA polymerase II holoenzyme. *Mol Cell* **1**: 895–904.

Kokubo, T., Swanson, M.J., Nishikawa, J.I., Hinnebusch, A.J. and Nakatani, Y. (1998) The yeast TAF145 inhibitory domain and TFIIA competitively bind to TATA-binding protein. *Mol. Cell. Biol.* **18**: 1003–1012.

Koleske, A.J. and Young, R.A. (1994) An RNA polymerase II holoenzyme responsive to activators. *Nature* **368**: 466–469.

Kuras, L. and Struhl, K. (1999) Binding of TBP to promoters *in vivo* is stimulated by activators and requires Pol II holoenzyme. *Nature* **399**: 609–613.

Lagrange, T., Kapanidis, A.N., Tang, H., Reinberg, D. and Ebright, R.H. (1998) New core promoter element in RNA polymerase II-dependent transcription: sequence-specific DNA binding by transcription factor IIB. *Genes Dev.* **12**: 34–44.

Laybourn, P.J. and Dahmus, M.E. (1989) Transcription-dependent structural changes in the *C*-terminal domain of mammalian RNA polymerase subunit IIa/o. *J. Biol. Chem.* **264**: 6693–6698.

Li, X.-Y., Virbasius, A., Zhu, X. and Green, M. (1999) Enhancement of TBP binding by activators and general transcription factors. *Nature* **399**: 605–609.

Li, Y., Bjorklund, S., Jiang, Y.W., *et al.* (1995) Yeast global transcriptional regulators Sin4 and Rgrl are components of mediator complex/RNA polymerase II holoenzyme. *Proc. Natl Acad. Sci. USA* **92**: 10864–10868.

Liao, S.M., Zhang, J., Jeffery, D.A., *et al.* (1995) A kinase-cyclin pair in the RNA polymerase II holoenzyme. *Nature* **374**: 193–196.

Liu, D., Ishima, R., Tong, K.I., *et al.* (1998) Solution structure of a TBP-TAF(II)230 complex: protein mimicry of the minor groove surface of the TATA box unwound by TBP. *Cell* **94**: 573–583.

Lu, H., Zawel, L., Fisher, L., Egly, J.M. and Reinberg, D. (1992) Human general transcription factor IIH phosphorylates the *C*-terminal domain of RNA polymerase II. *Nature* **358**: 641–645.

Lu, H., Flores, O., Weinmann, R. and Reinberg, D. (1991) The nonphosphorylated form of RNA polymerase II preferentially associates with the preinitiation complex. *Proc. Natl Acad. Sci. USA* **88**: 10004–10008.

Martinez, E., Kundu, T.K., Fu, J. and Roeder, R.G. (1998) A human SPT3-TAFII31-GCN5-L-acetylase complex distinct from transcription factor IID. *J. Biol. Chem.* **273**: 23781–23785.

Maxon, M.E. and Tjian, R. (1994) Transcriptional activity of transcription factor IIE is dependent on zinc binding. *Proc. Natl Acad. Sci. USA* **91**: 9529–9533.

McCracken, S. and Greenblatt, J. (1991) Related RNA polymerase-binding regions in human RAP30/74 and *Escherichia coli* sigma 70. *Science* **253**: 900–902. [Published erratum appears in *Science* (1992); 255(5049): 1195].

Means, A.L., Slansky, J.E., McMahon, S.L., Knuth, M.W. and Farnham, P.J. (1992) The HIP1 binding site is required for growth regulation of the dihydrofolate reductase gene promoter. *Mol. Cell. Biol.* **12**: 1054–1063.

Mizzen, C.A., Yang, X.J., Kokubo, T., *et al.* (1996) The TAF(II)250 subunit of TFIID has histone acetyltransferase activity. *Cell* **87**: 1261–1270.

Moreland, R.J., Hanas, J.S., Conaway, J.W. and Conaway, R.C. (1998) Mechanism of action of RNA polymerase II elongation factor Elongin. Maximal stimulation of elongation requires conversion of the early elongation complex to an Elongin-activable form. *J. Biol. Chem.* **273**: 26610–26617.

Myers, L.C., Gustafsson, C.M., Bushnell, D.A., *et al.* (1998) The Med proteins of yeast and their function through the RNA polymerase II carboxy-terminal domain. *Genes Dev.* **12**: 45–54.

Nikolov, D.B., Chen, H., Halay, E.D., *et al.* (1995) Crystal structure of a TFIIB-TBP-TATA-element ternary complex. *Nature* **377**: 119–128.

Nonet, M.L. and Young, R.A. (1989) Intragenic and extragenic suppressors of mutations in the heptapeptide repeat domain of *Saccharomyces cerevisiae* RNA polymerase II. *Genetics* **123**: 715–724.

Ohkuma, Y., Sumimoto, H., Horikoshi, M. and Roeder, R.G. (1990) Factors involved in specific transcription by mammalian RNA polymerase II: purification and characterization of general transcription factor TFIIE. *Proc. Natl Acad. Sci. USA* **87**: 9163–9167.

Orphanides, G., Lagrange, T. and Reinberg, D. (1996) The general transcription factors of RNA polymerase II. *Genes Dev.* **10**: 2657–2683.

Ozer, J., Mitsouras, K., Zerby, D., Carey, M. and Lieberman, P.M. (1998) Transcription factor IIA derepresses TATA-binding protein (TBP)-associated factor inhibition of TBP–DNA binding. *J. Biol. Chem.* **273**: 14293–14300.

Parvin, J.D. and Sharp, P.A. (1993) DNA topology and a minimal set of basal factors for transcription by RNA polymerase II. *Cell* **73**: 533–540.

Prelich, G. (1997) *Saccharomyces cerevisiae* BUR6 encodes a DRAP1/NC2alpha homolog that has both positive and negative roles in transcription *in vivo*. *Mol. Cell. Biol.* **17**: 2057–2065.

Prelich, G. and Winston, F. (1993) Mutations that suppress the deletion of an upstream activating sequence in yeast – involvement of a protein kinase and histone-H3 in repressing transcription *in vivo*. *Genetics* **135**: 665–676.

Pugh, B.F. (1996) Mechanisms of transcription complex assembly. *Curr. Opin. Cell Biol.* **8**: 303–311.

Pugh, B.F. and Tjian, R. (1991) Transcription from a TATA-less promoter requires a multisubunit TFIID complex. *Genes Dev.* **5**: 1935–1945.

Qureshi, S.A., Khoo, B., Baumann, P. and Jackson, S.P. (1995) Molecular cloning of the transcription factor TFIIB homolog from *Sulfolobus shibatae*. *Proc. Natl Acad. Sci. USA* **92**: 6077–6081.

Reines, D., Conaway, J.W. and Conaway, R.C. (1996) The RNA polymerase II general elongation factors. *Trends Biochem. Sci.* **21**: 351–355.

Reines, D., Ghanouni, P., Gu, W., Mote, Jr., J. and Powell, W. (1993) Transcription elongation by RNA polymerase II: mechanism of SII activation. *Cell Mol. Biol. Res.* **39**: 331–338.

Roberts, S.G. and Green, M.R. (1994) Activator-induced conformational change in general transcription factor TFIIB. *Nature* **371**: 717–720.

Roy, R., Adamczewski, J.P., Seroz, T., *et al.* (1994) The MO15 cell cycle kinase is associated with the TFIIH transcription-DNA repair factor. *Cell* **79**: 1093–1101.

Roy, A.L., Malik, S., Meisterernst, M. and Roeder, R.G. (1993) An alternative pathway for transcription initiation involving TFII-I. *Nature* **365**: 355–359.

Ruppert, S. and Tjian, R. (1995) Human TAFII250 interacts with RAP74: implications for RNA polymerase II initiation. *Genes Dev.* **9**: 2747–2755.

Selleck, S.B. and Majors, J. (1987) Photofootprinting *in vivo* detects transcription-dependent changes in yeast TATA boxes. *Nature* **325**: 173–177.

Serizawa, H., Makela, T.P., Conaway, J.W., Conaway, R.C., Weinberg, R.A. and Young, R.A. (1995) Association of Cdk-activating kinase subunits with transcription factor TFIIH. *Nature* **374**: 280–282.

Serizawa, H., Conaway, J.W. and Conaway, R.C. (1994) An oligomeric form of the large subunit of transcription factor (TF) IIE activates phosphorylation of the RNA polymerase II carboxyl-terminal domain by TFIIH. *J. Biol. Chem.* **269**: 20750–20756.

Shilatifard, A., Conaway, J.W. and Conaway, R.C. (1997) Mechanism and regulation of transcriptional elongation and termination by RNA polymerase II. *Curr. Opin. Genet. Dev.* **7**: 199–204.

Smale, S.T. and Baltimore, D. (1989) The 'initiator' as a transcription control element. *Cell* **57**: 103–113.

Sopta, M., Burton, Z.F. and Greenblatt, J. (1989) Structure and associated DNA-helicase activity of a general transcription initiation factor that binds to RNA polymerase II. *Nature* **341**: 410–414.

Thompson, C.M. and Young, R.A. (1995) General requirement for RNA polymerase II holoenzymes *in vivo*. *Proc. Natl Acad. Sci. USA* **92**: 4587–4590.

Timmers, H.T. (1994) Transcription initiation by RNA polymerase II does not require hydrolysis of the beta-gamma phosphoanhydride bond of ATP. *EMBO J.* **13**: 391–399.

Usheva, A. and Shenk, T. (1994) TATA-binding protein-independent initiation: YY1, TFIIB, and RNA polymerase II direct basal transcription on supercoiled template DNA. *Cell* **76**: 1115–1121.

Usheva, A., Maldonado, E., Goldring, A., *et al.* (1992) Specific interaction between the nonphosphorylated form of RNA polymerase II and the TATA-binding protein. *Cell* **69**: 871–881.

Verrijzer, C.P., Yokomori, K., Chen, J.L. and Tjian, R. (1994) *Drosophila* $TAF_{II}150$: similarity to yeast gene TSM-1 and specific binding to core promoter DNA. *Science* **264**: 933–941.

Wyrick, J.J., Holstege, F.C., Jennings, E.G., *et al.* (1999) Chromosomal landscape of nucleosome-dependent gene expression and silencing in yeast. *Nature* **402**: 418–421.
Xie, X., Kokubo, T., Cohen, S.L., *et al.* (1996) Structural similarity between TAFs and the heterotetrameric core of the histone octamer. *Nature* **380**: 316–322.
Yokomori, K., Verrijzer, C.P. and Tjian, R. (1998) An interplay between TATA box-binding protein and transcription factors IIE and IIA modulates DNA binding and transcription. *Proc. Natl Acad. Sci. USA* **95**: 6722–6727.
Zenzie-Gregory, B., Khachi, A., Garraway, I.P. and Smale, S.T. (1993) Mechanism of initiator-mediated transcription: evidence for a functional interaction between the TATA-binding protein and DNA in the absence of a specific recognition sequence. *Mol. Cell. Biol.* **13**: 3841–3849.

2

Regulatory transcription factors and *cis*-regulatory regions

Albert J. Courey

1. Introduction

In eukaryotes, the transcription of protein-encoding genes is directed by two classes of factors, the basal transcriptional apparatus and regulatory factors. The basal apparatus (reviewed in Chapter 1) is operationally defined as that set of factors required for accurately initiated basal transcription in a cell-free transcription system. The basal apparatus, which includes pol II and a number of general transcription factors, is directed to the 5′ end of a transcription unit by the core promoter. Although core promoters can direct accurate initiation *in vitro*, they do not generally direct significant levels of transcription *in vivo*, especially in the context of an intact eukaryotic chromosome. Measurable transcription *in vivo* generally requires the action of regulatory factors, which interact either directly or indirectly with *cis*-regulatory elements and modulate the efficiency of transcription from linked core promoters.

Regulatory transcription factors can be grouped into two broad categories, the sequence-specific regulators and the coregulators (Berk, 1999; Mitchell and Tjian, 1989; Tjian and Maniatis, 1994; Triezenberg, 1995). As their name implies, sequence-specific regulators interact directly with *cis*-regulatory elements in a sequence-specific manner. In eukaryotic cells, these factors are generally modular in nature. In other words, they consist of independent structural domains that are separately responsible for the various biochemical functions of the factors. Sequence-specific regulators can be further subdivided into activators, which stimulate core promoter utilization, and repressors, which inhibit core promoter utilization. It is not always straightforward to classify a sequence-specific regulator as an activator or a repressor, since many such factors can mediate either activation or repression depending upon the context of their binding sites, or upon the array of interacting ligands available to modulate their conformation and/or activity. In contrast to sequence-specific regulators, coregulators do not interact with DNA directly. Rather,

Transcription Factors, edited by J. Locker.

they are directed to regulatory targets by protein–protein interactions with sequence-specific regulators. Like sequence-specific regulators, coregulators can be further categorized according to whether their role is in gene activation, in which case they are termed coactivators, or in gene repression, in which case they are termed corepressors.

The *cis*-regulatory regions with which regulatory factors interact can be extremely complex in nature. Regulatory transcription factor binding sites are usually grouped together into regulatory modules. Such modules are referred to as enhancers, if their major role is in transcriptional activation, or as silencers, if their major role is in transcriptional repression. Many genes contain multiple enhancers and/or silencers, which can be spread out over extremely large regions (100 kilobases or more). In any given complex locus, different regulatory modules can be responsible for controlling transcription at different times during development, at different locations in a complex organism, or in response to different environmental stimuli.

2. Regulatory transcription factors

2.1 The modular nature of sequence-specific regulators

The modular nature of eukaryotic sequence-specific regulators was first recognized through studies of yeast activators such as Gcn4 (Hope and Struhl, 1986) and Gal4 (Ma and Ptashne, 1987a), and mammalian activators such as glucocorticoid receptor (Miesfeld *et al.*, 1987) and Sp1 (Courey and Tjian, 1988). These studies showed that the factors contained separable DNA-binding and activation domains, which could be mixed and matched in various domain-swapping experiments to generate functional chimeric transcription factors. Although we now take the modularity of these factors for granted, the initial discovery of this phenomenon came as a surprise. This was perhaps because of the impression left from studies of enzyme structure suggesting that proteins generally consisted of single cooperatively folding units that were stabilized by long-distance tertiary interactions. In addition, the modularity of eukaryotic factors appeared to stand in contrast to what had been learned from studies of certain paradigmatic prokaryotic transcriptional regulators, such as λ-repressor, in which residues responsible for transcriptional activation were found to be intimately linked to residues responsible for sequence-specific DNA recognition (Ptashne, 1986). The modularity of eukaryotic regulatory factors is likely a manifestation of the phenomenon of exon shuffling, whereby complex multifunctional proteins are thought to have evolved by the joining together of genes encoding more primitive monofunctional proteins (de Souza *et al.*, 1996).

As the foregoing discussion implies, most eukaryotic sequence-specific regulators consist of a single, independently folding DNA-binding domain as well as one or more independently folding regulatory domains. Regulatory domains can be further categorized as activation domains or repression domains.

2.2 DNA-binding domains

In contrast to prokaryotic transcription factors, which generally recognize *cis*-elements using a highly conserved helix-turn-helix motif, eukaryotic sequence-specific regulators utilize a large variety of structural motifs to interact with specific DNA sequences. Several of the most important examples of these motifs are discussed in detail in Chapter 4, and so the structure of DNA-binding domains will not be covered comprehensively here. Suffice to say that, although there are important exceptions, the majority of these motifs represent solutions to the problem of how to position an α-helix in the major groove of the recognition element in such a way that enables stereo-specific contacts to be made, including hydrogen-bond contacts. While these protein–DNA interactions can be of moderately high affinity, the specificity of the interaction between a single factor and its recognition element is not sufficient to account for the specificity with which genes are selected for regulation. Rather, as will be discussed further shortly, the specificity of target gene selection appears to depend upon the need to assemble a large DNA-bound protein complex (an enhanceosome) to modulate gene activity.

Although a multitude of different DNA-binding motifs (see Chapter 4) are used in recognition of *cis*-regulatory elements, certain structural motifs, which can be recognized on the basis of sequence homology, occur with high frequency, e.g., the Cys_2His_2 zinc-finger domain, the $(Cys_4)_2$ zinc-finger domain, the homeodomain, the basic-region-leucine-zipper (bZip) domain, and the basic-helix-loop-helix (bHLH) domain. Sequence-specific factors are usually grouped according to the class of DNA-binding motif they contain. This is justified, in part, because the DNA-binding motifs are usually the only parts of the factors that show significant sequence conservation across phyla. Thus, for example, the products of the *Drosophila* homeotic gene complex (*HOM-C*) comprise a set of homeodomain-containing transcription factors. These factors contain *bona fide* orthologs among the products of the vertebrate *HOX* genes. However, outside the homeodomain, there is very little homology between the *Drosophila HOM*-C encoded transcription factors and the vertebrate *HOX* gene products (Krumlauf, 1994).

2.3 Activation domains

As mentioned earlier, regulatory domains are of two varieties, activation domains and repression domains. A wealth of complex and sometimes contradictory information is available about activation domains, and this will be the topic of this section. Little information is available about the structure of repression domains, which will therefore not be discussed.

Although the exact mechanisms of activation are still hotly debated, it is generally agreed that activation domains usually provide interfaces for protein–protein interactions. (See Sections 2.4 and 2.5 for activation domain targets.) Given this role in protein–protein interactions, one might expect activation domains, like DNA-binding domains, to contain a relatively small number of conserved structural motifs that could easily be recognized on the basis of sequence homology, with different structural motifs interacting with different target proteins. However, this does not appear to be the case. As previously discussed, even when one compares orthologous transcription factors from different phyla, the extent of sequence conservation among the regulatory domains is extremely low.

There are many possible reasons for this lack of sequence and therefore of structural conservation. As we will see, activation domains are often involved in bringing ('recruiting') various target proteins (such as coactivators) to the template. It is possible that high-affinity, high-specificity interactions are simply not required for coactivator recruitment. Coactivators are often very large multi-polypeptide complexes with many protein–protein interaction surfaces. Recruitment of coactivators may involve multiple, synergistic, low-affinity interactions between sequence-specific activators and diverse surfaces on the coactivators. Given this possibility, high-affinity coactivator recruitment by a single sequence-specific factor may actually be undesirable since it might abrogate the possibility of synergy.

Although there is little sequence homology *per se* among activation domains, they are often characterized by a high content of a particular type of amino acid (Mitchell and Tjian, 1989; Triezenberg, 1995). Many activation domains, for example, tend to be rich in the negatively charged amino acids glutamate and aspartate. Such negatively charged activation domains are referred to as 'acidic activation domains'. Early experiments in which random amino acid sequences were tested for activation potential in yeast cells originally led to the conclusion that negative charge might be a feature of all activation domains (Ma and Ptashne, 1987b). However, since that time a wide variety of activation domains have been discovered that are not particularly rich in glutamate or aspartate. Instead, these domains are often characterized by a preponderance of some other residue such as glutamine or proline. The significance of this richness in a particular amino acid remains totally obscure. In fact, the mutational analyses of these domains suggests that the most important amino acids are not the acidic amino acids in acidic activation domains or the glutamines in glutamine-rich activation domains. Rather, in each case, it appears to be specific bulky hydrophobic amino acids that are most sensitive to mutagenesis (Gill *et al.*, 1994; Regier *et al.*, 1993; Triezenberg, 1995). The hydrophobic nature of the side-chains of these residues seems to be their most important characteristic since they can be replaced with other bulky hydrophobic residues with little effect on function, but not with polar amino acids. It is not clear if these hydrophobic amino acids are important because they directly mediate contacts with activation domain targets or because they help to stabilize the structure of the domain.

A number of attempts to characterize activation domains by biophysical and genetic approaches have led to the conclusion that these domains may lack a well-defined secondary and tertiary structure. For example, the absence of an essential long-range tertiary structure is supported by the results of deletion analyses, which show that many activation domains contain multiple short segments with activating potential that work together in a more or less additive manner (Courey and Tjian, 1988; Hope and Struhl, 1986; Ma and Ptashne, 1987a; Regier *et al.*, 1993; Walker *et al.*, 1993). Such findings suggest that activation domains are not like typical globular proteins in which a well defined tertiary structure is critical for function. Biophysical analyses of the acidic activation domains of yeast Gcn4, yeast Gal4 and the viral transactivator VP16 using approaches such as nuclear magnetic resonance (NMR), circular dichroism and fluorescence spectroscopy, suggest that these domains are essentially disordered in aqueous solution at physiological pH, although, in a number of cases, the formation of a secondary structure can be induced by lowering the pH or by introducing organic solvents into the media (Donaldson and Capone, 1992; Shen *et al.*, 1996a; van Hoy *et al.*, 1993).

How are we to reconcile the ability of activation domains to bind their targets specifically with this apparent lack of structure? The most likely explanation is that activation domains do adopt ordered structures, but only when liganded to specific protein targets. This idea fits with the ability of these domains to assume ordered structures in hydrophobic solvents. These solvents might mimic the environment an activation domain experiences when it interfaces with a hydrophobic surface on another protein.

Direct evidence that activation domains assume ordered conformations upon binding to other proteins comes from a number of biophysical analyses. For example, the cAMP response element-binding protein (CREB) contains a 60-residue-long protein kinase A-inducible activation domain (KID). Phosphorylation of this domain at serine 133 promotes binding to the coactivator, CREB-binding protein (CBP). In the absence of CBP, NMR analysis reveals that phosphorylated KID (pKID) is a random coil. Upon binding to CBP, a central ~25-residue-long region of pKID assumes an ordered conformation consisting of two α-helices with about a 90° turn between them. The second of these helices is amphipathic and its hydrophobic face packs into a hydrophobic groove on the surface of CBP (Radhakrishnan *et al.*, 1997) (*Figure 1*). How does phosphorylation of serine 133 in KID stabilize the complex? It was originally speculated that the negative charge on the phosphate group might convert KID into an acidic activation domain. However, this is unlikely to be the case since substitution of serine 133 with aspartate or glutamate does not lead to a constitutively active domain. Rather, the NMR structure suggests that the most important role of the phosphate group is to accept a hydrogen bond from a tyrosine residue in CBP, thereby stabilizing the pKID/CBP complex.

Other activation domains also appear to form amphipathic α-helices that pack against hydrophobic surfaces on effector proteins. For example, a portion of the activation domain of the p53 tumor suppressor protein has been crystallized in a complex with the murine double minute 2 (Mdm2) oncoprotein, a negative effector of p53 (Kussie *et al.*, 1996). In the absence of Mdm2, this region of p53 is probably unstructured, while in the presence of Mdm2 it forms an amphipathic α-helix that packs into a deep hydrophobic trough in the oncoprotein. Similarly, analysis of the VP16 acidic activation domain reveals that this domain, while unstructured in the absence of a ligand, assumes an α-helical conformation upon binding to certain subunits of TFIID (Shen *et al.*, 1996b; Uesugi *et al.*, 1997).

Thus, in the limited number of instances where structural information about activation domains is available, an effector protein is found to induce the formation of an amphipathic α-helix that packs against an apolar surface on the effector protein. Given the diversity of activation domains, and given that an activation domain is really, at heart, nothing more or less than a protein–protein interaction domain, it is likely that other structural motifs besides amphipathic α-helices will be found to function as activation domains.

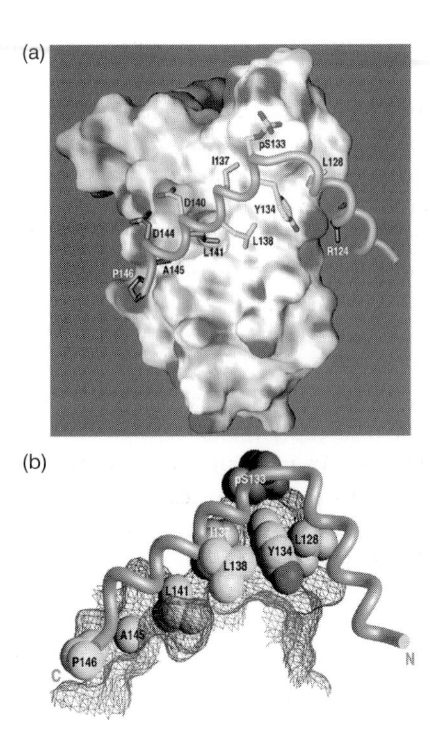

Figure 1. Structure of the complex between the phosphorylated protein kinase A inducible activation domain (pKID) of cAMP response element-binding protein (CREB) and the pKID interaction domain (KIX) in the coactivator, CREB-binding protein (CBP). Taken from (Radhakrishnan *et al.*, 1997). (a) The diagram shows the molecular surface of KIX, the pKID backbone, and side-chains of pKID that interact with KIX. (b) Hydrophobic interactions between the second α-helix of pKID and the shallow hydrophobic groove on KIX. The hydrophobic side-chains (CPK spheres) and the backbone of pKID are shown. The side-chains of phosphoserine 133 (pS133 – the protein kinase A phosphorylation target) is also depicted. The mesh indicates the part of the KIX surface in close proximity to pKID. Note the complementarity between the interacting surfaces. Reproduced from Radhakrishnan, I. *et al.* (1997) Solution structure of the KIX domain of CBP bound to the transactivation domain of CREB: a model for activator:coactivator interactions. *Cell* 91: 741–752, with permission from Cell Press.

2.4 General transcription factors as activation domain targets

The target proteins with which activation domains interact seem to be extremely variable. For example, interactions have been reported between activation domains and nearly every general transcription factor including TBP, TFIIB, TFIIA, TFIIH, TFIIF and pol II (Triezenberg, 1995). The interactions between activation domains and general transcription factors appear to lead to transcriptional activation, at least in some cases, by directly recruiting the general transcription factors to the template thereby facilitating the formation of a preinitiation complex. A number of experiments suggest that the act of recruitment is, at least sometimes, sufficient for transcriptional activation. For example, when components of the general transcription machinery are artificially fused to DNA-binding domains to allow their direct recruitment to the template via a sequence-specific protein–DNA interaction, the activation domain becomes dispensable (Hampsey and Reinberg, 1999; Ptashne and Gann, 1997). Experiments looking at the interactions between the Gal4 activation domain and TFIIB suggest that the stimulation of recruitment may involve more than the simple binding of activation domains to general factors. In these experiments, protease footprinting suggests that an interaction between the Gal4 activation domain and TFIIB can, in addition to facilitating the recruitment of TFIIB, induce a hinge motion between the *N*- and *C*-terminal domains of TFIIB. This change in conformation may, in turn, facilitate the recruitment of the PolII/TFIIF by TFIIB (Roberts and Green, 1994; Roberts *et al.*, 1993).

While recruitment of the pre-initiation complex is clearly a critical step in promoter activation, a number of experiments suggest that there may be other rate-limiting steps in this process. In particular, certain mutations in the TATA box and in TFIIB have much more dramatic effects on transcriptional initiation than on pre-initiation complex assembly (Cho and Buratowski, 1999; Hoopes *et al.*, 1998; Jacob *et al.*, 1994; Ranish *et al.*, 1999). This suggests the existence of post-recruitment steps in promoter activation that could well be regulated by activation domain interactions.

2.5 Coactivators as activation domain targets

In addition to interacting with components of the general transcription machinery, activation domains interact with coactivators, thereby recruiting them to the template (Berk, 1999). The coactivators recruited in this way can then activate transcription by at least two means. First, they can serve to mediate interactions between activators and the general transcription machinery, thereby helping to recruit or activate the pre-initiation complex. Second, they can serve to catalyze covalent or non-covalent changes in chromatin structure that somehow alter the accessibility of the template to the general transcription machinery.

The first group of proteins that were described as coactivators, and which are thought to mediate interactions between activators and the general transcription factor TBP, are the TBP-associated factors found in TFIID (the TAF_{II}s) (Verrijzer and Tjian, 1996). It is somewhat arbitrary to refer to these proteins as coactivators rather than as components of the general transcription machinery. According to the operational definition of general transcription factors given at the beginning of this chapter, TAF_{II}s are not general transcription factors since they are not usually required for optimal basal transcription in cell-free transcription systems. However, they are tightly associated with TBP, a basal transcription factor, and recent evidence suggests that they may play roles in core promoter recognition, a function normally associated with the general transcription machinery.

The role of TAF_{II}s as coactivators is suggested by a large number of *in vitro* transcription experiments in which the ability of recombinant TBP to substitute for TFIID (consisting of TBP plus about eight TAF_{II}s) in mediating activated transcription was examined (Verrijzer and Tjian, 1996). In these experiments, it was generally observed that, while TBP was capable of supporting basal transcription, TAF_{II}s, in addition to TBP, were required for activated transcription. Different activators were found to bind to different TAF_{II}s, and the TAF_{II}s bound by an activator were found to be required for activation by that activator. For example, the *Drosophila* activator NTF1 was found to bind TAF_{II}150, which is anchored to TBP by TAF_{II}250. In accord with these binding interactions, it was discovered that activation by NTF1 requires, at a minimum, TBP, TAF_{II}250 and TAF_{II}150. In contrast, human Sp1, which binds to TAF_{II}110, was found to additionally require TAF_{II}110 for activation (Chen *et al.*, 1994). In accord with the idea that TAF_{II}s are coactivators, recent genetic analysis indicates that TAF_{II}s are required *in vivo* for the activated transcription of a wide variety of genes in yeast (Berk, 1999) and *Drosophila* (Pham *et al.*, 1999; Zhou *et al.*, 1998).

Another protein complex that has the properties of a coactivator and that is thought to work by mediating interactions between the general machinery and activators is the Mediator complex (Berk, 1999; Hampsey and Reinberg, 1999). Once again, it is not completely clear if this complex should be considered a coactivator or a component of the general machinery since all or some of the polypeptides in the Mediator complex may associate with pol II as components of the pol II holoenzyme (see Chapter 1). The Mediator was first discovered during fractionation of a yeast transcription extract to look for factors that could support transcriptional activation by the VP16 activation domain (Flanagan *et al.*, 1991). This large complex contains a multitude of polypeptides including a subset of the SRB proteins (Kim *et al.*, 1994; Myers *et al.*, 1998), which are subunits of the Pol II holoenzyme (Koleske and Young, 1994). Like the SRB complex, the Mediator binds to Pol II through the C-terminal domain of the largest pol II subunit. Evidence that the Mediator complex is a *bona fide* coactivator *in vivo* comes from genetic analysis in yeast. This showed that mutations in certain mediator subunits resulted in defects in activated transcription of many genes.

Recently, a variety of multi-subunit coactivator complexes have been discovered in mammalian systems that contain subunits homologous to those found in the yeast mediator. The various mammalian complexes contain subunits in common with one another as well as subunits that are specifically associated with an individual complex (Berk, 1999; Hampsey and Reinberg, 1999). Like TAF_{II}s, subunits of these complexes can interact directly with activation domains. For example, human Sur2, a component of a human mediator complex, was purified from a crude nuclear protein fraction on the basis of its affinity for the activation domain in the adenovirus E1A protein (Boyer *et al.*, 1999). It thus appears that the mediator can be recruited by protein–protein interactions with activators.

What is the functional relationship between the TAF_{II} complex and the Mediator complex? Genetic analysis suggests that both complexes are required for the transcription of large, probably overlapping, subsets of genes (Berk, 1999; Hampsey and Reinberg, 1999). Biochemical analysis also suggests that many activators (e.g., Sp1) require both TAF_{II}s and a Mediator-like complex in order to activate transcription (Ryu *et al.*, 1999). Perhaps providing activators with multiple alternative targets ensures that factors interacting with regulatory elements organized in many different ways will be able to cooperate with one another to regulate transcription. In other words, these multiple interaction surfaces may provide a myriad of possibilities for intricately integrating the multiple regulatory inputs that determine the transcriptional state of a gene.

As previously mentioned, while some coactivators such as TAF_{II}s and the Mediator are thought to serve as adapters between the general machinery and activators, other coactivators work to recruit chromatin-modifying activities to the template. Principal among these chromatin-modifying coactivators are the complexes containing histone acetyltransferase activity as well as the ATP-dependent SWI/SNF-type chromatin remodeling complexes. This family of coactivators will not be discussed further here as they are extensively reviewed in Chapters 1 and 3. It is, however, worth mentioning that the line between mediator-type and chromatin-modifying-type coactivators is not always easy to draw. For example, the coactivator CBP has histone acetyltransferase activity and may also serve as an adapter between activators and the general machinery. TAF_{II}250, which, as a component of TFIID, is thought to mediate interactions with the general machinery, has also been found to possess histone acetyltransferase activity.

3. Organization of *cis*-acting regulatory regions and combinatorial control

Eukaryotic DNA-binding regulatory factors bind to a variety of specific recognition elements with varying affinities. Many of these would be expected to occur by chance with high frequency. The sites for families of factors can usually be distinguished by characteristic central motifs. For example, many bHLH domain-containing factors bind to sites containing the motif CANNTG (Garrell and Campuzano, 1991), although the full binding sites for specific bHLH dimers are somewhat larger (Ellenberger *et al.*, 1994). Similarly, ETS domain factors bind to fairly degenerate sequences based on a GGAA/T motif (Sharrocks *et al.*, 1997), and most homeodomains recognize a very similar set of recognition elements in which the most important feature is a core TAAT sequence (Gehring *et al.*, 1994). Other types of factors have larger, more specific binding sites. These include zinc-finger factors such as Sp1, which binds to a 10-base-pair site with high affinity (Kriwacki *et al.*, 1992), and nuclear receptor dimers which bind 12-base-pair sites (Luisi *et al.*, 1991). Even in the case of these longer high affinity sites, considerable base substitution is tolerated. On the basis of simple probability, short binding sites and degenerate weak sites can probably be found in virtually every gene in a eukaryotic genome. So how can regulatory factors have gene-specific effects on rates of transcription? The answer to this question almost certainly lies in the fact that transcription factors rarely act in isolation. A single factor binding to a single site in a gene is almost never sufficient to result in the activation of that gene. Rather, the transcriptional activity of any given gene seems to be dictated by combinations of factors acting together. This combinatorial control provides the means to ensure specificity of gene regulation, and, in addition, provides mechanisms to integrate multiple inputs into the decision to turn a gene on or off.

3.1 Combinatorial interactions and developmental complexity

In multicellular organisms, developmental complexity arises, for the most part, via the integration of initially simple spatially restricted patterns of transcription factor activity to generate ever more complex patterns of gene activity. Perhaps the best studied examples of this process are the mechanisms by which segmentally repeating stripes of gene activity are generated during the development of the *Drosophila* embryo. For example, the *even-skipped* (*eve*) gene is expressed in the early embryo in the classic seven transverse stripe pattern characteristic of most pair-rule genes. The expression of pair-rule genes such as *eve* in this striped pattern begins even before the cellular blastoderm has completely formed and is the first evidence of the metameric nature of the organism (Ingham, 1988). To a first approximation, each *eve* stripe is generated by the activity of a separate stripe enhancer. Each such enhancer is a composite of binding sites for multiple, spatially restricted transcription factors that work in combination to generate the stripe of *eve* expression (Gray and Levine, 1996a).

The best characterized stripe enhancer is the one that directs the second stripe of *eve* expression (Gray and Levine, 1996a; Small *et al.*, 1991). This regulatory module contains binding sites for two activators, Bicoid and Hunchback, both of which are required to activate transcription of the *eve* stripe. Since Bicoid and Hunchback are both distributed throughout the anterior half of the embryo, they are, in principal, capable of activating transcription throughout this region. However, the second stripe enhancer also contains binding sites for at least two repressor proteins that are capable of overriding activation by Bicoid and Hunchback. One of these, Giant, is present at the anterior end of the embryo and sets of the anterior boundary of the *eve* stripe, while the other, Krüppel, is present in a broad central domain, and sets the posterior border of the *eve* stripe. Thus, through combinatorial control mechanisms, four broadly expressed regulatory factors can generate a narrow stripe of gene expression (*Figure 2*).

The action of the *eve* second stripe enhancer is dependent upon at least two distinct forms of combinatorial control, transcriptional synergy and quenching. In transcriptional synergy, two factors (in the case of *eve*, Bicoid and Hunchback) work together in a greater than additive manner to activate a promoter. Requiring multiple factors for activation helps to guard against inappropriate gene activation that could occur due to the chance occurrence of binding sites for various factors in the genome. Possible molecular mechanisms of synergy will be discussed in the section on enhanceosomes. Quenching is a form of transcriptional repression in which a repressor protein is able to block activation by activators bound to elements located within perhaps 100 base-pairs of the DNA-bound repressor. For example, in the *eve* stripe 2 enhancer, all the known activation elements are within 100 base-pairs of both Krüppel and Giant sites, thus ensuring that the activators will be unable to activate transcription in the domains of Krüppel and Giant activity (Arnosti *et al.*, 1996; Gray and Levine, 1996b). The short-range nature of this repression is essential to the proper functioning of the extremely complex *eve* control region, which contains multiple stripe enhancers. If Krüppel and Giant were able to repress transcription at large distances, they would interfere with the function of other stripe enhancers in the locus (Gray *et al.*, 1994; Small *et al.*, 1993). While the phenomenon of quenching is well established, the molecular mechanism of quenching remains a mystery.

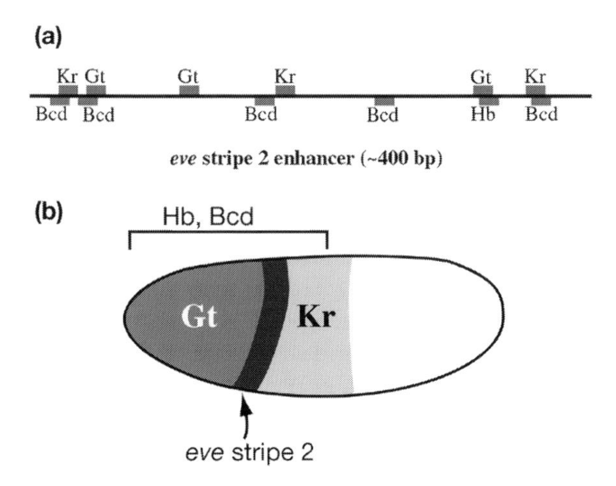

Figure 2. Combinatorial control of *eve* transcription. (a) A diagram of the *eve* stripe 2 enhancer showing the approximate position of binding sites for the four sequence-specific regulators, Krüppel (Kr), Giant (Gt), Bicoid (Bcd), and Hunchback (Hb). Bcd and Hb are activators that work together to synergistically activate transcription through the enhancer. Kr and Gt are repressors that are able to quench activation by Bcd and Hb. Note that each activator-binding site is within ~100 base-pairs of binding sites for each of the repressors. This is essential to allow quenching. (b) Expression patterns of Kr, Gt, Bcd and Hb relative to the position of the *eve* second stripe. In this diagram of an early *Drosophila* embryo, anterior is to the left. The activators Bcd and Hb are present throughout the anterior half of the embryo. By quenching Bcd/Hb activated transcription at the anterior end of the embryo, Gt sets the anterior border of *eve* stripe 2. By quenching transcription in the center of the embryo, Kr sets the posterior border of *eve* stripe 2.

3.2 Combinatorial interactions and signal transduction

Just as complex regulatory control regions provide a mechanism for integrating spatial information, so too do they provide a means of integrating information about the cellular environment. For example, the interferon-β enhancer contains single, closely spaced binding sites for many transcription factors, including the cAMP inducible factor, activating transcription factor 2 (ATF-2)/c-Jun, the interferon-γ inducible factor, interferon regulatory factor 1 (IRF-1), and the tumor necrosis factor-α inducible factor NF-κB (Maniatis *et al.*, 1992; Thanos and Maniatis, 1995). This enhancer responds only very weakly to cAMP, interferon-γ, and tumor-necrosis factor-α because the transcriptional response to these agents normally requires enhancers containing multiple binding sites for the appropriate inducible factor. All three of the factors just mentioned can also activate gene expression in response to viral infection, perhaps because viral infection somehow mimics the intracellular environment that results from the physiological signaling agents. The multiple virus inducible factors bound to the interferon-β enhancer direct a robust transcriptional response to viral infection due to their ability to interact with one another in a synergistic fashion. Thus, by combining binding sites for factors normally involved in diverse signaling processes, it is possible to generate an enhancer with a completely different pattern of responses.

Another means by which combinatorial interactions result in the integration of information about the environment is through the formation of transcription factor heterodimers. Many DNA-binding domains include surfaces for homo- or heterodimerization in addition to surfaces for making stereospecific contacts with DNA. Excellent examples of this are provided by the nuclear hormone receptors, which are discussed extensively in Chapter 8. Another good example is provided by the AP-1 family of bZip domain-containing transcription factors. The AP-1 family consists of Jun subfamily proteins and Fos subfamily proteins (Karin *et al.*, 1997). Jun subfamily members heterodimerize with high affinity to Fos subfamily members and the heterodimers bind with high specificity to a specific hexanucleotide sequence. In the *Drosophila* embryo, the Jun/Fos heterodimer is involved in the spatial and temporal regulation of *decapentaplegic* (*dpp*) expression in the so-called 'leading-edge cells' at the edge of the dorsal epidermal cell sheet (Riesgo-Escovar and Hafen, 1997; Zeitlinger *et al.*, 1997). Activation of *dpp* in these cells results in the migration of this epithelial sheet over the adjacent amnioserosa thereby leading to dorsal closure. Jun and Fos are expressed in broad patterns that overlap in the leading-edge cells ensuring that both factors are present in these cells. Furthermore, activation of Jun depends upon a phosphorylation cascade that is activated only at the correct time during development, perhaps by a signal emanating from the adjacent amnioserosa cells. Thus, a combinatorial interaction between Jun and Fos helps to integrate spatial and temporal information and to coordinate a complex cellular process. The role of Jun/Fos heterodimers in epithelial sheet movement may be widespread. For example, Fos is thought to play a role in wound healing, and both Jun and Fos are required for thorax closure during *Drosophila* metamorphosis (Zeitlinger and Bohmann, 1999).

3.3 Enhanceosomes and the mechanism of transcriptional synergy

Recent studies on a number of enhancers are beginning to suggest a mechanism by which multiple factors can act together to activate promoters synergistically. In particular, it appears that many enhancers may serve to nucleate the formation of a large nucleoprotein complex called an enhanceosome consisting of the enhancer DNA, multiple DNA-bound sequence-specific regulatory factors, and additional 'architectural factors' (Carey, 1998). Enhanceosomes are thought to assemble as a single cooperative unit stabilized by multiple protein–protein and protein–DNA interactions. As a result of this cooperativity, the absence of a single factor greatly destabilizes the enhanceosome and thus partially explains the mechanism of transcriptional synergy.

Perhaps the best characterized enhanceosome is the one that directs viral induction of the interferon-β gene. It was mentioned earlier that the enhancer of this gene, which is about 100 base-pairs in length, contains binding sites for at least three sequence-specific factors (Maniatis *et al.*, 1992). Mutagenesis of the binding site for any one of these factors dramatically reduces enhancer activity. Furthermore, insertion of a half-helical turn of DNA between the binding sites for any two factors greatly reduces transcriptional synergy, while insertion of a full helical turn does not (Thanos and Maniatis, 1995). Similarly, the cooperative binding of transcription factors (measured *in vitro*) to the enhancer is negatively impacted by the insertion of a half-helical turn of DNA between transcription factor-binding sites. These findings suggest that transcriptional synergy requires the stereo-specific alignment of multiple factors on the same face of the enhancer and that this allows adjacently bound factors to interact with one another to promote enhanceosome formation (*Figure 3*).

Both transcriptional synergy (measured *in vitro* and *in vivo*) and cooperative assembly of the interferon-β enhanceosome (measured *in vitro*) require the high mobility group I(Y) (HMGI(Y)) protein, an architectural factor (Kim and Maniatis, 1997; Thanos and Maniatis, 1995). This small basic protein appears to lack an activation domain and so probably does not participate in transcriptional activation directly. Rather, it facilitates assembly of the enhanceosome, at least in part, by binding to the minor groove of the DNA at certain sites within the enhancer and altering the curvature of the DNA. Specifically, it binds to the NF-

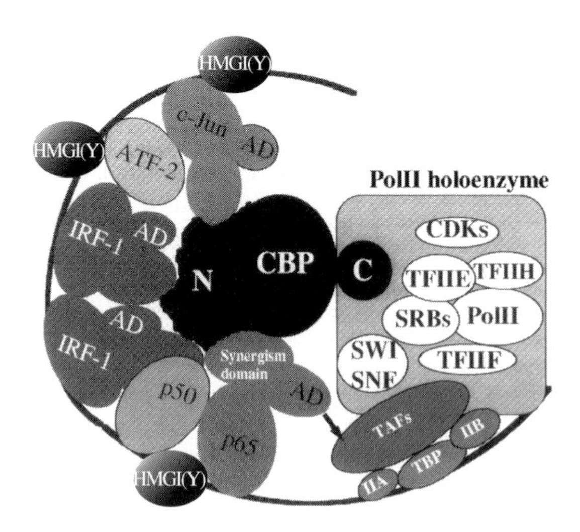

Figure 3. A model of the interferon-β (IFNβ) enhanceosome. Adapted from Merika *et al.*, 1998. Virus infection leads to the cooperative assembly of an enhanceosome containing the three activators, NF-κB (p50/p65 heterodimer), IRF-1, and the ATF2/c-Jun heterodimer as well as the architectural factor HMGI(Y). Regions of the three activators contact an N-terminal domain (N) of the coactivator, CREB-binding protein (CBP), which in turn interacts through its C-terminal domain (C) with the RNA polymerase II (Pol II) holoenzyme. NF-κB contains a region required for CBP recruitment (the synergism domain) that is distinct from another activation domain (AD) thought to play a role in transcription factor TFIID recruitment through its interaction with the TBP-associated factors (TAFs). The regulators also contact one another. HMGI(Y) probably facilitates the formation of this complex by altering the curvature of the DNA thereby increasing the affinity of the activators for DNA. By altering the curvature of the DNA, HMGI(Y) may also promote cooperative interactions between activators. The cooperative assembly of this large stereospecific complex probably accounts for the ability of the regulatory factors to synergistically activate transcription. In addition to polII and SRBs, other potential components of the pol II holoenzyme include the general factors TFIIE, H and F, the SWI/SNF complex and cyclin-dependent kinases (CDKs). Reproduced from Merika, M. *et al.* (1998) Recruitment of CBP/p300 by the IFN beta enhanceosome is required for synergistic activation of transcription. *Mol.Cell.* 1: 277–287, with permission from Cell Press.

κB and ATF-2/c-Jun sites and reduces or eliminates intrinsic bends in these sites that would otherwise disfavor transcription factor binding (Falvo *et al.*, 1995). In addition to favoring the binding of individual factors to their sites within the enhancer in this way, HMGI(Y) favors cooperative binding, perhaps by inducing subtle changes in the structure of the enhancer that favor interactions between adjacent DNA-bound factors. A different family of architectural factors, the so-called high motility group (HMG) box factors (note that HMGI(Y) is not an HMG box factor despite the similar name), facilitate enhanceosome formation at other enhancers such as the *TCRα* gene enhancer by inducing dramatic ~120° bends in the DNA that facilitate interactions between factors that are bound to sites separated by several turns of the DNA double helix (Giese *et al.*, 1995).

Recent studies on the interferon-β enhancer suggest that there may be mechanisms of synergistic activation beyond the cooperative assembly of the enhanceosome. In particular, the assembly of the enhanceosome may result in the formation of high-affinity protein surfaces for interaction with coactivators and/or components of the general machinery. For example, the coactivator CBP interacts independently with the three factors that bind to the enhancer, NF-κB, IRF-1, and ATF-2/c-Jun. Binding of these three factors to the enhancer appears to result in the formation of a high-affinity landing pad for the cooperative recruitment of CBP (Merika *et al.*, 1998). In addition, stereo-specific formation of the enhanceosome appears to be required for the recruitment of TFIIB and the pol II holoenzyme to the pre-initiation complex. The recruitment of the holoenzyme may be mediated by CBP (Kim *et al.*, 1998).

Just as interactions between the enhanceosome and the pre-initiation complex stabilize the pre-initiation complex, so too do these interactions appear to stabilize the enhanceosome. This reciprocity increases the size of the cooperative binding unit and thus the overall cooperativity of the transcriptional response (Carey, 1998; Ellwood *et al.*, 1999; Kim and Maniatis, 1997).

Transcriptional silencing also appears to involve the formation of multiprotein stereospecific complexes. This has perhaps been best studied in the case of the *zerknüllt* (*zen*) ventral repression region, a silencer that serves to mediate repression by the regulatory transcription factor, Dorsal, in the early embryo (Courey and Huang, 1995; Mannervik *et al.*, 1999). Dorsal can function as both an activator and a repressor of transcription (Rusch and Levine, 1996). Its ability to do both is essential for the correct dorsal/ventral patterning of the embryo. Multimerized Dorsal-binding sites direct transcriptional activation. Dorsal bound to these sites appears to activate transcription by recruiting multiple coactivators, including CBP and certain TAFs. In the *zen* ventral repression region, the Dorsal-binding sites are found in the proximity of a series of AT-rich elements (Jiang *et al.*, 1993; Kirov *et al.*, 1993). Proteins that bind to these AT-rich sites, including Dead-ringer, have the ability to convert Dorsal bound to the Dorsal sites from an activator to a repressor (Valentine *et al.*, 1998). This repression is largely dependent upon the corepressor, Groucho (Dubnicoff *et al.*, 1997). Both Dorsal and Dead-ringer bind to the corepressor Groucho, and, at least *in vitro*, are able to cooperatively recruit Groucho to the template. When Groucho or Dead-ringer are removed from the embryo, or when AT-rich sites are mutagenized, Dorsal bound to its nearby sites directs activation rather than repression.

Similarly, the insertion of a half-helical repeat of DNA between Dorsal sites and AT-rich sites interferes with transcriptional repression (Cai *et al.*, 1996). Thus, it appears that Dorsal and additional DNA-bound factors, including Dead-ringer, contribute to the formation of a high-affinity surface for the recruitment of Groucho. The ability of Groucho to repress transcription appears to depend upon protein–protein interactions with additional corepressor proteins, in particular the histone deacetylase Rpd3 (Chen *et al.*, 1999). These findings suggest that, just as transcriptional activation is often mediated by enhanceosomes, transcriptional repression can be mediated by 'silenceosomes'. Whether or not architectural factors are also required for silenceosome assembly remains to be determined. However, genetic analysis strongly suggests that Capicua is required for function of the *zen* silenceosome (Jiminez *et al.*, 2000). As an HMG box-containing protein, Capicua could well be serving an architectural role.

4. Conclusions

Eukaryotic cells contain thousands of protein-encoding genes, the transcription of which must be intricately coordinated in both time and space. It is thus not surprising that the regulatory machinery is extremely complex – 5–10% of the genes in eukaryotic genomes encode regulatory factors that are dedicated to controlling the rate at which protein-encoding genes are transcribed. These factors work by an array of mechanisms, including direct and indirect recruitment of the pre-initiation complex, activation of the pre-initiation complex, and modulation of the structure of the chromatin template. The effects of these regulatory factors are mediated by large coregulator complexes that appear to be designed to provide multiple protein–protein interaction surfaces to allow communication between a myriad of sequence-specific DNA-binding regulators and the general machinery. The *cis*-regulatory modules, with which regulatory factors interact, are equally diverse and complex. They are apparently designed to allow the integration of inputs from multiple regulatory factors thereby allowing for intricate spatially regulated gene expression in development as well as intricate responses to changes in the cellular environment.

References

Arnosti, D.N., Barolo, S., Levine, M. and Small, S. (1996) The *eve* stripe 2 enhancer employs multiple modes of transcriptional synergy. *Development* **122**: 205–214.

Berk, A.J. (1999) Activation of RNA polymerase II transcription. *Curr. Opin. Cell Biol.* **11**: 330–335.

Boyer, T.G., Martin, M.E., Lees, E., Ricciardi, R.P. and Berk, A.J. (1999) Mammalian Srb/Mediator complex is targeted by adenovirus E1A protein [see comments]. *Nature* **399**: 276–279.

Cai, H.N., Arnosti, D.N. and Levine, M. (1996) Long-range repression in the *Drosophila* embryo. *Proc. Natl Acad. Sci. USA* **93**: 9309–9314.

Carey, M. (1998) The enhanceosome and transcriptional synergy. *Cell* **92**: 5–8.

Chen, G., Fernandez, J., Mische, S. and Courey, A.J. (1999) A functional interaction between the histone deacetylase Rpd3 and the co-repressor Groucho in *Drosophila* development. *Genes Dev.* **13**: 2218–2230.

Chen, J.L., Attardi, L.D., Verrijzer, C.P., Yokomori, K. and Tjian, R. (1994) Assembly of recombinant TFIID reveals differential coactivator requirements for distinct transcriptional activators. *Cell* **79**: 93–105.

Cho, E.J. and Buratowski, S. (1999) Evidence that transcription factor IIB is required for a post-assembly step in transcription initiation. *J. Biol. Chem.* **274**: 25807–25813.

Courey, A.J. and Huang, J.D. (1995) The establishment and interpretation of transcription factor gradients in the *Drosophila* embryo. *Biochim. Biophys. Acta* **1261**: 1–18.

Courey, A.J. and Tjian, R. (1988) Analysis of Sp1 *in vivo* reveals multiple transcriptional domains, including a novel glutamine-rich activation motif. *Cell* **55**: 887–898.

de Souza, S.J., Long, M. and Gilbert, W. (1996) Introns and gene evolution. *Genes Cells* **1**: 493–505.

Donaldson, L. and Capone, J.P. (1992) Purification and characterization of the carboxyl-terminal transactivation domain of Vmw65 from herpes simplex virus type 1. *J. Biol. Chem.* **267**: 1411–1414.

Dubnicoff, T., Valentine, S.A., Chen, G., *et al.* (1997) Conversion of dorsal from an activator to a repressor by the global corepressor Groucho. *Genes Dev.* **11**: 2952–2957.

Ellenberger, T., Fass, D., Arnaud, M. and Harrison, S.C. (1994) Crystal structure of transcription factor E47: E-box recognition by a basic region helix-loop-helix dimer. *Genes Dev.* **8**: 970–980.

Ellwood, K., Huang, W., Johnson, R. and Carey, M. (1999) Multiple layers of cooperativity regulate enhanceosome-responsive RNA polymerase II transcription complex assembly. *Mol. Cell. Biol.* **19**: 2613–2623.

Falvo, J.V., Thanos, D. and Maniatis, T. (1995) Reversal of intrinsic DNA bends in the IFN beta gene enhancer by transcription factors and the architectural protein HMG I(Y). *Cell* **83**: 1101–1111.

Flanagan, P.M., Kelleher, R.D., Sayre, M.H., Tschochner, H. and Kornberg, R.D. (1991) A mediator required for activation of RNA polymerase II transcription *in vitro*. *Nature* **350**: 436–438.

Garrell, J. and Campuzano, S. (1991) The helix-loop-helix domain: a common motif for bristles, muscles and sex. *Bioessays* **13**: 493–498.

Gehring, W.J., Affolter, M. and Bürglin, T. (1994) Homeodomain proteins. *Annu. Rev. Biochem.* **63**: 487–526.

Giese, K., Kingsley, C., Kirshner, J.R. and Grosschedl, R. (1995) Assembly and function of a TCR alpha enhancer complex is dependent on LEF-1-induced DNA bending and multiple protein–protein interactions. *Genes Dev.* **9**: 995–1008.

Gill, G., Pascal, E., Tseng, Z.H. and Tjian, R. (1994) A glutamine-rich hydrophobic patch in transcription factor Sp1 contacts the dTAFII110 component of the *Drosophila* TFIID complex and mediates transcriptional activation. *Proc. Natl Acad. Sci. USA* **91**: 192–196.

Gray, S. and Levine, M. (1996a) Transcriptional repression in development. *Curr. Opin. Cell Biol.* **8**: 358–364.

Gray, S. and Levine, M. (1996b) Short-range transcriptional repressors mediate both quenching and direct repression within complex loci in *Drosophila*. *Genes Dev.* **10**: 700–710.

Gray, S., Szymanski, P. and Levine, M. (1994) Short-range repression permits multiple enhancers to function autonomously within a complex promoter. *Genes Dev.* **8**: 1829–1838.

Hampsey, M. and Reinberg, D. (1999) RNA polymerase II as a control panel for multiple coactivator complexes. *Curr. Opin. Genet. Dev.* **9**: 132–139.

Hoopes, B.C., LeBlanc, J.F. and Hawley, D.K. (1998) Contributions of the TATA box sequence to rate-limiting steps in transcription initiation by RNA polymerase II. *J. Mol. Biol.* **277**: 1015–1031.

Hope, I.A. and Struhl, K. (1986) Functional dissection of a eukaryotic transcriptional activator protein, GCN4 of yeast. *Cell* **46**: 885–894.

Ingham, P.W. (1988) The molecular genetics of embryonic pattern formation in *Drosophila*. *Nature* **335**: 25–34.

Jacob, G.A., Kitzmiller, J.A. and Luse, D.S. (1994) RNA polymerase II promoter strength *in vitro* may be reduced by defects at initiation or promoter clearance. *J. Biol. Chem.* **269**: 3655–3663.

Jiang, J., Cai, H., Zhou, Q. and Levine, M. (1993) Conversion of a dorsal-dependent silencer into an enhancer – evidence for dorsal corepressors. *EMBO J.* **12**: 3201–3209.

Jiminez, G., Guichet, A., Ephrussi, A. and Casanova, J. (2000) Relief of gene repression by *touso* RTK signaling: role of *Capicua* in *Drosophila* terminal and dorsoventral patterning. *Genes Dev.* **14**: 224–231.

Karin, M., Liu, Z. and Zandi, E. (1997) AP-1 function and regulation. *Curr. Opin. Cell Biol.* **9**: 240–246.

Kim, T.K., Kim, T.H. and Maniatis, T. (1998) Efficient recruitment of TFIIB and CBP–RNA polymerase II holoenzyme by an interferon-beta enhanceosome *in vitro*. *Proc. Natl Acad. Sci. USA* **95**: 12191–12196.

Kim, T.K. and Maniatis, T. (1997) The mechanism of transcriptional synergy of an *in vitro* assembled interferon-beta enhanceosome. *Mol. Cell* **1**: 119–129.

Kim, Y.J., Bjorklund, S., Li, Y., Sayre, M.H. and Kornberg, R.D. (1994) A multiprotein mediator of transcriptional activation and its interaction with the *C*-terminal repeat domain of RNA polymerase II. *Cell* **77**: 599–608.

Kirov, N., Zhelnin, L., Shah, J. and Rushlow, C. (1993) Conversion of a silencer into an enhancer – evidence for a co-repressor in dorsal-mediated repression in *Drosophila*. *EMBO J.* **12**: 3193–3199.

Koleske, A.J. and Young, R.A. (1994) An RNA polymerase II holoenzyme responsive to activators [see comments]. *Nature* **368**: 466–469.

Kriwacki, R.W., Schultz, S.C., Steitz, T.A. and Caradonna, J.P. (1992) Sequence-specific recognition of DNA by zinc-finger peptides derived from the transcription factor Sp1. *Proc. Natl Acad. Sci. USA* **89**: 9759–9763.

Krumlauf, R. (1994) *Hox* genes in vertebrate development. *Cell* **78**: 191–201.

Kussie, P.H., Gorina, S., Marechal, V., *et al.* (1996) Structure of the MDM2 oncoprotein bound to the p53 tumor suppressor transactivation domain. *Science* **274**: 948–953.

Luisi, B.F., Xu, W.X., Otwinowski, Z., Freedman, L.P., Yamamoto, K.R. and Sigler, P.B. (1991) Crystallographic analysis of the interaction of the glucocorticoid receptor with DNA. *Nature* **352**: 497–505.

Ma, J. and Ptashne, M. (1987a) Deletion analysis of Gal4 defines two transcriptional activating segments. *Cell* **48**: 847–853.

Ma, J. and Ptashne, M. (1987b) A new class of yeast transcriptional activators. *Cell* **51**: 113–119.

Maniatis, T., Whittemore, L., Du, W., *et al.* (1992) Positive and negative control of human interferon-β gene expression. In: *Transcriptional Regulation, Vol. 2,* (eds. S. L. McKnight and K. R. Yamamoto). Cold Spring Harbor Laboratory Press, Cold Spring Harbor, New York, pp. 1193–1220.

Mannervik, M., Nibu, Y., Zhang, H. and Levine, M. (1999) Transcriptional coregulators in development. *Science* **284**: 606–609.

Merika, M., Williams, A.J., Chen, G., Collins, T. and Thanos, D. (1998) Recruitment of CBP/p300 by the IFN beta enhanceosome is required for synergistic activation of transcription. *Mol. Cell* **1**: 277–287.

Miesfeld, R., Godowski, P.J., Maler, B.A. and Yamamoto, K.R. (1987) Glucocorticoid receptor mutants that define a small region sufficient for enhancer activation. *Science* **236**: 423–427.

Mitchell, P.J. and Tjian, R. (1989) Transcriptional regulation in mammalian cells by sequence-specific DNA binding proteins. *Science* **245**: 371–378.

Myers, L.C., Gustafsson, C.M., Bushnell, D.A., *et al.* (1998) The Med proteins of yeast and their function through the RNA polymerase II carboxy-terminal domain. *Genes Dev.* **12**: 45–54.

Pham, A.D., Müller, S. and Sauer, F. (1999) Mesoderm-determining transcription in *Drosophila* is alleviated by mutations in TAF(II)60 and TAF(II)110. *Mechan. Dev.* **84**: 3–16.

Ptashne, M. (1986) *A Genetic Switch: Gene Control and Phage Lambda.* Blackwell Scientific Publications, Cambridge, MA, pp. 128.

Ptashne, M. and Gann, A. (1997) Transcriptional activation by recruitment. *Nature* **386**: 569–577.

Radhakrishnan, I., Pérez-Alvarado, G.C., Parker, D., Dyson, H.J., Montminy, M.R. and Wright, P.E. (1997) Solution structure of the KIX domain of CBP bound to the transactivation domain of CREB: a model for activator: coactivator interactions. *Cell* **91**: 741–752.

Ranish, J.A., Yudkovsky, N. and Hahn, S. (1999) Intermediates in formation and activity of the RNA polymerase II preinitiation complex: holoenzyme recruitment and a postrecruitment role for the TATA box and TFIIB. *Genes Dev.* **13**: 49–63.

Regier, J.L., Shen, F. and Triezenberg, S.J. (1993) Pattern of aromatic and hydrophobic amino acids critical for one of two subdomains of the VP16 transcriptional activator. *Proc. Natl Acad. Sci. USA* **90**: 883–887.

Riesgo-Escovar, J. R. and Hafen, E. (1997) Common and distinct roles of dFos and dJun during *Drosophila* development. *Science* **278**: 669–672.

Roberts, S.G. and Green, M.R. (1994) Activator-induced conformational change in general transcription factor TFIIB. *Nature* **371**: 717–720.

Roberts, S.G.E., Ha, I., Maldonado, E., Reinberg, D. and Green, M.R. (1993) Interaction between an acidic activator and transcription factor-TFIIB is required for transcriptional activation. *Nature* **363**: 741–744.

Rusch, J. and Levine, M. (1996) Threshold responses to the dorsal regulatory gradient and the subdivision of primary tissue territories in the *Drosophila* embryo. *Curr. Opin. Genet. Dev.* **6**: 416–423.

Ryu, S., Zhou, S., Ladurner, A.G. and Tjian, R. (1999) The transcriptional cofactor complex CRSP is required for activity of the enhancer-binding protein Sp1. *Nature* **397**: 446–450.
Sharrocks, A.D., Brown, A.L., Ling, Y. and Yates, P.R. (1997) The ETS-domain transcription factor family. *Inter J. Biochem. Cell Biol.* **29**: 1371–1387.
Shen, F., Triezenberg, S.J., Hensley, P., Porter, D. and Knutson, J. R. (1996a) Critical amino acids in the transcriptional activation domain of the herpesvirus protein VP16 are solvent-exposed in highly mobile protein segments. An intrinsic fluorescence study. *J. Biol. Chem.* **271**: 4819–4826.
Shen, F., Triezenberg, S.J., Hensley, P., Porter, D. and Knutson, J.R. (1996b) Transcriptional activation domain of the herpesvirus protein VP16 becomes conformationally constrained upon interaction with basal transcription factors. *J. Biol. Chem.* **271**: 4827–4837.
Small, S., Arnosti, D.N. and Levine, M. (1993) Spacing ensures autonomous expression of different stripe enhancers in the even-skipped promoter. *Develop.* **119**: 762–772.
Small, S., Kraut, R., Hoey, T., Warrior, R. and Levine, M. (1991) Transcriptional regulation of a pair-rule stripe in *Drosophila*. *Genes Dev.* **5**: 827–839.
Thanos, D. and Maniatis, T. (1995) Virus induction of human IFN beta gene expression requires the assembly of an enhanceosome. *Cell* **83**: 1091–1100.
Tjian, R. and Maniatis, T. (1994) Transcriptional activation: a complex puzzle with few easy pieces. *Cell* **77**: 5–8.
Triezenberg, S.J. (1995) Structure and function of transcriptional activation domains. *Curr. Opin. Genet. Dev.* **5**: 190–196.
Uesugi, M., Nyanguile, O., Lu, H., Levine, A.J. and Verdine, G.L. (1997) Induced alpha helix in the VP16 activation domain upon binding to a human TAF. *Science* **277**: 1310–1313.
Valentine, S.A., Chen, G., Shandala, T., *et al.* (1998) Dorsal-mediated repression requires the formation of a multiprotein repression complex at the ventral silencer. *Mol. Cell. Biol.* **18**: 6571–6583.
Van Hoy, M., Leuther, K.K., Kodadek, T. and Johnston, S.A. (1993) The acidic activation domains of the GCN4 and GAL4 proteins are not alpha helical but form beta sheets. *Cell* **72**: 587–594.
Verrijzer, C.P. and Tjian, R. (1996) TAFs mediate transcriptional activation and promoter selectivity. *Trends Biochem. Sci.* **21**: 338–342.
Walker, S., Greaves, R. and O'Hare, P. (1993) Transcriptional activation by the acidic domain of Vmw65 requires the integrity of the domain and involves additional determinants distinct from those necessary for TFIIB binding. *Mol. Cell. Biol.* **13**: 5233–5244.
Zeitlinger, J. and Bohmann, D. (1999) Thorax closure in *Drosophila*: involvement of *fos* and the JNK pathway [in process citation]. *Development* **126**: 3947–3956.
Zeitlinger, J., Kockel, L., Peverali, F.A., Jackson, D.B., Mlodzik, M. and Bohmann, D. (1997) Defective dorsal closure and loss of epidermal decapentaplegic expression in *Drosophila fos* mutants. *EMBO J.* **16**: 7393–7401.
Zhou, J., Zwicker, J., Szymanski, P., Levine, M. and Tjian, R. (1998) TAFII mutations disrupt Dorsal activation in the *Drosophila* embryo. *Proc. Natl Acad. Sci. USA* **95**: 13483–13488.

3

Chromatin structure and the regulation of transcription

Alan P. Wolffe

1. Introduction

Chromatin appears to be an inhospitable environment for the molecular machines that use it as a substrate for transcription. Nucleosomes are remarkably stable to physical perturbation and under physiological conditions, nucleosomal arrays fold into higher-order structures that achieve concentrations in excess of 50 mg/ml within the nucleus. In spite of this apparent stability and compaction, complex metabolic processes involving DNA occur very efficiently *in vivo*. This contrasting requirement between storage and functional utility is met through the use of an effective filing system whereby regulatory elements and genes remain retrievable within the chromosome, and through the application of molecular machines that remodel chromatin. The nature of functional chromatin domains and the dynamic properties of chromatin are the focus of the Chapter. Particular importance is given to the activators and repressors that control transcription through chromatin disruption and modification.

This chapter briefly reviews the organization of the chromosomal domain and the roles of individual histones in nucleosome structure and stability. It also summarizes their known relevance in the control of gene expression. Also covered are the structural and functional consequences of covalently modifying the histones through acetylation and phosphorylation in light of the evidence that histone tail domains are key arbiters of chromatin function. These issues have special relevance owing to increasing evidence for the control of transcription by histone acetyltransferases and deacetylases. Finally the mechanisms of chromatin disruption and reassembly in response to the action of the SWI/SNF family of complexes will be discussed.

Transcription Factors, edited by J. Locker.

2. Chromosome architecture and transcription

2.1 *The radial loop and helical-folding models of chromosome structure*

Our most thorough understanding of chromosomal organization is for the most condensed and hence most visible of chromosomes – those at metaphase (Rattner and Lin, 1985). Although folding of DNA into nucleosomes leads to a seven-fold compaction in length, and the subsequent folding of arrays of nucleosomes into the chromatin fiber to a further seven-fold compaction, a massive 250-fold compaction of DNA follows the organization of the chromatin fiber into a metaphase chromosome (Earnshaw, 1991; 1988). Two principal models have been proposed to account for this compaction. The first suggests an organization of the fiber into loops that are radially arranged along the axis of the chromosome (Boy de la Tour and Laemmli, 1988; Gasser and Laemmli, 1986; Paulson and Laemmli, 1977). The second suggests a helical folding of the chromatin fiber (Sedat and Manuelidis, 1978).

The evidence for the organization of the chromatin fiber into loops attached to a central axis in normal cells comes from several experimental approaches. Long-standing observations on the morphology of lampbrush chromosomes in amphibian oocytes show a succession of loops emerging from a single chromosomal axis (Callan, 1986). Each contains very transcriptionally active DNA that is coated with RNA polymerase and associated ribonucleoprotein complexes. This generates a visible 'brush-like' appearance. The axes of lampbrush chromosomes, from which the loops project, consist visually of linear arrays of compacted beads, known as chromomeres. DNA is concentrated in the chromomeres, which represent compacted regions of chromatin. Distinctive loops can be recognized at invariant positions of the chromosomes and depend upon the DNA sequence contained within each loop (Callan *et al.*, 1987). Each loop may contain several transcription units and range in size up to 100 kilobases.

Chromosomal loops can also be detected in the chromatin of interphase cells using less direct methodologies. When intact chromosomes are subjected to very mild digestion with DNase I, so as to produce single-strand nicks, the sizes of the resulting chromosomal fragments decrease gradually until a plateau value is reached in which each fragment contains a single chromosomal loop of approximately 85 000 base-pairs of DNA complexed with protein. Electron microscopy, or sedimentation of nuclei that are extracted from histones by exposure to a high salt solution, allows direct measurement of the length of DNA on the microscope grid from where it exits a residual nuclear structure to where it re-enters this structure. Estimates of loop size from between 40 and 90 kilobase-pairs were obtained that are consistent with the biochemical measurements (Cook and Brazell, 1975; Jackson *et al.*, 1990; Paulson and Laemmli, 1977).

The development of pulsed-field gel electrophoresis (Schwartz and Cantor, 1984) allowed a more systematic analysis of the separation of cleavage sites following a mild nuclease digestion of nuclei (Filipski *et al.*, 1990). This technique allows the resolution of DNA molecules of a very large size using agarose electrophoresis. Preferential cleavage sites in the nuclei, spaced 50–300 kilobases apart, were detected. Taken together, these observations establish a strong case for the large independent loops (50–100 kilobases) of the chromatin fiber representing a unit of chromosome structure.

Support for the second model, proposing the folding of the chromatin fiber without specific attachments to an undisrupted chromosome axis, is based on sophisticated high voltage and conventional transmission electron microscopy combined with extensive computer analysis. The 'scaffolding' hypothesis would suggest that, although the arrangement of loops of the chromatin fiber about the chromosome axis might form distinct patterns, no discrete higher-order organization exists above the chromatin fiber. Belmont *et al.* (1987, 1989) demonstrated convincingly that such an order does in fact exist. Early work on large plant chromosomes during the meiotic cycle suggested that coils of chromatin fibers existed in the chromosome (White, 1973). Under certain fixation conditions chromosomes from animal cells could also appear as a spiral or zig-zag fiber (Onnuki, 1968). Belmont *et al.* (1987, 1989) examined native *Drosophila* mitotic chromosomes, observing a hierarchy of higher-order chromatin-folding patterns for 30 nm fibers and structures of 50, 100 and 130 nm in diameter were clearly discernible. Although a looping architecture for the 30 and 50 nm fibers could be detected under certain circumstances, the loops were not observed to be consistently orientated radially in three dimensions about any given axis, and no evidence for a central scaffolding was apparent. These observations have now been extended to mammalian cells (Belmont and Bruce, 1994). These results are not inconsistent with an important role for non-histone scaffolding proteins in the anchoring of local loops or domains of chromatin structure, but they do appear to rule out a strict radial symmetry or undisrupted central axis for such loops. Most likely, a diffuse organization of loops anchored to non-histone proteins actually exists *in vivo*.

2.2 The nuclear matrix, scaffold and skeleton: protein components and their function

Many studies have focused on the non-histone proteins present in the nucleus that generate the infrastructure on chromosomes and on the nucleus itself and the DNA sequences associated with them. The nature of the nuclear skeleton, the nuclear scaffold and the nuclear matrix has been the subject of much debate. The metaphase scaffold of a chromosome initially had a morphological definition as the complex structure at the axis of a mitotic chromosome visualized after swelling and extraction of the histones (Paulson and Laemmli, 1977). Biochemical extraction with high salt (2 M NaCl) or the detergent-like molecule, lithium diiodosalicylate (LIS), was used to define the residual nucleoprotein complex at which DNA was attached to the chromosome during interphase (Mirkovitch *et al.*, 1984; Paulson and Laemmli, 1977). This nuclear 'matrix' (after high-salt extraction) or 'scaffold' (after LIS extraction) is now known to contain a substantially more complex group of proteins than the metaphase scaffold itself (Gasser *et al.*, 1989). The use of these non-physiological extraction procedures was criticized because of their capacity to cause rearrangement of protein–DNA interactions and non-specific aggregation (Cook, 1988). An alternate strategy to study nuclear infrastructure is to encapsulate cells in agarose and extract most of the chromatin, thus leaving the nucleoprotein complexes essential for nuclear integrity under physiological conditions (Jackson *et al.*, 1988). This last methodology generates a nuclear 'skeleton' that retains the capacity to transcribe and replicate DNA. Recent comparative experiments are consistent with the nuclear matrix or scaffold interacting with gene-poor regions of the genome, whereas the nuclear skeleton interacts with gene-rich regions (Craig *et al.*, 1997). It is now clear that the function and composition of the nuclear skeleton is very different from that of the nuclear matrix or scaffold. All define chromosomal domains, yet the reasons for generating attachments to the chromatin fiber will differ (*Figure 1*).

The nuclear skeleton appears to represent the sites at which the large enzymatic complexes necessary to utilize chromatin as a substrate are localized. These include coactivators, corepressors and pol II itself. There is substantial evidence for these enzymes being associated with the nuclear infrastructure. In view of their size and complexity, these enzymes and their associated regulatory factors might be anticipated to have a structural role in the nucleus (Davie *et al.*, 1998). The actual DNA sequences associated with the nuclear scaffold are likely to vary, being dependent on the particular genes being transcribed.

In contrast to this dynamic quality found in the nuclear skeleton, it has been suggested that the nuclear matrix or scaffold contains specific DNA sequences that represent fixed sites of chromosome organization known as matrix or scaffold

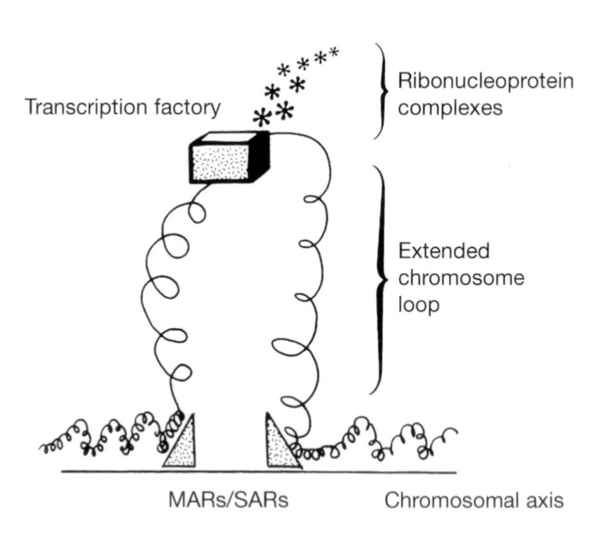

Figure 1. The location of the nuclear skeleton is compared to that of the nuclear matrix (scaffold). The nuclear skeleton contains gene-rich genomic segments. These will be assembled into transcription factories that will produce mRNA and other transcripts packaged into ribonucleoprotein particles. In contrast, the nuclear matrix (scaffold) contains fixed sites of DNA attachment known as matrix scaffold attachment regions (MARs or SARs). These are often modeled so that they are associated with the chromosomal axis. The extended loop of chromatin is digested by nucleases during the preparation of the nuclear skeleton or matrix.

attachment regions (MARs and SARs). Evidence for a function for the sequences in chromosomal dynamics comes from the synthesis of an artificial protein that preferentially binds to these AT-rich sequences (Strick and Laemmli, 1995). The presence of this protein interfered with the chromosomal dynamics that are normally observed during nuclear decondensation or chromosomal condensation in Xenopus egg extracts. However, in the absence of the identification of *bona fide* scaffold attachment sequence-binding proteins *in vivo* and examination of their function, an active role for scaffold or matrix attachments in chromosomal condensation remains speculative. In plants there is compelling evidence that AT-rich segments of the genome that have been biochemically defined as scaffold attachment regions promote the activity of transgenes (Allen *et al.*, 1996; 1993). However it should be noted that, in mammalian cells, biochemically defined SARS or MARS are not preferentially found in gene-rich portions of the genome (Craig *et al.*, 1997). Topoisomerase II and the histone, H1, are believed to bind preferentially to these AT-rich DNA sequences (Adachi *et al.*, 1989; Izaurralde *et al.*, 1989). It has also been suggested that proteins such as HMGI-Y might displace histone H1 selectively from SARs, thereby contributing to the local control of transcriptional activity (Zhao *et al.*, 1993). This type of selective association of proteins with SARs remains to be proven *in vivo*. Lysine-rich proteins like histone H1 have long been known to interact with AT-rich DNA (Leng and Felsenfeld, 1966). Hence the significance of *in vitro* binding experiments that enrich lysine-rich proteins bound to AT-rich SARs remains questionable.

2.3 Chromosome domains, position effect and locus control regions

Although the biochemical and structural definition of higher-order chromatin organization is far from complete, there is excellent genetic evidence for domains of influence within chromosomes (*Figure 2*). Early studies by cytologists led to the realization that some chromosomal regions have properties distinct from the rest of the chromosome (Henikoff, 1994). Large segments of chromatin were found to be highly condensed and to replicate late in the S phase. Geneticists determined that these chromosomal regions, which they called heterochromatin, have a significant effect on chromosomal function. The most common observed consequence of heterochromatin formation is the repression of transcription either in heterochromatin itself or in regions of chromatin that lie adjacent to the heterochromatin domain. The variability in gene expression at the border of the heterochromatin is described as 'position effect variegation'. Two explanations have been offered for this phenomenon. The first is that special proteins exist that cause heterochromatin to adopt its distinct structure, and these proteins can 'spill over' into regions of normal chromatin. The second is that heterochromatin represents the sequestration of chromosomal domains in specialized nuclear compartments from which the transcriptional machinery is excluded.

Position effects in mammalian chromosomes have been a recurrent problem for transgenic research since highly variable levels of transcriptional activity follow from the random introduction of reporter genes into the genome (Forrester *et al.*, 1987; Grosveld *et al.*, 1987; Stief *et al.*, 1989). These effects can be relieved by the introduction of a locus control region that exerts a dominant transcriptional activation function over

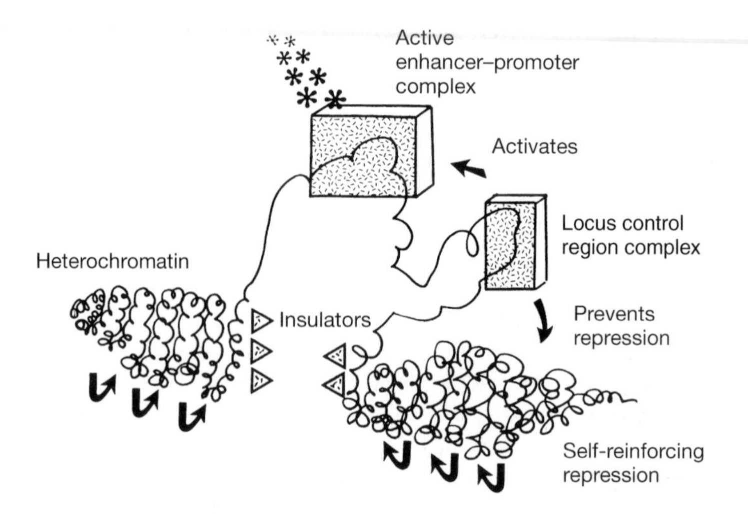

Figure 2. Model for an active versus a repressed chromosomal domain. The active chromosomal domain contains a locus control region complex that activates an enhancer–promoter complex to produce mRNA (asterisks). The locus control region also acts to prevent repressive influences spreading from adjacent heterochromatin. The heterochromatin exerts a self-reinforcing repression influence on transcription that can potentially spread to adjacent chromatin. This propagation of repression can be stopped by insulator elements.

a chromatin domain (10–100 kilobases). The mechanism of this activation function remains to be determined; however, communication between locus control regions, enhancers and promoters, either directly or through modifications of chromatin structural components, are favored hypotheses (Felsenfeld, 1992; Wijgerde *et al.*, 1995). Recent evidence suggests that both the locus control regions and enhancers act in *cis* to actively suppress position effect variegation (Festenstein *et al.*, 1996; Walters *et al.*, 1996). In this regard, locus control regions basically function as operationally defined 'powerful' enhancers. The coexistence of heterochromatin domains that can transmit repressive effects and the definition of the extensive long-range activation function of local control regions emphasize the necessary compartmentalization of the chromosome into discrete functional units which are prevented from influencing each other in a natural chromosomal context (Fiering *et al.*, 1995). This is due, in part, to the existence of special chromosomal regions which prevent the transmission of the chromatin-structural features that are associated with the boundaries of repressive or active domains. These specialized chromosomal regions are known as 'insulators' (Chung *et al.*, 1993; Wolffe, 1994). How insulators exert their function remains unknown.

3. Nucleosome structure

3.1 Features

Histone–histone and histone–DNA interactions are now understood in considerable structural detail (Arents *et al.*, 1991). The assembly of a stable nucleosome core depends on the initial heterodimerization of H3 with H4, and the subsequent dimerization of H3 to form the $(H3, H4)^2$ tetramer (Eickbusch and Moudrianakis, 1978). The $(H3, H4)^2$ tetramer can form a stable complex with more than 120 base-pairs of DNA (Hayes *et al.*, 1991a). Histones H2A and H2B form a stable heterodimer in a manner structurally homologous to H3/H4, but do not self-assemble into stable tetramer complexes (Arents *et al.*, 1991). Rather, dimers of (H2A, H2B) bind to either side of the $(H3, H4)^2$ tetramer and extend the wrapping of DNA within the nucleosome to over 160 base-pairs (Hayes *et al.*, 1990; 1991a). This creates a left-handed super-helical ramp of protein onto which the DNA is wrapped. It is essentially comprised of the four histone dimers linked end-to-end: (H2A,H2B)-(H4,H3)-(H3,H4)-(H2B,H2A) (Arents *et al.*, 1991). The (H3,H3) and (H2B,H4) dimer–dimer interfaces are comprised of a structurally similar four helix bundle; however, the latter does not remain stably associated in the absence of DNA in solutions containing physiological concentrations of salt. Given the stability of the individual heterodimers (Karantza *et al.*, 1996), the (H2B,H4) interface is a likely site for initial disruption of histone–histone interactions upon unfolding of the nucleosome core *in vivo* (*Figure 3*).

In order to follow the left-handed spiral formed by the histone-fold domains, the nucleosomal DNA is severely distorted into roughly two, 80 base-pairs, super-

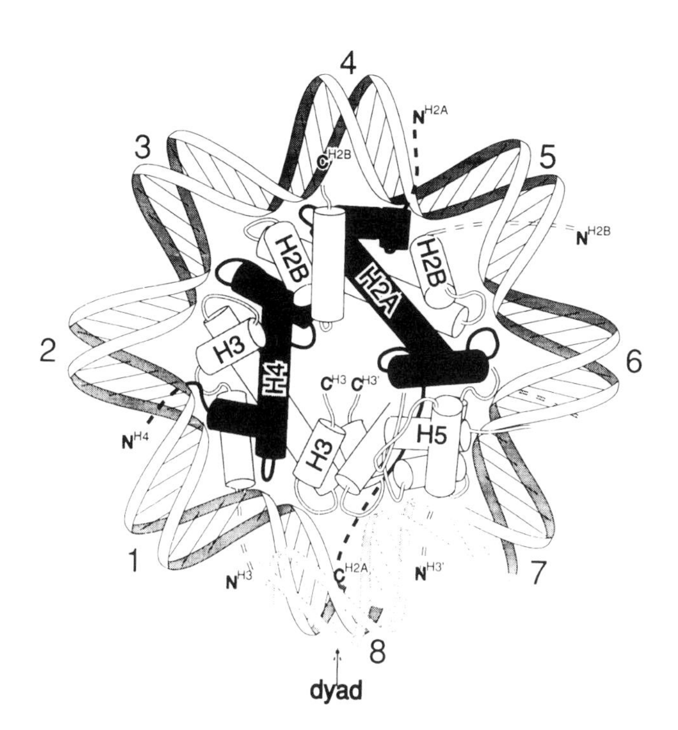

Figure 3. A model for the nucleosome. The approximate structure of the histone $(H3, H4)^2$ tetramer, the histone (H2A, H2B) dimer and H5 bound to DNA are shown. The dyad axis of the nucleosome is indicated. Numbers refer to integral turns of DNA away from the dyad axis. The approximate positions of the amino (*N*) and carboxyl (*C*) terminal tails of the core histones are indicated (dashed lines).

helical loops. Extended α-helical structures allow the histone-fold domains within each heterodimer of the octamer structure to contact approximately three double helical turns (~30 base-pairs) of DNA. Each contact involves an arginine residue penetrating the minor groove, several main polypeptide chain–amide interactions with two consecutive phosphates on each DNA strand, and, surprisingly, substantial hydrophobic interactions with the faces of the deoxyribose sugars in the DNA (Luger *et al.*, 1997). These precise histone–DNA interactions constrain all DNA sequences, regardless of inherent sequence-dependent structure, to adopt a relatively similar conformation in the nucleosome (Hayes *et al.*, 1991b). Because of the inherent anisotropic bending moments of most unique DNA sequences, a small number of preferred rotational orientations are found for most nucleosomal DNAs. However, precise sequence-dependent translational positioning of the nucleosome has been observed for only a small number of DNA sequences (Simpson, 1991; Wolffe and Kurumizaka, 1998). Although still poorly understood, translational positioning probably depends on how the inherent DNA structure matches the local variations in DNA curvature and helical periodicity found in the nucleosome.

External to the histone-fold domains, about 25% of the mass of the core histones is contained within the 'tail' domains. These domains, located at the N-termini of all four core histone proteins and at the C-terminus of histone H2A, were initially defined by their sensitivity to proteases (Böhm and Crane-Robinson, 1984). Proteolytic removal of the tail domains does not drastically alter the conformation or hydrodynamic properties of individual nucleosomes, and the tails do not play a role in nucleosome positioning or the correct assembly of nucleosomes *in vitro* (Hayes *et al.*, 1991). These N-termini, if fully extended, can project well beyond the super-helical turns of DNA in the nucleosome (Wolffe and Hayes, 1999). Consistent with their length, centrifugation studies with nucleosomal arrays lacking linker histones indicate that the histone tails mediate internucleosomal contacts as extended chains of nucleosomes are compacted to form the 30 nm chromatin fiber (Hansen, 1997). Further, the tails are critical for the self-assembly of condensed fibers into higher-order structures. Interestingly, histone tail interactions with DNA and protein change as the chromatin fiber undergoes folding or compaction (Fletcher and Hansen, 1996). Thus, certain post-translational modifications of the tail domains may evoke specific functional and/or conformational states of the chromatin fiber by inducing a defined alteration in the array of histone tail interactions (Hansen *et al.*, 1999).

3.2 Histone H1 and higher-order chromatin structure

Incorporation of linker histones into chromatin stabilizes nucleosomes and facilitates the assembly of higher-order chromatin structures. However, whereas core histones are essential for chromatin and chromosome assembly, linker histones are not required (Dasso *et al.*, 1994; Shen *et al.*, 1995). Metazoan linker histones have a three domain structure, a central globular domain, flanked by N- and C-terminal tails. The globular domain has a winged-helix domain structure (Ramakrishnan *et al.*, 1993) and can associate with the nucleosome core in a number of distinct ways (Pruss *et al.*, 1996; Zhou *et al.*, 1998). The N and C-terminal tails of the linker histones bind to DNA within the nucleosome core and in the linker DNA between nucleosome cores. The preponderance of basic residues within these tail domains serves to neutralize the polyanionic backbone of DNA, thus facilitating the folding of nucleosomal arrays into higher-order structures (Clark and Kimura, 1990). Inclusion of the linker histone into the nucleosome requires the presence of an octamer of core histones and restricts the translational mobility of histone octamers with respect to the DNA sequence (Pennings *et al.*, 1994; Ura *et al.*, 1995). Under physiological conditions the association of histone H1 with chromatin is much less stable than that of the core histones. Removal of histone H1 is therefore likely to represent a relatively simple means of destabilizing both local and higher-order chromatin structures and altering core histone–DNA interactions.

4. Chromatin remodeling and histone modification in transcription

4.1 Genetic evidence for individual histones and their domains in transcriptional control

Genetic experiments in *Saccharomyces cerevisiae* provide compelling evidence for general and specific roles for the histones in transcriptional control (Grunstein, 1990, 1997). Nucleosome depletion leads to the widespread activation of yeast promoters, and all four core histone N-termini are required for the repression of basal transcription. Acetylatable lysines in the N-termini of H3 and H4 have roles in transcriptional activation and repression. Interestingly, a region in the N-terminal tail of H4 that is known to be critical for silencing in yeast is observed to make protein–protein contacts with the surface of a (H2A,H2B) dimer in an adjacent core in the crystal structure of a nucleosome core particle (Luger *et al.*, 1997). Certain mutations of lysine to glutamine in the N-termini of H3 and H4 relieve the requirement for histone acetyltransferase activity in transcriptional activation (Zhang *et al.*, 1998). This suggests that histone acetylation is a major function of particular coactivators. Mutation of the histone-fold domains of the core histones can also lead to activation of certain yeast genes by relieving the requirement for the SWI/SNF family of molecular machines known to disrupt chromatin.

S. cerevisiae has an unusual non-essential linker histone that contains two globular domains, and when deleted has no detectable effects on gene expression (Patterton *et al.*, 1998). Deletion of *Tetrahymena* histone H1, which lacks the globular domain, does not influence transcription of the majority of genes; however, a subset of genes are either activated or repressed in H1-deficient strains (Shen and Gorovsky, 1996). Ablation of H1 during *Xenopus laevis* development leads to constitutive activation of certain oocyte specific *5S rRNA* genes and mesodermal specific genes (Bouvet *et al.*, 1994; Steinbach *et al.*, 1997). Repression can be restored by expression of the globular domain lacking N and C-terminal tails (Vermaak *et al.*, 1998). The molecular mechanism involved is now understood in some detail for one type of developmentally regulated gene. The globular domain of histone H1 has a precise architectural role for selective repression of the oocyte *5S rRNA* genes compared to somatic *5S DNA* in *X. laevis*. It binds to the 5S nucleosome asymmetrically, serving to position the histone octamer to repress certain genes while allowing continued activity of others (Chipev and Wolffe, 1992; Sera and Wolffe, 1998). Taken together, the histones can be seen as integral components of the transcriptional machinery with highly specific roles in gene control.

4.2 Structural and functional consequences of acetylation of the core histones

It has been known for some time that histone acetylation is intimately connected to transcriptional regulation. However, a direct link between chromatin function and acetylation was established by the discovery that the coactivator complexes which are required for transcriptional activation function as histone acetyltransferases (Brownell *et al.*, 1996), while corepressors containing histone deacetylases confer transcriptional repression (Taunton *et al.*, 1996). Histones are locally modified on target promoters (Kuo *et al.*, 1998) and specific lysines in particular histones are functional targets for acetyltransferases and deacetylases (Zhang *et al.*, 1998). Activator-dependent targeting of histone acetylase activity has recently been recapitulated *in vitro* (Utley *et al.*, 1998). Histone acetylation states are dynamic, with the acetylated lysines of hyper-acetylated histones turning over rapidly (with half-lives of minutes) within transcriptionally active chromatin, but much less rapidly for the hypo-acetylated histones of transcriptionally silent regions (Covault and Chalkley, 1980). The dynamics of histone acetylation provide an attractive mechanistic foundation for the reversible activation and repression of transcription (*Figure 4*).

Although the exact mechanism by which acetylation affects the biophysical properties of chromatin remains somewhat undefined, it is clear that acetylation of the core histone N-termini affects the transcriptional properties of chromatin at several levels of chromatin structure. Acetylation can facilitate the binding of

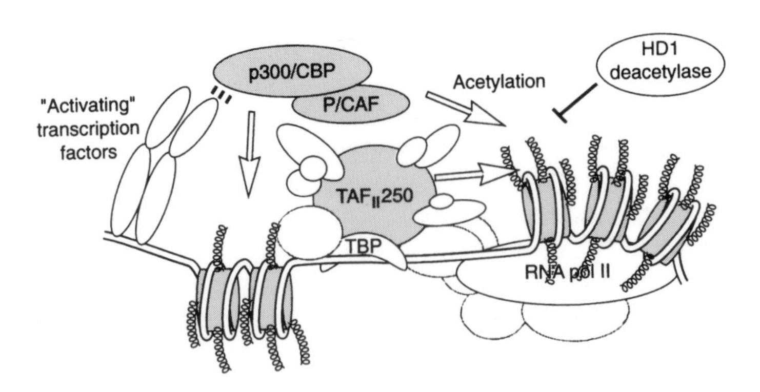

Figure 4. Histone acetylation states are dynamic. Activating transcription factors recruit transcriptional coactivators that are acetyltransferases. These enzymes modify chromatin continually to maintain gene activity in opposition to deacetylase enzymes. Repressive transcription factors will recruit transcriptional corepressors that maintain a deacetylated silent chromatin. CBP, CREB-binding protein; PCAF, p300/CBP-associated factor; TBP, TATA-binding protein; TAF, TBP-associated factor.

transcription factors to their recognition elements within isolated nucleosomes (Lee *et al.*, 1993; Vettese-Dadey *et al.*, 1996). Proteolytic removal of the N-termini of the core histones leads to comparable increases in transcription factor access to nucleosomal DNA and transcription of chromatin templates as histone acetylation (Lee *et al.*, 1993; Vettese-Dadey *et al.*, 1994). This is consistent with acetylation reducing the stability of interaction of the histone tails with nucleosomal DNA. It should nevertheless be noted that the *N*-termini of the core histones always make at least transient contacts with DNA in spite of acetylation (Mutskov *et al.*, 1998). Acetylated histones wrap DNA less tightly in mononucleosomes, which may result in a decrease in the amount of DNA super-helical writhe constrained by the nucleosome (Bauer *et al.*, 1994; Krajewski and Becker, 1998). These changes might be due to the fact that the acetylated *N*-terminal histone tails bind DNA with reduced affinity (Hong *et al.*, 1993) and are more mobile with respect to the DNA surface than unmodified tails (Cary *et al.*, 1982). Another interesting possibility is that acetylation disrupts the secondary structures which are known to exist within the H3 and H4 N-termini when they are bound to nucleosomal DNA (Baneres *et al.*, 1997). This might further destabilize interactions with DNA and the nucleosome itself.

Beyond effects on individual nucleosomes, acetylation facilitates factor access and transcription from nucleosomal arrays by decreasing the stability of the completely compacted 30 nm fiber (Tse *et al.*, 1998). It is also likely that acetylation leads to the destabilization of long range structures through which the chromatin fiber is folded into the chromosome itself (Annunziato *et al.*, 1988). Interactions between adjacent nucleosomal arrays are reduced when they are reconstituted with acetylated histones and chromatin solubility is increased (Perry and Chalkley, 1982). *In vivo*, the region of DNAse I sensitivity within the active β-globin locus also correlates with a region of increased histone acetylation (Hebbes *et al.*, 1994).

Interestingly, the level of histone modification required to facilitate the transcription process is relatively low, and a total of 12 acetylated lysines per histone octamer (out of 28 potential acetylated lysines) will promote *in vitro* transcription more than 15 fold. This level of modification reduces chromatin compaction to the same extent as proteolytic removal of the N-termini (Tse *et al.*, 1998), again suggesting that the primary consequence of hyper-acetylation is to reduce the interaction of the tails with the other components of chromatin, including nucleosomal DNA, linker DNA and the histones of adjacent nucleosomes. However, the level of charge neutralization necessary to facilitate the destabilization of chromatin higher-order structure is so low that other structural features must amplify the consequences of acetylation. As discussed earlier these might include alterations to secondary structure in the tail domains and/or changes in the association of the tails with other non-histone proteins. Acetylation of the histones probably serves to illuminate particular nucleosomes and/or segments of chromatin for interaction with other chromatin remodeling factors or components of the transcriptional machinery. The potential combination of direct chromatin structural transitions and modulation of protein–protein interactions following acetylation or deacetylation of the histone tails provides a powerful means of regulating transcription.

4.3 Structural and functional consequences of phosphorylation and ubiquitination of the core histones

In contrast to the many studies on the structural and functional consequences of histone acetylation, the impact of other post-translational modifications of the core histones is relatively unexplored. Significant future opportunities undoubtedly lie in this research area. Histone H3 is rapidly phosphorylated on serine residues within its basic amino terminal domain, when extracellular signals such as growth factors or phorbol esters stimulate quiescent cells to proliferate (Mahadevan *et al.*, 1991). Global phosphorylation of serine 10 in H3 occurs in pericentromeric chromatin in late G2 phase, completely spreads throughout the chromosome just before the prophase of mitosis, and is rapidly lost during anaphase (Hendzel *et al.*, 1997). This modification is spatially and temporally correlated with mitotic and meiotic chromatin condensation (Wei *et al.*, 1998). H3 serine 10 is located within the basic amino terminal domain of histone H3, and like the N-terminal domain of H4, may interact with the ends of DNA in the nucleosomal core particle and therefore, perhaps, with histone H1 (Glotov *et al.*, 1978). Indeed, based on charge effects, phosphorylation of histone H3 might be expected to have structural consequences comparable to acetylation. A change in either nucleosomal conformation or higher-order structure concomitant with phosphorylation of H3 within the chromatin of the proto-oncogenes *c-fos* and *c-jun* occurs following their rapid induction to high levels of transcriptional activity by phorbol esters (Chen and Allfrey, 1987). DNase I sensitivity of chromatin rapidly increases and proteins with exposed sulfhydryl groups accumulate on the proto-oncogene chromatin. The proteins that contain exposed sulfhydryl groups include both non-histone proteins, such as RNA polymerase, and molecules of H3 with exposed cysteine residues. The histone H3 cysteine residues, the only ones in the nucleosome, are normally buried within the particle. Exposure of the sulfhydryl groups implies that a major disruption of nucleosome structure occurs, which could involve the dissociation of an (H2A,H2B) dimer. Phosphorylation and acetylation of histone H3 might act in concert to cause these changes. There is likely to be an important link between cellular signal transduction pathways and chromatin targets for post-translational modification.

Ubiquitin is a 76-amino-acid peptide that is attached to the *C*-terminal tail of histone H2A and perhaps H2B. Ubiquitinated H2A is incorporated into nucleosomes, without major changes in the organization of nucleosome cores (Levinger and Varshavsky, 1980). Ubiquitination of histone H2A is associated with transcriptional activity. Only 1 nucleosome in 25 contains ubiquitinated histone H2A within non-transcribed chromatin. This increases to one nucleosome in two for the transcriptionally active *hsp70* genes (Levinger and Varshavsky, 1982). Enrichment in ubiquitinated H2A is especially prevalent at the 5′ end of transcriptionally active genes. Since the *C*-terminus of histone H2A contacts nucleosomal DNA at the dyad axis of the nucleosome (Usachenko *et al.*, 1994), ubiquitination of this tail domain might be anticipated to disrupt the interaction of linker histones with nucleosomal DNA. The bulky ubiquitin adduct might also be anticipated to disrupt higher-order chromatin structures by impeding internucleosomal interactions.

4.4 Phosphorylation of linker histones

Phosphorylation of histone H1 has been shown directly to weaken the interaction between the basic tails of the protein and DNA. Surprisingly, these changes influence the binding of the protein to chromatin even more than to DNA and thereby potentially destabilize the chromatin fiber (Hill *et al.*, 1991). Phosphorylation of the histone H1 tails occurs predominantly at conserved (S/T P-X- K/R, serine/threonine, proline, any amino acid, lysine/arginine) motifs of which several exist along the charged tail regions. Linker histone becomes heavily phosphorylated on transcriptional activation of the micronucleus of *Tetrahymena* during the sexual cycle (Sweet *et al.*, 1996). Transcriptional competence of the mouse mammary tumor virus (MMTV) promoter depends on the phosphorylation of histone H1 (Lee and Archer, 1998) and the active MMTV promoter is known to be selectively depleted in H1 (Bresnick *et al.*, 1992). In these examples, it seems probable that the transcriptional machinery will target the phosphorylation of linker histones as a component of activation pathways in order to alleviate the repressive influence of linker histones.

5. Transcriptional activators that remodel chromatin

5.1 The SAGA (PCAF) activator complex

A Gcn5p containing coactivator complex was identified through a genetic screen carried out by Berger *et al.* (1992) to identify mutations in genes that confer resistance to the toxic chimeric transcriptional activator, Gal4/VP16. Genes identified by this screen might be anticipated to be involved in facilitating gene activation by the VP16 acidic activation domain. In this way two 'adaptor' proteins, Ada2p and Ada3p were identified that were proposed to bridge interactions between activation domains and the basal transcriptional machinery (Guarente, 1995). A comparable mutation in the gene *Gcn5* impaired the activation of transcription by the transcription factor, Gcn4p (Georgakopoulos and Thireos, 1992). Subsequent genetic and biochemical experiments established that Gcn5p/Ada2p/Ada3p exist as a coactivator complex in yeast (Georgakopoulos *et al.*, 1995; Marcus *et al.*, 1994; Horiuchi *et al.*, 1995) and that the Ada2p interacts with both acidic activation domains and TBP (Barlev *et al.*, 1995). Recent experiments indicate that the Gcn5p/Ada2p/Ada3p complex is a component of an even more elaborate coactivator known as SAGA (Spt/Ada/Gcn5 acetyltransferase) complex (Grant *et al.*, 1998). The SAGA complex contains components of the basal transcriptional machinery, such as TAFs. Exactly how much Gcn5p is free, bound to Ada2p and Ada3p, or in the SAGA complex has not been resolved.

The SAGA coactivator is a histone acetyltransferase (Brownell *et al.*, 1996) that selectively modifies lysine 16 in the N-terminal tail domain of histone H4 (Kuo *et al.*, 1996). This property suggested, for the first time, that coactivators have the capacity to directly modify the chromatin template in order to facilitate transcription. *Gcn5* is not an essential gene in yeast; however, the capacity to induce gene expression by Gcn4p is reduced by 60% if *Gcn5* is not functional. This suggests that the individual histone acetyltransferases may not be essential in yeast.

This might reflect the presence of numerous genes with overlapping functions, and/or merely that the modification of chromatin structure is only one contributor to transcriptional regulation. The discovery that *S. cerevisiae* Gcn5p had histone acetyltransferase activity (Brownell *et al.*, 1996) led to the recognition that comparable regulatory mechanisms exist in metazoans (Yang *et al.*, 1996b). A human homolog of Gcn5p known as PCAF, acetylates histones (Yang *et al.*, 1996b) and has recently been shown to be part of a PCAF complex that is the human equivalent of SAGA (Martinez, 1998). Thus, the existence of multiple potentially redundant histone acetyltransferases in yeast is substantiated by recent observations in metazoans.

5.2 Histone acetyltransferases p300 and TAF_{II}*250*

p300/CBP itself binds the PCAF histone acetyltransferase complex and is also an independent histone actyl transferase (Ogryzko *et al.*, 1996). p300/CBP serves as an integrator to mediate regulation by a wide variety of sequence-specific transcription factors (Kamei *et al.*, 1996), including the steroid and nuclear hormone receptors, c-Jun/v-Jun, c-Myb/v-Myb, c-Fos and MyoD (Janknecht and Hunter, 1996). To strengthen the analogy with the Gcn5p/Ada2p/Ada3p complex, p300/CBP has a domain highly similar to part of Ada2p and associates with PCAF, the homolog of Gcn5p (Yang *et al.*, 1996b) (*Figure 4*). A component of the DNA-binding basal transcription factor, TFIID, has also been shown to have histone acetyltransferase activity (Mizzen *et al.*, 1996). TAF_{II}250 is the architectural core of TFIID interacting with the other TAFs as well as with TBP. TAF_{II}250 is required for the activation of particular genes that are indicative of coactivator function, and associates with components of the basal transcriptional machinery such as TFIIA, TFIIE and TFIIF (Dikstein *et al.*, 1996). In addition, TAF_{II}250 functions as both a kinase and a histone acetyltransferase (Dikstein *et al.*, 1996; Mizzen *et al.*, 1996). Thus diverse proteins in metazoans (and potentially in *S. cerevisiae*) possess histone acetyltransferase activity. In an interesting link with the mammalian SWI/SNF (Switch–Sucrose non-fermenting) activator complex, monoclonal antibodies against p300 immunoprecipitate a complex of p300/CBP together with at least seven other cellular proteins. Within this complex are TBP, TAF_{II}250 and SWI/SNF components including Brg1, suggesting that functions of histone acetyltransferases might be linked to those of other activators that contend with chromatin (Dallas *et al.*, 1998).

5.3 The SWI/SNF complex

SWI and *SNF* genes have been found to encode proteins that together assemble a large multi-subunit complex required for the regulation of a specific group of inducible genes in yeast (Cairns *et al.*, 1994; Peterson *et al.*, 1994). A major clue to the molecular mechanisms by which the SWI/SNF activator complex functions came from a genetic screen for mutations of genes that would allow transcription from the *HO* gene in the absence of specific *SWI* genes (Herskowitz *et al.*, 1992). These studies identified the *sin* genes (SWI independent). *Sin* 1–4 have been found, or inferred, to have a direct impact on chromatin structure and function. A simple model would predict that the SWI/SNF activator complex functions by overcoming the repressive effects of the *Sin* gene products on transcription. Indeed *in vivo* experiments in *S. cerevisiae* establish that the SWI/SNF activator complex activates transcription by altering chromatin structure (Hirschhorn *et al.*, 1992); and *in vitro* experiments using purified SWI/SNF complex indicate that stoichiometric amounts of SWI/SNF complex can alter histone–DNA interactions in the nucleosome (Côté *et al.*, 1994).

There is excellent precedent for pioneering experimental work in *S. cerevisiae*, which led to the recognition of comparable regulatory mechanisms in metazoans. The identification of the SWI/SNF complex (Peterson and Herskowitz, 1992) offered insight into potential regulatory roles for related proteins in *Drosophila* (Tamkun *et al.*, 1992). It was also shown that metazoan regulatory proteins, including the glucocorticoid receptor introduced into yeast, could make use of the SWI/SNF complex to activate synthetic promoters containing their recognition elements (Laurent and Carlson, 1992; Yoshinaga *et al.*, 1992). Animal homologs of the SWI/SNF complex have also been characterized (Chiba *et al.*, 1994; Khavari *et al.*, 1993; Muchardt *et al.*, 1993; Tsukiyama and Wu, 1995). The human complex includes *human brahma* (*hBrm*) and *brahma*-related gene product 1 (brg1), and possesses amino terminal proline and glutamine-rich regions that resemble transcriptional activation domains. Their capacity to interact with other components of the transcriptional machinery, including the glucocorticoid and estrogen receptor, is shown by their capacity to activate transcription in transient cotransfection assays that are largely independent of chromatin-mediated effects (Chiba *et al.*, 1994; Muchardt and Yaniv, 1993).

Evidence for the targeted disruption of chromatin by the mammalian SWI/SNF complex has remained elusive. A 100-fold molar excess of the 2×10^6 Dalton SWI/SNF complex can disrupt a synthetic nucleosome core (containing 0.1×10^6 Dalton of histone) *in vitro* (Imbalzano *et al.*, 1994). It has also been suggested that the pol II holoenzyme contains SWI/SNF and might remodel chromatin. However, recent experiments suggest that the yeast pol II holoenzyme might, under certain circumstances, disrupt chromatin independent of the presence of SWI/SNF (Gaudreau *et al.*, 1997).

How does the SWI/SNF complex disrupt nucleosomes? So far, no covalent modifications of the core histones have been shown to be conferred by SWI/SNF components. One model for disruption is that the complex tracks along DNA, rather like RNA and DNA polymerase, and displaces nucleosomes in a comparable way (Cairns, 1998). However this is difficult to reconcile with the continued wrapping of DNA on the surface of the histones in SWI/SNF disrupted nucleosomes, and with the recovery of normal histone stoichiometries from SWI/SNF treated nucleosomes (Côté *et al.*, 1998). An alternative idea is that histones H2A and H2B are displaced or destabilized within the nucleosome. Removal of H2A and H2B facilitates access of transcription factors to nucleosomal DNA (Hayes and Wolffe, 1992) and facilitates transcription (Hansen and Wolffe, 1994). Although complete displacement of (H2A, H2B) dimers seems unlikely, destabilization of (H2A, H2B) association would be consistent with genetic and structural data. This disruption might generate a structure prone to homologous dimerization (Schnitzler *et al.*, 1998).

Mutation of the core histone-fold domains can generate yeast strains that are Sin⁻. These Sin mutations lie either in regions of the core histones that mediate interaction between the (H3, H4)$_2$ tetramer and the (H2A, H2B) dimers (Santisteban *et al.*, 1997), or at sites that destabilize histone–DNA interactions (Kruger *et al.*, 1995). The boundaries of the nucleosome core are known to be mainly defined by the (H2A, H2B) dimers. Destabilization of (H2A, H2B) interactions in the nucleosome alone are insufficient to explain all of the features of SWI/SNF-mediated nucleosomal disruption, because removal of (H2A, H2B) dimers will not eliminate rotational constraint of DNA in the nucleosome cores. Such loss of rotational constraint, as assayed by DNase I cleavage, is seen in the presence of SWI/SNF (Côté *et al.*, 1994) and in nucleosome cores containing Sin2-mutant histones (Kurumizaka and Wolffe, 1997). The (H3, H4)$_2$ tetramer rotationally constrains DNA as efficiently as the histone octamer, thus the interaction of the (H3, H4)$_2$ tetramer with DNA must also be destabilized during SWI/SNF-mediated nucleosome disruption. It is possible that the binding of the SWI/SNF complex to the nucleosome destabilizes both the (H2A, H2B) dimer and the (H3, H4)$_2$ tetramer interactions with DNA and that this is accomplished by protein–protein interactions with the SWI/SNF complex on the face of the nucleosome. This interaction may require contact with the core histone tails (Georgel *et al.*, 1997) and may resemble the interaction of other nucleosome core-binding proteins, such as the globular domain of linker histones, HNF3 and NF1 (Alevizopoulos *et al.*, 1995; Cirillo *et al.*, 1998; Pruss *et al.*, 1996). Binding of SWI/SNF to the face of the nucleosome would allow contact with all four core histones and might be predicted to alter the contacts with DNA as has been observed following binding of linker histones (Guschin *et al.*, 1998; Usachenko *et al.*, 1996). Replacement of histone H1 might also facilitate nucleosome mobility, as reflected in the loss of clearly defined spacing (Tsukiyama *et al.*, 1994), and destabilize higher-order chromatin structure. Protein compositional analysis within nucleosome arrays containing H1 and SWI/SNF cross-linking to nucleosomal substrates will be necessary to test this hypothesis.

6. Transcriptional repressors that remodel chromatin

6.1 Histone deacetylase and mammalian Sin3

The purification of the mammalian histone deacetylase and the recognition of the similarities to *S. cerevisiae* Rpd3p (Repressor of Potassium Deficiency) (Taunton *et al.*, 1996) has provided considerable insight into transcriptional repression in metazoans. The first direct evidence for mammalian homologs of Rpd3p being involved in transcriptional repression came from two hybrid screens indicating that the transcriptional regulator, YY1, interacted with mouse and human Rpd3p (Yang *et al.*, 1996a). The fusion of mammalian Rpd3p to a targeted DNA-binding domain directed transcriptional repression by more than 10 fold. Mutations in a glycine-rich domain of YY1, which directs binding to Rpd3p, could abolish transcriptional repression by YY1 suggesting that it negatively regulates transcription by tethering Rpd3. YY1 is a mammalian zinc finger transcription factor (Shi *et al.*, 1991), which is thought to regulate cell growth and differentiation (Shrivastava and Calame, 1994).

A second well defined protein complex that influences cell growth and differentiation in mammalian cells is the Mad/Max heterodimer (Chen *et al.*, 1995; Hurlin *et al.*, 1995; Lahoz *et al.*, 1994). Max is a widely expressed sequence-specific transcriptional regulator of the basic region-helix-loop-helix-leucine zipper family (bHLH-Zip). Max heterodimerizes with the Myc family of bHLH-Zip proteins including Myc, Mad and Mxi1 (Ayer *et al.*, 1993; Zervos *et al.*, 1993). While the Myc/Max complex activates transcription and transformation, the Mad/Max complex represses these events. Ayer *et al* (1995) identified two mammalian proteins mSin3A and mSin3B that interact with Mad and that have striking homology to *S. cerevisiae* Sin3p, including the four-paired amphipathic helix (PAH) domains. The association between Mad/Max and mSin3A and B requires the second PAH domain. Mutations in this domain eliminate the interaction with mSin3A and prevent the Mad/Max complex from repressing transcription. The next step was to establish that the mSin3 proteins interact with the mammalian histone deacetylases. Mad, mSin3 and the mammalian histone deacetylases co-immunoprecipitate (Alland *et al.*, 1997; Laherty *et al.*, 1997). The third PAH domain of mSin3 interacts with the mammalian Rpd3p homologs and can confer transcriptional repression when attached to a DNA-binding domain. More subtle mutational analysis suggests that the cell transformation and transcriptional repression suppressed by the Mad/Max complex depend on distinct domains of the mSin3 proteins (Alland *et al.*, 1997). However an active role for histone deacetylation in transcriptional control in demonstrated by the use of deacetylase inhibitors such as Trichostatin A (Yoshida *et al.*, 1990) that abolish Mad's ability to repress transcription. The existence of a conserved transcriptional repression mechanism that utilizes Sin3p and histone deacetylase emphasizes the significance of the chromatin environment for transcriptional control. Histone deacetylation directs the assembly of a stable, repressive, chromatin structure.

6.2 Nuclear hormone receptors and histone deacetylase

A role for chromatin had already been established in the control of transcription by the thyroid hormone receptor (Wong *et al.*, 1995; 1997a; 1997b). These studies provide a useful example of how the histones can contribute to gene regulation. The assembly of mini-chromosomes within the *Xenopus* oocyte nucleus has been used to examine the role of chromatin in both transcriptional silencing and activation of the *Xenopus TRβA* promoter. Transcription from this promoter is under the control of thyroid hormone and the thyroid hormone receptor (Ranjan *et al.*, 1994), which exists as a heterodimer of the thyroid (TR) and retinoid X receptors (RXR). Micro-injection of either single-stranded or double-stranded DNA templates into the *Xenopus* oocyte nucleus offers the opportunity to examine the influence on gene regulation of chromatin-assembly pathways that are either coupled or uncoupled to DNA synthesis (Almouzni and Wolffe, 1993a). The staged injection of both mRNA that encodes the transcriptional regulatory proteins and of template DNA offers the potential for examining how and when transcription factors work within a chromatin environment. In particular, it is possible to discriminate between pre-emptive mechanisms, in which transcription factors bind during chromatin assembly to activate transcription, and post-replicative mechanisms, in which transcription factors gain access to their recognition elements after they have been assembled into mature chromatin structures. TR/RXR heterodimers bind constitutively within the mini-chromosome, independently of whether the receptor is synthesized before or after chromatin assembly. Rotational positioning of the thyroid hormone response element (TRE) on the surface of the histone octamer allows the specific association of the TR/RXR heterodimer *in vitro*. The coupling of chromatin assembly to the replication process augments transcriptional repression by unliganded TR/RXR, without influencing the final level of transcriptional activity, in the presence of thyroid hormone.

The molecular mechanisms by which the unliganded thyroid hormone receptor makes use of chromatin in order to augment transcriptional repression also involve mSin3 and histone deacetylase (Alland *et al.*, 1997; Heinzel *et al.*, 1997). The unliganded thyroid hormone receptor and retinoic acid receptor (RAR) bind a corepressor called nuclear corepressor (NCoR) (Horlein *et al.*, 1995). NCoR interacts with Sin3 and recruits the histone deacetylase (Alland *et al.*, 1997; Heinzel *et al.*, 1997). All of the transcriptional repression conferred by the unliganded thyroid hormone receptor in *Xenopus* oocytes (Wong *et al.*, 1997, 1995a) can be alleviated by the inhibition of histone deacetylase using Trichostatin A (Wong *et al.*, 1998), which is indicative of an essential role for deacetylation in establishing transcriptional repression in a chromatin environment.

The addition of thyroid hormone to the chromatin-bound receptor leads to the disruption of chromatin structure (Wong *et al.*, 1997a; 1997b; 1995). Chromatin disruption is not restricted to the receptor-binding site and involves the reorganization of chromatin structure in which targetted histone acetylation by the PCAF and p300/CBP activators have a contributory role (Ogryzko *et al.*, 1996; Yang *et al.*, 1996) (*Figure 4*). It is possible to separate chromatin disruption from productive recruitment of the basal transcription machinery *in vivo* by deletion of the regulatory elements that are essential for transcription initiation at the start site, and by using transcriptional inhibitors (Wong *et al.*, 1997a; 1995). Therefore chromatin disruption is an independent hormone-regulated function targetted by DNA-bound thyroid hormone receptor. It is remarkable just how effectively the various functions of the thyroid hormone receptor are mediated through the recruitment of enzyme complexes that modify chromatin. These results provide compelling evidence for the productive utilization of structural transitions in chromatin as a regulatory principle in gene control.

6.3 DNA methylation and transcriptional control

The covalent modification of DNA provides a direct and powerful mechanism to regulate gene expression (Kass *et al.*, 1997a). Considerable experimental evidence supports the existence of such a mechanism in the majority of plants and animals (Bird, 1995, 1986; Szyf, 1996; Yoder *et al.*, 1997). The genome of an adult vertebrate cell has 60–90% of the cytosines in CpG dinucleotides that are methylated by DNA methyltransferase (Riggs and Porter, 1996). This modification can alter the recognition of the double helix by the transcriptional machinery and alter the structural proteins that assemble chromatin (Kass *et al.*, 1997; Nan *et al.*, 1997).

DNA methylation could control gene activity either at a local level, through effects at a single promoter and enhancer, or through global mechanisms, which influence many genes within an entire chromosome or genome (Tate and Bird, 1993). An attractive suggestion is that DNA methylation evolved as a host defense mechanism in metazoans to protect the genome against genomic parasites such as transposable elements (Yoder *et al.*, 1997). An increase in methyl-CpG correlates with transcriptional silencing for whole chromosomes and transgenes, and particular developmentally regulated genes and human disease genes (Li *et al.*, 1993; Szyf, 1996). All of these systems exhibit epigenetic effects on transcriptional regulation, where identical DNA sequences are differentially utilized within the same cell nucleus. These patterns of differential gene activity are clonally inherited through cell division. Because specific methyl-CpG dinucleotides are maintained through DNA replication, DNA methylation states also provide an attractive mechanism (epigenetic mark) to maintain a particular state of gene activity through cell division and, thus, to contribute to the maintenance of the differentiated state (Holliday, 1987).

The most direct mechanism by which DNA methylation could interfere with transcription would be to prevent the binding of the basal transcriptional machinery and ubiquitous transcription factors to promoters. This is not a generally applicable mechanism because some promoters are transcribed effectively as naked DNA templates that are independent of DNA methylation (Busslinger *et al.*, 1983; Iguchi-Ariga and Schaffner, 1989; Kass *et al.*, 1997). Certain transcription factors (e.g. the cyclic AMP-dependent activator, CREB) bind less well to methylated recognition elements; however, the reduction in affinity is often insufficient to account for the inactivity of promoters *in vivo* (Hoeller *et al.*, 1988; Weih *et al.*, 1991). It seems unlikely that DNA methylation would function to repress transcription globally by modifying the majority of cytosine–guanosine (CpGs) in a chromosome if the only sites of action are to be a limited set of recognition elements for individual transcription factors.

The second possibility is that specific transcriptional repressors exist that recognize methyl-CpG and, either independently or together with other components of chromatin, turn off transcription. This mechanism would have the advantage of being substantially independent of the DNA sequence itself, thereby offering a simple means of global transcriptional control. It would be especially attractive if the methylation-dependent repressors worked in a chromatin context because then DNA could maintain the nucleosomal and chromatin-fiber architecture necessary to compact DNA (Jost and Hofsteenge, 1992; McArthur and Thomas, 1996). Moreover, because chromatin assembly also represses transcription, methylation-dependent repression mechanisms would add to those already in place.

Bird and colleagues have identified two repressors, methyl CpG dinucleotide binding proteins 1 and 2 (MeCP1 and MeCP2), that bind to methyl-CpG without apparent sequence specificity (Lewis *et al.*, 1992; Meehan *et al.*, 1989, 1992). Like DNA methylation itself, MeCP2 is dispensable for the viability of embryonic stem cells. However, it is essential for normal embryonic development. Consistent with the capacity of methylation-dependent repressors to operate in chromatin, recent studies indicate that MeCP2 is a chromosomal protein with the capacity to displace histone H1 from the nucleosome (Chandler *et al.*, 1999; Nan *et al.*, 1996). Moreover, MeCP2 contains a methyl-CpG DNA-binding domain, which might alter chromatin structure directly, and a repressor domain, which might function indirectly to confer long-range repression *in vivo* (Nan *et al.*, 1997, 1993). The capacity for MeCP2 to function in chromatin explains several phenomena which are connected to the unique aspects of the chromatin that is assembled on the methylated DNA.

A role for specialized chromatin structures in mediating transcriptional silencing by methylated DNA has been suggested by several investigators. High levels of methyl-CpG correlate with transcriptional inactivity and nuclease resistance in endogenous chromosomes (Antequera *et al.*, 1990; 1989). Methylated DNA transfected into mammalian cells is also assembled into a nuclease-resistant structure containing unusual nucleosomal particles (Keshet *et al.*, 1986). These unusual nucleosomes migrate as large nucleoprotein complexes on agarose gels, and the complexes are held together by higher-order protein–DNA interactions even though there is an abundance of micrococcal nuclease cleavage points within the DNA. Individual nucleosomes assembled on methylated DNA appear to interact together in a more stable manner than they do when assembled on unmethylated templates (Keshet *et al.*, 1986). The replacement of histone H1 with MeCP2 is a possible explanation for the assembly of a distinct chromatin structure on methylated DNA (Nan *et al.*, 1997).

Early experiments using the micro-injection of templates into the nuclei of mammalian cells suggested that the prior assembly of methylated, but not unmethylated, DNA into chromatin represses transcription (Buschhausen *et al.*, 1987). The importance of a nucleosomal infrastructure for transcriptional repression that is dependent on DNA methylation was reinforced by the observation that immediately after injection into *Xenopus* oocyte nuclei, both methylated and unmethylated templates have equivalent activity (Kass *et al.*, 1997). However, as chromatin is assembled, the methylated DNA is repressed, with the loss of both DNase I hypersensitivity and engaged RNA polymerase. The requirement for nucleosomes to exert efficient repression can be explained in several ways. The repression domain of MeCP2 recruits a corepressor complex containing Sin3 and histone deacetylase that directs the modification of the chromatin template into a more stable and transcriptionally inert state (Jones *et al.*, 1998; Nan *et al.*, 1998) (*Figure 5*). In addition, MeCP2 might bind more efficiently to nucleosomal rather than to naked DNA (Chandler *et al.*, 1999). Any cooperative interactions between molecules could propagate the association of MeCP2 along the nucleosomal array even into unmethylated DNA segments. This latter mechanism is analogous to the nucleation of heterochromatin assembly at the yeast telomeres by the DNA-binding protein, Rap1, which then recruits the repressors Sir3p and Sir4p to organize chromatin into a repressive structure (Grunstein *et al.*, 1995; Hecht *et al.*, 1996). All of these potential mechanisms could individually, or together, contribute to the assembly of a repressive chromatin domain.

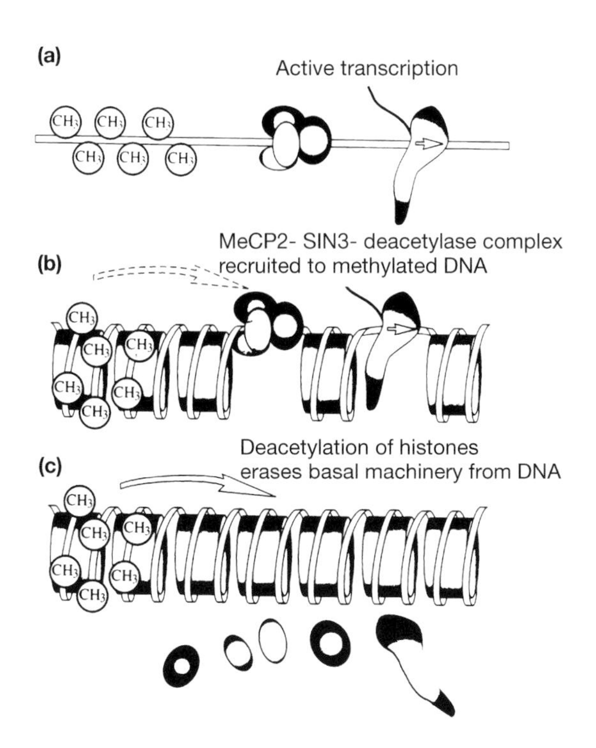

Figure 5. DNA methylation represses transcription by chromatin modification. (a) methylated naked DNA is transcriptionally active. (b) Methylated DNA assembled into nucleosomes is bound by MeCP2 which recruits a corepressor complex containing Sin3 and histone deacetylase. (c) Deacetylation of the histones leads to the assembly of a repressive chromatin structure that erases the basal transcriptional machinery from DNA.

If methylated DNA directs the assembly of a specialized repressive chromatin structure, it might be anticipated that the transcriptional machinery will have less access to such a structure than the orthodox chromatin assembled on unmethylated promoters and genes. Activators such as Gal4/VP16 can normally penetrate a pre-assembled chromatin template to activate transcription, even in the presence of histone Hl (Laybourn and Kadonaga, 1992). However once chromatin has been assembled on methylated DNA, Gal4/VP16 can no longer gain access to its binding sites and activate transcription (Kass *et al.*, 1997). This suggests that the specialized features of chromatin assembly on methylated DNA provide a molecular lock to silence the transcription process permanently (Siegfied and Cedar, 1997). This capacity of DNA methylation to strengthen transcriptional silencing in a chromatin context could be an important contributor to the separation of the genome into active and inactive compartments in a differentiated cell.

7. Conclusion

Chromatin and chromosomes represent the true environment for transcriptional control. Experiments have now established that chromatin and chromosomal architecture have essential functions in transcriptional control. Genetic and biochemical approaches have defined numerous chromatin remodeling machines that control nucleosome structure and transcriptional activity or repression. The challenge for the future is to understand how these machines are targeted, what they do in mechanistic terms, and how they contribute to the organization of the nucleus itself.

References

Adachi, Y., Kas, E. and Laemmli, U.K. (1989) Preferential, cooperative binding of topoisomerase II to scaffold associated regions. *EMBO J.* **8**: 3997–4006.

Alevizopoulos, A., Dusserre, Y., Tsai-Pflulgfelderm, von der Weid, T., Wahli, W. and Mermod, N. (1995) A proline-rich TGF-β-responsive transcriptional activator interacts with histone H3. *Genes Dev.* **9**: 3051–3066.

Almouzni, G. and Wolffe, A.P. (1993a) Replication coupled chromatin assembly is required for the repression of basal transcription *in vivo*. *Genes Dev.* **7**: 2033–2047.

Almouzni, G. and Wolffe, A.P. (1993b) Nuclear assembly, structure and function: the use of *Xenopus in vitro* systems. *Exp. Cell Res.* **205**: 1–15.

Alland, L., Muhle, R., Hou H. Jr., *et al.* (1997) Role of NCoR and histone deacetylase in Sin3-mediated transcriptional and oncogenic repression. *Nature* **387**: 49–55.

Allen, G.C., Hall, G. Jr., Michalowski, S. *et al.* (1996) High level transgene expression in plant cells: effects of a strong scaffold attachment region from tobacco. *Plant Cell* **8**: 899–913.

Allen, G.C., Hall, G.E. Jr., Childs, L.C., Wissinger, A.K., Spiker, S. and Thompson, W.F. (1993) Scaffold attachment regions increase reporter gene expression in stably transformed plant cells. *Plant Cell* 5: 603–613.

Annunziato, A.A., Frado, L.-L.Y., Seale, R.L. and Woodcock, C.L.F. (1988) Treatment with sodium butyrate inhibits the complete condensation of interphase chromatin. *Chromosoma* **96**: 132–138.

Antequera, F., Macleod, D. and Bird, A. (1989) Specific protection of methylated CpGs in mammalian nuclei. *Cell* **58**: 509–517.

Antequera, F., Boyes, J. and Bird, A. (1990) High levels of *de novo* methylation and altered chromatin structure at CpG islands in cell lines. *Cell* **62**: 503–514.

Arents, G., Burlingame, R.W., Wang, B.W., Love, W.E. and Moudrianakis, E.N. (1991) The nucleosomal core histone octamer at 3.1 Å resolution: a tripartite protein assembly and a left-handed superhelix. *Proc. Natl Acad. Sci. USA* **88**: 10148–10152.

Ayer, D.E., Kretzner, L. and Eisenman, R.N. (1993) Mad: a heterodimeric partner for Max that antogonizes Myc transcriptional activity. *Cell* **72**: 211–222.

Ayer, D.E., Lawrence, Q.A. and Eisenman, R.N. (1995) Mad–Max transcriptional repression is mediated by ternary complex formation with mammalian homologs of yeast repressor Sin3. *Cell* **80**: 767–776.

Baneres, J.L., Martin, A. and Parello, J. (1997) The *N* tails of histones H3 and H4 adopt a highly structured conformation in the nucleosome. *J. Mol. Biol.* **273**: 503–508.

Barlev, N.A., Candau, R., Wang, L., Darpino, P., Silverman, N. and Berger, S.L. (1995) Characterization of physical interactions of the putative transcriptional adaptor ADA2 with acidic activation domains and TATA-binding-protein. *J. Biol. Chem.* **270**: 19334–19337.

Bauer, W.R., Hayes, J.J., White, J.H. and Wolffe, A.P. (1994) Nucleosome structural changes due to acetylation. *J. Mol. Biol.* **236**: 685–690.

Belmont, A.S. and Bruce, K. (1994) Visualization of G1 chromosomes: a folded, twisted, supercoiled chromonema model of interphase chromatid structure. *J. Cell Biol.* **127**: 287–302.

Belmont, A.S., Sedat, J.W. and Agard, D.A. (1987) A three dimensional approach to mitotic chromosome structure: evidence for a complex hierarchical organization. *J. Cell Biol.* **105**: 77–92.

Belmont, A.S., Braunfeld, M.B., Sedat, J.W. and Agard, D.A. (1989) Large scale chromatin structural domains within mitotic and interphase chromosomes *in vivo* and *in vitro*. *Chromosoma* **98**: 129–43.

Berger, S.L., Pina, B., Silverman, N. *et al.* (1992) Genetic isolation of ADA2: a potential transcriptional adaptor required for function of certain acidic activation domains. *Cell* **70**: 251–265.

Bird, A. (1995) Gene number, noise reduction and biological complexity. *Trends Genet.* **11**: 94–100.

Bird, A.P. (1986) CpG-rich islands and the function of DNA methylation. *Nature* **321**: 209–213.

Böhm, L. and Crane-Robinson, C. (1984) Proteases as structural probes for chromatin: the domain structure of histones. *Biosci. Rep.* **4**: 365–386.

Bouvet, P., Dimitrov, S. and Wolffe, A.P. (1994) Specific regulation of chromosomal 5S rRNA gene transcription *in vivo* by histone H1. *Genes Dev.* **8**: 1147–1159.

Boy de la Tour, E. and Laemmli, U.K. (1988) The metaphase scaffold is helically folded: sister chromatids have predominantly opposite helical handedness. *Cell* **55**: 937–944.

Bresnick, E.H., Bustin, M., Marsaud, V., Richard-Foy, H. and Hager, G.L. (1992) The transcriptionally-active MMTV promoter is depleted of histone H1. *Nucl. Acids Res.* **20**: 273–278.

Brownell, J.E. and Allis, C.D. (1996) Special HATs for special occasions: linking histone acetylation to chromatin assembly and gene activation. *Curr. Opin. Genet. Dev.* **6**: 176–184.

Brownell, J.E., Zhou, J., Ranalli, T. *et al.* (1996) Tetrahymena histone acetyltransferase A: a homolog to yeast Gcn5p linking histone acetylation to gene activation. *Cell* **84**: 843–851.

Buschhausen, G., Wittig, B., Graessmann, M. and Graessman, A. (1987) Chromatin structure is required to block transcription of the methylated herpes simplex virus thymidine kinase gene. *Proc. Natl Acad. Sci. USA* **84**: 1177–1181.

Busslinger, M., Hurst, J. and Flavell, R.A. (1983) DNA methylation and the regulation of globin gene expression. *Cell* **34**: 197–206.

Cairns, B.R., Kim, Y.J., Sayre, M.H., Laurent, B.C. and Kornberg, R.D. (1994) A multisubunit complex containing the SWIA/ADR6, SWI2/SNF2, SWI3/SNF5 and SNF6 gene products isolated from yeast. *Proc. Natl Acad. Sci. USA* **91**: 1950–1954.

Cairns, B.R., Erdjument-Bromage, H., Tempst, P., Winston, F. and Kornberg, R.D. (1998) Two actin-related proteins are shared functional components of the chromatin-remodeling complexes RSC and SWI/SNF. *Mol. Cell* **2**: 639–651.

Callan, H.G. (1986) *Lampbrush Chromosomes*. Springer Verlag, Berlin.

Callan, H.G., Gall, J.G. and Berg, C.A. (1987) The lampbrush chromosomes of *Xenopus laevis*: preparation, identification and distribution of 5S DNA sequences. *Chromosoma* **95**: 236–250.

Cary, P.D., Crane-Robinson, C., Bradbury, E.M. and Dixon, G.H. (1982) Effect of acetylation on the binding of *N*-terminal peptides of histone H4 to DNA. *Eur. J. Biochem.* **127**: 137–143.

Chandler, S.P., Guschin, D., Landsberger, N. and Wolffe, A.P. (1999) The methyl CpG binding transcriptional repressor MeCP2 stably associates with nucleosomal DNA. *Biochemistry* **38**: 7008–7018.

Chen, T.A. and Allfrey, V.G. (1987) Rapid and reversible changes in nucleosome structure accompany the activation, repression and superinduction of the murine proto-oncogenes *c-fos* and *c-myc*. *Proc. Natl Acad. Sci. USA* **84**: 5252–5256.

Chen, J., Willingham, T., Margraf, L.R, Schreiber-Agus, N., DePinho, R.A. and Nisen, P.D. (1995) Effects of the MYC oncogene antagonist, MAD, on proliferation, cell cycling and the malignant phenotype of human brain tumor cells. *Nat. Med.* **1**: 638–643.

Chiba, H., Muramatsu, M., Nomoto, A. and Kato, H. (1994) Two human homologs of *Saccharomyces cerevisiae* SWI2/SNF2 and *Drosophila brahma* are transcriptional coactivators cooperating with the estrogen receptor and the retinoic acid receptor. *Nucl. Acids Res.* **22**: 1815–1820.

Chipev, C.C. and Wolffe, A.P. (1992) Chromosomal organization of *Xenopus laevis* oocyte and somatic 5S rRNA genes *in vivo*. *Mol. Cell. Biol.* **12**: 45–55.

Chung, J.H., Whiteley, M. and Felsenfeld, G. (1993) A 5′ element of the chicken β-globin domain serves as an insulator in human erythroid cells and protects against position effect in *Drosophila*. *Cell* **74**: 505–514.

Cirillo, L.A., McPherson, C.E., Bossard, P. *et al.* (1998) Binding of the winged-helix transcription factor HNF3 to a linker histone site on the nucleosome. *EMBO J.* **17**: 244–254.

Clark, D.J. and Kimura, T. (1990) Electrostatic mechanism of chromatin folding. *J. Mol. Biol.* **211**: 883–896.

Cook, P.R. (1988) The nucleoskeleton: artefact, passive framework or active site? *J. Cell Sci.* **90**: 1–6.

Cook, P.R. and Brazell, I.A. (1975) Supercoils in human DNA. *J. Cell Sci.* **19**: 261–279.

Côté, J., Quinn, J., Workman, J.L. and Peterson, C.L. (1994) Stimulation of GAL4 derivative binding to nucleosomal DNA by the yeast SWI/SNF complex. *Science* **265**: 53–60.

Côté, J., Peterson, C.L., Workman, J.L. (1998) Perturbation of nucleosome core structure by the SWI/SNF complex persists after its detachment, enhancing subsequent transcription factor binding. *Proc. Natl Acad. Sci. USA* **95**: 4947–4952.

Covault, J. and Chalkley, R. (1980) The identification of distinct populations of acetylated histone. *J. Biol. Chem.* **255**: 9110–9116.

Craig, J.M., Boyle, S., Perry, P. and Bickmore, W.A. (1997) Scaffold attachments within the human genome. *J. Cell Sci.* **110**: 2673–2682.

Dallas, P.B., Cheney, I.W., Liao, D.-W. *et al.* (1998) p300/CREB binding protein-related p270 is a component of mammalian SWI/SNF complexes. *Mol. Cell. Biol.* **18**: 3596–3603.

Dasso, M., Dimitrov, S. and Wolffe, A.P. (1994) Nuclear assembly is independent of linker histones. *Proc. Natl Acad. Sci. USA* **91**: 12477–12481.
Davie, J.R., Samuel S., Spencer, V. *et al.* (1998) Nuclear matrix: application to diagnosis of cancer and role in transcription and modulation of chromatin structure. *Gene Ther. Mol. Biol.* **1**: 509–528.
Dikstein, R., Ruppert, S. and Tjian, R. (1996) $TAF_{II}250$ is a bipartite protein kinase that phosphorylates the basal transcription factor RAP74. *Cell* **84**: 781–790.
Earnshaw, W.C. (1988) Mitotic chromosome structure. *Bioessays* **9**: 147–150.
Earnshaw, W.C. (1991) Large scale chromosome structure and organization. *Curr. Opin. Struct. Biol.* **1**: 237–244.
Eickbusch, T.H. and Moudrianakis, E.N. (1978) The histone core complex: an octamer assembled by two sets of protein–protein interactions. *Biochemistry* **17**: 4955–4967.
Felsenfeld, G. (1992) Chromatin: an essential part of the transcriptional apparatus. *Nature* **355**: 219–224.
Festenstein, R., Tolaini, M., Corbella, P. *et al.* (1996) Locus control region function and heterochromatin-induced position effect variegation. *Science* **271**: 1123–1125.
Fiering, S., Epner, E., Robinson, K. *et al.* (1995) Targeted deletion of 5′ HS2 of the murine β-globin locus reveals that it is not essential for proper regulation of the β-globin locus. *Genes Dev.* **9**: 2203–2213.
Filipski, J., Leblanc, J., Youdale, T., Sikorska, M. and Walker, P.R. (1990) Periodicity of DNA folding in higher-order chromatin structures. *EMBO J.* **9**: 1319–1327.
Fletcher, T.M. and Hansen, J.C. (1996) The nucleosomal array: structure/function relationships. *Crit. Rev. Eukaryot. Gene Exp.* **6**: 149–188.
Forrester, W., Takagawa, S., Papayannopoulou, T., Stamatoyannopoulos, G. and Groudine, M. (1987) Evidence for a locus activation region: the formation of developmentally stable hypersensitive sites in globin-expressing hybrids. *Nucl. Acids Res.* **15**: 10159–10177.
Gasser, S.M. and Laemmli, U.K. (1986) The organization of chromatin loops: characterization of a scaffold attachment site. *EMBO J.* 5: 511–518.
Gasser, S.M., Amati, B.B., Cardenas, M.E. and Hofmann, J.F. (1989) Studies on scaffold attachment sites and their relationship to genome function. *Int. Rev. Cytol.* **119**: 57–96.
Gaudreau, L., Schmid, A., Blaschke, D., Ptashne, M. and Horz, W. (1997) RNA polymerase II holoenzyme recruitment is sufficient to remodel chromatin at the yeast *PHO5* promoter. *Cell* **89**: 55–62.
Georgakopoulos, T. and Thireos, G. (1992) Two distinct yeast transcriptional activators require the function of the GCN5 protein to promote normal levels of transcription. *EMBO J.* **11**: 4145–4152.
Georgakopoulos, T., Gounalaki, N. and Thireos, G. (1995) Genetic evidence for the interaction of the yeast transcriptional co-activator proteins GCN5 and ADA2. *Mol. Gen. Genet.* **246**: 723–728.
Georgel, P.T., Tsukiyama, T. and Wu, C. (1997) Role of histone tails in nucleosome remodeling by *Drosophila* NURF. *EMBO J.* **16**: 4717–4726.
Glotov, B.O., Itkes, A.V., Nikolaev, L.G. and Severin, E.S. (1978) Evidence for close proximity between histones H1 and H3 in chromatin of intact nuclei. *FEBS Lett* **91**: 149–152.
Grant, P.A., Schieltz, D., Pray-Grant, M.G. *et al.* (1998) A subset of $TAF_{II}S$ are integral components of the SAGA complex required for nucleosome acetylation and transcriptional stimulation. *Cell* **94**: 45–53.
Grosveld, F., van Assendelft, G.B., Greaves, D.R. and Kollias, G. (1987) Position independent, high level expression of the human β-globin gene in transgenic mice. *Cell* **51**: 975–985.
Grunstein, M. (1990) Histone function in transcription. *Ann. Rev. Cell Biol.* **6**: 643–678.
Grunstein, M., Hecht, A., Fisher-Adams, G. *et al.* (1995) The regulation of enchromatin and heterochromatin by histones in yeast. *J. Cell Sci.* **519**: 29–36.
Grunstein, M. (1997) Histone acetylation in chromatin structure and transcription. *Nature* **389**: 349–352.
Guarente, L. (1995) Transcriptional coactivators in yeast and beyond. *Trends Biochem. Sci.* **20**: 517–527.
Guschin, D., Chandler, S. and Wolffe, A.P. (1998) Asymmetric linker histone association directs the asymmetric rearrangement of core histone interactions in a positioned nucleosome containing a thyroid hromone response element. *Biochemistry* **37**: 8629–8636.
Hansen, J.C. (1997) The core histone amino-termini:combinatorial interaction domains that link chromatin structure with function. *Chemtracts: biochem. mol. biol.* **10**: 737–750.
Hansen, J.C. and Wolffe, A.P. (1994) A role for histones H2A/H2B in chromatin folding and transcriptional repression. *Proc. Natl Acad. Sci. USA* **91**: 2339–2343.

Hansen, J.C., Tse, C. and Wolffe, A.P. (1998) Structure and function of the core histone *N*-termini: more than meets the eye. *Biochemistry* **37**: 17637–17641.

Hayes, J.J. and Wolffe, A.P. (1992) Histones H2A/H2B inhibit the interactions of transcription factor IIIA with the *Xenopus borealis* somatic 5S RNA gene in a nucleosome. *Proc. Natl Acad. Sci. USA* **89**: 1229–1233.

Hayes, J.J., Clark, D.J. and Wolffe, A.P. (1991a) Histone contributions to the structure of DNA in the nucleosome. *Proc. Natl Acad. Sci. USA* **88**: 6829–6833.

Hayes, J.J., Bashkin, J., Tullius, T.D. and Wolffe, A.P. (1991b) The histone core exerts a dominant constraint on the structure of DNA in a nucleosome. *Biochemistry* **30**: 8434–8440.

Hayes, J.J., Tullius, T.D. and Wolffe, A.P. (1990) The structure of DNA in a nucleosome. *Proc. Natl Acad. Sci. USA* **87**: 7405–7409.

Hebbes, T.R., Clayton, A.L., Thorne, A.W. and Crane-Robison, C. (1994) Core histone hyperacetylation comaps with generalized DNase I sensitivity in the chicken β-globin chromosomal domain. *EMBO J.* **13**: 1823–1830.

Hecht, A., Strahl-Bolsinger, S. and Grunstein, M. (1996) Spreading of the transcriptional repressor SIR3 from telomeric chromatin. *Nature* **383**: 92–96.

Heinzel, T., Laviusky, R.M., Mullen, T.M. *et al.* (1997) N-CoR, mSIN3, and histone deacetylase in a complex required for repression by nuclear receptors and Mad. *Nature* **387**: 43–48.

Hendzel, M.J., Wei, Y., Mancini, M.A. *et al.* (1997) Mitosis specific phosphorylation of H3 initiates primarily within pericentric heterochromatin during GZ and spreads in an ordered fashion coincident with mitotic chromosome condensation. *Chromosoma* **106**: 348–360.

Henikoff, S. (1994) A reconsideration of the mechanisms of position effect. *Genetics* **138**: 1–5.

Herskowitz, I., Andrews, B., Kruger, W. *et al.* (1992) Integration of multiple regulatory inputs in the control of *HO* expression in yeast. In: *Transcriptional Regulation* (eds S. McKnight and K. Yamamoto). Cold Spring Harbor Press, New York, Vol. 2, pp. 949–974.

Hill, C.S., Rimmer, J.M., Green, B.N., Finch, J.T. and Thomas, J.O. (1991) Histone–DNA interactions and their modulation by phosphorylation of Ser-Pro-X-Lys/Arg-motifs. *EMBO J.* **10**: 1939–1948.

Hirschhorn, J.N., Brown, S.A., Clark, C.D. and Winston, F. (1992) Evidence that SNF2/SWI2 and SNF activate transcription in yeast by altering chromatin structure. *Genes Dev.* **6**: 2288–2298.

Hoeller, M., Westin, G., Jiricny, J. and Schaffner, W. (1988) Spl transcription factor binds DNA and activates transcription even when the binding site is CpG methylated. *Genes Dev.* **2**: 1127–1135.

Holliday, R. (1987) Inheritance of epigenetic defects. *Science* **238**: 163–170.

Hong, L., Schroth, G.P., Matthews, H.R., Yau, P. and Bradbury, E.M. (1993) Studies of the DNA binding properties of the histone H4 amino terminus. *J. Biol. Chem.* **268**: 305–314.

Horiuchi, J., Silverman, N., Marcus, G.A., and Guarente, L. (1995) ADA3, a putative transcriptional adaptor, consists of two separable domains and interacts with ADA2 and GCN5 in a trimeric complex. *Mol. Cell. Biol.* **15**: 1203–1209.

Horlein, A.J., Naar, A.M., Heinzel, T. *et al.* (1995) Ligand-independent repression by the thyroid hormone receptor mediated by a nuclear receptor co-repressor. *Nature* **377**: 397–404.

Hurlin, P.J., Queva, C., Koskinen, P.J. *et al.* (1995) Mad3 and Mad4: novel Max-interacting transcriptional repressors that suppress c-Myc-dependent transformation and are expressed during neural and epidermal differentiation. *EMBO J.* **14**: 5646–5659.

Iguchi-Ariga, S.M.M. and Schaffner, W. (1989) CpG methylation of the cAMP responsive enhancer/promoter sequence TGACGTCA abolishes specific factor binding as well as transcriptional activation. *Genes Dev.* **3**: 612–619.

Imbalzano, A.M., Kwon, H., Green, M.R. and Kingston, R.E. (1994) Facilitated binding of TATA-binding protein to nucleosomal DNA. *Nature* **370**: 481–485.

Izaurralde, E., Kas, E. and Laemmli, U.K. (1989) Highly preferential nucleation of histone H1 assembly on scaffold associated regions. *J. Mol. Biol.* **210**: 573–585.

Jackson, D.A., Dickinson, P. and Cook, P.R. (1990) The size of chromatin loops in HeLa cells. *EMBO J.* **9**: 567–571.

Jackson, D.A., Yuan, J. and Cook, P.R. (1988) A gentle method for preparing cyto- and nucleoskeletons and associated chromatin. *J. Cell Sci.* **90**: 365–378.

Janknecht, R. and Hunter, T. (1996) A growing coactivator network. *Nature* **383**: 22–23.

Jones, P.L., Veenstra, G.J.C., Wade, P.A. *et al.* (1998) Methylated DNA and MeCP2 recruit histone deacetylase to repress transcription. *Nature Genet.* **19**: 187–191.

Jost, J.-P. and Hofsteenge, J. (1992) The repressor MDBP-2 is a member of the histone H1 family that binds preferentially *in vitro* and *in vivo* to methylated non specific DNA sequences. *Proc. Natl Acad. Sci. USA* **89**: 9499–9503.

Kamei, Y., Xu, L., Heinzel, T. *et al.* (1996) A CBP integrator complex mediates transcriptional activation and AP-1 inhibition by nuclear receptors. *Cell* **85**: 403–414.

Karantza, V., Freire, E. and Moudrianakis, E.N. (1996) Thermodynamic studies of the core histones: pH and ionic strength effects on the stability of the (H3-H4)/(H3-H4)^2 system. *Biochemistry* **35**: 2037–2046.

Kass, S.U., Landsberger, N. and Wolffe, A.P. (1997) DNA methylation directs a time-dependent repression of transcription initiation. *Curr. Biol.* **7**: 157–165.

Keshet, I., Lieman-Hurwitz, J. and Cedar, H. (1986) DNA methylation affects the formation of active chromatin. *Cell* **44**: 535–543.

Khavari, P.A., Peterson, C.L., Tamkun, J.W., Mendel, D.B. and Crabtree, G.R. (1993) BRG1 contains a conserved domain of the SWI2/SNF2 family necessary for normal mitotic growth and transcription. *Nature* **366**: 170–174.

Krajewski, W.A. and Becker, P.B. (1998) Reconstitution of hyperacetylated, DNase I-sensitive chromatin characterized by high conformational flexibility of nucleosomal DNA. *Proc. Natl Acad. Sci. USA* **95**: 1540–1545.

Kruger, W., Peterson, C.L., Sil, A. *et al.* (1995) Amino acid substitutions in the structured domains of histones H3 and H4 partially relieve the requirement of the yeast SWI/SNF complex for transcription. *Genes Dev.* **9**: 2770–2779.

Kuo, M.-H., Brownell, J.E., Sobal, R.E. *et al.* (1996) Transcription-linked acetylation by Gcn5p of histones H3 and H4 at specific lysines. *Nature* **383**: 269–272.

Kuo, M.-H., Zhou, J., Jambeck, P., Churchill, M.E.A. and Allis, C.D. (1998) Histone acetyltransferase activity of yeast Gcn5p is required for the activation of target genes *in vivo*. *Genes Dev.* **12**: 627–639.

Kurumizaka, H. and Wolffe, A.P. (1997) Sin mutations of histone H3: influence on nucleosome core structure and function. *Mol. Cell. Biol.* **17**: 6953–6969.

Laherty, C.D., Yang, W.M., Sun, J.M., Davie, J.R., Seto, E. and Eisenman, R.M. (1997) Histone deacetylase associated with the mSin3 corepressor mediate Mad transcriptional repression. *Cell* **89**: 349–356.

Lahoz, E.G., Xu, L., Schreiber-Agus, N. and DePinho, R.A. (1994) Suppression of Myc, but not E1a, transformation activity by Max-associated proteins, Mad and Mxi1. *Proc. Natl Acad. Sci. USA* **91**: 5503–5507.

Laurent, B.C. and Carlson, M. (1992) Yeast SNF2/SWI2, SNF5 and SNF6 proteins function coordinately with the gene-specific transcriptional activators Gal4 and Bicoid. *Genes Dev.* **6**: 1707–1715.

Lee, H.L. and Archer, T.K. (1998) Prolonged glucocorticoid exposure dephosphorylates histone H1 and inactivates the MMTV promoter. *EMBO J.* **17**: 1454–1466.

Lee, D.Y., Hayes, J.J., Pruss, D. and Wolffe, A.P. (1993) A positive role for histone acetylation in transcription factor binding to nucleosomal DNA. *Cell* **72**: 73–84.

Laybourn, P.J. and Kadonaga, J.T. (1992) Threshold phenomena and long distance activation of transcription by RNA polymerase II. *Science* **257**: 1682–1685.

Leng, M. and Felsenfeld, G. (1966) The preferential interaction of polylysine and polyarginine with specific base sequences in DNA. *Proc. Natl Acad. Sci. USA* **56**: 1325–1332.

Levinger, L. and Varshavsky, A. (1980) High-resolution fractionation of nucleosomes: minor particles 'whiskers' and separation of nucleosomes containing and lacking A24 semihistone. *Proc. Natl Acad. Sci. USA* **77**: 3244–3248.

Levinger, L. and Varshavsky, A. (1982) Selective arrangement of ubiquitinated and D1 protein containing nucleosomes within the *Drosophila* genome. *Cell* **28**: 375–385.

Lewis, J.D., Meehan, R.R., Henzel, W.J. *et al.* (1992) Purification, sequence and cellular localization of a novel chromosomal protein that binds to methylated DNA. *Cell* **69**: 905–914.

Li, E., Beard, C. and Jaenisch, R. (1993) Role of DNA methylation in genomic imprinting. *Nature* **366**: 362–365.

Luger, K., Mader, A.W., Richmond, R.K., Sargent, D.F. and Richmond, T.J. (1997) X-ray structure of the nucleosome core particle at 2.8Å resolution. *Nature* **389**: 251–259.

McArthur, M. and Thomas, J.O. (1996) A preference of histone H1 for methylated DNA. *EMBO J.* **15**: 1705–1714.

Mahadevan, L.C., Willis, A.C. and Barrah, M.J. (1991) Rapid histone H3 phosphorylation in response to growth factors, phorbol esters, okadaic acid and protein synthesis inhibitors. *Cell* **65**: 775–783.

Marcus, G.A., Silverman, N., Berger, S.L., Horiuchi, J. and Guarente, L. (1994). Functional similarity and physical association between GCN5 and ADA2: putative transcriptional adaptors. *EMBO J.* **1354**: 4807–4815.

Martinez, E., Kundu, T.K., Fu, J. and Roeder, R.G. (1998). A human SPT3-TAFII31-GCN5-L acetylase complex distinct from transcription factor IID. *J. Biol. Chem.* **273**: 23781–23785.

Meehan, R.R., Lewis, J.D., McKay, S., Kleiner, E.L. and Bird, A.P. (1989) Identification of a mammalian protein that binds specifically to DNA containing methylated CpGs. *Cell* **58**: 499–507.

Meehan, R.R., Lewis, J.D. and Bird, A.P. (1992) Characterization of MeCP2, a vertebrate DNA binding protein with affinity for methylated DNA. *Nucl. Acids Res.* **20**: 5085–5092.

Mirkovitch, J., Mirault, M.E. and Laemmli, U. (1984) Organization of the higher-order chromatin loop: specific DNA attachment sites on nuclear scaffold. *Cell* **39**: 223–232.

Mizzen, C.A., Yang, X.J., Kobuko, T. ***et al.*** (1996). The $TAF_{II}250$ subunit of TFIID has histone acetyltransferase activity. *Cell* **87**: 1261.

Muchardt, C. and Yaniv, M. (1993) A human homolog of *Saccharomyces cerevisiae* SNF2/SWI2 and *Drosophila brm* genes potentiates transcriptional activation by the glucocorticoid receptor. *EMBO J.* **12**: 4279–4290.

Mutskov, V., Gerber, D., Angelov, D., Ausio, J., Workman, J. and Dimitrov, S. (1998) Persistant interactions of core histone tails with nucleosomal DNA following acetylation and transcription factor binding. *Mol. Cell. Biol.* **18**: 6293–6304.

Nan, X., Meehan, R.R. and Bird, A.P. (1993) Dissection of the methyl-CpG binding domain from the chromosal protein MeCP2. *Nucl. Acids Res.* **21**: 4886–4892.

Nan, X. Tate, P., Li, E. and Bird, A.P. (1996) MeCP2 is a transcriptional repressor with abundant binding sites in genomic chromatin. *Mol. Cell. Biol.* **16**: 414–421.

Nan, X., Campoy, J. and Bird, A. (1997) MeCP2 is a transcriptional repressor with abundant binding sites in genomic chromatin. *Cell* **88**: 1–11.

Nan, X., Ng, H.H., Johnson, C.A. ***et al.*** (1998) Transcriptional repression by the methyl-CpG binding protein MeCP2 involves a histone deacetylase complex. *Nature* **393**: 386–389.

Onnuki, Y. (1968) Structure of chromosomes: morphological studies on the spiral structure of human somatic chromosomes. *Chromosoma* **25**: 402–408.

Ogryzko, V.V., Schiltz, R.L., Russanova, V., Howard, B.H. and Nakatani, Y. (1996) The transcriptional coactivators p300 and CBP are histone acetyltransferases. *Cell* **87**: 953–959.

Patterton, H.G., Landel, C.C., Landsman, D., Peterson, C.L. and Simpson, R.T. (1998) The biochemical and phenotypic characterization of *Hho1p*, the putative linker histone H1 of *Saccharomyces cerevisiae*. *J. Biol. Chem.* **273**: 7268–7276.

Paulson, J.R. and Laemmli, U.K. (1977) The structure of histone-depleted metaphase chromosomes. *Cell* **12**: 817–828.

Pennings, S., Meersseman, G. and Bradbury, E.M. (1994) Linker histones H1 and H5 prevent the mobility of positioned nucleosomes. *Proc. Natl Acad. Sci. USA* **91**: 10275–10279.

Perry, M. and Chalkley, R. (1982) Histone acetylation increases the solubility of chromatin and occurs sequentially over most of the chromatin. *J. Biol. Chem.* **257**: 7336–7347.

Peterson, C.L. and Herskowitz, I. (1992) Characterization of the yeast *SWI1*, *SWI2* and *SWI3* genes which encode a global activator of transcription. *Cell* **68**: 573–584.

Peterson, C.L., Dingwall, A. and Scott, M.P. (1994) SWI/SNF gene products are components of a large multisubunit complex required for transcriptional enhancement. *Proc. Natl Acad. Sci. USA* **91**: 2905–2908.

Pruss, D., Bartholomew, B., Persinger, J. (1996) An asymmetric model for the nucleosome: a binding site for linker histones inside the DNA gyres. *Science* **274**: 614–617.

Ramakrishnan, V., Finch, J.T., Graziano, V., Lee, P.L. and Sweet, R.M. (1993) Crystal structure of globular domain of histone H5 and its implications for nucleosome binding. *Nature* **362**: 219–224.

Ranjan, M., Wong, J. and Shi, Y.-B. (1994) Transcriptional repression of *Xenopus* TRβ gene is mediated by a thyroid hormone response element located near the start site. *J. Biol. Chem.* **269**: 24699–24705.

Rattner, J.B. and Lin, C.C. (1985) Radical loops and helical coils coexist in metaphase chromosomes. *Cell* **42**: 291–296.

Riggs, A.D. and Porter, T.N. (1996) in *Epigenetic Mechanisms of Gene Regulation* (eds. X. Russo, R.A. Martienssen and A.D. Riggs). Cold Spring Harbor Laboratory Press, New York; pp. 29–45.

Santisleban, M.S., Arents, G., Moudrianakis, E.N. and Smith, M.M. (1997) Histone octamer function *in vivo*: mutations in the dimer–tetramer interfaces disrupt both gene activation and repression. *EMBO J.* **16**: 2493–2506.

Schnitzler, G., Sif, S. and Kingston, R.E. (1998) Human SWI/SNF interconverts a nucleosome between its base state and a stable remodeled state. *Cell* **94**: 17–27.

Schwartz, D.C. and Cantor, C.R. (1984) Separation of yeast chromosome-sized DNAs by pulsed field gradient gel electrophoresis. *Cell* **37**: 67–75.

Sedat, J. and Manuelidis, L. (1978) A direct approach to the structure of mitotic chromosomes. *Cold Spring Harbor Symp. Quant. Biol.* **42**: 331–350.

Sera, T. and Wolffe, A.P. (1998) The role of histone H1 as an architectural determinant of chromatin structure and as a specific repressor of transcription on the *Xenopus* oocyte 5S rRNA genes. *Mol. Cell. Biol.* **18**: 3668–3680.

Shen, X., Yu, L., Weir, J.W. and Gorovsky, M.A. (1995) Linker histones are not essential and affect chromatin condensation *in vivo*. *Cell* **82**: 47–56.

Shen, X. and Gorovsky, M.A. (1996) Linker histone H1 regulates specific gene expression but not global transcription *in vivo*. *Cell* **86**: 475–483.

Shi, Y., Seto, E., Chang, L.-S. and Shenk, T. (1991) Transcriptional repression by YY1, a human GL1-Krüppel-related protein, and relief of repression by adenovirus E1A protein. *Cell* **67**: 377–388.

Shrivastava, A. and Calame, K. (1994) An analysis of genes regulated by the multi-functional transcriptional regulatory Yin yang 1. *Nucl. Acids Res.* **22**: 5151–5155.

Siegfied, Z. and Cedar, H. (1997) DNA methylation: a molecular lock. *Curr. Biol.* **7**: R305–R307.

Simpson, R.T. (1991) Nucleosome positioning: occurrence, mechanisms and functional consequences. *Prog. Nucl. Acids Res. Mol. Biol.* **40**: 143–184.

Stief, A., Winter, D.M., Stratting, W.E.H. and Sippel, A.E. (1989) A nuclear DNA attachment element mediates elevated and position independent gene activity. *Nature* 341: 343–345.

Steinbach, O.C., Wolffe, A.P. and Rupp, R. (1997) Somatic linker histones cause loss of mesodermal competence in *Xenopus*. *Nature* **389**: 395–399.

Strick, R. and Laemmli, U.K. (1995) SARs are *cis* DNA elements of chromosome dynamics: synthesis of a SAR repressor protein. *Cell* **83**: 1137–1148.

Sweet, M.T., Jones, K. and Allis, C.D. (1996) Phosphorylation of linker histones is associated with transcriptional activation in a normally silent nucleus. *J. Cell Biol.* **135**: 1219–1228.

Szyf, M. (1996) The DNA methylation machinery as a target for anticancer therapy. *Pharmacol. Ther.* **70**: 1–37.

Tamkun, J.W., Deuring, R., Scott, M.P. *et al.* (1992) Brahma: a regulator of *Drosophila* homeotic genes structurally related to the yeast transcriptional activator SNF2/SWI2. *Cell* **68**: 561–572.

Tate, P.H. and Bird, A.P. (1993) Effects of DNA methylation on DNA-binding proteins and gene expression. *Curr. Opin. Genet. Dev.* **3**: 226–231.

Taunton, J., Hassig, C.A. and Schreiber, S.L. (1996) A mammalian histone deacetylase related to a yeast transcriptional regulator Rpd3. *Science* **272**: 408–411.

Tse, C., Sera, T., Wolffe, A.P. and Hansen, J.C. (1998) Disruption of higher order folding by core histone acetylation dramatically enhances transcription of nucleosomal arrays by RNA polymerase III. *Mol. Cell. Biol.* **18**: 4629–4638.

Tsukiyama, T., Becker, P.B. and Wu, C. (1994) ATP-dependent nucleosome disruption at a heat-shock promoter mediated by binding of GAGA transcription factor. *Nature* **367**: 525–532.

Tsukiyama, T. and Wu, C. (1995) Purification and properties of an ATP-dependent nucleosome remodeling factor. *Cell* **83**: 1011–1020.

Ura, K., Hayes, J.J. and Wolffe, A.P. (1995) A positive role for nucleosome mobility in the transcriptional activity of chromatin templates: restriction by linker histones. *EMBO J.* **14**: 752–765.

Usachenko, S.I., Bavykin, S.G., Gavin, I.M. and Bradbury, E.M. (1994) Rearrangement of the histone H2A *C*-terminal domain in the nucleosome. *Proc. Natl Acad. Sci. USA* **91**: 6845–6849.

Usachenko, S.I., Gavin, I.M. and Bavykin, S.G. (1996) Alterations in nucleosome core structure in linker histone-depleted chromatin. *J. Biol. Chem.* **271**: 3831–3836.

Utley, R.T., Ikeda, K., Grant, P.A. *et al.* (1998) Transcriptional activators direct histone acetyltransferase complexes to nucleosomes. *Nature* **394**: 498–502.

Vermaak, D., Steinbach, O.C., Dimitrov, S., Rupp, R.A.W. and Wolffe, A.P. (1998) The globular domain of histone H1 is sufficient to direct specific gene repression in early *Xenopus* embryos. *Curr. Biol.* **8**: 533–536.

Vettese-Dadey, M., Grant, P.A., Hebbes, T.R., Crane-Robinson, C., Allis, C.D. and Workman, J.L. (1996) Acetylation of histone H4 plays a primary role in enhancing transcription factor binding to nucleosomal DNA *in vitro*. *EMBO J.* **15**: 2508–2518.

Vettese-Dadey, M., Walter, P., Chen, H., Juan, L-J. and Workman, J.L. (1994) Role of the histone amino termini in facilitated binding of a transcription factor, GAL4-AH to nucleosome cores. *Mol. Cell. Biol.* **14**: 970–981.

Walters, M.C., Magis, W., Fiering, S. (1996) Transcription enhancers act in *cis* to suppress position-effect variegation. *Genes Dev.* **10**: 185–195.

Wei, Y., Mizzen, C.A., Cook, R.G., Gorovsky, M.A. and Allis, C.D. (1998) Phosphorylation of histone and meiosis in *Tetrahymena*. *Proc. Natl. Acad. Sci. USA* **95**: 7480–7484.

Weih, F. *et al.* (1991) Analysis of CpG methylation and genomic footprinting at the tyrosine amino transferase gene: DNA methylation alone is not sufficient to prevent protein binding *in vivo*. *EMBO J.* **10**: 2559–2567.

White, M.J.D. (1973) *Animal Cytology and Evolution*. Cambridge University Press, Cambridge, pp. 1–58.

Wijgerde, M., Grosveld, F. and Fraser, P. (1995) Transcription complex stability and chromatin dynamics *in vivo*. *Nature* **377**: 209–213.

Wolffe, A.P. (1994) Insulating chromatin. *Curr. Biol.* **4**: 85–87.

Wolffe, A.P. and Hayes, J.J. (1999) Chromatin disruption and modification. *Nucl. Acids Res.* **27**: 711–720.

Wolffe, A.P. and Kurumizaka, H. (1998) The nucleosome – a powerful regulator of transcription. *Prog. Nucl. Acids Res. Mol. Biol.* **61**: 379–422.

Wong, J., Shi, Y.-B. and Wolffe, A.P. (1995) A role for nucleosome assembly in both silencing and activation of the *Xenopus TRβA* gene by the thyroid hormone receptor. *Genes Dev.* **9**: 2696–2711.

Wong, J., Shi, Y.-B. and Wolffe, A.P. (1997a) Determinants of chromatin disruption and transcriptional regulation instigated by the thyroid hormone receptor: hormone regulated chromatin disruption is not sufficient for transcriptional activation. *EMBO J.* **16**: 3158–3171.

Wong, J., Li, Q., Levin, B.-Z., Shi, Y.-B. and Wolffe, A.P. (1997b) Structural and functional features of a specific nucleosome containing a recognition element for the thyroid hormone receptor. *EMBO J.* **16**: 7130–7145.

Wong, J., Patterton, D., Imhof, A., Shi, Y.-B. and Wolffe, A.P. (1998) Distinct requirements for chromatin assembly in transcriptional repression by thyroid hormone receptor and histone deacetylase. *EMBO J.* **17**: 520–534.

Yang, W.M., Inouye, C., Zeng, Y., Bearss, D. and Soto, E. (1996a) Transcriptional repression by YY1 is mediated by interaction with a mammalian homolog of the yeast global regulator RPD3. *Proc. Natl Acad. Sci. USA* **93**: 12845–12850.

Yang, X.-J., Ogryzko, V.V., Nishikawa, J.-I., Howard, B. and Nakatani, Y. (1996b) A p300/CBP-associated factor that competes with the adenoviral E1A oncoprotein. *Nature* **382**: 319–324.

Yoder, J.A., Walsh, C.P. and Bestor, T.H. (1997) Cytosine methylation and the ecology of intragenomic parasites. *Trends Genet.* **13**: 335–340.

Yoshida, M., Kijima, M., Akita, M. and Beppu, T. (1990) Potent and specific inhibition of mammalian histone deacetylase both *in vivo* and *in vitro* by Trichostatin A. *J. Biol. Chem.* **265**: 17174–17179.

Yoshinaga, S.K., Peterson, S.L., Herskowitz, I. and Yamamoto, K.R. (1992) Roles of SWI1, SWI2 and SWI3 proteins for transcriptional enhancement by steroid receptors. *Science* **258**: 1598–1604.

Zervos, A.S., Gyuris, J. and Brent, R. (1993) Mxi1, a protein that specifically interacts with Max to bind Myc–Max recognition sites. *Cell* **72**: 223–232.

Zhang, W., Bone, J.R., Edmondson, D.G., Turner, B.M. and Roth, S.Y. (1998) Essential and redundant functions of histone acetylation revealed by mutation of target lysines and loss of the Gcn5p acetyltransferase. *EMBO J.* **17**: 3155–3167.

Zhao, K., Kas, E., Gonzalez, E. and Laemmli, U.K. (1993) SAR-dependent mobilization of histone H1 by HMG-I/Y *in vitro*: HMG-I/Y is enriched in H1-depleted chromatin. *EMBO J.* **12**: 3237–3247.

Zhou, Y.-B., Gerchman, S.E., Ramakrishnan, V., Travers, A. and Muyldermans, S. (1998) Position and orientation of the globular domain of linker histone H5 on the nucleosome. *Nature* **395**: 402–405.

4

DNA binding by transcription factors

Louise Fairall and John W.R. Schwabe

Structural biology has made a tremendous impact on our understanding of how transcription factors bind to and recognize their specific DNA targets. Over the last 15 years the structures of several hundred protein–DNA complexes have been determined by X-ray crystallography or NMR spectroscopy. These include many DNA-binding domains from both prokaryotic and eukaryotic transcriptional regulators (Lilley, 1995) and more recently, the structures of an increasing number of multiprotein–DNA complexes (Chen, 1999). As a result of these structures we are beginning to understand the myriad of protein–DNA and protein–protein interactions that bring about the precise regulation of gene transcription.

In this chapter we address protein–DNA recognition from two directions. First we discuss some of the general theoretical principles of the recognition of DNA by proteins. Second, we present a classification and overview of the architectural domains used to recognize specific DNA sequences.

1. Principles of protein–DNA recognition

All DNA-binding proteins must bind to their DNA targets with 'appropriate' affinity, specificity and kinetics. Obviously the appropriate values will depend wholly on the biological function and context of the interaction. For example elongation factors are likely to have very different 'needs' from transcriptional repressors. Lack of understanding of these needs often presents a problem in relating biology to its three-dimensional structures.

Despite the diverse thermodynamic and kinetic properties of different protein–DNA complexes, the three-dimensional structures reveal that DNA-binding proteins have many features in common:

- they all have a shape that allows them to form a significant surface area of contact with their DNA partner;
- they have a surface chemistry that allows them to interact favorably with the charged phosphate backbone of the DNA as well as make sequence-specific hydrogen bonds or van der Waals interactions with the bases;

Transcription Factors, edited by J. Locker.

- they make use of their aqueous environment to specifically exclude or include solvent at the intermolecular interface so as to modulate the affinity of binding or allow flexibility of recognition;
- they exploit the structural flexibility of their DNA substrate either to enhance specificity through recognizing the dynamic properties of their target site or for some biologically important function;
- many transcription factors use multiple recognition modules to increase the specificity of the resulting complex;
- most transcription factors recognize a family of related DNA targets.

These concepts are discussed next.

1.1 Shape recognition

The DNA-binding domains of transcription factors need to have a complementary shape to that of DNA for two reasons:

(i) they need to be able to form many individual hydrogen bonds, electrostatic and hydrophobic interactions with the DNA;
(ii) they need to be able to exclude surface waters from the interface.

Both of these drive the interaction between the two molecules through favorable enthalpic or entropic contributions to the Gibbs free energy of interaction.

Since DNA has a helical structure, most DNA-binding proteins need to have a structure that allows them to interact with both the phosphate backbone and the bases in the helical grooves. A great many DNA-binding proteins employ an α-helical element to reach into the major groove-excluding surface waters and which allows interaction with the edges of the bases (see section 2). However, the use of α-helices is certainly not a single solution; not only does the orientation of these helices vary greatly from one DNA-binding motif to another, but also many other proteins use β-strands, or simply protruding loops and tails, to reach into the helical grooves of DNA.

Despite this wide variation, it is still possible to classify the majority of DNA-binding proteins into a limited number of different groups, each with a different structural solution to the problem of shape recognition. Whether this implies that there can only be a limited number of solutions or whether this just reflects the nature of evolution is open to debate.

So, how large an area of surface complementarity is needed? If we survey the range of protein–DNA complexes we see that the typical recognition modules from transcription factors have an area of contact with DNA of $1600Å^2$ +/– $400Å^2$ (Jones *et al.*, 1999; Nadassy *et al.*, 1999). This translates to excluding water from the major groove for between approximately 3–6 base-pairs. Many proteins have a much larger interaction surface. This is achieved by using more than one recognition module, either through binding as a dimer, or through using tandemly repeated recognition modules (see section 1.5).

Interestingly, although there are rather few structures of non-specific complexes in those that are available, we see that the surface area of interaction is quite similar to the specific complexes. The difference, however, is that the interaction is less tight, with many trapped water molecules sandwiched between the protein and DNA (for a good example see Gewirth and Sigler, 1995). This emphasizes that complementarity of shape plays a major role in specificity.

1.2 Chemical recognition

High-resolution crystal structures of protein–DNA complexes reveal an intricate network of interactions that tie the two molecules together. These interactions include hydrogen bonds, van der Waals forces, hydrophobic interactions, global electrostatic interactions and local salt bridges. These interactions are possible because of the large area of closely interacting surfaces and are major contributors to both specificity and affinity.

The majority of proteins bind to DNA in the major groove with a near B-form structure. In general terms, in DNA that is close to the B-form structure, the major groove is wider and better suited to accommodating protein secondary structure than the minor groove. Furthermore, in the major groove, the pattern of hydrogen-bond donors and acceptors is unique for each base-pair, whereas in the minor groove it is less easy to distinguish between AT and TA base-pairs and between GC and CG base-pairs. Consequently, a major groove with B-like properties is best suited for allowing direct, sequence-specific interactions.

It has not been possible to formulate any global rules or recognition code that govern the specificity of the interactions between a protein side-chain and the DNA bases. However, certain recurring patterns have emerged; in particular, arginine and glutamine residues have long side-chains that are able to make bidentate contacts to individual bases. Indeed, the most commonly occurring interaction is of arginine with the N7 and O6 of guanine. Glutamine (and asparagine) can interact similarly with the N7 and N6 of adenine. However, both these amino acids are seen to make a variety of other contacts to bases. Finally, the bulky 5′-methyl group of thymine is suited to making van der Waals contacts with the methyls of several amino acids, although it may also play an important negative role in sterically preventing incorrect binding.

In this section we must not neglect the contacts involving the phosphate backbone of the DNA. Typically the interaction surfaces of DNA-binding proteins have many exposed, positively charged lysine and arginine residues along with other hydrogen-bond donors. These contacts may appear to be non-specific, but they play a critical role in orienting the protein so that specific contacts can be made in the major groove. Moreover, the precise arrangement of these side-chains can allow the protein to measure precisely the local structural parameters of the DNA, such as the width of the major groove. Through this, phosphate contacts can directly contribute to specificity.

1.3 Role of water

It is generally accepted that water plays a critical role in both the affinity and specificity of protein–DNA interactions (Schwabe, 1997). In most of the protein–DNA crystal structures determined to date, there are ordered water molecules present at the protein–DNA interface. It is clear that, in many cases, these water molecules mediate a network of hydrogen bonds between the protein and

DNA which contributes (enthalpically) to the affinity of interaction. Good examples are the *trp* repressor DNA complex (Otwinowski *et al.*, 1988) and the E2 DNA-binding domain (Hegde *et al.*, 1992).

However, arguably the most important water molecules are those that are not seen in crystal structures but those that are displaced on formation of a complex. These contribute to the affinity of interaction through the favorable entropic contribution of their being released into the bulk solvent pool. They also contribute to the specificity of interaction, since highly complementary specific interfaces displace more solvent molecules than non-specific interfaces. Although these water molecules cannot be seen in crystal structures, their contribution can be measured by observing the effect of osmotic pressure and temperature on the thermodynamics of protein–DNA interactions (Hyre and Spicer, 1995; Ladbury *et al.*, 1994; Lundbäck and Härd, 1996). Such studies show that the displacement of solvent in some complexes provides the dominant driving force for interaction. A good example of this is the DNA-binding domain from the glucocorticoid receptor (Gewirth and Sigler, 1995; Luisi *et al.*, 1991).

When considering the role of water in these interactions it is also important to note that *in vitro* experimental conditions probably bear little resemblance to the environment *in vivo*. Inside the cell there is an extremely high concentration of dissolved macromolecules — maybe as high as 300 mg/ml (Garner and Burg, 1994). This macromolecular crowding enhances the affinity of intermolecular interactions and also reduces the macromolecular diffusion rate. Recent experimental data suggest that the change in hydration when proteins bind to DNA is a key thermodynamic variable in protein–DNA interactions, and that in general, as the two surfaces come together, the interaction energies associated with the hydration forces increase exponentially with the number of water molecules displaced (Garner and Rau, 1995).

1.4 Role of DNA structure

DNA is often the forgotten partner in protein–DNA complexes. However, even the earliest X-ray fiber diffraction studies of DNA revealed that the double helix could adopt discrete conformations depending on the sequence and upon the degree of hydration (Arnott and Selsing, 1974a, 1974b) with the extreme conformations A and B forms (Fuller *et al.*, 1965; Langridge *et al.*, 1960). It is now generally accepted that the structure and flexibility of the double helix is continuously variable and that both local structure and deformability play a major role in protein–DNA recognition.

The 434 repressor protein provides a striking example of how bases that are not in contact with protein can significantly affect affinity. Mutation of the central base-pairs of the 434 repressor binding site (Koudelka *et al.*, 1987) from TA to AT had no effect, whereas mutation to GC or CG decreased the affinity of binding by 50-fold. It is clear from the crystal structure that the dinucleotides, TA or AT, can adopt the required narrow minor groove needed to allow the 434 repressor dimer to bind (Aggarwal *et al.*, 1988). GC or CG does not readily adopt such a conformation.

In other complexes it is apparent that the flexibility of DNA is important for protein–DNA recognition. Indeed, there are now many examples of protein–DNA complexes in which the DNA is significantly distorted. The catabolite activator protein (CAP) and E2 proteins (Hegde *et al.*, 1992; Schultz *et al.*, 1991) bound to DNA result in significant local and global bending of the DNA double helix so that it takes up the shape of the protein surface.

Perhaps the most extreme example of DNA distortion is seen in the structure of the TATA-binding protein bound to its DNA target in which the DNA has two 90° bends (Kim *et al.*, 1993a; 1993b). In this case the protein binds lengthways in the minor groove which is splayed apart. Clearly, the structural properties of the TATAAAA sequence are an important factor in facilitating this distortion.

1.5 Multiple recognition modules

One problem faced by the generally small and compact DNA-binding domains is that one such domain is not able to make a sufficient number of contacts with the DNA to specify a unique target site or bind with reasonable affinity. It seems that several strategies have been employed to overcome this problem. The first is simply to add on arms or tails that recognize additional features of the DNA, particularly in the minor groove, e.g. the homeodomains (Kissinger *et al.*, 1990; Wolberger *et al.*, 1991). The second is to double up on the recognition by forming either homo- or heterodimers and so specifying a longer DNA sequence; e.g. the estrogen receptor DNA-binding domain or Fos/Jun (Glover and Harrison, 1995; Schwabe *et al.*, 1993). In the case of Fos/Jun this also vastly increases the recognition possibilities through a combinatorial strategy. The third method of increasing specificity is to employ multiple DNA-binding domains, either by using tandem repeats of the same type of DNA-binding motif; e.g. the zinc-finger motif (Nolte *et al.*, 1998), or by linking together different types of motif (Klemm *et al.*, 1994).

1.6 Multiple recognition sites

Most sequence-specific DNA-binding proteins do not recognize a unique DNA target. Rather, they recognize a family of related DNA sequences. Whilst in most structural analyses proteins are crystallized initially with their consensus targets, in a few cases there is structural information for the protein bound to more than one DNA sequence. These studies show that there are three ways in which proteins can interact with a non-consensus DNA target. In the simplest case it appears that the protein can rearrange the conformation of surface side-chains to create a slightly different network of hydrogen bonds with the alternative sequence. This is exemplified by the estrogen receptor DNA-binding domain bound to a non-consensus target in which a GC base-pair is replaced by an AT base-pair (Schwabe *et al.*, 1995). In this case a lysine side-chain moves to allow alternative hydrogen bonds to be made. However this mutation results in a reduced binding affinity.

The phage 434 repressor illustrates a similar but more intricate situation. A comprehensive comparison of complexes with several different DNA targets reveals that, in addition to side-chain rearrangements, the DNA structure and relative orientation of the protein is different in each of the complexes and that this readjustment plays an essential role in facilitating flexibility of specific recognition (Rodgers and Harrison, 1993).

Interestingly, for certain dimeric proteins the spacing between half-sites is invariant, whereas for others the spacing can vary by one or more nucleotides without compromising specific DNA-binding, e.g. Gcn4 and NF-κB. In these cases the protein adapts to the different spacing by distorting the DNA and protein structure on the different sequences, so that the interactions of the individual monomers with the DNA are essentially identical (Ellenberger *et al.*, 1992; Ghosh *et al.*, 1995; König and Richmond, 1993; Müller *et al.*, 1995). Similarly, flexible linkers between DNA-binding modules and dimerization domain allows certain dimeric proteins to bind without much constraint on the arrangement of half-sites. This is true of certain nuclear receptors as well as certain fungal Zn_2Cys_6 transcription factors.

The third and final way in which proteins seem to overcome the problem of non-consensus DNA targets is through part of the protein (e.g. one monomer in a dimer) binding non-specifically to DNA. This type of binding appears to be important for the nuclear hormone receptors for which one half-site of the palindromic binding site frequently bears little resemblance to the consensus. The structure of the glucocorticoid receptor DNA-binding domain/DNA complex shows very clearly how one half of the dimer adjusts to non-specific interaction with DNA (Luisi *et al.*, 1991).

2. Architecture of DNA-binding domains

There is some confusion in the literature regarding the terms used to describe DNA-binding proteins. In this section we use the terms 'domain', 'module', and 'motif'. A DNA-binding domain is that region of a DNA-binding protein that is necessary and sufficient to direct sequence-specific DNA recognition. This sometimes, but not always, corresponds to a single 'structural domain'. A recognition module is a single structural domain that is part of a DNA-binding protein but needs additional modules to enable recognition of the DNA-target site. A motif is a characteristic pattern of amino acid types that is diagnostic for a specific type of DNA-binding domain or module.

The DNA-binding domain structures are illustrated, all on the same scale, using molscript figures. While these show the protein architecture very clearly, they do not succeed in illustrating the intimate nature of the complexes. To illustrate this, *Figure 1* shows a surface and van de Waals sphere representation of a protein–DNA complex alongside the equivalent molscript view.

2.1 Helix-turn-helix containing DNA-binding domains

The term helix-turn-helix (HTH) was first used to describe the structural elements responsible for DNA-binding observed in the crystal structures of the prokaryotic proteins λ-Cro, *E. coli* CAP and λ-C1 (Anderson *et al.*, 1981; McKay and Steiz, 1981; Pabo and Lewis, 1982). As a sequence motif the HTH is poorly conserved. Structurally, the HTH is made up of two α-helices linked by a sharp

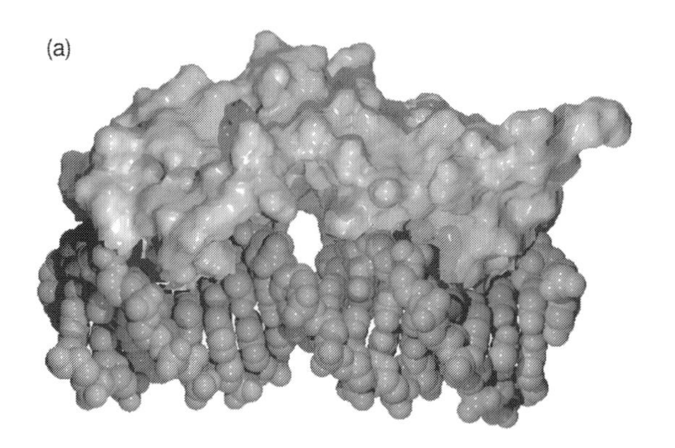

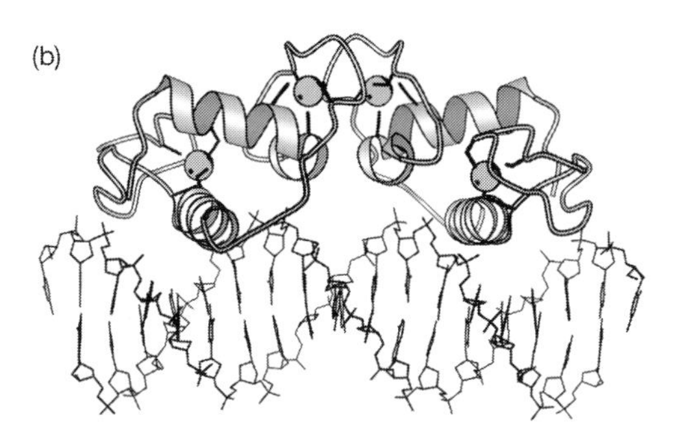

Figure 1. A comparison of (a) the molscript representation (Kraulis, 1991) with (b) a molecular surface/van der Waals representation of the same protein–DNA complex. The aim of this figure is to set the remaining figures in context. Note the extensive complementary surface between the protein and DNA.

β-turn. Mutagenesis experiments suggested that the carboxyl-terminal helix was used for sequence-specific DNA-binding and was thus termed the DNA-recognition helix (Wharton *et al.*, 1984). Co-crystal structures of various proteins containing the HTH motif show that the recognition helix lies in the major groove of the DNA as shown in *Figure 2a* for the 434 repressor (Aggarwal *et al.*, 1988; Anderson *et al.*, 1987; Jordan and Pabo, 1988; Schultz *et al.*, 1991; Wolberger *et al.*, 1988). The exact orientation of the recognition helix in the major groove is variable in the different complexes and appears to be dependent upon the structure of the rest of the protein and the contacts made to the phosphate backbone of the DNA.

Homeodomain and Myb domain. The eukaryotic homeodomain and the Myb domain contain eukaryotic versions of the HTH motif. The lack of sequence conservation indicates that the eukaryotic motifs may not be evolutionarily related to the prokaryotic motifs, but are structurally analogous. Homeodomains and Myb domains share a conserved pattern of hydrophobic amino acids (König and Rhodes, 1997) that form a hydrophobic core between three α-helices (two of which form the HTH motif). Unlike the prokaryotic HTH-containing proteins, the eukaryotic homeodomain proteins are able to bind DNA as monomers (*Figure 2b*). This may be due, in part, to the involvement of an additional N-terminal tail that wraps around the DNA and makes base contacts in the minor groove (Kissinger *et al.*, 1990; Otting *et al.*, 1990; Wolberger *et al.*, 1991). The Myb protein also binds to DNA as a monomer, but this protein has three tandem HTH

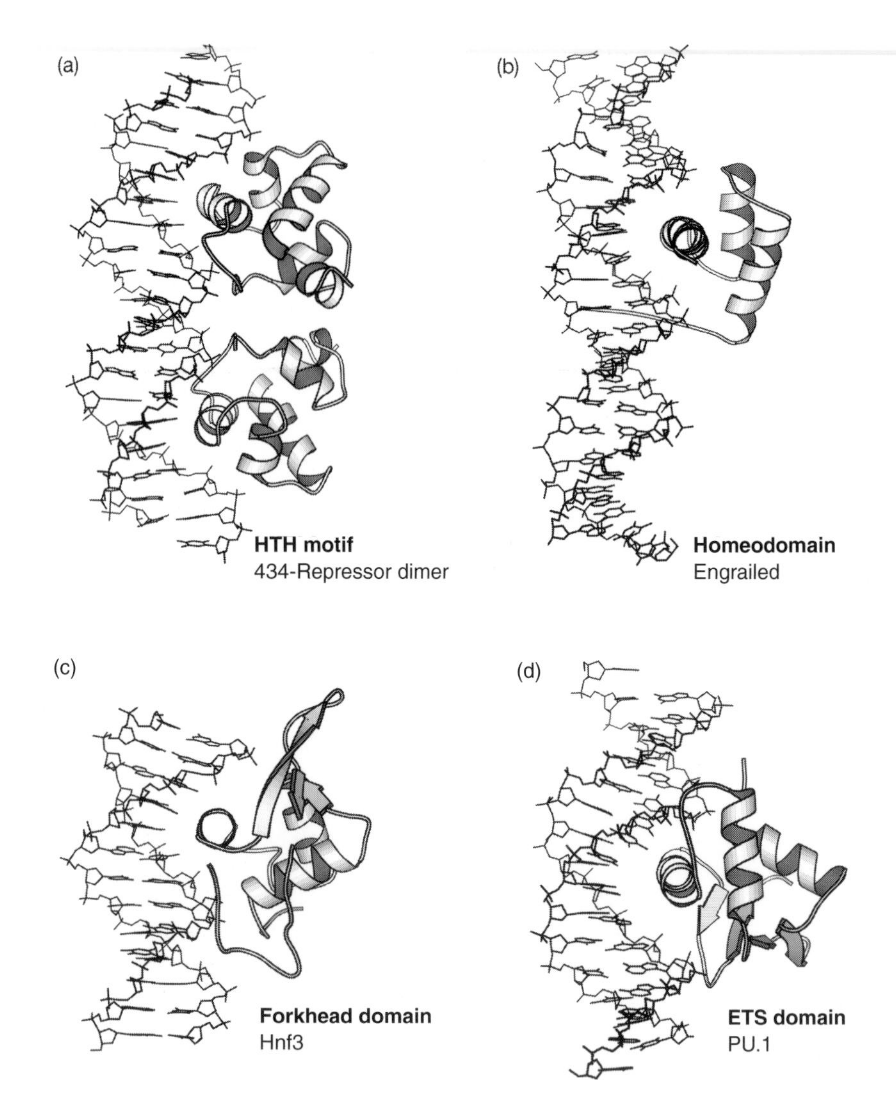

Figure 2. Helix-turn-helix motifs. α-helices are shown as coils and β-strands as arrows. Molscript representations of the crystal structures of: (a) 434-repressor dimer–DNA complex (Aggarwal *et al.*, 1988); (b) engrailed homeodomain–DNA complex (Kissinger *et al.*, 1990); (c) Hnf3 (forkhead)–DNA complex (Clark *et al.*, 1993); and (d) PU.1 ETS domain–DNA complex (Kodandapani *et al.*, 1996).

repeats. There is direct interaction between the recognition helices from each motif (Ogata *et al.*, 1994).

α/β HTH motifs. A variation on the HTH motif, these are found in the Forkhead or 'winged helix' and ETS-type DNA-binding domains. The structure of the HNF3 (hepatocyte nuclear factor)/forkhead DNA-binding domain consists of three helices containing a HTH motif, a twisted, antiparallel three β-sheet and a C-terminal random coil (*Figure 2c*) (Clark *et al.*, 1993). The HTH motif interacts in the major groove of the DNA. In addition, the random coil at the C-terminus makes a base contact in the major groove. Other parts of the protein make phosphate contacts, particularly the loop between the second and third β-strands, which has become known as a 'wing'. Strikingly, the structure of the globular domain of histone H5 (Ramakrishnan *et al.*, 1993) closely resembles that of Hnf3, indicating that the requirements of chromatin packaging and transcription factor recruitment can be met by the same structural domain.

The ETS domain is composed of three helices containing a HTH motif and a four-stranded antiparallel β-sheet. The structures of the Ets1 and PU.1 ETS/domain DNA complexes illustrated in *Figure 2d* (Kodandapani *et al.*, 1996; Werner *et al.*, 1997) indicate that these domains interact with DNA using not only the HTH motif, but also the loop between the third and fourth β-strands (the wing), and also the 'turn' within the HTH motif (best described as a loop in these structures). Intriguingly the structure of the heat-shock transcription factor is very similar to that of the ETS domain, except that the loop between the third and fourth β-strands forms a dimer interface between two adjacent DNA-binding domains when bound to DNA (Littlefield and Nelson, 1999).

2.2 Zinc-containing DNA-binding domains

Classical zinc fingers (TFIIIA type). The classical (Cys_2His_2) zinc-finger motif was first observed in the protein transcription factor, TFIIIA, from the frog *Xenopus laevis* (Miller *et al.*, 1985). The motif is found as a tandemly repeated unit from 2 to as many as 37 times. The zinc finger consists of a two-stranded β-sheet and a α-helix that is held together by a zinc ion ligated to two cysteine and two histidine residues. The structures of zinc finger–DNA complexes illustrate a consensus-binding mode, but there are also many exceptions to that mode. The consensus mode is found in Zif268, which is illustrated by *Figure 3a* (Pavletich and Pabo, 1991) and Tramtrack (Fairall *et al.*, 1993) complexes. These proteins bind by winding around the DNA major groove with each zinc finger using three amino acids in the α-helix to recognize three adjacent bases on one strand of DNA. The zinc fingers of YY1 interact in the major groove and use similar amino acids to make DNA contacts, but make more extensive contacts to the other DNA strand (Houbaviy *et al.*, 1996). The consensus zinc finger–DNA complex represents the simplest kind of protein–DNA interaction that has been observed to date in any complex, and has enabled novel proteins to be designed that can recognize almost any DNA sequence (Choo and Klug, 1997).

Other zinc-finger proteins exhibit rather different uses of the (Cys_2His_2) zinc-finger modules. For instance, GLI has five zinc fingers of which the *N*-terminal zinc finger makes no contact with the DNA; the second finger makes only one base contact and a few phosphate contacts, while the third makes only phosphate contact (Pavletich and Pabo, 1993). The fourth and the fifth fingers make extensive base contacts using, between them, most of the positions in the α-helix. Similarly the structure of the first six fingers of TFIIIA in *Figure 3b* (Nolte *et al.*, 1998) also shows differences from the consensus mode of binding. Zinc finger 1 binds in the major groove but the helix is oriented differently; fingers 2, 3 and 5 bind like the consensus, and fingers 4 and 6 are used to span the minor groove.

Nuclear hormone receptors. These form a super-family of ligand-activated transcription factors that share a highly conserved DNA-binding domain which is characterized by eight cysteine residues that nucleate two zinc-binding clusters. These two zinc-finger-like motifs fold to form a single structural domain and hence are quite distinct from the classical modular zinc fingers. The dominant feature of this structural domain is an arrangement of two helices that are oriented perpendicularly to each other and crossing at their mid-points. The N-terminal helix binds in the major groove of DNA.

Many nuclear hormone receptor DNA-binding domains bind cooperatively to DNA as either homo or heterodimers. This allows them to recognize not only the DNA sequence of their binding sites, but also the spacing and orientation of the two half-sites. *Figures 3c* and *3d* show homodimeric (ER) and heterodimeric (RXR/TR) nuclear receptor DNA-binding domains linking to their DNA targets. In each case the protein binds in adjacent major grooves on one side of the DNA double helix. Some nuclear receptors bind to DNA as monomers where additional residues at the C-terminus of the DNA-binding domain make extended contacts to the DNA.

GATA factors. GATA factors, so-called because they recognize a GATA sequence, also contain a Cys_2Cys_2 motif. The NMR structures of erythoid Gata1 (*Figure 3e*) (Omichinski *et al.*, 1993) and fungal AREA (Starich *et al.*, 1998) GATA-binding domains show that, in addition to the Cys_2Cys_2 motif, a *C*-terminal tail is also required for DNA binding. The zinc ion bound to its four cysteine ligands form the core of the domain, which is composed of two small, irregular, two-stranded antiparallel β-sheets and an α-helix followed by a long loop that leads into the carboxyl terminal tail. The helix and the loop that joins the two β-sheets interact with the major groove of the DNA. The carboxyl terminal tail, which contributes to the sequence-specific binding of the domain, behaves differently in Gata1 and AREA; in the former it wraps around and interacts in the minor groove while in the latter it interacts with the phosphate backbone. The core structure of Gata1 and AREA is strikingly similar, in both sequence and structure, to the *N*-terminal zinc-binding motif of the nuclear hormone receptors described earlier. This similarity strongly suggests that the nuclear receptor DNA-binding domains arose as an evolutionary duplication of this motif.

Zn_2Cys_6 binuclear cluster. This describes the DNA-binding domains found in a family of fungal transcription factors of which Gal4 is the best studied example (*Figure 3f*). Each of these proteins binds to DNA as a dimer with a coiled-coil dimerization domain joined by an extended linker to a DNA-binding module. The DNA-binding module has six cysteines that ligate two zinc ions in a single cluster, with the first and fourth cysteines shared between the two zinc ions. It has two short α-helices each capped at their N-terminus by a pair of cysteine ligands: the first and second cysteines in the first helix and the fourth and fifth in the second helix. Each helix is followed by an extended peptide strand forming a helix-strand motif.

Each DNA-binding module recognizes the sequence CGG. The dimeric proteins therefore bind to two CGG sequences. The spacing and orientation of these confers specificity between the different proteins. For instance, Gal4 and Ppr1

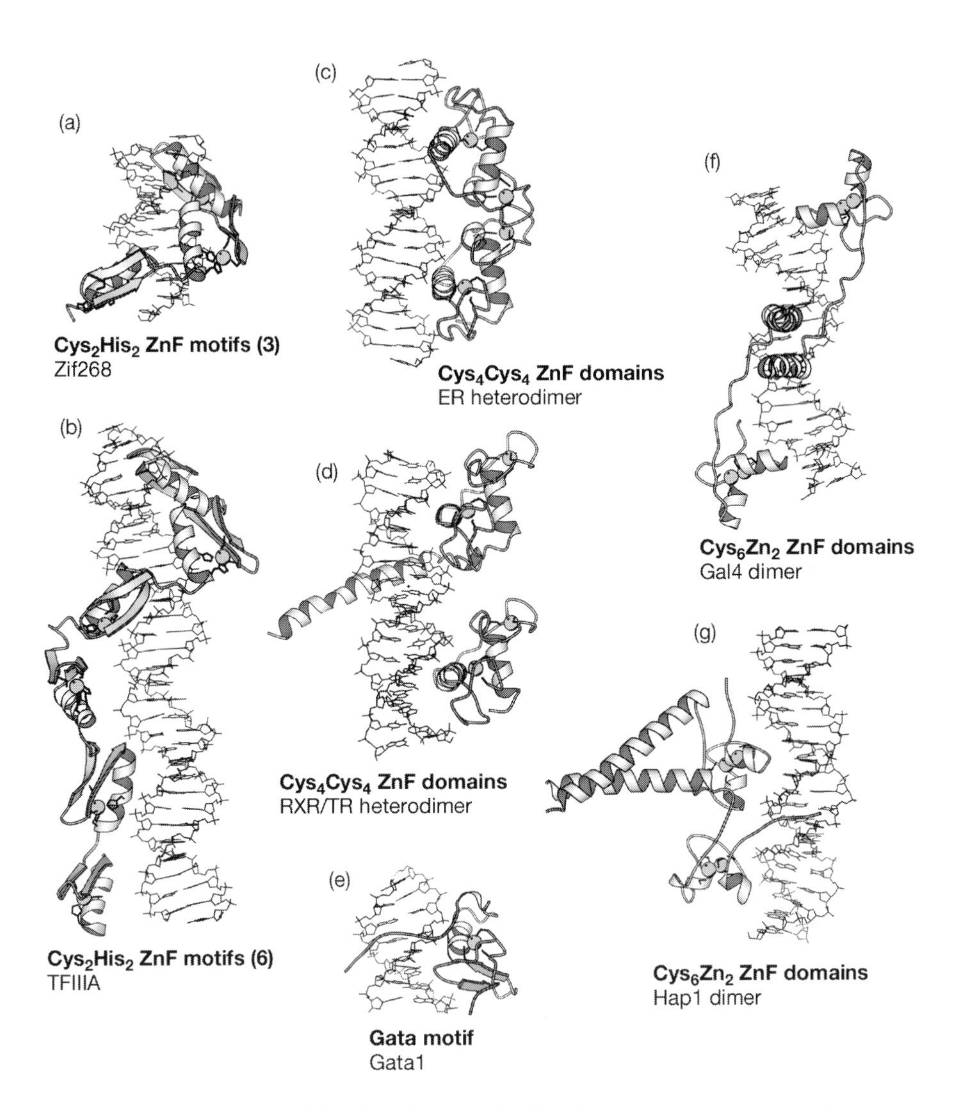

Figure 3. Zinc-containing DNA binding motifs. Zinc ions are shown as grey spheres. Molscript representations of the crystal structures of: (a) three zinc fingers of Zif268 complexed with DNA (Pavletich and Pabo, 1991); (b) first six zinc fingers of TFIIIA complexed with DNA (Nolte *et al.*, 1998); (c) homodimer of the estrogen receptor (ER) DNA-binding domain complexed with DNA (Schwabe *et al.*, 1993); (d) heterodimer of the retinoid X receptor (RXR)/thyroid receptor (TR) DNA-binding domains complexed with DNA (Rastinejad *et al.*, 1995), (e) NMR structure of Gata1 DNA-binding domain complexed with DNA (Omichinski *et al.*, 1993); (f) crystal structures of the DNA-binding domains of Gal4 (Marmorstein *et al.*, 1992); and (g) Hap1 (King *et al.*, 1999).

bind inverted repeats with a separation of 11 and 6 base-pairs respectively (Marmorstein *et al.*, 1992; Marmorstein and Harrison, 1994). Hapl, on the other hand (*Figure 3g*), binds a direct repeat with a spacing of six base-pairs (King *et al.*, 1999). The specificity for this spacing and orientation appears to reside in the structure and properties of the linker region between the DNA-binding module and the coiled-coil dimerization domain.

2.3 Other DNA-binding domains that use α-helices to bind DNA

MADS. This DNA-binding domain (named after the transcription factors Mcml, AG, DEFA and SRF in which it was first discovered) from the serum response factor (SRF) binds DNA as a dimer and is organized in three layers as illustrated in *Figure 4a* (Pellegrini *et al.*, 1995). The N-terminus is the primary DNA-binding element and consists of an antiparallel coiled coil of two amphipathic α-helices, one from each monomer of the dimer. The coiled-coil is oriented parallel to the minor groove of the binding site and the DNA is bent around the protein with each α-helix binding in adjacent major grooves. The next layer of the domain is a four-stranded antiparallel β-sheet that is packed against the coiled-coil, followed by an irregular coiled region and a short α-helix. The MADS domain is also seen in the structure of Mcml in complex with Matα2 and DNA (Tan and Richmond, 1998).

bZip motif. The basic-region-leucine-zipper (bZip) DNA-binding domain perhaps represents the simplest possible protein structure for DNA recognition. It consists of a basic region that binds to DNA and a leucine-rich region which is involved in the dimerization of bZip proteins. Crystal structures of a Gcn4 homodimer (Ellenberger *et al.*, 1992; König and Richmond, 1993) and c-Fos-c-Jun heterodimer (*Figure 4b*) (Glover and Harrison, 1995) show that the bZip dimer is a pair of continuous α-helices that dimerize at their *C*-termini, using a parallel coiled-coil with the *N*-termini binding in the major groove of the DNA binding site. The protein grips the DNA like a pair of forceps with no sharp kinks or bends in the α-helices. The basic region is disordered in solution before binding to DNA and makes many base and phosphate contacts upon binding.

bHLH and bHLHZ motifs. The HLH motif was initially identified as a region of extensive identity between immunoglobulin enhancer-binding proteins, *Drosophila daughterless*, *achaete-scute* and *twist* gene products, the *myc* family and *MyoD* (Murre *et al.*, 1989). In the bHLH proteins MyoD (*Figure 4c*) (Ma *et al.*, 1994), E47 (Ellenberger *et al.*, 1994) and Pho4 (Shimizu *et al.*, 1997), the basic region forms a α-helix that interacts in the major groove of DNA. The HLH motif serves as a dimerization domain that forms a four-helix bundle in the dimer. In the bHLHZ proteins, Max (*Figure 4d*) (Ferré-D'Amaré *et al.*, 1993) and upstream stimulatory factor (Ferré-D'Amaré *et al.*, 1994), the four-helix bundle is extended away from the DNA to form a coiled-coil leucine zipper. Interestingly it has been shown that upstream stimulatory factor is a tetramer in the absence of DNA. Tetramer formation requires the leucine zipper region. The tetramer dissociates upon DNA-binding.

High mobility group domain. The high mobility group (HMG) proteins consist of a family of sequence-specific DNA-binding proteins, for example Lefl

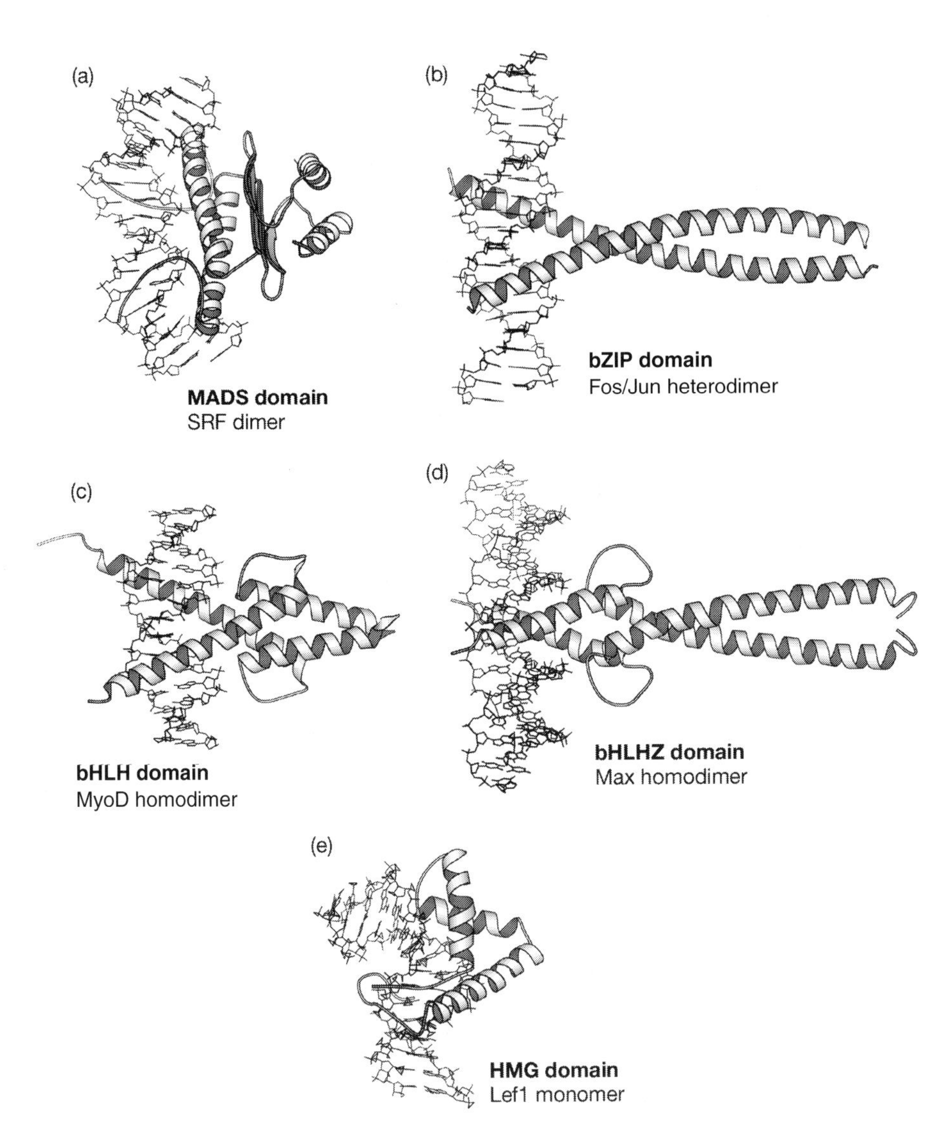

Figure 4. Other motifs that use α-helices. Molscript representations of the crystal structures of the dimer of (a) MADS DNA-binding domain of serum response factor (SRF) complexed with DNA (Pellegrini *et al.*, 1995); (b) heterodimer of the basic-region-leucine-zipper (bZip) DNA-binding domains of c-Fos-c-Jun complexed with DNA (Glover and Harrison, 1995); (c) homodimer of the basic helix-loop-helix bHLH DNA-binding domain of MyoD complexed with DNA (Ma *et al.*, 1994); (d) homodimer of the bHLHZ DNA-binding domain of Max complexed with DNA (Ferré-D'Amaré *et al.*, 1993); and (e) NMR structure of the high motility group (HMG) domain of Lef1 complexed with DNA (Love *et al.*, 1995).

(Love *et al.*, 1995) and a family of non-sequence specific proteins, for example HMG-D (Murphy *et al.*, 1999). Both HMG families bind to DNA using an 80 amino acid HMG domain. The structures are very similar and consist of a characteristic L-shaped fold, which is comprised of three α-helices held together by two hydrophobic cores (*Figure 4e*). The DNA in both of the protein–DNA complexes is severely distorted. The DNA specificity in Lef1 seems to reside in a few amino acids (Murphy *et al.*, 1999) and in the Lef1 structure, a C-terminal basic region contributes to DNA-binding by linking across the narrow minor groove.

2.4 DNA-binding domains that use β-strands to bind DNA

Met repressor. The *E. coli* Met repressor and related proteins use a β-ribbon-helix-helix motif to bind to DNA (*Figure 5a*). The Met repressor, itself a dimer, binds cooperatively to tandem binding sites. Each dimer interacts with the major groove of DNA via a short antiparallel two-stranded β-ribbon (Somers and Phillips, 1992). Each monomer contributes one strand to this β-ribbon. C-terminal to the β-strand are two helices: the first mediates cooperative contacts to the next repressor dimer; the second forms the dimer interface and binds to the phosphate backbone. The structure of the Arc repressor of *Salmonella* bacteriophage P22 determined by NMR methods is very similar to that of the Met repressor and the Arc repressor binds to DNA in the same way (Raumann *et al.*, 1994).

Integration host factor and HU. The prokaryotic integration host factor (IHF) has been shown to bind as a dimer to the minor groove of DNA and phosphate backbone using β-ribbons (*Figure 5b*) (Nash and Granston, 1991). The protein dimer kinks the DNA in two places by intercalating a proline from each monomer between the base-pairs. It is proposed that the prokaryotic DNA-packaging protein, HU, binds to DNA in a similar way (White *et al.*, 1989).

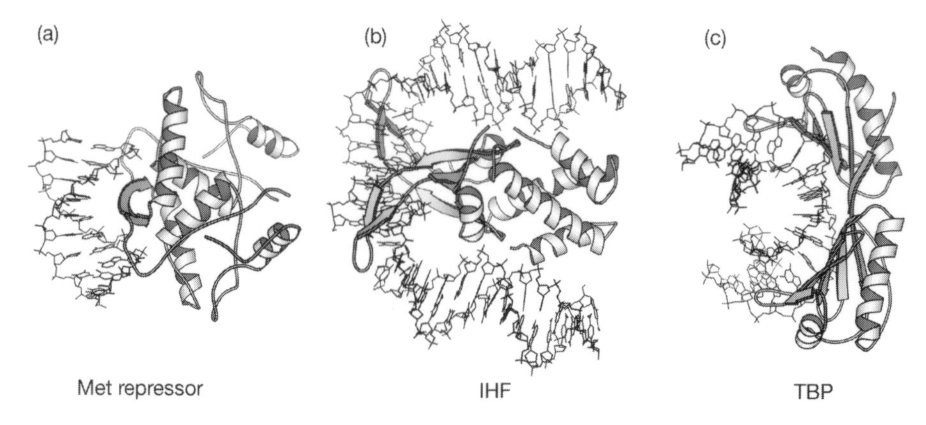

Figure 5. β-strands. Molscript representations of the crystal structures of: (a) homodimer of the Met-repressor bound to DNA (Somers and Phillips, 1992); (b) heterodimer of the integration host factor (IHF) bound to DNA (Rice *et al.*, 1996); and (c) TATA-box binding protein (TBP) bound to DNA (Kim *et al.*, 1993a; 1993b).

TATA box-binding protein. The DNA-binding domain of the TATA box-binding protein (TBP) is a curved antiparallel β-sheet. The concave surface of the β-sheet binds to the minor groove of the TATA box, and the DNA is kinked in two places with an overall bend of 100° (*Figure 5c*). The main interactions involve: two pairs of phenylalanine residues inserted between adjacent base-pairs that produce the kinks; polar side-chains making contacts with the base-pairs in the minor groove; and van der Waals and hydrophobic interactions, also in the minor groove (Kim *et al.*, 1993a; 1993b).

2.5 β-barrels and β-sandwiches

β-barrel. The dominant transcriptional regulator of the papillomavirus, E2, binds to DNA using a dimeric antiparallel β-barrel with two helices that make specific contacts in successive major grooves (*Figure 6a*) (Hegde *et al.*, 1992). The DNA is smoothly bent around the barrel. The general transcription factor, TFIIA, also contains a β-barrel that interacts with TBP and extends the TBP β-sheet (Geiger *et al.*, 1996; Tan *et al.*, 1996). It spans the major groove immediately upstream of the TATA box making contacts with the phosphate backbone.

Immunoglobulin β-sandwich. The use of the immunoglobulin-type fold (a sandwich of two β-sheets) to provide a stable structure on which to build DNA-binding activity was first seen in the tumor suppressor protein, p53, (*Figure 6b*) (Cho *et al.*, 1994). Since then many proteins, including human transcription factor, NF-κB, (*Figure 6c*) (Ghosh *et al.*, 1995; Müller *et al.*, 1995), STAT (signal transducers and activators of transcription) proteins (*Figure 6d*) (Becker *et al.*, 1998; Chen *et al.*, 1998), NFATC (Zhou *et al.*, 1998) and Brachyury T-domain (Müller and Herrmann, 1997) have been found to have immunoglobulin-type folds. All of these proteins bind to DNA by using loops and structures that protrude from the same face of the immunoglobulin fold. For example, the p53 (Cho *et al.*, 1994) monomer has a loop-sheet-helix motif that packs against one end of the β-sandwich. Of this motif the helix and the loop bind in the major groove (*Figure 6b*). The p53 DNA-binding domain structure also has two large loops (one of which is part of the DNA-binding interface) which are held together by a tetrahedrally coordinated zinc atom. This large loop uses an arginine to bind to the minor groove of the DNA. The whole p53 protein binds to DNA as a tetramer, and binds with enhanced affinity (McLure and Lee, 1998).

Another example is NF-κB (Ghosh *et al.*, 1995; Müller *et al.*, 1995), which is a heterodimer of two subunits, p50 and p65. The structure of the p50 subunit contains two β-sandwich domains of which the N-terminal domain binds to the DNA bases using well-defined loops between the β-strands (*Figure 6c*). The C-terminal domain is the dimerization domain and is involved in binding to the phosphate backbone.

3. Conclusions

Although brief, we hope that this survey of DNA-binding proteins gives a flavor of the enormous amount that has been learnt using structural techniques to study protein–DNA complexes. At the beginning of the chapter we noted that, in

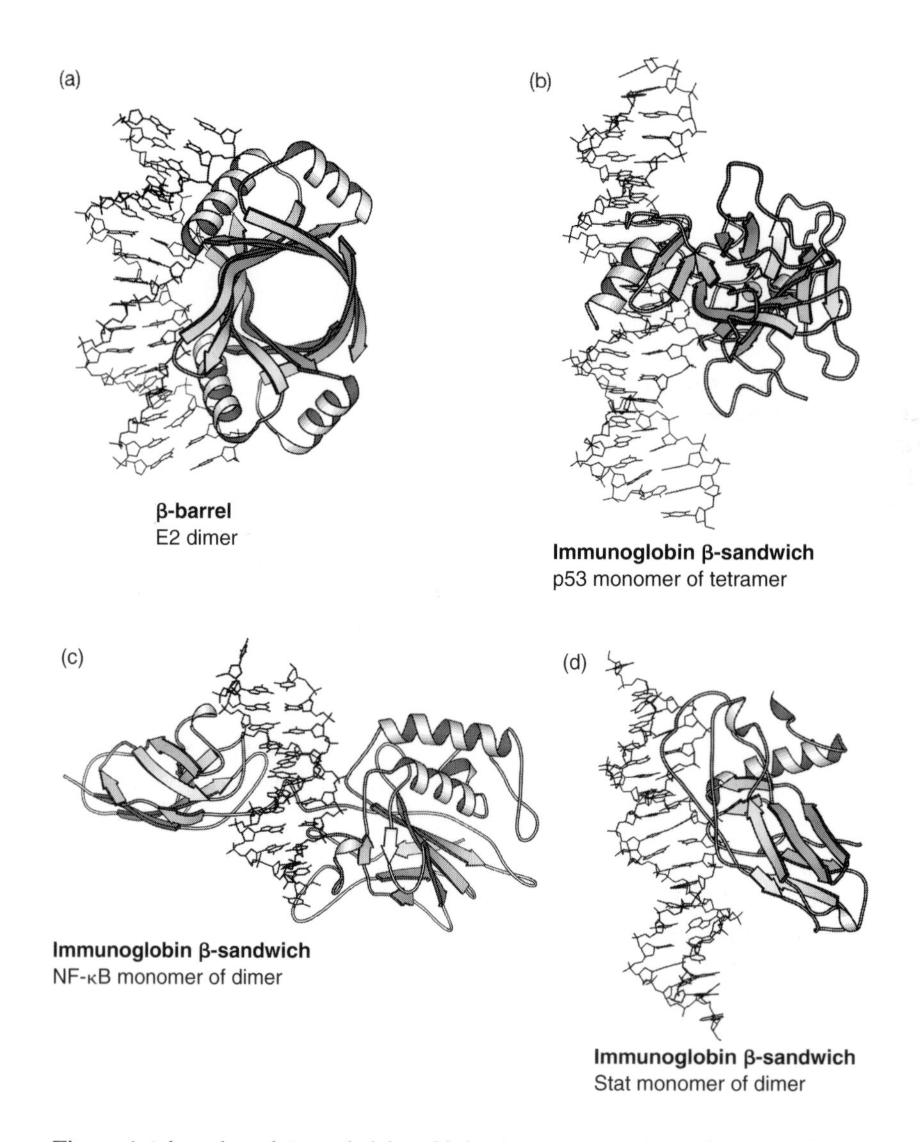

Figure 6. β-barrels and β-sandwiches. Molscript representations of the crystal structures of: (a) homodimer of the DNA-binding domain of the papillomavirus E2 protein–DNA complex (Hegde *et al.*, 1992); (b) core domain of p53–DNA complex (Cho *et al.*, 1994); (c) one monomer of the homodimer of NF-κB p50–DNA complex (Ghosh *et al.*, 1995; Müller *et al.*, 1995); and (d) DNA-binding domain of one monomer of the homodimer of Stat1 (Chen *et al.*, 1998).

biological terms, what is important is that DNA-binding proteins should bind to their DNA targets with 'appropriate' affinity, specificity and kinetics. A fair question is therefore: Do the structures explain these critical biological parameters? The answer is yes and no. We certainly understand the types of features in the structures that contribute to these parameters — many of these were discussed in the first half of the chapter. What we cannot do is extrapolate from known structures a quantitative description of the thermodynamic and kinetic properties of a complex (von Hippel, 1998). This is therefore an enormously important area for research in the future, but any general thermodynamic description is likely to present a formidable challenge.

In a more tractable time-frame, more and more attention is being paid to multiprotein complexes. There has been considerable success with looking at complexes containing more than one transcriptional regulator (Chen, 1999). It is likely that we will soon see structures of the much larger complexes that have been termed 'enhanceosomes'. Of course many other complexes also play a role in transcriptional regulation. These include the polymerase holocomplex, chromatin modifying complexes and the mediator complex. It seems likely that the next 15 years will see enormous progress in these directions, until finally we will have a complete structural description of transcription.

References

Aggarwal, A.K., Rodgers, D.W., Drottar, M., Ptashne, M. and Harrison, S.C. (1988) Recognition of a DNA operator by the repressor of phage 434: a view at high resolution. *Science* **242**: 899–907.

Anderson, J.W., Ptashne, M. and Harrison, S.C. (1987) Structure of the repressor–operator complex of bacteriophage 434. *Nature* **326**: 846–852.

Anderson, W.F., Ohlendorf, D.H., Takeda, Y. and Matthews, B.W. (1981) Structure of the cro repressor from bacteriophage λ and its interaction with DNA. *Nature* **290**: 754–758.

Arnott, S. and Selsing, E. (1974a) The structure of polydeoxyguanylic acid.polydeoxycytidylic acid. *J. Mol. Biol.* **88**: 551–552.

Arnott, S. and Selsing, E. (1974b) Structures for the polynucleotide complexes poly(dA).poly(dT) and poly(dT).poly(dA).poly(dT). *J. Mol. Biol.* **88**: 509–521.

Becker, S., Groner, B. and Müller, C.W. (1998) Three-dimensional structure of the Stat3β homodimer bound to DNA. *Nature* **394**: 145–151.

Chen, L. (1999) Combinatorial gene regulation by eukaryotic transcription factors. *Curr. Opin. Struct. Biol.* **9**: 48–55.

Chen, X., Vinkemeier, U., Zhao, Y., Jeruzalmi, D., Darnell, J.E.J. and Kuriyan, J. (1998) Crystal structure of a tyrosine phosphorylated STAT-1 dimer bound to DNA. *Cell* **93**: 827–839.

Cho, Y., Gorina, S., Jeffrey, P.D. and Pavletich, N.P. (1994) Crystal structure of a p53 tumor suppressor–DNA complex: understanding tumorigenic mutations. *Science* **265**: 346–355.

Choo, Y., and Klug, A. (1997) Physical basis of a protein-DNA recognition code. *Curr. Opin. Struct. Biol.* **7**: 117–125.

Clark, K.L., Halay, E.D., Lai, E. and Burley, S.K. (1993) Co-crystal structure of the HNF-3/fork head DNA-recognition motif resembles histone H5. *Nature* **364**: 412–420.

Ellenberger, T., Fass, D., Arnaud, M. and Harrison, S.C. (1994) Crystal structure of transcription factor E47: E-box recognition by a basic region helix-loop-helix dimer. *Genes Dev.* **8**: 970–980.

Ellenberger, T.E., Brandl, C.J., Struh, K. and Harrison, S.C. (1992) The GCN4 basic region leucine zipper binds to DNA as a dimer of uninterrupted α-helices: crystal structure of the protein–DNA complex. *Cell* **71**: 1223–1237.

Fairall, L., Schwabe, J.W.R., Chapman, L., Finch, J.T. and Rhodes, D. (1993) The crystal structure of a two zinc-finger peptide reveals an extension to the rules for zinc-finger/DNA recognition. *Nature* **366**: 483–487.

Ferré-D'Amaré, A.R., Pognonec, P., Roeder, R.G. and Burley, S.K. (1994) Structure and function of the b/HLH/Z domain of USF. *EMBO J.* **13**: 180–189.

Ferré-D'Amaré, A.R., Prendergast, G.C., Ziff, E.B. and Burley, S.K. (1993) Recognition by Max of its cognate DNA through a dimeric b/HLH/Z domain. *Nature* **363**: 38–45.

Fuller, W., Wilkins, M.H.F., Wilson, H.R. and Hamilton, L.D. (1965) The molecular configuration of deoxyribonucleic acid. IV. X-ray diffraction study of the A-form. *J. Mol. Biol.* **12**: 60–80.

Garner, M.M. and Burg, M.B. (1994) Macromolecular crowding and confinement in cells exposed to hypertonicity. *Am. J. Physiol. Cell Physiol.* **266**: C877–C892.

Garner, M.M. and Rau, D.C. (1995) Water release associated with specific binding of *gal* repressor. *EMBO J.* **14**: 1257–1263.

Geiger, J.H., Hahn, S., Lee, S. and Sigler, P.B. (1996) Crystal structure of the yeast TFIIA/TBP/DNA complex. *Science* **272**: 830–836.

Gewirth, D.T. and Sigler, P.B. (1995) The basis for half-site specificity explored through a non-cognate steroid receptor–DNA complex. *Nat. Struct. Biol.* **2**: 386–394.

Ghosh, G., van Duyne, G., Ghosh, S. and Sigler, P.B. (1995) Structure of NF-κB p50 homodimer bound to a κB site. *Nature* **373**: 303–310.

Glover, J.N. and Harrison, S.C. (1995) Crystal structure of the heterodimeric bZIP transcription factor c-*Fos*-c-*Jun* bound to DNA. *Nature* **373**: 257–261.

Hegde, R.S., Grossman, S.R., Laimins, L.A. and Sigler, P.B. (1992) Crystal-structure at 1.7 Å of the bovine papillomavirus-1 E2 DNA-binding domain bound to its DNA target. *Nature* **359**: 505–512.

Houbaviy, H.B., Usheva, A., Shenk, T. and Burley, S.K. (1996) Cocrystal structure of YY1 bound to the adeno-associated virus P5 initiator. *Proc. Natl Acad. Sci.* **93**: 13577–13582.

Hyre, D.E. and Spicer, L.D. (1995) Thermodynamic evaluation of binding interactions in the methionine repressor system of *Escherichia coli* using isothermal titration calorimetry. *Biochemistry* **34**: 3212–3221.

Jones, S., van Heyningen, P., Berman, H.M. and Thornton, J.M. (1999) Protein–DNA interactions: a structural analysis. *J. Mol. Biol.* **287**: 877–896.

Jordan, S.R. and Pabo, C.O. (1988). Structure of the lambda complex at 2.5 Å resolution: details of the repressor–operator interactions. *Science* **242**: 893–899.

Kim, J.L., Nikolov, D.B. and Burley, S.K. (1993a). Co-crystal structure of TBP recognising the minor groove of a TATA element. *Nature* **365**: 520–527.

Kim, Y., Geiger, J.H., Hahn, S. and Sigler, P.B. (1993b) Crystal structure of a yeast TBP/TATA-box complex. *Nature* **365**: 512–520.

King, D.A., Zhang, L., Guarente, L. and Marmorstein, R. (1999) Structure of a HAP1–DNA complex reveals dramatically asymmetric DNA binding by a homodimeric protein. *Nat. Struct. Biol.* **6**: 64–71.

Kissinger, C.R., Liu, B., Martin-Blanco, E., Kornberg, T.B. and Pabo, C.O. (1990) Crystal structure of an engrailed homeodomain–DNA complex at 2.8 Å resolution: a framework for understanding homeodomain–DNA interactions. *Cell* **63**: 579–590.

Klemm, J.D., Rould, M.A., Aurora, R., Herr, W. and Pabo, C.O. (1994) Crystal structure of the Oct-1 POU domain bound to an octamer site: DNA recognition with tethered DNA-binding modules. *Cell* **77**: 21–32.

Kodandapani, R., Pio, F., Ni, C.Z., *et al.* (1996) A new pattern for helix-turn-helix recognition revealed by the PU.1 ETS-domain-DNA complex. *Nature* **380**: 456–460.

König, P. and Rhodes, D. (1997) Recognition of telomeric DNA. *Trends Biochem. Sci.* **22**: 43–47.

König, P. and Richmond, T.J. (1993) The X-ray structure of the GCN4-bZIP bound to ATF/CREB site DNA shows the complex depends on DNA flexibility. *J. Mol. Biol.* **233**: 139–154.

Koudelka, G.B., Harrison, S.C. and Ptashne, M. (1987) Effect of non-contacted bases on the affinity of 434 operator for 434 repressor and Cro. *Nature* **326**: 886–888.

Kraulis, P.J. (1991) MOLSCRIPT: a program to produce both detailed and schematic plots of protein structures. *J. Appl. Cryst.* **24**: 946–950.

Ladbury, J.E., Wright, J.G., Sturtevant, J.M. and Sigler, P.B. (1994) A thermodynamic study of the *trp* repressor–operator interaction. *J. Mol. Biol.* **238**: 669–681.

Langridge, R., Marvin, D.A., Seeds, W.E., *et al.* (1960) The molecular configuration of deoxyribonucleic acid. II. Molecular models and their fourier transforms. *J. Mol. Biol.* **2**: 38–64.

Lilley, D.M.J. (ed.) (1995) *DNA-Protein: Structural Interactions*. IRL Press, Oxford.

Littlefield, O. and Nelson, H.C. (1999) A new use for the 'wing' of the 'winged' helix-turn-helix motif in the HSF–DNA cocrystal. *Nat. Struct. Biol.* **6**: 464–470.

Love, J.J., Li, X., Case, D.A., Giese, K., Grosschedl, R. and Wright, P.E. (1995) Structural basis for DNA bending by the architectural transcription factor LEF-1. *Nature* **376**: 791–795.

Luisi, B.F., Xu, W.X., Otwinowski, Z., Freedman, L.P., Yamamoto, K.R. and Sigler, P.B. (1991) Crystallographic analysis of the interaction of the glucocorticoid receptor with DNA. *Nature* **352**: 497–505.

Lundbäck, T. and Härd, T. (1996) Sequence-specific DNA-binding dominated by dehydration. *Proc. Natl Acad. Sci.* **93**: 4754–4759.

Ma, P.C., Rould, M.A., Weintraub, H. and Pabo, C.O. (1994) Crystal structure of MyoD bHLH domain–DNA complex: perspectives on DNA recognition and implications for transcriptional activation. *Cell* **77**: 451–459.

Marmorstein, R., Carey, M., Ptashne, M. and Harrison, S.C. (1992) DNA recognition by GAL4: structure of a protein–DNA complex. *Nature* **356**: 408–414.

Marmorstein, R. and Harrison, S.C. (1994) Crystal structure of a PPR1–DNA complex: DNA recognition by proteins containing a Zn_2Cys_6 binuclear cluster. *Genes Dev.* **8**: 2504–2512.

McKay, D.B. and Steiz, T.A. (1981) Structure of catabolite gene activator protein at 2.9 Å resolution suggests binding to left-handed B-DNA. *Nature* **290**: 744–749.

McLure, K.G. and Lee, P.W. (1998) How p53 binds DNA as a tetramer. *EMBO J.* **17**: 3342–3350.

Miller, J., McLachlan, A.D. and Klug, A. (1985) Repetitive zinc-binding domains in the protein transcription factor IIIA from *Xenopus* oocytes. *EMBO J.* **4**: 1609–1614.

Müller, C.W. and Herrmann, B.G. (1997) Crystallographic structure of the T domain–DNA complex of the Brachyury transcription factor. *Nature* **389**: 884–888.

Müller, C.W., Rey, F.A., Sodeoka, M., Verdine, G.L. and Harrison, S.C. (1995) Structure of the NF-kappa B p50 homodimer bound to DNA. *Nature* **373**: 311–317.

Murphy, F.V., Sweet, R.M. and Churchill, M.E.A. (1999) The structure of a chromosomal high mobility group protein–DNA complex reveals sequence-neutral mechanisms important for non-sequence-specific DNA recognition. *EMBO J.* **18**: 6610–6618.

Murre, C., McCaw, P.S. and Baltimore, D. (1989) A new DNA binding and dimerization motif in immunoglobulin enhancer binding, *daughterless*, *MyoD*, and *myc* proteins. *Cell* **56**: 777–783.

Nadassy, K., Wodak, S.J. and Janin, J. (1999) Structural features of protein–nucleic acid recognition sites. *Biochemistry* **38**: 1999–2017.

Nash, H.A. and Granston, A.E. (1991) Similarity between the DNA-binding domains of IHF protein and TFIID protein. *Cell* **67**: 1037–1038.

Nolte, R.T., Conlin, R.M., Harrison, S.C. and Brown, R.S. (1998) Differing roles for zinc fingers in DNA recognition: structure of a six-finger transcription factor IIIA complex. *Proc. Natl Acad. Sci.* **95**: 2938–2943.

Ogata, K., Morikawa, S., Nakamura, H., *et al.* (1994) Solution structure of a specific DNA complex of the Myb DNA-binding domain with cooperative recognition helices. *Cell* **79**: 639–648.

Omichinski, J.G., Clore, G.M., Schaad, O., *et al.* (1993) NMR structure of a specific DNA complex of Zn-containing DNA binding domain of GATA-1. *Science* **261**: 438–446.

Otting, G., Qian, Y.Q., Billeter, M., *et al.* (1990) Protein–DNA contacts in the structure of a homeodomain–DNA complex determined by nuclear magnetic resonance spectroscopy in solution. *EMBO J.* **9**: 3085–3092.

Otwinowski, Z., Schevitz, R.W., Zhang, R.-G., *et al.* (1988) Crystal structure of *trp* repressor/operator complex at atomic resolution. *Nature* **335**: 321–329.

Pabo, C.O. and Lewis, M. (1982) The operator-binding domain of λ repressor: structure and DNA recognition. *Nature* **298**: 443–447.

Pavletich, N.P. and Pabo, C.O. (1991) Zinc finger-DNA recognition: Crystal structure of a Zif268–DNA complex at 2.1 Å. *Science* **252**: 809–817.

Pavletich, N.P. and Pabo, C.O. (1993) Crystal structure of a five-finger Gli–DNA complex: new perspectives on zinc fingers. *Science* **261**: 1701–1707.

Pellegrini, L., Tan, S. and Richmond, T.J. (1995) Structure of serum response factor core bound to DNA. *Nature* **376**: 490–498.

Ramakrishnan, V., Finch, J.T., Graziano, V., Lee, P.L. and Sweet, R.M. (1993) Crystal structure of globular domain of histone H5 and its implications for nucleosome binding. *Nature* **362**: 219–223.

Rastinejad, F., Perlmann, T., Evans, R.M. and Sigler, P.B. (1995) Structural determinants of nuclear receptor assembly on DNA direct repeats. *Nature* **375**: 203–211.

Raumann, B.E., Rould, M.A., Pabo, C.O. and Sauer, R.T. (1994) DNA recognition by beta-sheets in the Arc repressor–operator crystal structure. *Nature* **367**: 754–757.

Rice, P.A., Yang, S., Mizuuchi, K. and Nash, H.A. (1996) Crystal structure of an IHF–DNA complex: a protein-induced DNA U-turn. *Cell* **87**: 1295–1306.

Rodgers, D.W. and Harrison, S.C. (1993) The complex between the phage 434 repressor DNA-binding domain and operator site OR3: structural differences between consensus and non-consensus half sites. *Structure* **1**: 227–240.

Schultz, S.C., Shields, G.C. and Steitz, T.A. (1991) Crystal structure of a CAP–DNA complex: the DNA is bent by 90°. *Science* **253**: 1001–1007.

Schwabe, J.W.R. (1997) The role of water in protein–DNA interactions. *Curr. Opin. Struct. Biol.* **7**: 126–134.

Schwabe, J.W.R., Chapman, L., Finch, J.T. and Rhodes, D. (1993) The crystal structure of the estrogen receptor DNA-binding domain bound to DNA: how receptors discriminate between their response elements. *Cell* **75**: 567–578.

Schwabe, J.W.R., Chapman, L. and Rhodes, D. (1995) The oestrogen receptor recognizes an imperfectly palindromic response element through an alternative side-chain conformation. *Structure* **3**: 201–213.

Shimizu, T., Toumoto, A., Ihara, K., *et al.* (1997) Crystal structure of PHO4 bHLH domain–DNA complex: flanking base recognition. *EMBO J.* **16**: 4689–4697.

Somers, W.S. and Phillips, S.E.V. (1992) Crystal structure of the met repressor–operator complex at 2.8 Å resolution reveals DNA recognition by β-strands. *Nature* **359**: 387–393.

Starich, M.R., Wikström, M., Arst, H.N.J., Clore, G.M. and Gronenborn, A.M. (1998) The solution structure of a fungal AREA protein–DNA complex: an alternative binding mode for the basic carboxyl tail of GATA factors. *J. Mol. Biol.* **277**: 605–620.

Tan, S., Hunziker, Y., Sargent, D.F. and Richmond, T.J. (1996) Crystal structure of a yeast TFIIA/TBP/DNA complex. *Nature* **381**: 127–151.

Tan, S. and Richmond, T.J. (1998) Crystal structure of the yeast MATα2/MCM1/DNA ternary complex. *Nature* **391**: 660–666.

von Hippel, P.H. (1998) An integrated model of the transcription complex in elongation, termination, and editing. *Science* **281**: 660–665.

Werner, M.H., Clore, G.M., Fisher, C.L., *et al.* (1997) Correction of the NMR structure of the ETS1/DNA complex. *J. Biomol. NMR* **10**: 317–328.

Wharton, R.P., Brown, E.L. and Ptashne, M. (1984) Substituting an α-helix switches the sequence-specific DNA interactions of a repressor. *Cell* **38**: 361–369.

White, S.W., Appelt, K., Wilson, K.S. and Tanaka, I. (1989) A protein structural motif that bends DNA. *Proteins* 5: 281–288.

Wolberger, C., Dong, Y., Ptashne, M. and Harrison, S.C. (1988) Structure of a phage 434 Cro/DNA complex. *Nature* **335**: 789–795.

Wolberger, C., Vershon, A.K., Lui, B., Johnson, A.D. and Pabo, C.O. (1991) Crystal structure of a MATα2 homeodomain–operator complex suggests a general model for homeodomain–DNA interactions. *Cell* **67**: 517–528.

Zhou, P., Sun, L.J., Dötsch, V., Wagner, G. and Verdine, G.L. (1998) Solution structure of the core NFATC1/DNA complex. *Cell* **92**: 687–696.

5

Regulation of transcription during the cell cycle

Brian Dynlacht

1. Introduction

In the past 10 years there has been rapid progress in uncovering the links between gene expression and cell-cycle progression. An understanding of these connections has come about partly as a result of the biochemical characterization of cyclin-dependent kinases (CDK), which drive the cell cycle, and partly because of the detailed characterization of the multitude of factors that constitute the basal transcription machinery. In addition, the role of well-established transcription factors, including pRB, E2F, p53, and c-Myc, which play a prominent role in growth regulation, have undergone further clarification (Dang, 1999; Dyson, 1998; Evan and Littlewood, 1998; 1993; Giaccia and Kastan, 1998; Ko and Prives, 1996). This chapter will emphasize recent advances in our understanding of these and other factors during cell-cycle progression in mammalian cells. Studies in this system are providing a unified view of the regulatory controls based on the sequential activation of CDKs during G1 phase and S phase entry. These kinases progressively inhibit the retinoblastoma (pRB) tumor-suppressor protein and initially promote, and then inhibit, E2F, a transcription factor that displays the properties of an oncogene and tumor suppressor. This leads to the timely activation of genes required for cell-cycle progression. Similarly, Cyclin B/Cdc2 activity inhibits a number of transcriptional regulators during M phase. Two other CDK complexes that have less clear roles in cell-cycle regulation, cyclin H/Cdk7 and cyclin C/Cdk8, modify pol II thereby regulating the intiation and progression of transcription. Recent results further suggest intriguing parallels between the mechanisms that regulate transcription by all three RNA polymerases during the cell cycle and conservation of transcriptional controls between yeast and mammalian cells. Thus, an intricate balance of transcriptional and post-translational controls govern the activity of transcription factors that propel the cell cycle in a unidirectional manner.

Transcription Factors, edited by J. Locker.

2. An emerging role for mammalian and yeast CDKs in transcriptional regulation

Since the identification of Cdc2 as the catalytic component of maturation/mitosis promoting factor (MPF), an extensive family of kinases with homology to Cdc2 has been isolated, with several (Cdk2, Cdk4, Cdk6, Cdk7) having been implicated directly in cell-cycle control. Like Cyclin B/Cdc2, these kinases execute their functions by phosphorylating key substrates in a timely, cell-cycle-specific fashion. Periodic expression of Cyclins E and A during mid-to-late G1 drives the activation of Cdk2, and Cdk2 activity is essential for S phase entry (Hunter and Pines, 1994). Although pRB is an important physiological substrate of Cdk2 (see Section 5), recent data from several laboratories suggest that phosphorylation of this protein is not sufficient to account for the required role of Cdk2 in cell proliferation (Alevizopoulos *et al.*, 1997; Leng *et al.*, 1997; Lukas *et al.*, 1997). Indeed, it is apparent that several substrates of Cdk2 exist in addition to pRB and that phosphorylation of such substrates plays an important role in cell-cycle progression. The identity of these key Cdk2 substrates is not known.

A list of potential Cdk2 substrates is given in Table 1, and it is likely that additional targets of these kinases will be found in the near future. Included in this list are transcription factors with a known role in cell-cycle regulation (e.g., p53 and E2F, which are discussed later) as well as others where the role of phosphorylation is less well defined. The discussion in this chapter will be limited to the best-studied examples.

2.1 Transcription factors governing yeast cell-cycle progression

Yeast transcription factors, like their mammalian counterparts, are also targets of Cyclin/CDK complexes (Cross, 1995; Koch and Nasmyth, 1994). Budding yeast utilize at least nine distinct cyclin/CDK complexes that contain Cdc28, the yeast homolog of mammalian Cdc2, which is the only CDK that is strictly required for cell-cycle progression. Cdc28 is activated by G1 cyclins (Clns 1–3) and the S phase and mitotic cyclins (Clbs 1–6). Cell-cycle transitions in yeast are affected by transcriptional and post-transcriptional mechanisms that tightly regulate the abundance and activity of both CLNs and CLBs. Expression of these genes is controlled by both positive and negative factors. For example, transcription of CLN1 and CLN2 is activated by SBF, a heterodimeric transcription factor composed of Swi4 and Swi6, that binds to a G1/S-specific upstream activating sequence present in each of these genes. Clb5 and Clb6 are activated just prior to S phase by another factor, MBF, itself a heterodimer composed of Swi6 and MBP (McIntosh, 1993). A second regulatory mechanism imparts an even greater degree of control over SBF activity: near the onset of S phase, CLN/Cdc28 kinases phosphorylate and activate SBF bound to the promoter; later, during G2 and M phase, CLBs 1–4 combine with Cdc28 to inhibit SBF binding to the Cln2 promoter (Koch *et al.*, 1996). The net result is that CLNs are expressed only once per cell cycle during a defined period just prior to the DNA replicative phase. In this regard, yeast CDKs appear to modulate cell-cycle-dependent transcriptional activity of SBF in an elegant manner reminiscent of the Cyclin A/Cdk2 regulation of E2F transcription described later.

Table 1. Potential transcription factor substrates of cyclin-dependent kinases in mammalian cells and yeast

Substrate	Kinase pair/period of activity	Regulatory effect	Reference
Mammals			
pRB	Cyclin D/Cdk4, cyclin E/Cdk2 (G1)	Inhibitory	Weinberg, 1995 (Review)
UBF	Cyclin D/Cdk4, cyclin E/Cdk2 (G1)	Activation of transcription	Voit *et al.*, 1999
Id2	Cyclin E/Cdk2 (G1)	Altered DNA-binding activity	Hara *et al.*, 1997
B-Myb	Cyclin A/Cdk2 (S)	Stimulatory	Lane *et al.*, 1997;Bartsch *et al.*, 1999 Ziebold *et al.*, 1997
E2F	Cyclin A/Cdk2 (S/G2)	Inhibitory	Dynlacht *et al.*, 1994; 1997; Krek *et al.*, 1994; Xu *et al.*, 1994
p53	Cyclin A/Cdk2, cyclin B/Cdc2 (S/G2)	Stimulatory Altered DNA-binding specificity	Wang and Prives, 1995; Bischoff *et al.*, 1990 Sturzbecher *et al.*, 1990
c-Myb	Cyclin B/Cdc2 (G2/M)	Inhibitory	Luscher and Eisenman, 1992
TFIID	Cyclin B/Cdc2 (G2/M)	Inhibitory	Segil *et al.*, 1996
TFIIH	Cyclin B/Cdc2 (G2/M)	Inhibitory	Akoulitchev and Reinberg, 1998; Long *et al.*, 1998
TFIIIB	Cyclin B/Cdc2 (G2/M)	Inhibitory	Gottesfeld *et al.*, 1994; White *et al.*, 1995
RNA Pol II CTD	Cyclin H/Cdk7 (G1, S, G2)	Stimulation of pol II elongation	Dynlacht, 1997 (Review)
RNA Pol II CTD	Cyclin C/Cdk8 (all phases)	Inhibition of pol II initiation	Dynlacht, 1997 (Review)
Poly(A) polymerase	Cyclin B/Cdc2 (G2/M)	Inhibitory	Colgan *et al.*, 1996
SL1	Cyclin B/Cdc2 (G2/M)	Inhibition of pol I transcription	Heix *et al.*, 1998
Yeast			
SBF	Cln/Cdc28 (Start)	Stimulatory	Koch *et al.*, 1996
	Clb/Cdc28 (G2/M)	Inhibitory	Koch *et al.*, 1996
sw15	Clb/Cdc28 (G2/M)	Inhibition of nuclear localization	Moll *et al.*, 1991

M phase cyclins also regulate the activity of at least one other yeast transcription factor, Swi5, but in this case, regulation is at the level of nuclear localization. Localization of Swi5 is regulated by CLB/Cdc28 activity: CLB phosphorylation results in the cytoplasmic localization of Swi5 during anaphase, while inactivation of this kinase in late anaphase allows nuclear localization of Swi5 and transcription of target genes (Moll *et al.*, 1991; Nasmyth *et al.*, 1990). One important target of Swi5 is $p40^{Sic1}$, a CDK inhibitor that preferentially inactivates CLB/Cdc28 kinases. This could, in part, explain the decrease in CLB-associated kinase activity that results as cells exit mitosis and enter G1 (Knapp *et al.*, 1996; Toyn *et al.*, 1997). Subsequent ubiquitination and proteolytic destruction of $p40^{Sic1}$ allows the activation of CLB kinases and S phase entry.

3. Cdk7 and Cdk8: two components of the basal transcription machinery

One recent and unexpected finding was the demonstration that CDKs are an integral component of two basal transcription factors, TFIIH and RNA polymerase holoenzyme, suggesting an even more intimate linkage between cell cycle-control and transcriptional regulation.

TFIIH is a nine-subunit basal transcription factor complex with multiple enzymatic activities (Svestrup *et al.*, 1996). These include ATPase, DNA helicase, and kinase activities that are involved in transcription and DNA repair. The kinase activity associated with TFIIH is capable of phosphorylating the CTD of pol II, a heptapeptide of Tyr-Ser-Pro-Thr-Ser-Pro-Ser that is repeated 52 times in the human protein (Lu *et al.*, 1992). The kinase activity of TFIIH is required for transcription of certain promoters (DHFR, a TATA-less promoter), but not others (adenovirus major later promoter, a TATA-box containing promoter) (Akoulitchev *et al.*, 1995; Makela *et al.*, 1995). CTD kinase activity is provided by a Cyclin/CDK pair, Cyclin H/Cdk7/Mat1 (Feaver *et al.*, 1994; Makela *et al.*, 1994; Roy *et al.*, 1994; Serizawa *et al.*, 1995; Shiekhattar *et al.*, 1995). Phosphorylation of the CTD is thought to play a pivotal role in the transcription cycle, since it is associated with release of pol II from the pre-initiation complex, promoter clearance and efficient transcriptional elongation (Dahmus, 1995). Moreover, the hyperphosphorylated CTD is associated with an elongation-competent form of pol II (Bartholomew *et al.*, 1986; Laybourn and Dahmus, 1989; O'Brien *et al.*, 1994). The yeast counterpart of cyclin H/Cdk7, Kin28/Ccl1, is also a component of TFIIH, and it is generally required for transcription of the yeast genome (Feaver *et al.*, 1994; Holstege *et al.*, 1998). Interestingly, the mammalian kinase had been isolated originally as the CDK-activating kinase (CAK) an essential activator of Cdc2, Cdk2 and Cdk4 enzymes. Although experiments in yeast have identified an unrelated, monomeric kinase with properties of CAK, an important unresolved issue is whether higher eukaryotic cells contain a CAK enzyme distinct from Cyclin H/Cdk7 that is dedicated solely to the activation of CDKs. If no other CAK activity exists in higher eukaryotes, then an unknown connection between cell cycle and transcriptional control may soon be revealed. In this regard, a possible further connection between DNA repair and basal transcription may come from recent work demonstrating that p53 is phosphorylated by Cyclin H/Cdk7 in a manner dependent on a regulatory subunit, p36/Mat1, which suggests that Mat1 may function as a substrate specificity factor (Ko *et al.*, 1997; Yankulov and Bentley, 1997).

Additional findings have recently confirmed the connection between activity of the Cyclin H/Cdk7/Mat1 complex and cell-cycle progression. First, ablation of Mat1 RNA causes a pronounced block in the G1 phase followed by apoptosis, at least in some cell types (Wu *et al.*, 1999). In addition, the activity of Cyclin H/Cdk7 is dramatically reduced during M phase as a consequence of phosphorylation (see Section 8). Thus, expression from promoters dependent on this enzyme would be silenced by this mechanism during this stage of the cell cycle.

Shortly after the identification of Cyclin H/Cdk7 as a component of the TFIIH complex, a second Cyclin/CDK pair, composed of Srb10 and Srb11, was found to interact genetically and biochemically with the yeast pol II holoenzyme (Liao *et al.*, 1995). Loss of Srb10 function resulted in the activation of a small percentage of the yeast genome, suggesting that this kinase is a negative regulator of gene expression (Holstege *et al.*, 1998). The mammalian counterpart of Srb10/Srb11, Cyclin C/Cdk8, has been identified and found to co-purify with the pol II holoenzyme, and like Cyclin H/Cdk7, it is a potent CTD kinase (Maldonado *et al.*, 1996; Rickert *et al.*, 1996). Cyclin C levels do not oscillate during the cell cycle, suggesting that its activity may not change as a function of the cell cycle. It is not presently known why there are two Cyclin/CDK complexes associated with the basal transcription machinery that are able to phosphorylate the CTD. One possibility, based on biochemical studies, is that Srb10 is a negative regulator that functions prior to recruitment of the pre-initiation complex to the promoter, while Kin28 phosphorylates the CTD after pol II is recruited to the promoter (Hengartner *et al.*, 1998).

Because no fewer than 10 kinases have been shown to phosphorylate the CTD *in vitro*, a connection between CTD phosphorylation and the cell cycle remains elusive, and while it appears that Cdk2, Cdk7 and Cdk8 phosphorylate, in some cases, common transcription factor targets (*Table 2*), the physiological relevance of each remains to be determined. Further, it is possible that neither Cyclin H/Cdk7 nor Cyclin C/Cdk8 play a direct role in cell-cycle regulation: transcriptional control by cyclins and CDKs may simply represent another application of an efficient on–off regulatory switch that is provided by the cyclins.

4. The pRB family of proteins

In addition to post-translational modification as a mechanism of temporal modulation of transcription factor activity and cell-cycle progression, stable protein–protein interactions also govern cell-cycle-specific gene expression. One of the most thoroughly studied links of this type has emerged from studies of pRB. The prototypical tumor suppressor, pRB was first implicated as a tumor suppressor in genetic studies of children with retinoblastoma (Knudson, 1971). Molecular cloning of the *RB* gene and demonstration of its deregulation by DNA tumor viruses and inhibition by cell-cycle-dependent phosphorylation confirmed

Table 2. Transcription factors that bind to pRB family members

Target	Proposed function/ net effect on transcription	Interaction domain	*In vivo* association[a]	References
ATF2	Regulation of growth? (+)	Not determined	No	Kim *et al.*, 1992
hBrm/BRG	Chromatin remodeling (+)	LXCXE	Yes	Dunaief *et al.*, 1994; Singh *et al.*, 1995; Strober *et al.*, 1996
C/EBP family	Regulation of differentiation (+)	E2F-1-like domain	Yes	Chen *et al.*, 1996c
E2F family	Regulation of growth (–)	*C*-terminal domain	Yes	Dyson, 1998 (Review)
Elf1	Regulation of differentiation (–)	LXCXE	Yes	Wang *et al.*, 1993
HBP1	Regulation of growth/differentiation (–)	LXCXE	No	Tevosian *et al.*, 1997
HDAC1	Chromatin remodeling (–)	LXCXE	Yes	Brehm *et al.*, 1998; Luo *et al.*, 1998; Magnaghi-Jaulin *et al.*, 1998
Id-2	Regulation of growth	HLH domain	Yes	Iavarone *et al.*, 1994
c-Myc	Regulation of growth (–)	*N*-terminal domain	No	Beijersbergen *et al.*, 1994; Gu *et al.*, 1994; Rustgi *et al.*, 1991
MyoD/myogenin	Regulation of differentiation (+)	bHLH domain	Yes	Gu *et al.*, 1993
NF-IL6	Regulation of differentiation (+)	E2F-1-like domain	Yes	Chen *et al.*, 1996b
PU.1	Regulation of differentiation (–)	Acidic activation domain	No	Weintraub *et al.*, 1995; Hagemeier *et al.*, 1993
TAF250	Pol II regulation (–)	Multiple domains	No	Shao *et al.*, 1995
TFIIIB	Pol III repression; growth control (–)	Not determined	Yes	Larminie *et al.*, 1997
UBF	Pol I repression; growth control (–)	LXCXE	Yes	Cavanaugh *et al.*, 1995; Voit *et al.*, 1997

[a] *In vivo* association: yes and no indicate published detection of interaction between endogenous protein and pRB family member.

its pivotal role in proliferation and provoked a search for potential targets (Weinberg, 1995).

It is also well established that transcription factors are the principal targets of pRB action (*Table 2*). The phosphorylation state of pRB appears to determine its activity (*Figure 1*), since all targets known to interact with it involve the hypo-phosphorylated form of this protein. Inactivation of pRB is thought to occur through a series of sequential phosphorylation events. First, it is moderately phosphorylated upon activation of Cyclin D/Cdk4 during mid-G1 phase of the cell cycle (Lundberg and Weinberg, 1998). Later, upon activation of Cyclin E/Cdk2 and Cyclin A/Cdk2, it becomes hyper-phosphorylated and is rendered inactive both as a repressor of E2F and as a growth suppressor. Multiple CDK phosphorylation sites are also present in the p107 and p130 proteins, and phosphorylation of these proteins by G1 phase kinases results in their functional inactivation and, in the case of p130, degradation (Beijersbergen *et al*., 1995; Smith *et al*., 1996).

Shortly after the cloning of the *RB* gene, two highly related proteins, p107 and p130, which were first identified based on their binding to viral oncoproteins, were subsequently isolated and characterized. These proteins share the highest percent of homology with pRB in the so-called 'pocket domain' that includes the oncoprotein and E2F binding regions. They are distinguished from pRB by a region dedicated to binding Cyclin E/Cdk2 and Cyclin A/Cdk2 complexes. This 'spacer region' is distinct from the E2F binding region, enabling p107 and p130 to simultaneously interact with both E2F and Cyclin/CDK complexes. Indeed, p107/p130 complexes with E2F-4/Cyclin E/Cdk2 and E2F-4/Cyclin A/Cdk2 exist during G1 and S phase of the cell cycle, and although they are presumed to be transcriptionally inactive, the function of these complexes remains enigmatic.

A list of transcription factors known to bind the pRB family of proteins is shown in *Table 2*. In many, but not all, cases, pRB interacts with its binding partner through an LXCXE motif also found in several viral oncoproteins that interact with the tumor suppressor. Every protein thus far shown to interact with pRB does so exclusively with the hypo-phosphorylated form of this protein. While

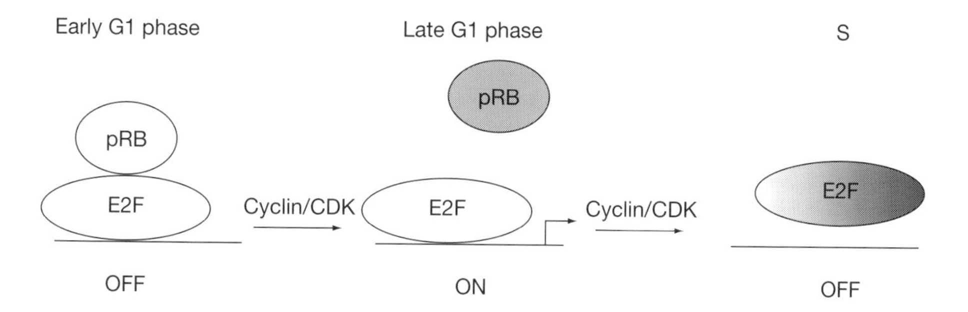

Figure 1. pRB regulation of E2F. pRB is in the unphosphorylated form during early G1 phase and is a tight-binding inhibitor of E2F. During late G1, retinoblastoma protein (pRB) is phosphorylated by Cyclin-dependent kinases (CDK) complexes and is no longer able to bind E2F. Active 'free' E2F is able to potentiate transcription from responsive genes. Later, other Cyclin/CDK complexes phosphorylate E2F, inhibiting its ability to bind DNA and activate transcription.

each of the factors listed in *Table 2* has been shown to interact with pRB or its relatives, several have not been shown to interact under physiological conditions, and in some cases, the transcriptional outcome induced by pRB remains to be determined. Thus, it is not always possible to assign to pRB a meaningful role in regulating their transcriptional activity. Further, since many of these proteins bind identical regions of pRB, it has been impossible to assess the relative contributions of any given interaction based on mutant alleles of pRB. However, recent studies suggest that the ability of pRB to restrain cell growth (by repressing E2F transcription) can be genetically separated from its ability to coactivate transcription (mediated by such factors as MyoD and C/EBPβ/NF-IL6) which promotes differentiation (Sellers *et al.*, 1998).

While it is clear that interaction with one transcription factor, E2F, cannot fully account for all of the growth-suppressive and differentiation-inducing properties of pRB (Qian *et al.*, 1992; Sellers *et al.*, 1995; Welch and Wang, 1993; Zhu *et al.*, 1993), it remains one of the most convincing transcription factor targets of pRB. The interactions between pRB and E2F will be discussed in Section 5.6. In addition to E2F, a cohort of factors that regulate transcription from all three classes of promoters appear to interact with pRB (*Table 2*). In many cases, association with pRB is thought to inhibit cell growth and promote differentiation. Not only does pRB regulate the activity of several upstream activator proteins, but it also modulates the activity of chromatin remodeling factors. Many new insights into the various ways that pRB regulates transcription are now emerging, and these recent advances are now discussed.

4.1 A common theme: regulation of RNA polymerase I and III

Although initially characterized by its ability to inhibit E2F activation of pol II transcription, it is now recognized that pRB may exert its growth-suppressive effect by globally inhibiting pol I and pol III transcription as well (*Table 2*). pRB represses pol I-mediated transcription through its interactions with upstream binding factor (UBF) (Cavanaugh *et al.*, 1995). UBF is known to potentiate transcription from pol I promoters by stimulating the recruitment of SL1, a factor required for class I gene expression. pRB could repress UBF, and therefore pol I activity, by preventing interactions between UBF and the basal machinery, or by inhibiting UBF DNA-binding activity (Voit *et al.*, 1997). Likewise, pol III transcription of 5S and tRNA synthesis is inhibited by interactions between pRB and the TFIIIB complex (Chu *et al.*, 1997; Larminie *et al.*, 1997). Transcription by pol I and pol III is thus finely tuned by a combination of cell-cycle controls that also govern pol II regulation, including phosphorylation by CDKs and repression by pRB.

4.2 Mechanisms of retinoblastoma protein

Our knowledge of the mechanism(s) used by pRB to repress transcription is still fairly limited. One model initially proposed that pRB could inhibit transcription by simply blocking interactions between the transactivation domain of the associated factor and the basal transcription machinery. This was an especially attractive way to explain pRB repression of E2F, since the pRB-binding domain of E2F overlaps with its *trans*-activation domain. This model further suggested that the activation domains of factors bound proximal to E2F (or any other factor that could recruit pRB) might be similarly inhibited.

However, the latest experiments suggest that the E2F/pRB complex can act like an active transcriptional repressor or silencer (Weintraub *et al.*, 1995), and recent detailed mechanistic studies of pRB (Ross *et al.*, 1999) suggest the possibility that it may not inhibit transcription solely by the simple masking of transactivation domains. Rather, pRB might interact with a basal factor in the pre-initiation complex. To test these possible mechanisms of pRB repression, it is necessary to carry out reconstituted transcription reactions with purified factors. Using an *in vitro* system, pRB repression of at least one pol II promoter has been shown to occur prior to, or at the initial stages of, assembly of the pre-initiation complex, although pRB had no obvious effect on E2F binding at a promoter (Ross *et al.*, 1999). E2F can recruit transcription factors TFIID and TFIIA, and pRB could function, in part, by preventing recruitment of these basal factors. However, once fully formed, the pre-initiation complex is completely resistant to pRB repression. Because pRB is unable to inhibit transcription directed by TBP, it is possible that a target of pRB may exist in the TFIID complex (Ross *et al.*, 1999). One such target might be $TAF_{II}250$ (Shao *et al.*, 1995). Intriguingly, repression of pol III (but not pol I) promoters by pRB can occur before or after a pre-initiation complex is formed (Larminie *et al.*, 1997; Voit *et al.*, 1997). In this respect, the mechanism of transcriptional repression of pol II promoters might resemble more closely that of the pol I promoters, although additional experiments will be required in the future to determine the identity of potential targets in the pre-initiation complex in both cases. Further work will also be required to understand whether transcriptional repression by p107 and p130 occurs by a similar or different mechanism. It will also be important to ultimately determine the extent to which each mechanism governs regulation of E2F-responsive promoters during cell-cycle progression.

Although it is clear that pRB can repress transcription in a reconstituted transcription system devoid of chromatin (Dynlacht *et al.*, 1994; Ross *et al.*, 1999), others have suggested that pRB may function as a repressor through the recruitment of the histone deacetylase, HDAC1 (Brehm *et al.*, 1998; Luo *et al.*, 1998; Magnaghi-Jaulin *et al.*, 1998). This evidence was based on an association between pRB and cellular histone deacetylase activity and the ability of histone deacetylase inhibitors, such as Trichostatin A, to reverse pRB-mediated repression of E2F. However, it was also apparent that an association between histone deacetylases and pRB was required to repress certain promoters but not others, and interestingly, the related protein p107 did not recruit histone deacetylase nor require histone deacetylation to repress transcription (Luo *et al.*, 1998). If histone deacetylase is required for promoter-specific repression by pRB, it will be both interesting and important to understand the basis of such specificity. One caveat with the above studies is that Trichostatin A appears to de-repress transcription in a general way, causing, in some cases, growth arrest and cellular differentiation (Yoshida *et al.*, 1995). Furthermore, these studies were performed with transfected reporter genes and ectopically expressed proteins rather than endogenous promoters and physiological levels of pRB. It will be important to study the role of pRB in facilitating histone deacetylation of promoters in their native context.

In addition, pRB and related proteins could alter transcription by interacting with two other chromatin remodeling factors, hBrm and Brgl (*Table 2*), which are components of the human SWI/SNF complex (Dunaief *et al.*, 1994; Singh *et al.*, 1995; Strober *et al.*, 1996). The association between pRB and hBrm/Brgl has a profound impact on cell growth as well, since there is a strict correlation between the ability of these proteins to interact and to induce a flat-cell, growth-arrested phenotype. pRB may also interact with the LXCXE motif of hBrm to play a positive role in potentiating transcription. In this setting, pRB was able to cooperate with the glucocorticoid receptor and hBrm to activate transcription from glucocorticoid receptor-responsive genes (Singh *et al.*, 1995). However, the precise functional relationship between pRB and the hBrm/Brgl remodeling activity has not been elucidated, and a general role for pRB in the regulation of chromatin remodeling during the cell cycle has not been demonstrated. Nevertheless, the above examples suggest the intriguing possibility that pRB could act to facilitate establishment of chromatin in either a repressed or de-repressed state.

5. The E2F transcription factor

Table 2 lists a fraction of the ~50 cellular polypeptides thought to bind the pRB family (Dyson, 1998), but as previously described, there is now compelling biochemical evidence to suggest that the E2F transcription factor family is a major downstream target of the pRB family *in vivo*. Furthermore, genetic data obtained from knockout mice have recently confirmed the *in vivo* importance of the pRB/E2F-1 interaction: loss of the *E2F1* gene reduced the frequency of tumors seen in an *RB*(+/-) mouse (Yamasaki *et al.*, 1998). The role of E2F in cell-cycle progression and the regulation of its activity will be the focus of this section; those seeking a more detailed review of the E2F literature are directed to a number of excellent reviews (Weinberg, 1995; Dyson, 1998).

Among the first indications that the E2F transcription factor was important in regulation of cell proliferation stemmed from the fact that when it was overexpressed, E2F-1 drove quiescent cells into a proliferative state (Johnson *et al.*, 1993), and subsequent experiments suggested that E2F-1 could function as an oncogene or as an inducer of apoptosis when ectopically expressed (Qin *et al.*, 1994; Singh *et al.*, 1994; Xu *et al.*, 1995). These studies prompted investigations into downstream targets of E2F. It was anticipated that a list of E2F-responsive genes would include a set that was required for entry into the DNA-replicative (S) phase, and indeed, this was shown to be the case. It is now clear that the list of downstream genes includes other transcription factors that drive cell-cycle progression (*E2F*, *RB*), cell-cycle regulatory proteins (*B-myb*, *c-Myc*, *cyclin A*, *cyclin E*, *CDC2*), gene products necessary for DNA synthesis (*DHFR*, *TK*, *TS*, *RNR*), and proteins necessary for DNA replication itself (*CDC6*, *DNA polymerase* α, *MCM*s, *Orc1*, *PCNA*) (Dyson, 1998). Since most of these targets are required for S phase entry or DNA replication it is clear how activation of expression could promote proliferation while inhibition of expression by the pRB family could restrain growth.

5.1 E2F family members

The active E2F transcription factor, referred to as 'free' E2F, is a heterodimer of one E2F and one DP family polypeptide (*Figure 2*). The mammalian E2F family consists of at least six genes (collectively referred to as *E2F* throughout this chapter), termed *E2F-1–E2F-6*, and the DP family is represented by two genes, *DP1* and *DP2*. Homology between E2F family members is largely restricted to DNA-binding and dimerization domains. In addition, all E2Fs except E2F-6 share a conserved domain that binds the pRB family of proteins and overlaps with its transactivation domain. Since E2F-6 lacks a transactivation domain, this protein is thought to be a constitutive repressor.

Despite these apparent structural similarities, E2F family members can be distinguished by additional regulatory layers. First, although all E2F proteins except E2F-6 are able to bind pRB-related proteins, only a subset (E2F-1, E2F-2, E2F-3, and E2F-4) bind to pRB *in vivo*. E2F-4 is also able to interact, not only with pRB, but also with the pRB-related proteins, p107 and p130 (discussed later), while E2F-5 interacts exclusively with p130. Second, three E2F family members (E2F-1, E2F-2 and E2F-3) are related by an additional amino-terminal domain that confers stable binding to a cyclin-dependent kinase, Cyclin A/Cdk2. Interestingly, similar kinase targeting sequences (variously termed the Cy or RXL motif) have been identified in p107, p130, and the p21/p27/p57 cyclin-dependent kinase inhibitor family (Adams *et al.*, 1996; Castano *et al.*, 1998; Chen *et al.*, 1996a; Dynlacht *et al.*, 1997; Zhu *et al.*, 1995), as well as an expanding list of proteins with diverse roles in the cell cycle.

E2F abundance and activity are regulated by several mechanisms during the course of the cell cycle. First, transcription of *E2F* genes occurs in 'waves': *E2F-4* and *E2F-5* are expressed during early G1, while *E2F-1*, *E2F-2* and *E2F-3* are expressed later, as cells approach the S phase (Leone *et al.*, 1998; Sardet *et al.*, 1995; Sears *et al.*, 1997). At least two E2F promoters (E2F-1 and E2F-2) are autoregulated by E2F and pRB, and E2F and c-Myc can collaborate to regulate expression of *E2F-2* (Sears *et al.*, 1997). Regulation of E2F activity occurs at the post-transcriptional level as well, since it has been shown that 'free' E2F is ubiquitinated and degraded by the proteasome, while the pRB family of proteins

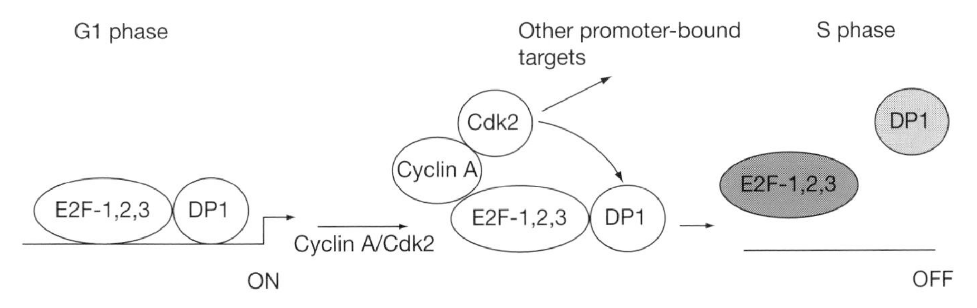

Figure 2. Specific E2F family members are regulated by cyclin/CDK complexes. E2F-1, -2, and -3 have cyclin-binding domains that stably bind Cyclin-dependent kinase (CDK) complexes. The net result is phosphorylation of the E2F and DP1 subunits, the latter of which appears to be important in regulating the activity of the heterodimer. It is not known whether other targets might be regulated by a promoter-bound Cyclin/CDK complex.

enhances the stability of E2F by blocking ubiquitination of E2F (Campanero and Flemington, 1997; Hateboer *et al.*, 1996; Hofmann *et al.*, 1996).

E2F is phosphorylated, in a temporally restricted manner, by cyclin-dependent kinases (*Figure 2* and *Table 1*). The cyclin-binding domain of E2F-1 (as well as E2F-2 and E2F-3) targets DP1 (and the associated E2F polypeptide) for phosphorylation, resulting in the downregulation of E2F DNA-binding and transcriptional activity (Dynlacht *et al.*, 1994; 1997; Krek *et al.*, 1994; Xu *et al.*, 1994). Phosphorylation of the E2F family by CDKs is remarkably specific in two ways. First, Cyclin A-dependent kinases are potent regulators of E2F activity, but the G1-specific kinases Cyclin D/Cdk4 and Cyclin E/Cdk2 are not (Dynlacht *et al.*, 1994; 1997). Second, only those E2F family members with a cyclin-binding domain are phosphorylated by CDKs (Dynlacht *et al.*, 1997). The net result of this regulation is that E2F activity may be restricted to a defined window during the cell cycle (Dynlacht *et al.*, 1994) (*Figure 2*). Indeed, downregulation of E2F activity may be essential for completion of the S phase, since expression of constitutively active E2F-1 heterodimers, which lacked either the cyclin-binding domain of E2F-1 or major phosphorylation sites of DP1, resulted in an S phase block and eventual apoptosis (Krek *et al.*, 1995).

5.2 E2F as a transcriptional activator and repressor

It is attractive to speculate that individual E2F/DP combinations will activate unique subsets of responsive genes based on the non-overlapping phenotypes of *E2F-1* and *E2F-5* null mice (Lindeman *et al.*, 1998; Yamasaki *et al.*, 1996) and the observation that ectopic expression of specific E2F family members activates distinct targets (DeGregori *et al.*, 1997). However, because the full complement of transcriptional targets of E2F is not known, it is not clear precisely how individual E2F family members contribute to cell cycle progression. Nor is any single downstream target able to replace E2F, with the exception of *cyclin E* in *Drosophila* dE2F mutants (Duronio and O'Farrell, 1995). It does seem likely, however, that targets of E2F will include genes unrelated to proliferation, based on both *dE2F* mutations in flies and *E2F-5* knockout mice (Brook *et al.*, 1996; Lindeman *et al.*, 1998).

In addition to the role of E2F as a transactivator of genes, such as *DHFR* required for S phase entry, recent data have emerged to suggest that E2F-binding sites function as transcriptional repressor elements. Genomic footprinting confirms that E2F sites in the *B-Myb*, *cyclin A*, *cyclin E* and *Cdc2* genes are occupied during quiescence, a period that coincides with the presence of p130/E2F complexes and low levels of promoter activity (Huet *et al.*, 1996; Le Cam *et al.*, 1999; Tommasi and Pfeifer, 1995; Zwicker *et al.*, 1996). Interestingly, the *cyclin E* promoter has an E2F recognition site able to bind a novel p130/E2F complex that confers repression during quiescence and early G1 (Le Cam *et al.*, 1999). This E2F recognition motif is actually a bipartite cell-cycle-regulated element that contains a variant E2F site adjacent to a TA-rich sequence conserved in the mouse and human promoters. Cyclin E induction at the G1/S transition results from the loss of p130/E2F binding (de-repression) and activation by a second E2F-binding site (Botz *et al.*, 1996; Le Cam *et al.*, 1999). It is possible that the *cyclin E* promoter may represent an example of a composite E2F site that juxtaposes two elements and confers binding by specific members of the E2F and pRB families.

The biological importance of E2F as a repressor was further suggested by the *E2F-1* knockout mouse, which, surprisingly, developed a range of tumors (Yamasaki *et al.*, 1996). The fact that E2F-1 resembles a tumor suppressor might imply that the repressive aspect of this transcription factor is at least as important as its role as an activator. Indeed, if it is true that E2F functions solely as a repressor on some promoters, such as those that control *E2F-1* and *B-myb* expression, loss of repression could give rise to inappropriate expression of these genes, both of which display oncogenic properties. Tumors in *E2F-1* knockout mice could result from inappropriate levels of these and/or other genes or inappropriate timing of their expression, which then induces S phase. Alternatively, since apoptosis was partially abrogated in the *E2F-1-/-* mouse, tumor suppression could result from the loss of an apoptotic function. It is intriguing to note that induction of apoptosis is not a general property of the E2F. That is, although E2F-1, E2F-2, and E2F-3 are equally capable of promoting S phase through the induction of responsive genes, E2F-1 was exclusively able to induce apoptosis (DeGregori *et al.*, 1997). Although E2F-1-mediated apoptosis appears to be p53-dependent in many cell types, the precise mechanisms governing this response are not known. One possibility is that E2F can inhibit activation of anti-apoptotic signals, including NF-κB expression (Phillips *et al.*, 1999). Alternatively, prolonged proliferative signals engendered by inappropriate levels of E2F activity, or the inability to downregulate E2F during S phase, could activate a checkpoint that causes cells to delay or arrest in S phase and ultimately undergo apoptosis (Krek *et al.*, 1995).

It will be important to understand the contributions of individual E2F family members to cell-cycle progression and apoptosis. To this end, methods to detect binding by E2F and pRB family members to endogenous promoters in living cells must be used to uncover physiologically relevant gene targets regulated by E2F, both positively and negatively. In addition, the systematic disruption of all mouse *E2F* genes will ultimately be of great importance to this effort.

6. c-Myc transcription factor and its partners

In addition to E2F and p53, c-Myc is one of the most extensively studied mammalian transcription factors known to play a role in cell proliferation and apoptosis and has been extensively reviewed in Evan and Littlewood (1993 and 1998) and in Dang (1999). As such, this section will focus on the more recent studies of c-Myc regulation of cell growth.

c-Myc activation or amplification has been implicated in a variety of human cancers, including lymphomas and carcinomas of the lung, breast, and colon (Cole, 1986). c-Myc function is required for the proliferation of many cell types, and conversely, its inactivation is essential for differentiation. c-Myc functions as a transcriptional activator by dimerizing with Max, allowing it to bind to an E-box recognition sequence. c-Myc, but not Max, contains an amino-terminal transactivation domain that stimulates transcription, perhaps by contacting TBP (Hateboer *et al.*, 1993; Maheswaran *et al.*, 1994). c-Myc function is antagonized by the Mad family of proteins, which forms heterodimers with Max (Ayer *et al.*, 1993; Zervos *et al.*, 1993). The Mad proteins repress transcription by recruiting Sin3, a component of a complex that contains the corepressor NCor and the histone deacetylase, HDAC1 (Hassig *et al.*, 1997; Heinzel *et al.*, 1997; Laherty *et al.*, 1997; Nagy *et al.*, 1997; Zhang *et al.*, 1997). Recruitment of histone deacetylases is thought to constitute an important mechanism for Mad-mediated transcriptional repression.

A number of protein–protein interactions have been shown to modulate c-Myc function, and interestingly, a subset of them appears to regulate E2F function as well. These include pRB and the related protein, p107 (Beijersbergen *et al.*, 1994; Gu *et al.*, 1994; Hoang *et al.*, 1995; Rustgi *et al.*, 1991), and TRRAP, a protein homologous to the ATM/PI3-kinase family and a component of the PCAF and SAGA complexes (Grant *et al.*, 1998; McMahon *et al.*, 1998; Saleh *et al.*, 1998; Vassilev *et al.*, 1998). Presumably, the pRB family represses c-Myc by binding and sterically blocking its amino terminal transactivation domain. On the other hand, complexes containing TRRAP could promote c-Myc and E2F oncogenesis by potentiating histone acetylation and transactivation of target genes. However, since many of these experiments were performed largely by ectopically expressing potential c-Myc regulators, it will be important to determine whether these proteins are *bona fide* regulators of c-Myc *in vivo* and to unravel the mechanisms through which these proteins regulate c-Myc transcriptional activity and cell proliferation.

How does the c-Myc family promote proliferation? At least 30 genes are proposed to be direct c-Myc targets (Dang, 1999). Although *c-Myc* is a well-known immediate–early response gene, the issue of c-Myc targets, especially those associated with cell cycle progression, remains a controversial one, due in part to differences in experimental systems. A multitude of genes with a direct role in cell-cycle progression have been implicated as c-Myc targets. These include many of the cyclin genes involved in G1 and S phase progression (*cyclins A*, *D1* and *E*), *Cdc2*, and *Cdc25A*, a phosphatase that activates Cdk2 and Cdk4 (Daksis *et al.*, 1994; Galaktionov *et al.*, 1996; Hanson *et al.*, 1994; Hoang *et al.*, 1994; Jansen-Durr *et al.*, 1993; Perez-Roger *et al.*, 1997; Philipp *et al.*, 1994). However, many of these published experiments rely on ectopic expression of c-Myc, yielding, in some cases, conflicting results (see Dang, 1999), and recent experiments with *c-myc* null cells appear to cast doubt on some of the putative targets while confirming the authenticity of others (Bush *et al.*, 1998). In this study, growth arrest and DNA damage inducible gene (*gadd45*) expression was elevated and carbamoyl-phosphate synthase/aspartate carbamoyltransferase/dihydroorotase (*CAD*) expression was reduced significantly. These genes were the only identified targets that appear to contribute to growth regulation by c-Myc, although caveats regarding compensatory changes necessarily apply to all studies of knockout cells. Such targets could play an important role in growth control, since *CAD* is involved in pyrimidine biosynthesis and would therefore be essential for G1/S phase progression. Similarly, *gadd45*, which is normally induced by DNA damage and repressed by c-Myc (Marhin *et al.*, 1997) is de-repressed in *c-Myc* null cells. Although not previously defined as a c-Myc target, recent ectopic expression experiments have identified the $p19^{ARF}$ tumor suppressor gene as such (Zindy *et al.*, 1998). $p19^{ARF}$ is also thought to be a transcriptional target of E2F (Bates *et al.*, 1998; DeGregori *et al.*, 1997). A number of recent observations suggest that cells protect themselves from the deleterious effects of activated oncogenes by promoting apoptosis or growth arrest (Weinberg, 1997). Therefore, c-Myc and E2F-mediated signaling through $p19^{ARF}$ and p53 (Section 7) could, in part, underlie oncogene-mediated apoptosis. Future *in vivo* experiments will be required to determine if $p19^{ARF}$ is a direct target of c-Myc and E2F and whether it is responsible for apoptosis mediated by these transcription factors.

c-Myc could promote S phase entry by transcriptional enhancement of cyclins and CDKs that promote proliferation, or by increasing or decreasing expression of genes that activate or inhibit CDKs, respectively (Alevizopoulos *et al.*, 1997; Galaktionov *et al.*, 1996; Leone *et al.*, 1997; Perez-Roger *et al.*, 1997). In addition to direct effects of c-Myc on transcription, this activator might play a less well-defined, and perhaps indirect, role in cell-cycle progression by modulating levels of key regulators, such as Cyclin E (Amati *et al.*, 1998). In some settings, c-Myc could promote proliferation by collaborative interactions with Ras: ectopic expression of c-Myc and Ras led to elevated levels of Cyclin E/Cdk2 activity and decreased expression of the $p27^{Kip1}$ cyclin-dependent kinase inhibitor (CKI) (Alevizopoulos *et al.*, 1997; Leone *et al.*, 1997; Perez-Roger *et al.*, 1997). The net result of decreased expression of this CKI and increased levels of Cyclin E/Cdk2 is hyper-phosphorylation of pRB family members and accumulation of E2F, provoking induction of S phase entry.

Recent experiments have been performed with a transactivation-deficient c-Myc mutant that retains an ability to repress certain promoters and stimulate cell proliferation (Xiao *et al.*, 1998). These experiments suggest that the role of c-Myc in proliferation is more complex than anticipated and imply that one role of c-Myc in growth control might involve a mechanism other than activation through canonical (E box)-binding sites, such as transcriptional repression or activation of genes through non-consensus binding sites.

7. Role of p53 as transcription factor and growth suppressor

p53 is the most frequently mutated gene in human cancers and is associated with leukemias, and lung, brain and breast tumors, among others (Levine *et al.*, 1991), attesting to the central role that the p53 protein plays in maintaining normal cell growth and genomic stability. This section focuses on some very recent studies regarding the role of p53 in growth suppression and in response to DNA damage. More extensive surveys on p53 can be found in a number of thorough reviews (Giaccia and Kastan, 1998; Ko and Prives, 1996).

In normal growing cells, p53 is present at very low levels in a latent form that can be activated through post-translational modification in response to stressful or oncogenic stimuli from internal and external sources (Giaccia and Kastan, 1998). p53 is not necessary for cell viability; rather, its central growth-suppressive and apoptotic functions stem from this role as a 'sensor'. As described earlier, a number of kinases, including the CDKs (*Table 1*), are thought to regulate p53

DNA-binding and transcriptional activity in a cell-cycle-dependent manner. CAK phosphorylates an amino terminal residue of p53 (Ser 33), while Cdk2 and Cdc2 phosphorylate a carboxyl terminal residue (Ser315) *in vitro*. Although these findings are intriguing, additional work is needed to tie such phosphorylation to normal cell-cycle progression.

In contrast, regulation of p53 in response to DNA damage has been studied more extensively. p53 plays a pivotal role in the cellular response to DNA damage, and p53 stability and activity are altered at several levels in response to DNA damage. First, p53 protein levels and activity are regulated post-translationally by diverse pathways in response to a number of genotoxic and stressful stimuli. Several kinases, including the DNA-dependent protein kinase (DNA-PK), ATM, and ATR kinases have been implicated in the activation of p53 in response to DNA damage, although there has been some recent dispute regarding the role of DNA-PK (Jimenez *et al.*, 1999). It has been proposed that DNA-PK, ATM and ATR kinases phosphorylate p53 on Ser15 in response to DNA damage, increasing its stability and activity, as a result of decreased binding to Mdm2 (Shieh *et al.*, 1997). Interestingly, ultraviolet but not ionizing radiation, irradiation induces the phosphorylation of a specific p53 residue that results in increased sequence-specific DNA-binding (Kapoor and Lozano, 1998; Lu *et al.*, 1998). This phosphorylation occurs in an ATM-independent manner, suggesting the involvement of a different kinase. Further complicating our understanding of p53 regulation is the recent observation that p53 is acetylated in response to either ionizing radiation or ultraviolet treatment, and the net result is the enhancement of sequence-specific DNA-binding activity (Gu and Roeder, 1997; Sakaguchi *et al.*, 1998). DNA-damage-induced acetylation of p53 at specific lysine residues may be dictated by prior phosphorylation of certain amino-terminal sites (Sakaguchi *et al.*, 1998). The p300/CBP and PCAF acetyltransferases are likely to be the physiologically relevant enzymes that modify p53 (Liu *et al.*, 1999).

In addition to phosphorylation and acetylation, a number of protein–protein interactions also regulate the stability and activity of the p53 protein (Sherr, 1998). In this category is the Mdm2 oncoprotein, which was originally characterized based on its ability to antagonize p53 transcriptional activation, perhaps by blocking interactions between p53 and the basal transcription machinery (Oliner *et al.*, 1993). Interestingly, Mdm2 apparently counteracts p53 function by a second mechanism; namely by catalyzing its destruction, probably through a nuclear–cytoplasmic shuttling function of Mdm2 (Haupt *et al.*, 1997; Honda *et al.*, 1997; Kubbutat *et al.*, 1997). The $p19^{ARF}$ protein on the other hand, functions as a tumor suppressor by stabilizing p53, most likely by antagonizing the nuclear export of Mdm2 that is necessary for p53 destruction (Tao and Levine, 1999; Weber *et al.*, 1999; Zhang and Xiong, 1999).

As for c-Myc, a cadre of potential p53-responsive genes have been identified (Ko and Prives, 1996). Of these, the CKI, $p21^{Waf1}$, is one of the most compelling candidates for a downstream target whose upregulation correlates with growth arrest in response to DNA damage (Brugarolas *et al.*, 1995; Deng *et al.*, 1995; El-Deiry *et al.*, 1993). These studies showed that p21 is markedly induced in response to p53 activation, and that loss of the *p21* gene impairs the p53-dependent G1 arrest caused by DNA damage. Induction of *p21* inhibits the activity of the CDKs, resulting in hypo-phosphorylated pRB, which is the growth suppressive form of this protein (see Section 5.4).

Recent work has shown that p53 belongs to a family of related proteins that includes p63 and p73. Such studies have also begun to show the similarities and differences between the properties of p53 and p73, which, unlike p53, is not inactivated by viral oncoproteins and is not induced by DNA damage (Kaelin, 1999). Moreover, a major distinction is that, in contrast with p53, p63 appears to play a pronounced role in development (Mills *et al.*, 1999). Further work will be needed to distinguish a role, if any, for these p53-related proteins in the DNA damage response, normal cell growth control, and tumorigenesis.

8. M phase repression of transcription

It has been known for several decades that gene expression in eukaryotic cells (with the exception of yeast) is confined to the first and second gap phases (G1 and G2) and to the DNA replicative phase (S), whereas transcription is for the most part silenced during mitosis (M phase) (Gottesfeld and Forbes, 1997; Johnson and Holland, 1965; Prescott and Bender, 1962; Taylor, 1960). Several general mechanisms have been proposed recently to explain M phase transcriptional silencing, and it now appears likely that a mitotic kinase, originally defined as the MPF and now known as Cyclin B/Cdc2, will play a major role in suppressing gene expression during this phase of the cell cycle. Three mitosis-specific mechanisms are now considered:

(i) phosphorylation of sequence-specific and basal transcription factors required for transcription of all three classes of promoters;
(ii) alterations in chromatin;
(iii) inhibition of events subsequent to initiation, including inactivation of a factor contributing to mRNA stability and abortion of elongation complexes.

8.1 Phosphorylation of basal transcription factors

In addition to its role in activating condensin and inactivating sequence-specific DNA-binding proteins (*Table 1*), recent *in vitro* studies have shown that cyclin B/Cdc2 can also phosphorylate and inhibit the basal transcription machinery of all three RNA polymerases during M phase. These targets include the pol I factor SL1, two pol II basal factors, TFIID and TFIIH, and the TFIIIB component of pol III (*Table 2*).

It is known that transactivation by a wide array of mammalian upstream activator proteins requires the recruitment and activity of the basal factor TFIID, which is a multi-protein complex composed of TBP and as many as 13 TAFs (Tjian and Maniatis, 1994). Mitosis-specific phosphorylation of one or more TAFs, possibly by Cyclin B/Cdc2, was shown to reduce the ability of mammalian TFIID to support activation of transcription by p53 and Sp1 (Segil *et al.*, 1996). Basal transcription levels, however, were not significantly affected in this study, suggesting that although TBP was phosphorylated in mitotic cells, TAFs were the relevant functional targets of phosphorylation.

Interestingly, an analogous mechanism is apparently used to regulate pol I and pol III transcription during mitosis. Like pol II, pol I and pol III utilize basal factors SL1 and TFIIIB that contain TBP and polymerase-specific TAFs. Mitotic suppression of transcription of 5S and tRNA genes by pol III could be recapitulated *in vitro* using interphase and M phase extracts (Hartl *et al.*, 1993). Biochemical fractionation experiments revealed that the TFIIIB component of pol III was the likely target of suppression and that it is phosphorylated by cyclin B/Cdc2 (Gottesfeld *et al.*, 1994; White *et al.*, 1995). Subsequent reconstitution experiments with isolated TFIIIB, which is composed of TBP and TAFs of 75 and 92 kDa, suggested that the latter component (known as Brf/TAF_{III}90/TFIIIB90) and TBP are the most likely targets of this mitotic kinase (Leresche *et al.*, 1996). Similarly, mitotic repression of rRNA transcription is most likely to be due to phosphorylation of TAFs that are present in the SL1 complex, which is composed of TBP and three pol I-specific TAFs. Human SL1 is recruited to ribosomal gene promoters by UBF, and mitotic suppression of transcription could occur through the impairment of UBF/SL1 interactions as a result of phosphorylation by Cyclin B/Cdc2 or another mitotic kinase (Heix *et al.*, 1998).

Cyclin B/Cdc2 may also facilitate transcriptional repression of pol II by phosphorylating and inactivating another mammalian basal transcription factor, TFIIH. TFIIH includes a CTD kinase that is composed of Cyclin H/Cdk7 (discussed in Section 3). Recently, it has been shown that the Cyclin H/Cdk7 component of TFIIH is itself phosphorylated and inhibited by Cyclin B/Cdc2 (Akoulitchev and Reinberg, 1998; Long *et al.*, 1998). In these studies, it was apparent that TFIIH that had been isolated from mitotic cells had significantly reduced activity compared with TFIIH purified from interphase cells. This reduced activity can be traced to the phosphorylation of Cdk7, which impaired CTD kinase activity as well as basal transcription. Thus, as for TFIID, Cyclin B/Cdc2 is a candidate for the mitotic kinase involved in Cyclin H/Cdk7 inactivation.

Cyclin B/Cdc2 is also able to phosphorylate the CTD *in vitro* (Cisek and Corden, 1989; Gebara *et al.*, 1997). Such phosphorylation is expected to have profound effects on transcription, as it is the unphosphorylated form of pol II that is recruited to the pre-initiation complex (Laybourn and Dahmus, 1989; Lu *et al.*, 1991). The direct inactivation of pol II by MPF *in vitro* could provide an explanation for mitotic inactivation (Zawel *et al.*, 1993), although others have not seen gross differences between the phosphorylation state of mitotic and cycling mammalian cells (Parsons and Spencer, 1997).

8.2 Cell-cycle-dependent changes in chromatin

One recently proposed mechanism that explains transcriptional silencing during M phase in higher eukaryotic cells involves the apparent exclusion of sequence-specific transcription factors from chromatin (Martinez-Balbas *et al.*, 1995). In this study, genomic footprinting of one promoter (*hsp70*) indicated that several sequence-specific transcription factors, including Oct1, were excluded from this promoter during mitosis. This is consistent with the notion that mitosis-specific phosphorylation of the Oct1 homeodomain abrogates its DNA-binding activity (Roberts *et al.*, 1991; Segil *et al.*, 1991). Although the kinase(s) responsible for mitotic phosphorylation of Oct1 and related POU-domain proteins must be identified, phosphorylation may not be due solely to the action of Cdc2 (Caelles *et al.*, 1995). The studies on mitotic exclusion of factors from chromatin have focused on a single pol II promoter, but it is conceivable that similar findings will be uncovered for other pol II promoters as well as promoters driving pol I and pol III transcription. Therefore, chromatin exclusion might represent a more general mechanism for reprogramming transcription of promoters prior to active synthesis during G1 (Martinez-Balbas *et al.*, 1995).

In addition to mitotic phosphorylation of specific transcription factors such as Oct1 and Myb (Luscher and Eisenman, 1992), changes in chromatin condensation and structure during M phase may also account for repression of transcription during this period of the cell cycle. Phosphorylation of histones H1 and H3 during mitosis has been associated with chromosome condensation (Roth and Allis, 1992). Furthermore, it is known that histone H3 is phosphorylated during mitosis on one particular residue, ser10 (Hendzel *et al.*, 1997). This suggests the intriguing possibility that phosphorylation of this histone might function as a signal for chromosome condensation, either by altering the structure of chromatin or by recruitment of additonal factors that are involved in condensation. Recently, components of the cellular machinery that are involved in chromosome condensation have been identified. This large multi-protein complex called Condensin compacts chromatin in an ATP-dependent manner specifically during mitosis (Hirano *et al.*, 1997). It will be important to determine how the activity of this complex is confined specifically to M phase, but phosphorylation by Cyclin B/Cdc2, itself active only during M phase, and perhaps another kinase(s), could promote periodic Condensin phosphorylation and activation (Kimura *et al.*, 1998). Future experiments will determine whether Condensin is recruited to mitotic chromosomes as a consequence of chromatin modification. Formation of repressed chromatin during mitosis also results from inactivation of the chromatin-remodeling factor, SWI/SNF (Sif *et al.*, 1998), or its exclusion from chromatin (Muchardt *et al.*, 1996). SWI/SNF inactivation correlated with phosphorylation of two subunits, hBrm and Brg1, while reversal of phosphorylation reactivated remodeling activity. The kinase(s) and phosphatase(s) responsible for phosphorylating and dephosphorylating SWI/SNF have not been identified, although Cyclin/CDK complexes do not appear to be involved (Sif *et al.*, 1998). In any event, reversible mitotic inactivation of a remodeling complex could further promote cell-cycle-dependent, global repression of transcription, in collaboration with other mechanisms.

8.3 Post-initiation regulatory mechanisms

Still other mechanisms may exist to reduce gene expression beyond the initiation phase of transcription. First, pools of poly(A)+ RNA are decreased during mitosis, concomitant with a decrease in protein synthesis, and mechanisms probably exist to regulate its abundance at appropriate cell-cycle stages. Indeed, this is likely to be the case, since it has been shown recently that poly(A) polymerase is found in the hyper-phosphorylated state in mitotic extracts, and such phosphorylation is inhibitory (Colgan *et al.*, 1996). Poly(A) polymerase is phosphorylated by Cyclin B/Cdc2 *in vitro*, suggesting again that MPF may be the kinase responsible for inactivating the enzyme. Another mechanism for mitotic regulation of transcriptional events subsequent to initiation is abortion of nascent transcripts, which has been postulated based on experiments in *Drosophila* (Shermoen and O'Farrell, 1991). In mammalian cells, elongation complexes were excluded from mitotic chromosomes from late prophase to late telophase, although arrested pol II complexes were detected on certain chromosomal regions (Parsons and Spencer, 1997).

In summary then, transcription directed by all three RNA polymerases can be completely shut down during mitosis as a result of phosphorylation by a single kinase, Cyclin B/Cdc2. It is possible that other kinases with similar substrate specificity, such as the mitogen-activated protein kinases (MAPK), may function analogously, and it will be important to ascribe cell cycle-regulated phosphorylation events to modification of particular residues in pol II to clarify the role of phosphorylation of this enzyme in mitotic transcriptional repression. That transcription by all polymerases is globally and completely inhibited attests to the importance of chromatin condensation during M phase and to a promoter reconfiguration prior to the subsequent G1 phase. Although a number of mechanisms underlying mitotic repression have already been revealed, it is likely that others will emerge in the future.

9. Conclusion

Although many intricate details have recently emerged regarding transcriptional control of the cell cycle, many surprises are likely to be uncovered in the near future. A major challenge will be to understand how the function of various combinations of factors is integrated at different promoters that exhibit cell-cycle periodicity, in much the same way that combinatorial mechanisms govern transcription during the course of development. In addition, it will be necessary to understand the functional relationships within complex, multi-protein assemblies—transcription factor complexes with the pRB family and Cyclin/CDK complexes as well as histone deacetylase—that modulate transcriptional activity during the cell cycle.

Acknowledgments

I apologize to my colleagues for the omission of suitable references due to space restrictions. Work in my laboratory was supported by the Pew Foundation, NIH and Department of Defense.

References

Adams, P.D., Sellers, W.R., Sharma, S.K., Wu, A.D., Nalin, C.M. and Kaelin, W.G. (1996) Identification of a cyclin-cdk2 recognition motif present in substrates and p21-like cyclin-dependent kinase inhibitors. *Mol. Cell. Biol.* **16**: 6623–6633.

Akoulitchev, S., Makela, T.P., Weinberg, R.A. and Reinberg, D. (1995) Requirement for TFIIH kinase activity in transcription by RNA polymerase II. *Nature* **377**: 557–560.

Akoulitchev, S. and Reinberg, D. (1998) The molecular mechanism of mitotic inhibition of TFIIH is mediated by phosphorylation of CDK7. *Genes Dev.* **12**: 3541–3550.

Alevizopoulos, K., Vlach, J., Hennecke, S. and Amati, B. (1997) Cyclin E and c-Myc promote cell proliferation in the presence of p16INK4a and of hypophosphorylated retinoblastoma family proteins. *EMBO J.* **16**: 5322–5323.

Amati, B., Alevizopoulos, K. and Vlach, J. (1998) Myc and the cell cycle. *Front. Biosci.* **3**: D250–D268.

Ayer, D.E., Kretzner, L. and Eisenman, R.N. (1993) Mad: a heterodimeric partner for Max that antagonizes Myc transcriptional activity. *Cell* **72**: 211–222.

Bartholomew, B., Dahmus, M. E. and Meares, M. E. (1986) RNA contacts subunits IIo and IIc in HeLa RNA polymerase II transcription complexes. *J. Biol. Chem.* **261**: 14226–14231.

Bartsch, O., Horstmann, S., Toprak, K., Klempnauer, K. H. and Ferrari, S. (1999) Identification of cyclin A/Cdk2 phosphorylation sites in B-Myb. *Eur. J. Biochem.* **260**: 384–391.

Bates, S., Phillips, A.C., Clark, P.A. *et al.* (1998) p14ARF links the tumour suppressors RB and p53. *Nature* **395**: 124–125.

Beijersbergen, R.L., Carlee, L., Verkhoven, R.M. and Bernards, R. (1995) Regulation of the retinoblastoma protein-related p107 by G1 cyclin complexes. *Genes Dev.* **9**: 1340–1353.

Beijersbergen, R.L., Hijmans, E.M., Zhu, L. and Bernards, R. (1994) Interaction of c-Myc with the pRb-related protein p107 results in inhibition of c-Myc-mediated transactivation. *EMBO J.* **13**: 4080–4086.

Bischoff, J.R., Friedman, P.N., Marshak, D.R., Prives, C. and Beach, D. (1990) Human p53 is phosphorylated by p60–cdc2 and cyclin B–cdc2. *Proc. Natl Acad. Sci. USA* **87**: 4766–4770.

Botz, J., Zerfass-Thome, K., Spitkovsky, D. *et al.* (1996) Cell cycle regulation of the murine cyclin E gene depends on an E2F binding site in the promoter. *Mol. Cell. Biol.* **16**: 3401–3409.

Brehm, A., Miska, E.A., McCance, D.J., Reid, J.L., Bannister, A.J. and Kouzarides, T. (1998) Retinoblastoma protein recruits histone deacetylase to repress transcription. *Nature* **391**: 597–601.

Brook, A., Xie, J.-E., Du, W. and Dyson, N. (1996) Requirements for dE2F function in proliferating cells and in post-mitotic differentiating cells. *EMBO J.* **15**: 3676–3683.

Brugarolas, J., Chandrasekaran, C., Gordon, J.I., Beach, D., Jacks, T. and Hannon, G.J. (1995) Radiation-induced cell cycle arrest compromised by p21 deficiency. *Nature* **377**: 552–557.

Bush, A., Mateyak, M., Dugan, K. *et al.* (1998) *c-myc* null cells misregulate *cad* and *gadd45* but not other proposed c-Myc targets. *Genes Dev.* **12**: 3797–3802.

Caelles, C., Hennemann, H. and Karin, M. (1995) M-phase-specific phosphorylation of the POU transcription factor GHF-1 by a cell cycle-regulated protein kinase inhibits DNA binding. *Mol. Cell Biol.* **15**: 6694–6701.

Campanero, M.R. and Flemington, E.K. (1997) Regulation of E2F through ubiquitin-proteasome-dependent degradation: stabilization by the pRB tumor suppressor protein. *Proc. Natl Acad. Sci. USA* **94**: 2221–2226.

Castano, E., Kleyner, Y. and Dynlacht, B.D. (1998) Dual cyclin-binding domains are required for p107 to function as a kinase inhibitor. *Mol. Cell Biol.* **18**: 5380–5391.

Cavanaugh, A.H., Hempel, W.M., Taylor, L.J., Rogalsky, V., Todorov, G. and Rothblum, L.I. (1995) Activity of RNA polymerase I transcription factor UBF blocked by Rb gene product. *Nature* **374**: 177–180.

Chen, J., Saha, P., Kornbluth, S., Dynlacht, B.D. and Dutta, A. (1996a) Cyclin-binding motifs are essential for the function of p21^{cip1}. *Mol. Cell. Biol.* **16**: 4673–4682.

Chen, P.-L., Riley, D.J., Chen-Kiang, S. and Lee, W.-H. (1996b) Retinoblastoma protein directly interacts with and activates the transcription factor NF–IL6. *Proc. Natl Acad. Sci. USA* **93**: 465–469.

Chen, P.L., Riley, D.J., Chen, Y. and Lee, W.H. (1996c) Retinoblastoma protein positively regulates terminal adipocyte differentiation through direct interaction with C/EBPs. *Genes Dev.* **10**: 2794–2804.

Chu, W.M., Wang, Z., Roeder, R.G. and Schmid, C.W. (1997) RNA polymerase III transcription repressed by Rb through its interactions with TFIIIB and TFIIIC2. *J. Biol. Chem.* **272**: 14755–14761.

Cisek, L.J. and Corden, J.L. (1989) Phosphorylation of RNA polymerase by the murine homologue of the cell cycle control protein cdc2. *Nature* **339**: 679–684.

Cole, M.D. (1986) The *myc* oncogene: its role in transformation and differentiation. *Ann. Rev. Genet.* **20**: 361–384.

Colgan, D.F., Murthy, K.G.K., Prives, C. and Manley, J.L. (1996) Cell cycle related regulation of poly(A) polymerase by phosphorylation. *Nature* **384**: 282–285.

Cross, F. (1995) Transcriptional regulation by a cyclin-cdk. *Trends Genet.* **11**: 209–211.

Dahmus, M.E. (1995) Phosphorylation of the *C*-terminal domain of RNA polymerase II. *Biochim. Biophys. Acta* **1261**: 171–182.

Daksis, J.I., Lu, R.Y., Facchini, L.M., Marhin, W.W. and Penn, L.J. (1994) Myc induces cyclin D1 expression in the absence of *de novo* protein synthesis and links mitogen-stimulated signal transduction to the cell cycle. *Oncogene* **9**: 3635–3645.

Dang, C.V. (1999) c-Myc target genes involved in cell growth, apoptosis, and metabolism. *Mol. Cell. Biol.* **19**: 1–11.

DeGregori, J., Leone, G., Miron, A., Jakoi, L. and Nevins, J.R. (1997) Distinct roles for E2F proteins in cell growth control and apoptosis. *Proc. Natl Acad. Sci. USA* **94**: 7245–7250.

Deng, C., Zhang, P., Harper, J.W., Elledge, S. and Leder, P. (1995) Mice lacking p21/CIP1/WAF1 undergo normal development, but are defective in G1 checkpoint control. *Cell* **82**: 675–684.

Dunaief, J.L., Strober, B.E., Guha, S. *et al.* (1994) The retinoblastoma protein and BRG1 form a complex and cooperate to induce cell cycle arrest. *Cell* **79**: 119–130.

Duronio, R.J. and O'Farrell, P.H. (1995) Developmental control of the G1 to S transition in *Drosophila*; cyclin E is a limiting downstream target of E2F. *Genes Dev.* **9**: 1456–1468.

Dynlacht, B. (1997) Regulation of transcription by proteins that control the cell cycle. *Nature* **389**: 149–152.

Dynlacht, B.D., Flores, O., Lees, J.A. and Harlow, E. (1994) Differential regulation of E2F *trans*-activation by cyclin-cdk2 complexes. *Genes Dev.* **8**: 1772–1786.

Dynlacht, B.D., Moberg, K., Lees, J.A., Harlow, E. and Zhu, L. (1997) Specific regulation of E2F family members by cyclin-dependent kinases. *Mol. Cell. Biol.* **17**: 3867–3875.

Dyson, N. (1998) The regulation of E2F by pRB-family proteins. *Genes Dev.* **12**: 2245–2262.

El-Deiry, W.S., Tokino, T., Velculescu, V.E. *et al.* (1993) *WAF1*, a potential mediator of p53 tumor suppression. *Cell* **75**: 817–825.

Evan, G. and Littlewood, T. (1993) The role of c-myc in cell growth. *Curr. Opin. Genet. Dev.* **3**: 44–49.

Evan, G. and Littlewood, T. (1998) A matter of life and cell death. *Science* **281**: 1317–1322.

Feaver, W.J., Svejstrup, J.Q., Henry, N.L. and Kornberg, R.D. (1994) Relationship of CDK-activating kinase and RNA polymerase II CTD kinase TFIIH/TFIIK. *Cell* **79**: 1103–1109.

Galaktionov, K., Chen, X. and Beach, D. (1996) Cdc25 cell cycle phosphatase as a target of c-myc. *Nature* **382**: 511–517.

Gebara, M.M., Sayre, M.H. and Corden, J.L. (1997) Phosphorylation of the carboxy-terminal repeat domain in RNA polymerase II by cyclin-dependent kinases is sufficient to inhibit transcription. *J. Cell. Biochem.* **64**: 390–402.

Giaccia, A.J. and Kastan, M.B. (1998) The complexity of p53 modulation: emerging patterns from divergent signals. *Genes Dev.* **12**: 2973–2983.

Gottesfeld, J.M. and Forbes, D.J. (1997) Mitotic repression of the transcriptional machinery. *Trends Biochem. Sci.* **22**: 197–202.

Gottesfeld, J.M., Wolf, V.J., Dang, T., Forbes, D.J. and Hartl, P. (1994) Mitotic repression of RNA polymerase III transcription *in vitro* mediated by phosphorylation of a TFIIIB component. *Science* **263**: 81–84.

Grant, P.A., Schieltz, D., Pray-Grant, M.G., Yates, J.R. and Workman, J.L. (1998) The ATM-related cofactor Tra1 is a component of the purified SAGA complex. *Mol. Cell* **2**: 863–867.

Gu, W., Bhatia, K., Magrath, I.T., Dang, C.V. and Dalla-Favera, R. (1994) Binding and suppression of the Myc transcriptional activation domain by p107. *Science* **264**: 251–254.

Gu, W. and Roeder, R.G. (1997) Activation of p53 sequence-specific DNA binding by acetylation of the p53 *C*-terminal domain. *Cell* **90**: 595–606.

Gu, W., Schneider, J.W., Condorelli, G., Kaushai, S., Mahdavi, V. and Nadal-Ginard, B. (1993) Interaction of myogenic factors and the retinoblastoma protein mediates muscle cell commitment and differentiation. *Cell* **72**: 309–324.

Hagemeier, C., Bannister, A.J., Cook, A. and Kouzarides, T. (1993) The activation domain of transcription factor PU.1 binds the retinoblastoma (RB) protein and the transcription factor TFIID *in vitro*: RB shows sequence similarity to TFIID and TFIIB. *Proc. Natl Acad. Sci. USA* **90**: 1580–1584.

Hanson, K.D., Shichiri, M., Follansbee, M.R. and Sedivy, J.M. (1994) Effects of c-myc expression on cell cycle progression. *Mol. Cell. Biol.* **14**: 5748–5755.

Hara, E., Hall, M. and Peters, G. (1997) Cdk2-dependent phosphorylation of Id2 modulates activity of E2A-related transcription factors. *EMBO J.* **16**: 332–342.

Hartl, P., Gottesfeld, J. and Forbes, D.J. (1993) Mitotic repression of transcription *in vitro*. *J. Cell Biol.* **120**: 613–624.

Hassig, C.A., Fleischer, T.C., Billin, A.N., Schreiber, S.L. and Ayer, D.E. (1997) Histone deacetylase activity is required for full transcriptional repression by mSin3A. *Cell* **89**: 341–347.

Hateboer, G., Kerkhoven, R.M., Shvarts, A., Bernards, R. and Beijersbergen, R.L. (1996) Degradation of E2F by the ubiquitin-proteasome pathway: regulation by retinoblastoma family proteins and adenovirus transforming proteins. *Genes Dev.* **10**: 2960–2970.

Hateboer, G., Timmers, H.T., Rustgi, A.K., Billaud, M., van't Veer, L.J. and Bernards, R. (1993) TATA-binding protein and the retinoblastoma gene product bind to overlapping epitopes on c-Myc and adenovirus E1A protein. *Proc. Natl Acad. Sci. USA* **90**: 8489–8493.

Haupt, Y., Maya, R., Kazaz, A. and Oren, M. (1997) Mdm2 promotes the rapid degradation of p53. *Nature* **387**: 296–299.

Heinzel, T., Lavinsky, R.M., Mullen, T.M. *et al.* (1997) A complex containing N-CoR, mSin3 and histone deacetylase mediates transcriptional repression. *Nature* **387**: 43–48.

Heix, J., Vente, A., Voit, R., Budde, A., Michaelidis, T.M. and Grummt, I. (1998) Mitotic silencing of human rRNA synthesis: inactivation of the promoter selectivity factor SL1 by cdc2/cyclin B-mediated phosphorylation. *EMBO J.* **17**: 7373–7381.

Hendzel, M.J., Wei, Y., Mancini, M.A. *et al.* (1997) Mitosis-specific phosphorylation of histone H3 initiates primarily within pericentromeric heterochromatin during G_2 and spreads in an ordered fashion coincident with mitotic chromosome condensation. *Chromosoma* **106**: 348–360.

Hengartner, C.J., Myer, V.E., Liao, S.-M., Wilson, C.J., Koh, S.S. and Young, R.A. (1998) Temporal regulation of RNA polymerase II by Srb10 and Kin28 cyclin-dependent kinases. *Mol. Cell* **2**: 43–53.

Hirano, T., Kobayashi, R. and Hirano, M. (1997) Condensins, chromosome condensation protein complexes containing XCAP-C, XCAP-E, and *Xenopus* homolog of the *Drosophila* Barren protein. *Cell* **89**: 511–521.

Hoang, A.T., Cohen, K.J., Barrett, J.F., Bergstrom, D.A. and Dang, C.V. (1994) Participation of cyclin A in Myc-induced apoptosis. *Proc. Natl Acad. Sci. USA* **91**: 6875–6879.

Hoang, A. T., Lutterbach, B., Lewis, B. C. *et al.* (1995) A link between increased transforming activity of lymphoma-derived MYC mutant alleles, their defective regulation by p107, and altered phosphorylation of the c-Myc transactivation domain. *Mol. Cell Biol.* **15**: 4031–4042.

Hofmann, F., Martelli F, Livingston, D.M. and Wang, Z. (1996) The retinoblastoma gene product protects E2F-1 from degradation by the ubiquitin-proteasome pathway. *Genes Dev.* **10**: 2949–2959.

Holstege, F.C.P., Jennings, E.G., Wyrick, J.J., *et al.* (1998) Dissecting the regulatory circuitry of a eukaryotic genome. *Cell* **95**: 717–728.

Honda, R., Tanaka, H. and Yasuda, H. (1997) Oncoprotein MDM2 is a ubiquitin ligase E3 for tumor suppressor p53. *FEBS Lett.* **420**: 25–27.

Huet, X., Rech, J., Plet, A., Vie, A. and Blanchard, J.M. (1996) Cyclin A expression is under negative transcriptional control during the cell cycle. *Mol. Cell. Biol.* **16**: 3789–3798.

Hunter, T. and Pines, J. (1994) Cyclins and Cancer II: cyclin D and cdk inhibitors come of age. *Cell* **79**: 573–582.

Iavarone, A., Garg, P., Lasorella, A., Hsu, J. and Israel, M.A. (1994) The helix-loop-helix protein Id-2 enhances cell proliferation and binds to the retinoblastoma protein. *Genes Dev.* **8**: 1270–1284.

Jansen-Durr, P., Meichle, A., Steiner, P., *et al.* (1993) Differential modulation of cyclin gene expression by MYC. *Proc. Natl Acad. Sci. USA* **90**: 3685–3689.

Jimenez, G.S., Bryntesson, F., Torres-Arzayus, M.I., *et al.* (1999) DNA-dependent protein kinase is not required for the p53-dependent response to DNA damage. *Nature* **400**: 81–83.

Johnson, D.G., Schwartz, J.K., Cress, W.D. and Nevins, J.R. (1993) Expression of transcription factor E2F1 induces quiescent cells to enter S phase. *Nature* **365**: 349–352.

Johnson, L.H. and Holland, J.J. (1965) Ribonucleic acid and protein synthesis in mitotic HeLa cells. *J. Cell Biol.* **27**: 565–574.

Kaelin, J.W.G. (1999) The emerging p53 gene family. *J. Natl Cancer Inst.* **91**: 594–598.

Kapoor, M. and Lozano, G. (1998) Functional activation of p53 via phosphorylation following DNA damage by UV but not gamma radiation. *Proc. Natl Acad. Sci. USA* **95**: 2834–2837.

Kim, S.-J., Wagner, S., Liu, F., O'Reilly, M.A., Robbins, P.D. and Green, M.R. (1992) Retinoblastoma gene product activates expression of the human TGF-β2 gene through transcription factor ATF-2. *Nature* **358**: 331–334.

Kimura, K., Hirano, M., Kobayashi, R. and Hirano, T. (1998) Phosphorylation and activation of 13S condensin by Cdc2 *in vitro*. *Science* **282**: 487–490.

Knapp, D., Bhoite, L., Stillman, D.J. and Nasmyth, K. (1996) The transcription factor Swi5 regulates expression of the cyclin kinase inhibitor p40SIC1. *Mol. Cell. Biol.* **16**: 5701–5707.

Knudson, J. A.G. (1971) Mutation and cancer: statistical study of retinoblastomas. *Proc. Natl Acad. Sci. USA* **68**: 820–823.

Ko, L.J. and Prives, C. (1996) p53: puzzle and paradigm. *Genes Dev.* **10**: 1054–1072.

Ko, L.J., Shieh, S.Y., Chen, X. *et al.* (1997) p53 is phosphorylated by CDK7-cyclin H in a p36MAT1-dependent manner. *Mol. Cell. Biol.* **17**: 7220–7229.

Koch, C. and Nasmyth, K. (1994) Cell cycle regulated transcription in yeast. *Curr. Opin. Cell Biol.* **6**: 451–459.

Koch, C., Schleiffer, A., Ammerer, G. and Nasmyth, K. (1996) Switching transcription on and off during the yeast cell cycle: Cln/Cdc28 kinases activate bound transcription factor SBF (Swi4/Swi6) at start, whereas Clb/Cdc28 kinases displace it from the promoter in G2. *Genes Dev.* **10**: 129–141.

Krek, W., Ewen, M.E., Shirodkar, S., Arany, Z., Kaelin, W.G. and Livingston, D. (1994) Negative regulation of the growth-promoting transcription factor E2F-1 by a stably bound cyclin A-dependent protein kinase. *Cell* **78**: 161–172.

Krek, W., Xu, G. and Livingston, D.M. (1995) Cyclin A-kinase regulation of E2F-1 DNA binding function underlies suppression of an S phase checkpoint. *Cell* **83**: 1149–1158.

Kubbutat, M.H., Jones, S.N. and Vousden, K.H. (1997) Regulation of p53 stability by Mdm2. *Nature* **387**: 299–303.

Laherty, C.D., Yang, W.M., Sun, J.M., Davie, J.R., Seto, E. and Eisenman, R.N. (1997) Histone deacetylases associated with the mSin3 corepressor mediate mad transcriptional repression. *Cell* **89**: 349–356.

Lane, S., Farlie, P. and Watson, R. (1997) B-Myb function can be markedly enhanced by cyclin A-dependent kinase and protein truncation. *Oncogene* **14**: 2445–2453.

Larminie, C. G., Cairns, C. A., Mital, R. *et al.* (1997) Mechanistic analysis of RNA polymerase III regulation by the retinoblastoma protein. *EMBO J.* **16**: 2061–2071.

Laybourn, P.J. and Dahmus, M.E. (1989) Transcription-dependent changes in the *C*-terminal domain of mammalian RNA polymerase subunit IIa/o. *J. Biol. Chem.* **264**: 6693–6698.

Le Cam, L., Polanowska, J., Fabbrizio, E. *et al.* (1999) Timing of *cyclin E* gene expression depends on the regulated association of a bipartite repressor element with a novel E2F complex. *EMBO J.* **18**: 1878–1890.

Leng, X., Connell-Crowley, L., Goodrich, D. and Harper, J. W. (1997) S-Phase entry upon ectopic expression of G1 cyclin-dependent kinases in the absence of retinoblastoma protein phosphorylation. *Curr. Biol.* **7**: 709–712.

Leone, G., DeGregori, J., Sears, R., Jakoi, L. and Nevins, J.R. (1997) Myc and Ras collaborate in inducing accumulation of active cyclin E/Cdk2 and E2F. *Nature* **387**: 422–426.

Leone, G., DeGregori, J., Yan, Z. *et al.* (1998) E2F3 activity is regulated during the cell cycle and is required for the induction of S phase. *Genes Dev.* **12**: 2120–2130.

Leresche, A., Wolf, V.J. and Gottesfeld, J.M. (1996) Repression of RNA polymerase II and III transcription during M phase of the cell cycle. *Exp. Cell Res.* **229**: 282–288.

Levine, A.J., Momand, J. and Finlay, C.A. (1991) The p53 tumour suppressor gene. *Nature* **351**: 453–456.

Liao, S.-M., Zhang, J., Jeffery, D. A. *et al.* (1995) A kinase-cyclin pair in the RNA polymerase II holoenzyme. *Nature* **374**: 193–196.

Lindeman, G.J., Dagnino, L., Gaubatz, S. *et al.* (1998) A specific, nonproliferative role for E2F-5 in choroid plexus function revealed by gene targeting. *Genes Dev.* **12**: 1092–1098.

Liu, L., Scolnick, D.M., Trievel, R.C. *et al.* (1999) p53 sites acetylated *in vitro* by PCAF and p300 are acetylated *in vivo* in response to DNA damage. *Mol. Cell. Biol.* **19**: 1202–1209.

Long, J.J., Leresche, A., Kriwacki, R.W. and Gottesfeld, J.M. (1998) Repression of TFIIH transcriptional activity and TFIIH-associated cdk7 kinase activity at mitosis. *Mol. Cell. Biol.* **18**: 1467–1476.

Lu, H., Flores, O., Weinmann, R. and Reinberg, D. (1991) The non-phosphorylated form of RNA polymerase II preferentially associates with the pre-initiation complex. *Proc. Natl Acad. Sci. USA* **88**: 10004–10008.

Lu, H., Taya, Y., Ikeda, M. and Levine, A.J. (1998) Ultraviolet radiation, but not gamma radiation or etoposide-induced DNA damage, results in the phosphorylation of the murine p53 protein at serine-389. *Proc. Natl Acad. Sci. USA* **95**: 6399–6402.

Lu, H., Zawel, L., Fisher, L., Egly, J.-M. and Reinberg, D. (1992) Human general transcription factor IIH phosphorylates the *C*-terminal domain of RNA polymerase II. *Nature* **358**: 641–645.

Lukas, J., Herzinger, T., Hansen, K. *et al.* (1997) Cyclin E-induced S phase without activation of the pRb/E2F pathway. *Genes Dev.* **11**: 1479–1492.

Lundberg, A.S. and Weinberg, R.A. (1998) Functional inactivation of the retinoblastoma protein requires sequential modification by at least two distinct cyclin-cdk complexes. *Mol. Cell. Biol.* **18**: 753–761.

Luo, R.X., Postigo, A.A. and Dean, D.C. (1998) Rb interacts with histone deacetylase to repress transcription. *Cell* **92**: 463–473.

Luscher, B. and Eisenman, R.N. (1992) Mitosis-specific phosphorylation of the nuclear oncoproteins Myc and Myb. *J. Cell. Biol.* **118**: 775–784.

Magnaghi-Jaulin, L., Groisman, R., Naguibneva, I. *et al.* (1998) Retinoblastoma protein represses transcription by recruiting a histone deacetylase. *Nature* **391**: 601–605.

Maheswaran, S., Lee, H. and Sonenshein, G.E. (1994) Intracellular association of the protein product of the c-myc oncogene with the TATA-binding protein. *Mol. Cell Biol.* **14**: 1147–1152.

Makela, T.P., Parvin, J.D., Kim, J., Huber, L.J., Sharp, P.A. and Weinberg, R.A. (1995) A kinase-deficient transcription factor TFIIH is functional in basal and activated transcription. *Proc. Natl Acad. Sci. USA* **92**: 5174–5178.

Makela, T.P., Tassan, J.-P., Nigg, E.A., Frutiger, S., Hughes, G.J. and Weinberg, R.A. (1994) A cyclin associated with the CDK-activating kinase MO15. *Nature* **371**: 254–257.

Maldonado, E., Shiekhattar, R., Sheldon, M. *et al.* (1996) A human RNA polymerase II complex associated with SRB and DNA-repair proteins. *Nature* **381**: 86–89.

Marhin, W.W., Chen, S., Facchini, L.M., Fornace, J.A.J. and Penn, L.Z. (1997) Myc represses the growth arrest gene *gadd45*. *Oncogene* **14**: 2825–2834.

Martinez-Balbas, M.A., Dey, A., Rabindran, S.K., Ozato, K. and Wu, C. (1995) Displacement of sequence-specific transcription factors from mitotic chromatin. *Cell* **83**: 29–38.

McIntosh, E.M. (1993) MCB elements and the regulation of DNA replication in yeast. *Curr. Genet.* **24**: 185–192.

McMahon, S.B., Van Buskirk, H.A., Dugan, K.A., Copeland, T.D. and Cole, M.D. (1998) The novel ATM-related protein TRRAP is an essential cofactor for the c-Myc and E2F oncoproteins. *Cell* **94**: 363–374.

Mills, A.A., Zheng, B., Wang, X.J., Vogel, H., Roop, D.R. and Bradley, A. (1999) p63 is a p53 homologue required for limb and epidermal morphogenesis. *Nature* **398**: 708–713.

Moll, T., Tebb, G., Surana, U., Robitsch, H. and Nasmyth, K. (1991) The role of phosphorylation and the CDC28 protein kinase in cell cycle-regulated nuclear import of the *S. cerevisiae* transcription factor SWI5. *Cell* **66**: 743–758.

Muchardt, C., Reyes, J. C., Bourachot, B., Leguoy, E. and Yaniv, M. (1996) The hbrm and BRG-1 proteins, components of the human SNF/SWI complex, are phosphorylated and excluded from the condensed chromosomes during mitosis. *EMBO J.* **15**: 3394–3402.

Nagy, L., Kao, H. Y., Chakravarti, D. *et al.* (1997) Nuclear receptor repression mediated by a complex containing SMRT, mSin3A, and histone deacetylase. *Cell* **89**: 373–380.

Nasmyth, K., Adolf, G., Lydall, D. and Seddon, A. (1990) The identification of a second cell cycle control on the HO promoter in yeast: cell cycle regulation of SW15 nuclear entry. *Cell* **62**: 631–647.

O'Brien, T., Hardin, S., Greenleaf, A. and Lis, J.T. (1994) Phosphorylation of RNA polymerase II *C*-terminal domain and transcription elongation. *Nature* **370**: 75–77.

Oliner, J.D., Pietenpol, J.A., Thiagalingram, S., Gyuris, J., Kinzler, K.W. and Vogelstein, B. (1993) Oncoprotein MDM2 conceals the activation domain of tumor suppressor p53. *Nature* **362**: 857–860.

Parsons, G.G. and Spencer, C.A. (1997) Mitotic repression of RNA polymerase II transcription is accompanied by release of transcription elongation complexes. *Mol. Cell. Biol.* **17**: 5791–5802.

Perez-Roger, I., Solomon, D.L., Sewing, A. and L. and, H. (1997) Myc activation of cyclin E/Cdk2 kinase involves induction of cyclin E gene transcription and inhibition of p27(Kip1) binding to newly formed complexes. *Oncogene* **14**: 2373–2381.

Philipp, A., Schneider, A., Vasrik, I. *et al.* (1994) Repression of cyclin D1: a novel function of MYC. *Mol. Cell. Biol.* **14**: 4032–4043.

Phillips, A.C., Ernst, M.K., Bates, S., Rice, N.R. and Vousden, K.H. (1999) E2F-1 potentiates cell death by blocking antiapoptotic signaling pathways. *Mol. Cell* **4**: 771–781.

Prescott, D.M. and Bender, M.A. (1962) Synthesis of RNA and protein during mitosis in mammalian tissue culture cells. *Exp. Cell Res.* **26**: 260–268.
Qian, Y., Luckey, C., Horton, L., Esser, M. and Templeton, D.J. (1992) Biological function of the retinoblastoma protein requires distinct domains for hyperphosphorylation and transcription factor binding. *Mol. Cell. Biol.* **12**: 5363–5372.
Qin, X.-Q., Livingston, D. M., Kaelin, W. G. J. and Adams, P. (1994) Deregulated transcription factor E2F-1 expression leads to S-phase entry and p53-mediated apoptosis. *Proc. Natl Acad. Sci. USA.* **91**: 10918–10922.
Rickert, P., Seghezzi, W., Shanahan, F., Cho, H. and Lees, E. (1996) Cyclin C/CDK8 is a novel CTD kinase associated with RNA polymerase II. *Oncogene* **12**: 2631–2640.
Roberts, S.B., Segil, N. and Heintz, N. (1991) Differential phosphorylation of the transcription factor Octl during the cell cycle. *Science* **253**: 1022–1026.
Ross, J.F., Liu, X. and Dynlacht, B.D. (1999) Mechanism of transcriptional repression of E2F by the retinoblastoma tumor suppressor protein. *Mol. Cell* **3**: 195–205.
Roth, S.Y. and Allis, D. (1992) Chromatin condensation: does histone H1 dephosphorylation play a role? *Trends Biochem. Sci.* **17**: 93–98.
Roy, R., Adamczewski, J. P., Seroz, T. *et al.* (1994) The MO15 cell cycle kinase is associated with the TFIIH transcription-DNA repair factor. *Cell* **79**: 1093–1101.
Rustgi, A.K., Dyson, N.J. and Bernards, R. (1991) Amino-terminal domains of c-Myc and N-myc proteins mediate binding to the retinoblastoma gene product. *Nature* **352**: 541–544.
Sakaguchi, K., Herrera, J.E., Saito, S. *et al.* (1998) DNA damage activates p53 through a phosphorylation-acetylation cascade. *Genes Dev.* **12**: 2831–2841.
Saleh, A., Schieltz, D., Ting, N. *et al.* (1998) Tra1p is a component of the yeast Ada.Spt transcriptional regulatory complexes. *J. Biol. Chem.* **273**: 26559–26565.
Sardet, C., Vidal, M., Cobrinik, D. *et al.* (1995) E2F-4 and E2F-5, two novel members of the E2F family, are expressed in the early phases of the cell cycle. *Proc. Natl Acad. Sci. USA* **92**: 2403–2407.
Sears, R., Ohtani, K. and Nevins, J.R. (1997) Identification of positively and negatively acting elements regulating expression of the E2F2 gene in response to cell growth signals. *Mol. Cell. Biol.* **17**: 5227–5235.
Segil, N., Guermah, M., Hoffmann, A., Roeder, R.G. and Heintz, N. (1996) Mitotic regulation of TFIID: inhibition of activator-dependent transcription and changes in subcellular localization. *Genes Dev.* **10**: 2389–2400.
Segil, N., Roberts, S.B. and Heintz, N. (1991) Mitotic phosphorylation of the Oct-1 homeodomain and regulation of Oct-1 DNA binding activity. *Science* **254**: 1814–1816.
Sellers, W.R., Novitch, B.G., Miyake, S. *et al.* (1998) Stable binding to E2F is not required for the retinoblastoma protein to activate transcription, promote differentiation, and suppress tumor cell growth. *Genes Dev.* **12**: 95–106.
Sellers, W.R., Rodgers, J.W. and Kaelin, W.G. (1995) A potent transrepression domain in the retinoblastoma protein induces a cell cycle arrest when bound to E2F sites. *Proc. Natl. Acad. Sci. USA* **92**: 11544–11548.
Serizawa, H., Makela, T.P., Conaway, J.W., Conaway, R.C., Weinberg, R.A. and Young, R.A. (1995) Association of cdk-activating kinase subunits with transcription factor TFIIH. *Nature* **374**: 280–282.
Shao, Z., Ruppert, S. and Robbins, P.D. (1995) The retinoblastoma susceptibility gene product binds directly to the human TATA-binding protein-associated factor TAFII250. *Proc. Natl Acad. Sci. USA* **92**: 3115–3119.
Shermoen, A.W. and O'Farrell, P.H. (1991) Progression of the cell cycle through mitosis leads to abortion of nascent transcripts. *Cell* **67**: 303–310.
Sherr, C.J. (1998) Tumor surveillance via the ARF-p53 pathway. *Genes Dev.* **12**: 2984–2991.
Shieh, S.Y., Ikeda, M., Taya, Y. and Prives, C. (1997) DNA damage-induced phosphorylation of p53 alleviates inhibition by MDM2. *Cell* **91**: 325–334.
Shiekhattar, R., Mermelstein, F., Fisher, R. P. *et al.* (1995) Cdk-activating kinase complex is a component of human transcription factor TFIIH. *Nature* **374**: 283–287.
Sif, S., Stukenberg, P.T., Kirschner, M.W. and Kingston, R.E. (1998) Mitotic inactivation of a human SWI/SNF chromatin remodeling complex. *Genes Dev.* **12**: 2842–2851.
Singh, P., Coe, J. and Hong, W. (1995) A role for retinoblastoma protein in potentiating transcriptional activation by the glucocorticoid receptor. *Nature* **374**: 562–565.

Singh, P., Wong, S.H. and Hong, W. (1994) Overexpression of E2F-1 in rat embryo fibroblasts leads to neoplastic transformation. *EMBO J.* **13**: 3329–3338.

Smith, E.J., Leone, G., DeGregori, J., Jakoi, L. and Nevins, J.R. (1996) The accumulation of an E2F–p130 transcriptional repressor distinguishes a G0 cell state from a G1 cell state. *Mol. Cell Biol.* **16**: 6965–6976.

Strober, B.E., Dunaief, J.L., Guha, S. and Goff, S.P. (1996) Functional interactions between the hBRM/hBRG1 transcriptional activators and the pRB family of proteins. *Mol. Cell Biol.* **16**: 1576–1583.

Sturzbecher, H.-W., Maimets, T., Chumakov, P. *et al.* (1990) p53 interacts with p34cdc2 in mammalian cells: implications for cell cycle control and oncogenesis. *Oncogene* **5**: 795–801.

Svestrup, J.Q., Vichi, P. and Egly, J.M. (1996) The multiple roles of transcription/repair factor TFIIH. *Trends Biochem. Sci.* **21**: 346–350.

Tao, W. and Levine, A.J. (1999) P19(ARF) stabilizes p53 by blocking nucleo-cytoplasmic shuttling of Mdm2. *Proc. Natl. Acad. Sci. USA* **96**: 6937–6941.

Taylor, J. (1960) Nucleic acid synthesis in relation to the cell division cycle. *Ann. NY* Acad. Sci. **90**: 409–421.

Tevosian, S.G., Shih, H.H., Mendelson, K.G., Sheppard, K.-A., Paulson, K.E. and Yee, A.S. (1997) HBP1: a HMG box transcriptional repressor that is targeted by the retinoblastoma family. *Genes Dev.* **11**: 383–396.

Tjian, R. and Maniatis, T. (1994) Transcriptional activation: a complex puzzle with few easy pieces. *Cell* **77**: 5–8.

Tommasi, S. and Pfeifer, G.P. (1995) *In vivo* structure of the cdc2 promoter: release of a p130–E2F-4 complex from sequences immediately upstream of the transcription initiation site coincides with induction of cdc2 expression. *Mol. Cell. Biol.* **15**: 6901–6913.

Toyn, J.H., Johnson, A.L., Donovan, J.D., Toone, W.M. and Johnston, L.H. (1997) The Swi5 transcription factor of *Saccharomyces cerevisiae* has a role in exit from mitosis through induction of the cdk-inhibitor Sic1 in telophase. *Genetics* **145**: 85–96.

Vassilev, A., Yamauchi, J., Kotani, T. *et al.* (1998) The 400 kDa subunit of the PCAF histone acetylase complex belongs to the ATM superfamily. *Mol. Cell* **2**: 869–875.

Voit, R., Hoffmann, M. and Grummt, I. (1999) Phosphorylation by G1-specific cdk-cyclin complexes activates the nucleolar transcription factor UBF. *EMBO J.* **18**: 1891–1899.

Voit, R., Schafer, K. and Grummt, I. (1997) Mechanism of repression of RNA polymerase I transcription by the retinoblastoma protein. *Mol. Cell. Biol.* **17**: 4230–4237.

Wang, C.-J., Petryniak, B., Thompson, C.B., Kaelin, W.G. and Leiden, J.M. (1993) Regulation of the Ets-related transcription factor Elf-1 by binding to the retinoblastoma protein. *Science* **260**: 1130–1135.

Wang, Y. and Prives, C. (1995) Increased and altered DNA binding of p53 by S and G2/M but not G1 cyclin dependent kinases. *Nature* **376**: 88–91.

Weber, J.D., Taylor, L.J., Roussel, M.F., Sherr, C.J. and Bar-Sagi, D. (1999) Nucleolar Arf sequesters Mdm2 and activates p53. *Nat. Cell Biol.* **1**: 20–26.

Weinberg, R.A. (1995) The retinoblastoma protein and cell cycle control. *Cell* **81**: 323–330.

Weinberg, R.A. (1997) The cat and mouse games that genes, viruses, and cells play. *Cell* **88**: 573–575.

Weintraub, S.J., Chow, K.N.B., Luo, R.X., Zhang, S.H., He, S. and Dean, D.C. (1995) Mechanism of active transcriptional repression by the retinoblastoma protein. *Nature* **375**: 812–815.

Welch, P.J. and Wang, J.Y.J. (1993) A *C*-terminal protein-binding domain in the retinoblastoma protein regulates nuclear c-abl tyrosine kinase in the cell cycle. *Cell* **75**: 779–790.

White, R.J., Gottlieb, T.M., Downes, C.S. and Jackson, S.P. (1995) Mitotic regulation of a TATA-binding protein-containing complex. *Mol. Cell. Biol.* **15**: 1983–1992.

Wu, L., Chen, P., Hwang, J. J. *et al.* (1999) RNA antisense abrogation of MAT1 induces G1 phase arrest and triggers apoptosis in aortic smooth muscle cells. *J. Biol. Chem.* **274**: 5564–5572.

Xiao, Q., Claassen, G., Shi, J., Adachi, S., Sedivy, J. and Hann, S.R. (1998) Transactivation-defective c-MycS retains the ability to regulate proliferation and apoptosis. *Genes Dev.* **12**: 3803–3808.

Xu, G., Livingston, D.M. and Krek, W. (1995) Multiple members of the E2F transcription factor family are the products of oncogenes. *Proc. Natl Acad. Sci. USA* **92**: 1357–1361.

Xu, M., Sheppard, K.A., Peng, C.Y., Yee, A.S. and Piwnica-Worms, H. (1994) Cyclin A/CDK2 binds directly to E2F-1 and inhibits the DNA-binding activity of E2F-1/DP-1 by phosphorylation. *Mol. Cell Biol.* **14**: 8420–8431.

Yamasaki, L., Bronson, R., Williams, B.O., Dyson, N.J., Harlow, E. and Jacks, T. (1998) Loss of E2F-1 reduces tumorigenesis and extends the lifespan of Rb1(+/–) mice. *Nature Genet.* **18**: 360–364.

Yamasaki, L., Jacks, T., Bronson, R., Goillot, E., Harlow, E. and Dyson, N.J. (1996) Tumor induction and tissue atrophy in mice lacking E2F-1. *Cell* **85**: 537–548.

Yankulov, K.Y. and Bentley, D.L. (1997) Regulation of CDK7 substrate specificity by MAT1 and TFIIH. *EMBO J.* **16**: 1638–1646.

Yoshida, M., Horinouchi, S. and Beppu, T. (1995) Trichostatin A and trapoxin: novel chemical probes for the role of histone acetylation in chromatin structure and function. *Bioessays* **17**: 423–430.

Zawel, L., Lu, H., Cisek, L.J., Corden, J.L. and Reinberg, D. (1993) The cycling of RNA polymerase II during transcription. *Cold Spring Harb. Symp. Quant. Biol.* **58**: 187–198.

Zervos, A.S., Gyuris, J. and Brent, R. (1993) Mxi1, a protein that specifically interacts with Max to bind Myc–Max recognition sites. *Cell* **72**: 223–232.

Zhang, Y., Iratni, R., Erdjument-Bromage, H., Tempst, P. and Reinberg, D. (1997) Histone deacetylases and SAP18, a novel polypeptide, are components of a human Sin3 complex. *Cell* **89**: 357–364.

Zhang, Y. and Xiong, Y. (1999) Mutations in human ARF exon 2 disrupt its nucleolar localization and impair its ability to block nuclear export of MDM2 and p53. *Mol. Cell* **3**: 579–591.

Zhu, L., Harlow, E. and Dynlacht, B.D. (1995) p107 uses a p21^{CIP1}-related domain to bind cyclin/cdk2 and regulate interactions with E2F. *Genes Dev.* **9**: 1740–1752.

Zhu, L., van den Heuvel, S., Helin, K. *et al.* (1993) Inhibition of cell proliferation by p107, a relative of the retinoblastoma protein. *Genes Dev.* **7**: 1111–1125.

Ziebold, U., Bartsch, O., Marais, R., Ferrari, S. and Klempnauer, K.H. (1997) Phosphorylation and activation of B-Myb by cyclin A-Cdk2. *Curr. Biol.* **7**: 253–260.

Zindy, F., Eischen, C.M., Randle, D.H. *et al.* (1998) Myc signaling via the ARF tumor suppressor regulates p53-dependent apoptosis and immortalization. *Genes Dev.* **12**: 2424–2433.

Zwicker, J., Liu, N., Engeland, K., Lucibello, F.C. and Muller, R. (1996) Cell cycle regulation of E2F site occupation *in vivo*. *Science* **271**: 1595–1597.

6

Regulation of transcription by extracellular signals

Paul Shore and Andrew D. Sharrocks

1. Introduction

One of the major alterations in eukaryotic cells upon exposure to different extracellular environments is a change in their gene expression profiles. In simple eukaryotes such as yeast, these changes might be in response to changes in the nutrients available. In more complex multicellular organisms, cells respond to different types of stimuli, generated from either contact with neighboring cells, short range signals involving secreted molecules or long range signals transported by the circulatory system such as hormones. In addition, cells respond to various stresses that might either be mechanical or be due to external agents such as UV light. Over recent years, our understanding of how extracellular signals are sensed and transmitted into nuclear responses has increased substantially. Numerous different pathways and different types of mechanisms for transmitting these signals have been elucidated. These signaling pathways play critical roles both in the adult and during development and in many cases, the components of these pathways function in multiple different processes. Furthermore, in many diseases, notably cancer, some of the key genetic lesions are in genes encoding components of these signaling pathways which results in the perturbation of signal transmission to the nucleus. Ultimately, the signaling pathways often converge either directly or indirectly, on transcription factor substrates. Indirect effects might be via coregulatory proteins known as coactivator and corepressor proteins. For the purposes of this review, we define transcription factors as the DNA binding components of complexes which up- or down-regulate transcription. Coregulatory proteins are the non-DNA binding components of these complexes. We use the term transcriptional regulator when these two classes of proteins are referred to together.

The signaling pathways themselves are extremely complex with multiple points of regulation. However, in this review, we will concentrate on how extracellular signals are transmitted to the nucleus via transcription factors and the

Transcription Factors, edited by J. Locker.

changes that are elicited in these proteins following receipt of these signals. In addition, it is becoming clear that coregulatory proteins serve to modulate the activity of these proteins and their responses to signaling pathways. Recent progress in this area will also be reviewed. Furthermore, we also concentrate on the role of reversible protein phosphorylation in modifying the activities of transcriptional regulators as this is currently thought to be the major mechanism of signal transduction. We also touch on other recent studies that suggest that other direct mediators exist such as calcium ions. Further studies suggest a potential role for protein acetylation in modifying transcription factor function, which might also play a role in signal transduction. Many of the major advances in our understanding of signaling mechanisms have been achieved by studying a variety of different organisms. Due to space limitations, in this review we concentrate on the mechanisms of signaling to mammalian transcriptional regulatory proteins, although it should be emphasized that many of the principles we describe are applicable to a range of eukaryotic organisms, from yeasts, through worms and insects to lower and higher vertebrates. We also refer readers to earlier reviews on signaling to and phosphorylation of transcription factors for further background information (Hill and Treisman, 1995; Hunter and Karin, 1992; Karin and Hunter, 1995; Karin and Smeal, 1992).

2. Mechanisms of signal transmission to the nucleus

In order to gain a nuclear response to extracellular signals, these signals must be transmitted across the cell membrane and through the cytoplasm to reach the transcription factor (or coregulatory protein) targets. These target proteins might either be located in the nucleus or require subsequent translocation into the nucleus after modification by the pathways. Reversible phosphorylation is the major mechanism thought to be employed to transduce signals through intermediary proteins, to transcriptional regulators, although alternative mediators have been identified. The numerous different ways of transmitting signals to the nucleus which are currently known are illustrated in *Figure 1* and examples of each type of mechanism are discussed below (see *Table 1*).

2.1 Type I: protein kinase cascades to transcription factor substrates

One of the most commonly detected mechanisms of signal transduction to the nucleus is via protein kinase cascades that are triggered by the activation of membrane-bound receptors by extracellular signals. The mitogen-activated protein kinase (MAPK) pathways provide a paradigm for this type of signaling (Robinson and Cobb, 1997; Treisman, 1996; Whitmarsh and Davis, 1996). These pathways are initiated at receptor tyrosine kinases (RTKs) and the signals are transmitted by a series of membrane associated intermediates to a cascade of kinases which sequentially activate each other. The cascades terminate at MAPKs and these terminal kinases are responsible for either activating further downstream kinases or phosphorylating the transcription factor substrates. These substrates might be located either in the cytoplasm or in the nucleus. In the MAPK cascades, upon activation,

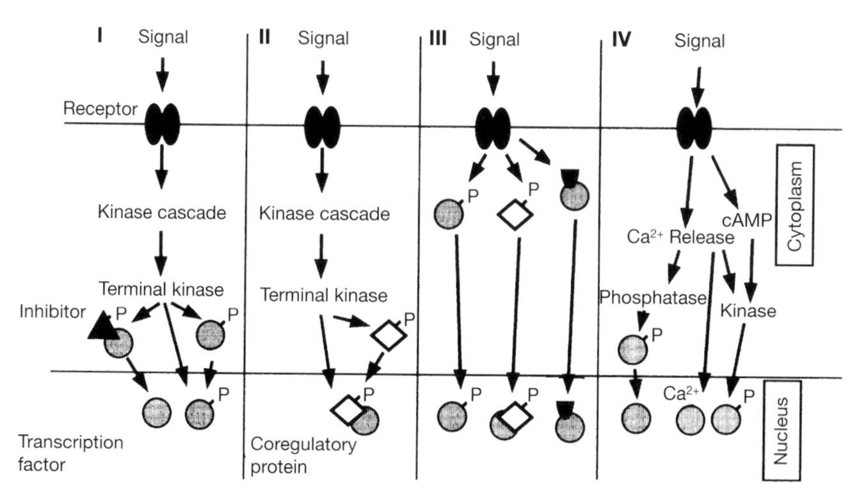

Figure 1. Mechanisms of signalling to the nucleus. Four different types of mechanisms for transmitting signals to nuclear targets are depicted (I–IV: see text for details). Transcription factor substrates are shown as grey circles, co-regulatory proteins as white diamonds and inhibitory proteins as black triangles. One pathway in mechanism III involves the translocation of a proteolytically cleaved cytoplasmic domain of the receptor (which subsequently acts as a coactivator) to the nucleus together with an associated transcription factor. The cellular location at which phosphorylation takes place is denoted by the addition of a phosphate group (P) to the substrates.

the terminal kinase translocates into the nucleus where it locates and phosphorylates its substrates. One of the best studied examples of this type of cascade is the extracellular signal regulated (Erk) MAPK pathway, which starts at RTKs such as the epidermal growth factor (EGF) receptor and transmits signals via the intermediates Grb/SOS to the small guanosine triphosphatase (GTPase), Ras. Activated Ras, then leads to the sequential activation of the kinases Raf-1, MEK and Erk1/2, and terminates in the phosphorylation of nuclear substrates such as the ETS-domain transcription factor Elk-1. Numerous other parallel MAPK pathways have been identified which respond to different signals and lead to the activation of different transcription factor substrates (*Table 1*) and it appears that in all cases, the general mechanism of signal transduction is identical.

A bifurcation in the MAPK pathway at the level of Ras has recently been elucidated which leads to a different pathway and a similar mechanism of signaling. In this case, a different series of kinases are activated (PI3 kinase and PkB/Akt) which leads to the activation of forkhead transcription factors (Brunet *et al.*, 1999; Downward, 1998; Kops *et al.*, 1999). Phosphorylation of the transcription factor is again thought to take place in the nucleus which then leads to export of the transcription factor into the cytoplasm (see Section 4.1).

The JAK-STAT pathway represents a further variation on this type of signaling (Ihle, 1996; Liu *et al.*, 1998; Stark *et al.*, 1998). Again, signaling is initiated at RTKs but in this case, the receptor directly phosphorylates and activates the terminal Janus kinase (JAKs) which in turn phosphorylates its transcription factor

Table 1. Components of signaling cascades

Type of signaling	Signals/stimuli	Receptors/kinase cascade	Terminal kinase	Target
I	Growth factors, stress inducers, cytokines	Receptor tyrosine kinases (RTKs), MAPK	Erk, Jnk, p38, Erk-5, MAPKs	ETS-domain (e.g. Elk-1), bZIP (e.g. cJun), STAT & MADS-box (e.g. MEF2C) transcription factors, nuclear hormone receptors (e.g. ER), Smads
	Insulin, growth factors, survival factors	RTKs, PI3K	PKB/Akt	Forkhead transcription factors (e.g. FKHRL1 and AFX).
	Stress inducers, cytokines	RTKs, IKK complex	IKK	IκB (NFκB).
	Cytokines	RTKs, JAKs	JAKs	STATs
II	Wnt	Frizzled receptor/Dsh	GSK-3β	β-catenins (TCF/LEF)
III	TGF-β family	TGF-β Receptor	TGF-β Receptor	Smads
	Delta/jagged	Notch receptor	–	CBF-1/Su(H)
IV	Neurotransmitters	CaMKK, Ca^{2+}	CaMKIV	CREB
	Neurotransmitters, hormones	G proteins, (cAMP)	PKA	CREB
	MHC–antigen complexes	T cell receptor, Ca^{2+}	Calcineurin[#] Ca^{2+}	NFAT DREAM

[#]Calcineurin is a protein phosphatase rather than a kinase.
Examples of each type of signaling mechanism are indicated. The direct targets of the pathways are indicated, and in the case of coactivators and inhibitory proteins, the partner transcription factor is shown in brackets. In the cascades involving calcium and cAMP (type IV), the role of these molecules as either direct effectors or components of the signaling cascades is indicated.

substrate (STATs). Another significant difference with this type of signalling is that the transcription factor is phosphorylated in the cytoplasm and the STAT, rather than the kinase, translocates into the nucleus.

The pathways that lead to NFκB activation (DiDonato *et al.*, 1997) exemplify another variation on this theme. Again, RTKs are responsible for initiating the cascade but a different kinase complex, IκB kinase (IKK), is responsible for the transduction of the signal. The key difference with this pathway is that it is an inhibitor protein rather than the transcription factor that is the target of the pathway. IκB acts as an inhibitor by sequestering the transcription factor NF-κB in the cytoplasm. Phosphorylation of IκB results in release of NF-κB which translocates into the nucleus (see Section 4.1).

2.2 Type II: protein kinase cascades to coactivator and corepressor proteins

In addition to activating transcription factors directly, signaling pathways can also target coregulatory proteins and hence up- or down-regulate the activity of transcription factors. The active component might be either the phosphorylated or unmodified form of the coregulatory protein (*Figure 1*). One of the best examples of this type of signaling is the coactivator protein β-catenin that translocates to the nucleus in the absence of phosphorylation, where it interacts with the HMG-box protein T-cell factor/lymphoid enhancer binding factor-1 (TCF/LEF-1) (Willert and Nusse, 1998). In this pathway, Wnt proteins signal to members of the seven transmembrane domain Frizzled receptor family. This signal is propogated via the Dishevelled (Dsh) protein and culminates in the inactivation of the protein kinase glycogen synthase kinase (GSK-3β). When active, GSK-3β directly phosphorylates β-catenin and leads to its destabilization and a block in the translocation of β-catenein to the nucleus. Wnt signaling therefore reverses this effect and permits β-catenin translocation to the nucleus.

Variations on this theme might include the phosphorylation of nuclear coregulatory proteins following kinase translocation or alternatively translocation of the phosphorylated transcriptional coregulator, although no clear examples of this type of signaling have been demonstrated so far.

2.3 Type III: direct signaling from receptors to transcription factors and coregulatory proteins

In addition to signaling via protein kinase cascades, receptors can directly phosphorylate their transcription factor/coregulatory protein substrates. A paradigm for this type of signaling is members of the transforming growth factor-β (TGF-β) receptor family which directly phosphorylate the Smad proteins (Kretzschmar and Massague, 1998). The Smads subsequently oligomerize and translocate to the nucleus where they can function in two modes, either as bone-fide DNA binding transcription factors or as coactivator proteins by binding to other transcription factors. In this pathway, the transmembrane receptors are composed of two subunits (type I and type II receptors) which upon association and activation by extracellular ligands (TGF-β and bone morphogenic proteins [BMPs]), phosphorylate the Smads on serine and threonine residues.

A modified version of this type of signaling mechanism is exemplified by Notch signaling where the receptor itself becomes a coactivator protein (reviewed in Baker, 2000; Osborne and Miele, 1999). Here the phosphorylation status of the transcription factor or coactivator does not appear to be altered, rather upon activation, a proteolytically cleaved intracellular domain of the receptor is released, which then moves into the nucleus with the associated transcription factor CBF-1/Su(H). Once in the nucleus, this domain acts as a coactivator for CBF-1/Su(H) (see Chapter 7).

2.4 Type IV: signaling by small molecules

In the examples described above, protein phosphorylation plays a major role in transmitting signals down protein kinase cascades. Other mechanisms exist for transducing these signals which involve the direct participation of small molecules such as cyclic adenosine monophosphate (cAMP) and calcium ions, and downstream effector kinases such as protein kinase A (PKA) and calmodulin-dependent kinase IV (CaMKIV) (reviewed in Brandon *et al.*, 1997; Hardingham and Bading, 1998; Soderling, 1999). For example, enhanced levels of cAMP caused by G-protein activation, leads to the activation of the protein kinase PKA and subsequent phosphorylation of the transcription factor cyclic AMP response element binding protein (CREB). Similarly, enhanced levels of intracellular calcium ions leads to the activation of CaMKIV and CREB phosphorylation. Furthermore, CREB is also activated indirectly by both cAMP and nuclear calcium as these molecules also lead to the direct activation of the CREB coactivator protein CBP (CREB binding protein) (Chawla *et al.*, 1998).

Enhanced levels of intracellular calcium ions can also lead to the activation of phosphatases that then regulate downstream transcription factors. A pertinent example is calcineurin that is activated by elevated levels of calcium in response to T cell receptor activation. This phosphatase then dephosphorylates NFAT (nuclear factor of activated T-cells) proteins, which subsequently translocate to the nucleus (reviewed in Rao *et al.*, 1997).

Recently, a more direct mode of action by calcium has been elucidated. In this case, calcium ions bind directly to the transcription factor, DREAM, which leads to a reduction in its ability to bind DNA and act as a transcriptional repressor (Carrion *et al.*, 1999).

3. Mechanisms of specificity and diversity generation in signaling pathways

One of the key outstanding questions is how signaling pathways maintain specificity and at the same time generate a diverse set of nuclear outcomes. For example, the MAP kinase cascades are known to play key roles in multiple different processes in various organisms including mating-type determination in *Saccharomyces cerevisiae*, vulval development in *C. elegans* and eye development in *Drosophila*. In mammals, MAP kinase pathways can regulate cellular proliferation and growth, cellular differentiation or nuclear responses to stress signals (reviewed in Treisman, 1996). However, within a cell at any given time, a specific response will be elicited to the activation of a signaling pathway. Mechanisms of generating specificity and diversity from a limited number of pathways are discussed below.

3.1 Linear, convergent and divergent signaling

Signals can be transmitted to transcription factors and their coregulatory proteins by the differential use of signaling pathways (*Figure 2*). For example, signals can make use of individual pathways (linear signaling) or two or more pathways that converge on either the same transcriptional regulator or different components of a regulatory complex (convergent signaling). Alternatively, signals can be transmitted to different regulatory proteins that control different genes (divergent signaling).

Examples of linear signaling (*Figure 2a*) are difficult to find, as more is known about which pathways are responsible for modifying the activities of particular transcriptional regulators. For example, c-Jun was identified as a specific target of

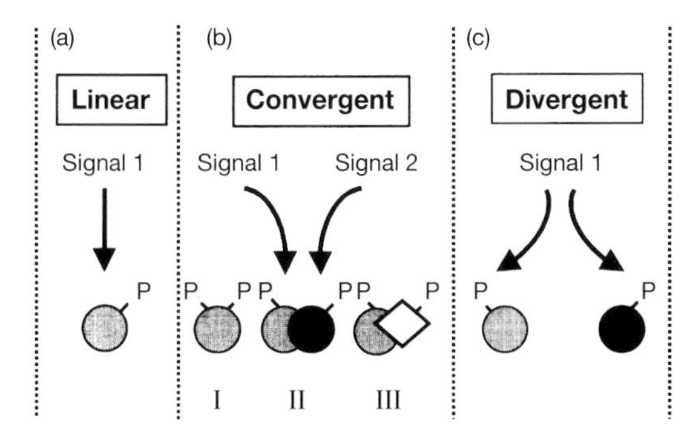

Figure 2. Types of signaling. Three different ways in which signals reach nuclear targets are depicted (a–c; see text for details). Different nuclear outcomes of convergent signaling are also indicated (I–III; see text for details). In this mode, two pathways can either converge on the same transcription factor (I), two different transcription factors in a complex (II) or a transcription factor and coactivator in the same complex (III). Transcription factor substrates are shown as grey and black circles and coregulatory proteins as white diamonds. For simplicity, the transcription factor/coregulatory protein modification is depicted as a single phosphorylation (P) event.

the Jnk MAP kinase pathway in response to UV stimulation (reviewed in Angel, 1995). However, it is now known that UV and the JNK pathway both result in multiple nuclear effects. In practice, although theoretically possible, it is unlikely that an individual signal elicits a response via one pathway and an individual nuclear target in higher eukaryotes.

The most common mode of signaling appears to be the convergence of two or more pathways. In this case, different pathways might converge on the same substrate by phosphorylating different residues (e.g. p38α/p38β_2 and Erk5 kinases and MEF2C; Han *et al.*, 1997; Kato *et al.*, 1997; Yang *et al.*, 1998c) (*Figure 2b*, mode I). Alternatively, the same residues in a substrate can be phosphorylated by different pathways, depending on which pathways are activated (e.g. Erk and Jnk MAPKs and Elk-1; reviewed in Treisman, 1996; Whitmarsh and Davis, 1996). Therefore, by using a limited number of transcription factors, different pathways can achieve a similar nuclear response. A further refinement of this type of signaling is the convergence of two pathways on two components of a transcriptional regulatory complex that forms on a promoter. This might either be two transcription factors (*Figure 2b*, mode II) or a transcription factor and its coregulatory partner (*Figure 2b*, mode III). An example of convergent signaling (mode II) is at the ternary complex that forms on the c-*fos* serum response element (SRE). The two transcription factors in this complex (serum response factor [SRF] and ternary complex factor [TCF]) respond to different pathways triggered by different serum components (*Figure 3a*) (Hill *et al.*, 1995; reviewed in Treisman, 1996). The TCF component is a direct target of the Ras/Erk pathway, whereas SRF is regulated by components downstream of RhoA. It is currently unclear what the components downstream from RhoA are, although both NF-κB and C/EBPβ have been implicated in this response (Montaner *et al.*, 1999). The CREB-CBP transcriptional complex represents another point of convergence of signaling (mode III). In this case, the same signal (increased levels of intracellular calcium) can lead to the independent activation of both parts of the complex, thereby permitting two levels of control. Cytoplasmic calcium appears to act via the Ras pathway to enhance CBP recruitment by the transcription factor CREB, whereas nuclear calcium acts via CaMKIV and the coactivator CBP directly (Chawla *et al.*, 1998). The convergence of several pathways on transcription factor complexes permits additional specificity in the signaling process as a single input is often insufficient to trigger a complete response on a promoter.

An alternative type of convergent signaling is when two pathways converge on a transcription factor but act in an antagonistic manner. A classical example of this is the interplay between the TGF-β and Erk signaling pathways (*Figure 3b*) (Kretzschmar *et al.*, 1999; reviewed in Kretzschmar and Massague, 1998). Stimulation of the TGF-β receptor leads to phosphorylation and activation of the Smad transcriptional regulators. However, simultaneous activation of the Erk pathway by growth factors, leads to phosphorylation of the Smads at different sites, nuclear exclusion and hence antagonism of TGF-β signaling.

Some of the best-studied examples of divergent signaling (*Figure 2c*) are the MAP kinase cascades (reviewed in Treisman, 1996). For example, an increasing number of diverse transcription factors and coregulatory proteins have been shown to be phosphorylated and regulated by the Erk pathway. These include several ETS-domain proteins (e.g. Elk-1; reviewed in Sharrocks *et al.*, 1997), Smads (reviewed in Kretzschmar and Massague, 1998), STATs (signal transducers and activators of transcription) (Ihle, 1996), heat shock factor, HSF-1 (Knauf *et al.*, 1996) and nuclear hormone receptors (Tremblay *et al.*, 1999). In all the examples quoted here, it is important to note that on these substrates, divergent signaling appears to operate in conjunction with convergent signaling as each individual target gene is subject to regulation by other transcription factors. Thus, whilst conceptually separable, divergent and convergent signaling combine in the cell to fine-tune the nuclear response to a particular signal.

3.2 Specificity-determining mechanisms

Part of the specificity determining mechanisms in signaling is the requirement for multiple pathways to impinge on a particular transcription factor complex that assembles on a promoter. However, in addition, each pathway has to initially reach its specific transcriptional regulator targets. There are several mechanisms by which signaling specificity is maintained in the cascades themselves (reviewed in Tan and Kim, 1999; Whitmarsh and Davis, 1998). However, it is usually the final step in the pathway, i.e. the terminal kinase-transcriptional regulator interaction, which ultimately determines the specificity of the signaling outcome. Several mechanisms exist for maintaining this specificity.

Consensus sites. Each protein kinase phosphorylates specific residues in its substrates, and the specificity is determined both by the nature of the residue itself (Ser, Thr or Tyr in higher eukaryotes) and the local context surrounding these residues (*Table 2*). However, although the consensus sequence is specific to a particular class of protein kinase and certain sites are favored by individual kinases, related kinases appear to phosphorylate similar sites with a significant

Table 2. Consensus phosphorylation sites for protein kinases

Protein kinase class	Phosphoacceptor residue	Consensus site
MAP kinases	Ser/Thr	PxS/TPx
PKB/Akt	Ser/Thr	RxRxxS/Tx
PKA	Ser/Thr	Rx_{1-2}S/Tx
GSK-3β	Ser/Thr	SxxxS/TPxxS
CaMKIV	Ser/Thr	*RRPSYRK
IKK	Ser	*RHDSGLDSMKDE
JAKs	Tyr	–

*In the case of CaMKIV and IKK, no consensus is available, and a representative sequence for each kinase is provided (from CREB and IkBα respectively).
The types of phosphoacceptor residues (Thr, Ser or Tyr) known to be phosphorylated by each class of kinase are indicated. The consensus sites derived from several known targets are indicated (MAPK, GSK-3β and PKA, Kreegipuu *et al.*, 1998; PKB, Kops *et al.*, 1999). The phosphoacceptor residues within these sites are underlined.

overlap in specificity. The summary of consensus sites can therefore only be taken as a guide to potential phosphoacceptor sites and their physiological significance must be determined experimentally. This is exemplified by the MAP kinases, where sites in transcription factors related to the consensus sequence PXS/TP, can be phosphorylated *in vitro* by all the different major classes of kinases (Erk, Jnk and p38). Moreover, physiologically relevant MAP kinase phosphorylation sites often conform to the more relaxed sequence S/TP, suggesting that further specificity determinants must exist *in vivo* (reviewed in Treisman, 1996). A further level of complexity is exhibited by CREB which can be phosphorylated on the same site by at least two different kinases, PKA and CaMKIV, indicating that a single site can be the site of integration of two different pathways.

Kinase-transcription factor docking. Direct binding of the kinase to the transcription factor substrate is often observed. It is proposed that this binding imparts further specificity on the phosphorylation process. A paradigm for this type of mechanism is the binding of the Jnk MAPKs to c-Jun (Kallunki *et al.*, 1996). In this case, a docking site (delta domain), which is distinct from the phosphoacceptor motifs, is required for binding of the kinase and maximal phosphorylation, in addition to the local context of the phosphoacceptor sites. Recently, this paradigm has been extended to other subclasses of MAP kinases and their substrates including Jnk-NFAT4, Erk/Jnk-Elk-1 and $p38\alpha/p38\beta_2$-MEF2A/C (Chow *et al.*, 1997; Yang *et al.*, 1998a; 1998b; 1999a). In the case of Erk–Elk-1 interactions, it has been proposed that two different types of docking motif exist in the transcription factor which provides at least three different components in the kinase–substrate recognition process and hence further specificity (Jacobs *et al.*, 1999). It is likely that similar mechanisms will exist to enhance the specificity of action of other kinase–transcription factor combinations. Moreover, a similar scenario has recently been reported for protein phosphatases. In this case, dephosphorylation of NFAT transcription factors by calcineurin requires the presence of a short docking domain on the transcription factor for maximal efficiency (Aramburu *et al.*, 1998).

Thus, the requirement of docking domains for kinases and phosphatases plays a pivotal role in regulating transcription factor phosphorylation status.

Cell-specific expression and subcellular compartmentalization. A simple way of maintaining specificity is to only express the signaling molecules or transcription factors at particular times or in certain cell types. Examples of cell-specific expression are provided in Chapters 7, 10, and 11.

An additional mechanism for maintaining specificity in signaling is subcellular compartmentalization. For example, several transcriptional regulators are phosphorylated in the cytoplasm (e.g. IκB) and in some cases in the vicinity of the cell membrane (e.g. Smads and STATs) rather than in the nucleus. Conversely, other transcription factors are phosphorylated in the nucleus and require the translocation of the kinase into the nucleus (e.g. c-Jun). The location of both the kinases and their substrates is of critical importance to permit phosphorylation of the correct substrates.

Strength and kinetics of signaling. Although signals induce a particular pathway or set of pathways, the eventual outcome can also be determined by the strength and duration of the signal. In mammalian cells, the classical example for this is growth factor signaling in neuronal PC12 cells (reviewed in Marshall, 1995). Here, transient activation of the Erk MAP kinase cascade by growth factors induces proliferation whereas sustained activation of this cascade induces cell-cycle exit and differentiation. In keeping with this observation, the repertoire of AP-1 transcription factor family members expressed in response to sustained signaling differs from that induced by transient signaling through the Erk pathway (Cook *et al.*, 1999).

3.3 Inactivation mechanisms

Protein phosphorylation by kinases and dephosphorylation by phosphatases is a pivotal switching mechanism used to control the activity of transcriptional regulatory proteins. In most cases, phosphorylation is thought to act as an activation signal for signaling cascades. Protein dephosphorylation acts in these cases to downregulate signaling. However, the converse situation also applies, where phosphorylation acts to block transcription factor activity and dephosphorylation is the activating signal. A pertinent example is the NFAT transcription factors where Jnk-mediated phosphorylation promotes its nuclear exclusion (Chow *et al.*, 1997), whereas calcineurin-mediated dephosphorylation promotes nuclear accumulation and hence activation of NFAT (reviewed in Rao *et al.*, 1997).

An alternative mechanism of inactivation is the antagonistic activities of two signaling pathways (see above and *Figure 3*). In this case, two different phosphorylation events at distinct sites on Smad2 and Smad3 lead to opposing effects on their nuclear localization (Kretzschmar *et al.*, 1999).

In addition to acting on the transcription factors themselves, downregulation of signaling can take place indirectly by turning off upstream components. A good example of this is the MAP kinase pathways where one consequence of their activation is to turn on the genes encoding dual-specificity phosphatases which in turn feed back to turn off the terminal MAP kinases in these cascades, thereby attenuating signaling (reviewed in Keyse, 1998).

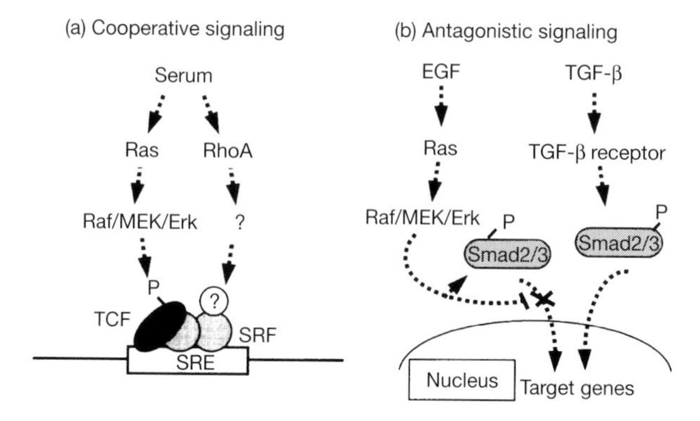

Figure 3. Cooperative and antagonistic signaling. (a) Cooperative signalling at the c-*fos* SRE by the convergence of two serum-induced pathways on a complex containing the TCF (black elipse) and SRF (grey circle) transcription factors. The effector kinases downstream of RhoA are unknown, as is the target of these putative kinases. A putative co-regulatory protein for SRF is indicated by a white circle. (b) Antagonistic signaling to Smad transcription factors by the Ras/Erk pathway (negative action) and the TGF-β pathway (positive action). Phosphorylation mediated by the two pathways takes place at different locations on Smad2 and Smad3, leading to either stimulation or inhibition of nuclear import of Smad2/3. Smad2/3 interact with Smad4 prior to entering the nucleus.

4. Modification of transcription factor activity

Phosphorylation of transcription factors occurs on serine, threonine and tyrosine residues. The phosphate group introduces a negative charge and thus alters the electrostatic properties of the protein. Changes in the electrostatic properties can result in changes in protein conformation, the creation of salt bridges and negatively charged surfaces, and disruption of hydrophobic pockets. The consequences of phosphorylation can thus have a profound effect upon the molecular interactions of transcription factors.

The phosphorylation status of transcription factors and their coregulatory proteins can influence their activities by two general mechanisms. The first mechanism involves regulation of nuclear translocation and the second involves regulation of transcription factor activity within the nucleus. However, it is important to emphasize that phosphorylation often affects both the intrinsic biochemical activities of transcription factors and their subcellular localization. Indeed, the complex interplay between the different effects of phosphorylation is illustrated by the Smads where phosphorylation affects multiple functions, including oligomerization, nuclear localization, DNA binding and coregulatory protein recruitment. Here, oligomerization is actually a prerequisite for triggering the other functions but is not sufficient to mediate all these effects.

Phosphorylation can take place at different locations within transcription factors (*Figure 4*) and this often leads to different effects on their activity. In general, phosphorylation takes place in or around the transcriptional activation domains although phosphorylation can occur within or in the vicinity of the DNA binding domains. When multiple phosphorylation events take place in response to different

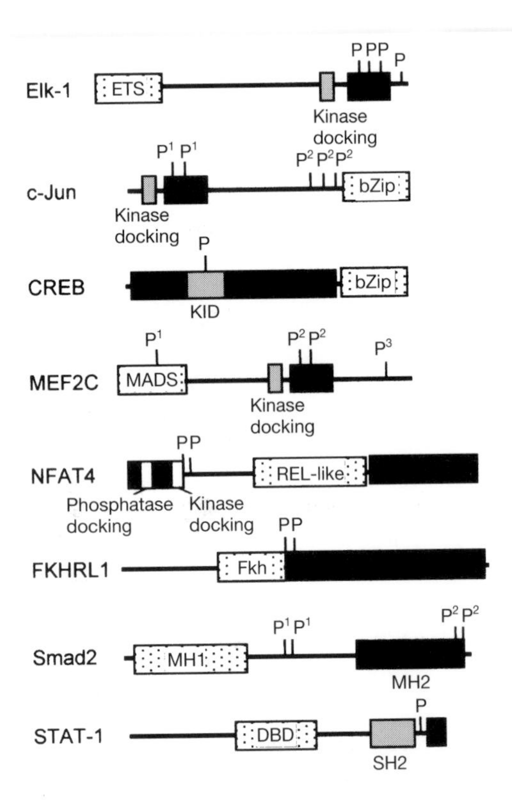

Figure 4. The location of phosphorylation sites within transcription factors. Schematic illustration of representative members of subclasses of transcription factors is shown. Transcriptional activation domains are depicted as black boxes, DNA binding domains as stippled boxes and additional domains relevant to this review are shown as grey or white boxes. The class of DNA-binding domains is indicated; Fkh is an abbreviation of Forkhead. The STAT-1 DNA-binding domain (DBD) has an overall immunoglobulin (Ig)-like fold. The sizes of the transcriptional activation domains reflect differences in the resolution of mapping. The overall size of the STAT and NFAT proteins relative to the other proteins is not to scale (50% reduced). The locations of phosphorylation sites are indicated (P) (see text for details). In the substrates which are phosphorylated by different classes of kinases, the different sites are: c-Jun, Jnk (P^1) and CKII (P^2); MEF2C, CKII (P^1), p38α/p38β_2 (P^2) and Erk5 (P^3); Smad2, Erk (P^1) and TGF-β receptor (P^2).

types of pathways, this typically takes place in different parts of the protein and results in different effects on transcription factor activity.

The following discussion refers to specific examples to illustrate the different ways in which signaling pathways regulate transcription factor activities. These different mechanisms for altering transcription factor function are illustrated in *Figure* 5.

4.1 Regulation of nuclear translocation

Nuclear import. Transcription factors are often rendered functionally inactive by virtue of their cytoplasmic location. Activation of transcription factors held in the

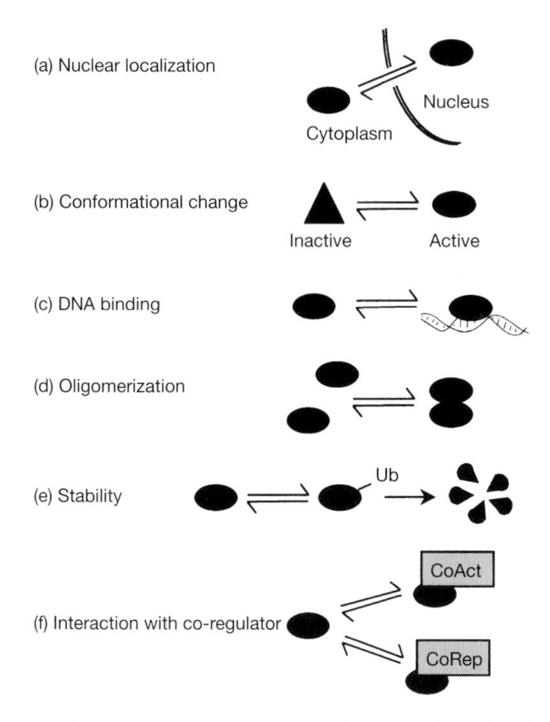

Figure 5. Mechanisms for regulating transcription factor activity by phosphorylation. The reversible arrows indicate that regulation of the particular activity by phosphorylation might occur in either direction. Phosphorylation might result in the production of active or inactive forms of the transcription factors. Transcription factors are indicated by black circles and different conformations are indicated by black triangles (part b). In part (e), proteins are first modified with ubiquitin (Ub) before degradation. In part (f), coactivators (CoAct) and corepressors (CoRep) are indicated by grey boxes.

cytoplasm can be achieved by signaling pathways that stimulate the translocation of activated transcription factor complexes to the nucleus. In a similar manner, inactivation of transcription factors can be achieved by stimulating their export from the nucleus (*Figure 5a*). Transcription factors often contain a nuclear localization signal (NLS) which enables them to be targeted to the nucleus. In some cases it has been shown that the translocation of transcription factors to the nucleus is achieved via specific nuclear import pathways and it is likely that most transcription factors enter the nucleus by specific mechanisms (reviewed in Jans and Hubner, 1996).

The TGF-β pathway is a well-defined example of a signaling pathway that stimulates activation of gene transcription by inducing nuclear translocation of transcription factors (reviewed in Kretzschmar and Massague, 1998). TGF-β signaling is mediated by the TGF-β receptor family of serine/threonine kinases and the Smad transcriptional regulators (see section 2.3). Stimulation with TGF-β leads to receptor activation and phosphorylation of members of the Smad family of proteins. The Smad proteins subsequently form an activated heteromeric complex which translocates to the nucleus and activates target genes. In the case of

TGF-β signaling, Smads 2 and 3 are substrates of the receptor complex and Smad4 is a 'co-Smad'. Phosphorylation of serine residues in the MH2 domain of Smad2 by the receptor enables it to mediate the interaction with Smad4 and the resultant complex translocates to the nucleus. The Smad2 subunit of the Smad2/Smad4 complex determines its nuclear translocation.

Interferon signaling pathways also act by regulating the nuclear translocation of a group of transcription factors referred to as STATs (Liu *et al.*, 1998; Stark *et al.*, 1998). This pathway functions in an analogous manner to the TGF-β pathway but in this case interferons induce the assembly of an active receptor complex which leads to the activation of receptor associated JAK kinases. The JAK kinases phosphorylate the receptor complex on tyrosine residues which subsequently allows recruitment of STAT monomers to the receptor via their phosphotyrosine-specific SH2 domains. Once bound to the receptor, the STAT monomers are themselves phosphorylated on tyrosine residues which promotes their subsequent dimerization. The active STAT dimers then dissociate from the receptor and translocate to the nucleus.

Another paradigm for regulation of transcription factor activity by nuclear translocation is that of the NF-κB family (reviewed in Ghosh *et al.*, 1998). Members of the NF-κB family of transcription factors are held as inactive dimers in the cytoplasm by interaction with inhibitor proteins of the IκB family. IκB interacts with the NLS of NF-κB and masks it from the nuclear import machinery. The release of NF-κB from the inhibitory complex and its subsequent translocation to the nucleus can be achieved by stimulation with inflammatory cytokines which lead to the activation of the IKK complex of serine/threonine kinases. Phosphorylation of IκB by the IKK complex targets IκB for proteolytic degradation by the ubiquitin pathway thus unmasking the NLS of NF-κB and enabling it to be translocated to the nucleus.

Nuclear import of transcription factors can also be achieved by dephosphorylation. This mechanism of regulation is illustrated by the NFAT family of transcription factors (reviewed in Rao *et al.*, 1997). The region N-terminal to the DNA-binding domain of NFAT contains an NLS and multiple phosphoserine residues. The protein phosphatase, calcineurin, interacts with this region and when activated, calcineurin dephosphorylates NFAT, thus enabling the NLS to mediate its translocation to the nucleus (Aramburu *et al.*, 1998).

Nuclear export. In addition to stimulating nuclear import, signaling pathways can also prevent the nuclear accumulation of transcription factors. For example, the accumulation of Smads in the nucleus can be antagonized by phosphorylation. In this case the Erk MAP kinases phosphorylate sites within the linker region of Smads (Kretszchmar *et al.*, 1999). In a similar manner, phosphorylation of NFAT by the JNK MAP kinase can also antagonize the nuclear import of NFAT (Chow *et al.*, 1997). Activation of the serine/threonine kinase, PKB/Akt, has also been shown to cause cytoplasmic retention of members of the Forkhead transcription factor family (Kops *et al.*, 1999; Brunet *et al.*, 1999). One recent example is that of the Forkhead transcription factor FKHRL1, where activation of Akt results in phosphorylation of FKHRL1 which enables it to be retained in the cytoplasm by 14–3–3 proteins (Brunet *et al.*, 1999). A further example of this type of regulation is during

TGF-β signaling in myogenic cells where activation of this pathway results in the translocation of the MADS-box transcription factor myocyte enhancer binding factor 2 (MEF2) from the nucleus to the cytoplasm (De Angelis *et al.*, 1998). However, the molecular mechanism underlying this process is currently not known.

4.2 Regulation of intrinsic transcription factor activity

Conformational changes. Conformational changes are one way that phosphorylation might affect the activities of transcription factors (*Figure 5b*). Indeed, phosphorylation can induce changes in the structure of proteins and it is often inferred that the phosphorylation status of eukaryotic transcription factors affects their conformation. However, whilst it is likely that the biochemical activities of many eukaryotic transcription factors are regulated by phosphorylation-dependent conformational changes this has not yet been confirmed by the elucidation of the three dimensional structures of the phosphorylated and non-phosphorylated forms of the same transcription factor. Moreover, there is also very little reported biochemical evidence for such changes.

One example where a conformational change has been strongly inferred, is that of the MADS-box transcription factor SRF, where changes in the partial proteolysis pattern indicate that the protein conformation in the vicinity of the DNA-binding domain changes upon serine phosphorylation by casein kinase II (CKII) (Manak and Prywes, 1991). Recently, a combination of biochemical and biophysical methods has been used to probe the structure of the ETS-domain transcription factor Elk-1. Upon phosphorylation by Erk MAP kinase, Elk-1 undergoes a complex conformational change in both its secondary and tertiary structural elements (Yang *et al.*, 1999b). Elk-1 is phosphorylated on serine and threonine residues in the C-terminal transactivation domain, which in turn induces a conformational change that accompanies the stimulation of its DNA-binding activity. The transmission of the phosphorylation signal through the Elk-1 molecule via intermediary domains can be viewed as a mechanism for intramolecular transmission of signals rather than the intermolecular transmission of the signal observed in the cascades themselves.

DNA binding. Phosphorylation can also regulate the DNA-binding activity of transcription factors (*Figure 5c*). This can be achieved indirectly by phosphorylation of regulatory subunits or directly by phosphorylation of the transcription factor. Indirect regulation is illustrated by the phosphorylation of the IκB/NF-κB complex. In addition to sequestering NF-κB in the cytoplasm, IκB also inhibits its DNA-binding activity. In the IκB/NFκB complex, the conformation of NFκB differs from that of the DNA-bound structure and it has been suggested that electrostatic interactions between IκB and NF-κB may lock NF-κB in a conformation that is incompatible with DNA binding (reviewed in Baeuerle, 1998). As discussed above, serine phosphorylation of IκB releases NF-κB which then translocates to the nucleus where, in the absence of IκB, it is free to interact with DNA.

Examples of transcription factors whose DNA-binding activity is directly stimulated by phosphorylation include MEF2C and Elk-1. Serine phosphorylation of MEF2C by CKII or p38 MAP kinases results in stimulation of its DNA binding activity. CKII-mediated phosphorylation occurs in the DNA-binding domain itself (*Figure 4*; Molkentin *et al.*, 1996) and might directly affect DNA contacts, whereas p38-mediated phosphorylation occurs in the transcriptional activation domain (*Figure 4*; Han *et al.*, 1997) and is therefore likely to act in an allosteric manner. Similarly, DNA-binding by Elk-1 is stimulated upon phosphorylation by MAP kinases on serine and threonine residues within its C-terminal region (*Figure 4*; reviewed by Treisman, 1996) and leads to allosteric regulation of DNA binding by the N-terminal ETS-domain (Yang *et al.*, 1999b).

Phosphorylation can also directly inhibit DNA-binding. For example, phosphorylation of c-Jun by CKII on serine and threonine residues adjacent to the DNA-binding domain inhibit its DNA-binding activity (*Figure 4*; Lin *et al.*, 1992). Conversely, dephosphorylation of c-Jun at these sites can potentiate DNA binding by c-Jun (reviewed in Hill and Treisman, 1995). In this case, the introduction of negatively charged phosphate groups close to the DNA binding region of c-Jun might act to repel the phosphate backbone of the DNA and hence reduce protein-DNA contacts. This also provides a further example of intramolecular transduction of phosphorylation signals as N-terminal phosphorylation of c-Jun leads to dephosphorylation of the C-terminal sites, presumably by eliciting a conformational change that alters the accessibility of the phosphorylated C-terminal sites (Papavassiliou *et al.*, 1995).

Oligomerization. The activity of transcription factors can be regulated by phosphorylation-dependent oligomerization (*Figure 5d*). Transcription factors that are regulated in this manner require specific obligatory oligomerization partners and are unable to function as monomers. For example, tyrosine phosphorylation of the STAT proteins induces their dimerization via a reciprocal interaction between the SH2 domains and the phosphotyrosine-containing motifs of each monomer (Chen *et al.*, 1998).

A second example of phosphorylation-induced oligomerization is illustrated by the Smad family. Association of the receptor phosphorylated Smads with the co-Smad, Smad4, is dependent upon the phosphorylation state of the C-terminal MH2 domain (*Figure 4*). Smads contain two domains; the N-terminal MH1 domain and the C-terminal MH2 domain separated by a linker region. The MH1 domain mediates DNA binding and serves to autoinhibit the function of the MH2 domain possibly by a direct interaction. The MH2 domain mediates oligomerization between Smad proteins and phosphorylation might relieve the autoinhibitory effect of the MH1 domain by disrupting their interaction, thus enabling the MH2 domain of the receptor-phosphorylated Smads to interact with the MH2 domain of Smad4 (reviewed in Kretzschmar and Massague, 1998).

Transactivation. Transcription factors often contain transcriptional activation domains whose activity can be stimulated by phosphorylation. Transcriptional activation by direct phosphorylation of activation domains can be achieved by the MAP kinases. For example, phosphorylation of the C-terminal transactivation domain of Elk-1 (*Figure 4*) directly stimulates its ability to activate transcription (reviewed in Treisman, 1996). The activation potential of c-Jun and ATF-2 can also be stimulated by MAP kinase-dependent phosphorylation of their N-terminal transactivation domains (*Figure 4*). In c-Jun the important residues are serines (reviewed in Whitmarsh and Davis, 1996) whereas threonine phosphorylation of activating transcription factor 2 (ATF-2) is essential for activation by Jnk MAP kinases (Livingstone *et al.*, 1995).

Protein stability. Phosphorylation can also affect the stability of proteins within the cell and hence indirectly affect the activities of transcription factors. In addition to regulating transcription factor activity by stimulating the proteolytic degradation of regulatory proteins like IκB, phosphorylation can also have a direct affect on the stability of transcription factors (reviewed in Fuchs *et al.*, 1998). One of the major mechanisms for altering protein stability is via the ubiquitin pathway, which acts to target proteins for destruction by the proteosome (*Figure 5e*). Again the MAPKs provide the best examples for controlling such a regulatory mechanism. In the case of c-Jun, association with inactive Jnk targets it for ubiquitination and subsequent proteolytic degradation. In contrast, phosphorylation by activated Jnk protects c-Jun from ubiquitination and thus prolongs its half life. In a similar manner, ubiquitination of p53 occurs when it is associated with Jnk. Phosphorylation of p53 by activated Jnk leads to the dissociation of the Jnk–p53 complex and a concomitant increase in p53 stability.

A second example of signal-regulated ubiquitination is the Wnt signaling pathway, where GSK-3-mediated phosphorylation leads to the ubiquitination and subsequent degradation of β-catenin (see Section 2.2; reviewed in Fuchs *et al.*, 1998; Willert and Nusse, 1998). This prevents the accumulation of β-catenin, which is thought to be important in the redistribution of its subcellular location from the membrane and cytoplasm to the nucleus.

A related pathway to ubiquitination has also been identified in which the small protein, small ubiquitin-like modifier protein (SUMO-1), is attached to protein substrates. IκB has been shown to be modified by SUMO-1 which acts antagonistically to ubiquitin and leads to enhanced IκB stability and hence NF-κB inactivation (Desterro *et al.*, 1999). Whilst ubiquitin modification is stimulated by phosphorylation, SUMO-1 modification of IκB is inhibited. It is currently unclear whether other transcription factors are targeted by SUMO-1, but this represents a further attractive mechanism for modification of their activities.

Proteolytic cleavage. In addition to affecting overall protein stability, signaling pathways can affect specific cleavage of transcription factors. In hedgehog signaling for example, the activation of the pathway results in blocking of cleavage of the Ci transcription factor. This results in the generation of an active form of this protein (Ci^{155}) rather than a repressive form (Ci^{75}) (reviewed in Ingham, 1998). The exact mechanisms by which hedgehog signaling affects Ci modification are unknown, although PKA-mediated phosphorylation has been implicated in the initiation of the proteolytic cleavage event.

A variation on this theme is in the Notch signaling pathway. In this case, the cleavage event occurs in the receptor protein that results in the generation of a coactivator derived from the intracellular domain (reviewed in Baker, 2000; Osborne and Miele, 1999). However, it is currently unclear whether ligand binding induces this cleavage event or merely results in the release of a precleaved domain.

4.3 Interaction with coregulatory proteins

In addition to regulating the intrinsic activities of transcription factors, phosphorylation can also affect their ability to interact with coregulators (*Figure 5f*). The best direct evidence for this is demonstrated by the interaction of CREB with the co-activator CBP which is dependent upon CREB phosphorylation. Phosphorylation of CREB within the kinase-inducible domain (KID) stimulates the recruitment of CBP by interaction with its KIX domain. A combination of structural and mutagenic studies have shown that the KID:KIX complex is stabilized by hydrogen bond interactions between the critical phosphoserine residue in the KID and the hydroxyl group of a tyrosine residue in the KIX domain (Radhakrishnan *et al.*, 1997).

A second example of this type of regulation is provided by nuclear receptor transcription factor, estrogen receptor beta (ERβ). MAP kinase pathways can stimulate the phosphorylation-dependent recruitment of the steroid receptor coactivator SRC-1 by ERβ. Phosphorylation of two serine residues in the N-terminal activation domain of ERβ plays a pivotal role in this process (Tremblay *et al.*, 1999).

Further links to coregulatory complexes are suggested by the observation that SRF-mediated transcriptional activation in response to activation of the RhoA pathway also requires histone hyperacetylation in the c-*fos* promoter region (Alberts *et al.*, 1998). This implies that coactivator complexes containing histone actetyltransferases (HATs) are recruited to this promoter following RhoA activation.

Finally, a more complex series of regulatory networks is suggested by studies on the POU-domain transcription factor Pit-1. In this case, the activation of various signaling pathways leads to the recruitment or release of different corepressor and coactivator complexes (Xu *et al.*, 1998). In the absence of stimulation, Pit-1 appears to recruit the N-coR/mSin3/HDAC complex and hence repress transcription. However, upon activation of either the cAMP pathway or growth factor regulated Erk MAP kinase pathway, coactivator complexes containing CBP and P/CAF are recruited. Moreover, upon recruitment, the HAT activities of these two coactivators are differentially required by the different pathways, with CBP being needed for the cAMP pathway but P/CAF being needed for response to the Erk pathway. It is currently unknown what the molecular basis is to this dynamic use of coregulatory protein complexes. However, it is clear that a fine balance between a variety of positively and negatively acting factors likely determines the transcriptional outcome mediated by individual transcription factors in response to the activation of different pathways.

5. Perspectives and future directions

To date, many of the major different classes of signaling pathways and transcription factors have been identified. An increasing number of different types of coregulatory proteins are being discovered and many of these are found to be part of larger coactivator or corepressor complexes. Many different principles in signaling to transcription factors have been uncovered, although different variations on the themes described here and novel types of signaling mechanisms are still likely to arise. There are several areas that are likely to yield significant insights into the mechanisms of signaling to the nucleus in the future.

5.1 Genome sequencing

Recent and future advances in the sequencing of the genomes of several eukaryotic organisms will provide us with a complete complement of the sequences of all the potential signaling and transcriptional regulatory molecules found in eukaryotic cells. The next major challenge is to then integrate this information and position the various proteins into the correct pathways. The frameworks derived so far will facilitate this process.

A second major advance made possible by genome projects is the development of microarrays that enable the simultaneous investigation of the expression of large numbers of genes. A recent application of this technology was to examine the range of different genes that were up- and down-regulated in response to serum stimulation (Iyer *et al.*, 1999). It should be possible in the future to refine this approach and to examine in a more global manner which genes are regulated by distinct signals, by components of signaling pathways or by their transcription factor targets.

5.2 Direct signaling to coactivators

It is becoming apparent that coactivator and coregulatory proteins play an integral part in transducing signals into a transcriptional response. To date, studies have mainly focused on how transcription factor modification determines the recruitment of different co-regulatory protein complexes. An additional component of the signaling process might be directly to the coregulatory proteins themselves as suggested by a recent report implicating Erks in phosphorylating and activating CBP in response to nerve growth factor (NGF) signaling (Liu *et al.*, 1999). A further example of this is the Smad proteins that can act as coactivator proteins and are directly phosphorylated by the Erk MAP kinases and TGF-β receptors (reviewed in Kretzschmar and Massague, 1998). Future studies are likely to uncover roles for direct coregulatory protein phosphorylation in response to extracellular signals, which will serve to complement or counteract signals reaching transcription factors and hence modify the transcriptional outcome from transcriptional complexes.

5.3 Specificity in signaling

To date, many pathways and links to transcription factors have been elucidated. However, as more similar signaling pathway components and transcription factor targets are identified from the various different families of proteins, it is becoming important to re-evaluate whether the correct components have been placed in each pathway. A clear example of this is the MAP kinase pathways where each tier of the pathways contains a different kinase responsible for activating downstream components. On each level, the kinases exhibit a large degree of similarity, both in their sequence, mechanism of activation and substrate specificity. The terminal MAP kinases provide pertinent examples. One way in which signaling specificity is maintained is by assembling these pathways into complexes by binding to scaffold proteins (reviewed in Whitmarsh and Davis, 1998). Docking with transcription factors represents an additional level of downstream specificity (see Section 3.2). Future studies are likely to uncover further specificity determining mechanisms that permit precise nuclear responses to be elicited in response to extracellular signals.

5.4 Signaling by acetylation

Reversible acetylation is a major regulator of transcriptional responses. This is clearly demonstrated by the number of coactivator complexes that contain acetylase activity and corepressor complexes that contain deacetylase activity. To date, the focus has been on histone modification, however, an increasing number of transcription factor targets have been identified (reviewed in Kouzarides, 1999). Moreover, phosphorylation of transcription factors often affects their ability to recruit complexes containing either acetylase or deacetylases. Acetylation acts in a conceptually similar way to phosphorylation by altering the local charge of the phosphoacceptor site (removal of a positive charge and replacement with a neutral hydrophobic side chain rather than replacement of a hydrophilic sidechain with a negative charge). It is therefore tempting to speculate that cascades might also exist where acetylation is used as the switch that sequentially turns on downstream components. Furthermore, direct links might be discovered where phosphorylation directly affects the activity of an acetylase/deacetylase or vice versa. Answers as to whether acetylation plays a wider role in signaling and how far the interplay between phosphorylation and acetylation extends are likely to result from future studies.

Acknowledgments

We would like to thank Shen-Hsi Yang and Alan Dickson for helpful discussions and critical comments on the manuscript. The work in the authors' laboratories is supported by the Wellcome Trust (ADS and PS), the BBSRC and the cancer research campaign [CRC] (ADS). PS holds a Wellcome Trust International Prize Research Fellowship. ADS is a Research Fellow of the Lister Institute of Preventative Medicine.

References

Alberts, A.S., Geneste, O. and Treisman, R. (1998) Activation of SRF-regulated chromosomal templates by Rho-family GTPases requires a signal that also induces H4 hyperacetylation. *Cell* **92**: 475–487.

Angel, P. (1995) The role and regulation of the Jun proteins in response to phorbol ester and UV light. In: *Inducible Gene Expression*. Vol. 1. (ed. P.A. Baeurle). Birkhauser, Boston, pp. 62–69.

Aramburu, J., Garcia-Cozar, F., Raghavan, A., Okamura, H., Rao, A. and Hogan, P.G. (1998) Selective inhibition of NFAT activation by a peptide spanning the calcineurin targeting site of NFAT. *Mol. Cell* **1**: 627–637.

Baker, N.E. (2000) Notch signaling in the nervous system. Pieces still missing from the puzzle. *Bioessays* **22**: 264–273.

Baeuerle, P.A. (1998) IkappaB-NF-kappaB structures: at the interface of inflammation control. *Cell* **95**: 729–731.

Brandon, E.P., Idzerda, R.L. and McKnight, G.S. (1997) PKA isoforms, neural pathways, and behaviour: making the connection. *Curr. Opin. Neurobiol.* **7**: 397–403.

Brunet, A., Bonni, A., Zigmond, M.J., Lin, M.Z., Juo, P., Hu, L.S., Anderson, M.J., Arden, K.C., Blenis, J. and Greenberg, M.E. (1999) Akt promotes cell survival by phosphorylating and inhibiting a Forkhead transcription factor. *Cell* **96**: 857–868.

Carrion, A.M., Link, W.A., Ledo, F., Mellstrom, B. and Naranjo, J.R. (1999) DREAM is a Ca^{2+}-regulated transcriptional repressor. *Nature* **398**: 80–84.

Chawla, S., Hardingham, G.E., Quinn, D.R. and Bading, H. (1998) CBP: a signal-regulated transcriptional coactivator controlled by nuclear calcium and CaM kinase IV. *Science* **281**: 1505–1509.

Chen, X., Vinkemeier, U., Zhao, Y., Jeruzalmi, D., Darnell, J.E. Jr. and Kuriyan, J. (1998) Crystal structure of a tyrosine phosphorylated STAT-1 dimer bound to DNA. *Cell* **93**: 827–839.

Chow, C.W., Rincon, M., Cavanagh, J., Dickens, M. and Davis, R.J. (1997) Nuclear accumulation of NFAT4 opposed by the JNK signal transduction pathway. *Science* **278**: 1638–1641.

Cook, S.J., Aziz, N. and McMahon, M. (1999) The repertoire of fos and jun proteins expressed during the G1 phase of the cell cycle is determined by the duration of mitogen-activated protein kinase activation. *Mol. Cell Biol.* **19**: 330–341.

De Angelis, L., Borghi, S., Melchionna, R. ***et al.*** (1998) Inhibition of myogenesis by transforming growth factor beta is density-dependent and related to the translocation of transcription factor MEF2 to the cytoplasm. *Proc. Natl. Acad. Sci. USA* **95**: 12358–12363.

Desterro, J.M., Rodriguez, M.S. and Hay, R.T. (1998) SUMO-1 modification of IkappaBalpha inhibits NF-kappaB activation. *Mol. Cell* **2**: 233–239.

DiDonato, J.A., Hayakawa, M., Rothwarf, D.M., Zandi, E. and Karin, M. (1997) A cytokine-responsive IkappaB kinase that activates the transcription factor NF-kappaB. *Nature* **388**: 548–554.

Downward, J. (1998) Mechanisms and consequences of activation of protein kinase B/Akt. *Curr. Opin. Cell Biol.* **10**: 262–267.

Fuchs, S.Y., Fried, V.A. and Ronai, Z. (1998) Stress-activated kinases regulate protein stability. *Oncogene* **17**: 1483–1490.

Ghosh, S., May, M.J. and Kopp, E.B. (1998) NF-kappaB and Rel proteins: evolutionarily conserved mediators of immune responses. *Annu. Rev. Immunol.* **16**: 225–260.

Han, J., Jiang, Y., Li., Z., Kravchenko, V.V. and Ulevitch, R.J. (1997) Activation of the transcription factor MEF2C by the MAP kinase p38 in inflammation. *Nature* **386:** 296–299.

Hardingham, G.E. and Bading, H. (1998) Nuclear calcium: a key regulator of gene expression. *Biometals* **11**: 345–358.

Hill, C.S. and Treisman, R. (1995) Transcriptional regulation by extracellular signals: mechanisms and specificity. *Cell* **80**: 199–211.

Hunter, T. and Karin, M. (1992) The regulation of transcription by phosphorylation. *Cell* **70**: 375–387.

Ihle, J.N. (1996) STATs and MAPKs: obligate or opportunistic partners in signaling. *Bioessays* **18**: 95–98.

Ingham, P.W. (1998) Transducing hedgehog: the story so far. *EMBO J.* **17**: 3505–3511.

Iyer, V.R., Eisen, M.B., Ross, D.T., ***et al.*** (1999) The transcriptional program in the response of human fibroblasts to serum. *Science* **283**: 83–87.

Jacobs, D., Glossip, D., Xing, H., Muslin, A.J. and Kornfeld, K. (1999) Multiple docking sites on substrate proteins form a modular system that mediates recognition by ERK MAP kinase. *Genes Dev.* **13**: 163–175.

Jans, D.A. and Hubner, S. (1996) Regulation of protein transport to the nucleus: central role of phosphorylation. *Physiol. Rev.* **76**: 651–685.

Kallunki, T., Deng, T., Hibi, M. and Karin, M. (1996) c-Jun can recruit JNK to phosphorylate dimerization partners via specific docking interactions. *Cell* **87**: 929–939.

Karin, M. and Smeal, T. (1992) Control of transcription factors by signal transduction pathways: the beginning of the end. *Trends Biochem Sci.* **17**: 418–422.

Karin, M. and Hunter, T. (1995) Transcriptional control by protein phosphorylation: signal transmission from the cell surface to the nucleus. *Curr. Biol.* **5**: 747–757.

Kato, Y., Kravchenko, V.V., Tapping, R.I., Han, J., Ulevitch, R.J. and Lee, J.-D. (1997) BMK1/ERK5 regulates serum-induced early gene expression through transcription factor MEF2C. *EMBO J.* **16**: 7054–7066.

Keyse, S.M. (1998) Protein phosphatases and the regulation of MAP kinase activity. *Semin. Cell. Dev. Biol.* **9**: 143–152.

Knauf, U., Newton, E.M., Kyriakis, J. and Kingston, R.E. (1996) Repression of human heat shock factor 1 activity at control temperature by phosphorylation. *Genes Dev.* **10**: 2782–2793.

Kops, G.J., de Ruiter, N.D., De Vries-Smits, A.M., Powell, D.R., Bos, J.L. and Burgering, B.M. (1999) Direct control of the Forkhead transcription factor AFX by protein kinase B. *Nature* **398**: 630–634.

Kouzarides, T. (1999) Histone acetylases and deacetylases in cell proliferation. *Curr. Opin. Genet. Dev.* **9**: 40–48.

Kreegipuu, A., Blom, N., Brunak, S. and Jarv, J. (1998) Statistical analysis of protein kinase specificity determinants. *FEBS Lett.* **430**: 45–50.

Kretzschmar, M. and Massague, J. (1998) SMADs: mediators and regulators of TGF-beta signaling. *Curr. Opin. Genet. Dev.* **8**: 103–111.

Kretzschmar, M., Doody, J., Timokhina, I. and Massague, J. (1999) A mechanism of repression of TGFbeta/Smad signaling by oncogenic Ras. *Genes Dev.* **13**: 804–816.

Lin, A., Frost, J., Deng, T., Smeal, T., al-Alawi, N., Kikkawa, U., Hunter, T., Brenner, D. and Karin, M. (1992) Casein kinase II is a negative regulator of c-Jun DNA binding and AP-1 activity. *Cell* **70**: 777–789.

Liu, K.D., Gaffen, S.L. and Goldsmith, M.A. (1998) JAK/STAT signaling by cytokine receptors. *Curr. Opin. Immunol.* **10**: 271–278.

Liu, Y.Z., Thomas, N.S. and Latchman, D.S. (1999) CBP associates with the p42/p44 MAPK enzymes and is phosphorylated following NGF treatment. *Neuroreport* **10**: 1239–1243.

Livingstone, C., Patel, G. and Jones, N. (1995) ATF-2 contains a phosphorylation-dependent transcriptional activation domain. *EMBO J.* **14**: 1785–1797.

Manak, J.R. and Prywes, R. (1991) Mutation of serum response factor phosphorylation sites and the mechanism by which its DNA-binding activity is increased by casein kinase II. *Mol. Cell. Biol.* **11**: 3652–3659.

Marshall, C.J. (1995) Specificity of receptor tyrosine kinase signaling: transient versus sustained extracellular signal-regulated kinase activation. *Cell* **80**: 179–185.

Molkentin, J.D., Li, L. and Olson, E.N. (1996) Phosphorylation of the MADS-Box transcription factor MEF2C enhances its DNA binding activity. *J. Biol. Chem.* **271**: 17199–17204.

Montaner, S., Perona, R., Saniger, L. and Lacal, J.C. (1999) Activation of serum response factor by RhoA is mediated by the nuclear factor-kappaB and C/EBP transcription factors. *J. Biol. Chem.* **274**: 8506–8015.

Osborne, B. and Miele, L. (1999) Notch and the immune system. *Immunity* **11**: 653–663.

Papavassiliou, A.G., Treier, M. and Bohmann, D. (1995) Intramolecular signal transduction in c-Jun. *EMBO J.* **14**: 2014–2019.

Radhakrishnan, I., Perez-Alvarado, G.C., Parker, D., Dyson, H.J., Montminy, M.R. and Wright, P.E. (1997) Solution structure of the KIX domain of CBP bound to the transactivation domain of CREB: a model for activator: coactivator interactions. *Cell* **91**: 741–752.

Rao, A., Luo, C. and Hogan, P.G. (1997) Transcription factors of the NFAT family: regulation and function. *Annu. Rev. Immunol.* **15**: 707–747.

Robinson, M.J. and Cobb, M.H. (1997) Mitogen-activated protein kinase pathways. *Curr. Opin. Cell. Biol.* **9**: 180–186.

Sharrocks, A.D., Brown, A.L., Ling, Y. and Yates, P.R. (1997) The ETS-domain transcription factor family. *Int. J. Biochem.* **29**: 1371–1387.

Soderling, T.R. (1999) The Ca-calmodulin-dependent protein kinase cascade. *Trends Biochem. Sci.* **24**: 232–236.

Stark, G.R., Kerr, I.M., Williams, B.R., Silverman, R.H. and Schreiber, R.D. (1998) How cells respond to interferons. *Annu. Rev. Biochem.* **67**: 227–264.

Tan, P.B. and Kim, S.K. (1999) Signaling specificity: the RTK/RAS/MAP kinase pathway in metazoans. *Trends Genet.* **15**: 145–149.

Treisman, R. (1996) Regulation of transcription by MAP kinase cascades. *Curr. Opin. Cell Biol.* **8**: 205–215.

Tremblay, A., Tremblay, G.B., Labrie, F. and Giguere, V. (1999) Ligand-independent recruitment of SRC-1 to estrogen receptor beta through phosphorylation of activation function AF-1. *Mol. Cell* **3**: 513–519.

Whitmarsh, A.J. and Davis, R.J. (1996) Transcription factor AP-1 regulation by mitogen-activated protein kinase signal transduction pathways. *J. Mol. Med.* **74**: 589–607.

Whitmarsh, A.J. and Davis, R.J. (1998) Structural organization of MAP-kinase signaling modules by scaffold proteins in yeast and mammals. *Trends Biochem. Sci.* **23**: 481–485.

Willert, K. and Nusse, R. (1998) Beta-catenin: a key mediator of Wnt signaling. *Curr. Opin. Genet. Dev.* **8**: 95–102.

Xu, L., Lavinsky, R.M., Dasen, J.S., *et al.* (1998) Signal-specific co-activator domain requirements for Pit-1 activation. *Nature* **395**: 301–306.

Yang, S.-H., Yates, P., Whitmarsh, A.J., Davis, R.J. and Sharrocks, A.D. (1998a) The Elk-1 ETS-domain transcription factor contains a mitogen-activated protein kinase targeting motif. *Mol. Cell. Biol.* **18**: 710–720.

Yang, S.-H., Whitmarsh, A.J., Davis, R.J. and Sharrocks, A.D. (1998b) Differential targeting of MAP kinases to the ETS-domain transcription factor Elk-1. *EMBO J.* 1740–1749.

Yang, C.-C., Ornatsky, O.I., McDermott, J.C., Cruz, T.F. and Prody, C.A. (1998c) Interaction of myocyte enhancer factor 2 (MEF2) with a mitogen-activated protein kinase, ERK5/BMK1. *Nucleic Acids Res.* **26**: 4771–4777.

Yang, S.-H., Galanis, A. and Sharrocks, A.D. (1999a) Targeting of p38 MAPKs to MEF2 transcription factors. *Mol. Cell. Biol.* **19**: 4028–4038.

Yang, S.-H., Shore, P., Willingham, N., Lakey, J.H. and Sharrocks, A.D. (1999b) The mechanism of phosphorylation-inducible activation of the ETS-domain transcription factor Elk-1. *EMBO J.* **18**: 5666–5674.

7

Transcriptional regulation of early lymphocyte development

James Hagman and Gerald Siu

1. Introduction

B and T lymphocytes comprise the cellular components of the antigen-specific immune system. The major role of B lymphocytes is the synthesis and secretion of immunoglobulin (Ig) proteins in response to the activation of individual cells by specific antigen. The roles of T lymphocytes are more complex. T cells elicit soluble factors following their activation by antigens in the context of major histocompatibility complex (MHC) molecules on antigen-presenting cells (including B cells). These factors augment immune responses by increasing the proliferation and survival of other cells, including B cells. Functions of T cells also include the direct killing of target cells that present stimulating antigens.

Lymphocytes provide important model systems for studying molecular pathways that control cellular differentiation. Notably, both B and T cells progress through multiple stages of differentiation characterized by the expression of distinct sets of genes. In this chapter, we review transcriptional control mechanisms in early B and T lymphocytes that are important for their early development. In the first section, we focus on transcription factors that are essential for the expression of genes at early stages of B-cell differentiation. In addition to new insights from mice that are deficient for these factors, abilities of these factors to activate quiescent target genes in non-lymphoid cells suggest mechanisms for the establishment of early patterns of gene expression. Other studies defined a key role for one of these factors, Pax-5/BSAP, for B-cell lineage commitment and progression: it activates B-cell-specific transcription while repressing genes that promote differentiation of progenitor cells to other cell lineages. We then review control of the T-cell-specific *CD4* gene, which encodes a co-receptor molecule that is critical for early thymocyte development and mature T-cell function, as a paradigm for

Transcription Factors, edited by J. Locker.

developmentally regulated gene expression in T lymphocytes. CD4 is regulated by positively acting enhancer and negatively acting silencer modules at different stages of T-cell differentiation.

2. Regulation of early B-cell commitment and differentiation

2.1 B-cell development

The hallmark of early B-cell development (*Figure 1*) is the stepwise expression and assembly of components of the functional receptor for antigen, the BCR (B-cell receptor). Prior to the initiation of Ig gene rearrangements, transcripts are detectable from promoters adjacent to (μ0) and upstream of (Iμ) the μ intronic enhancer region of the heavy chain locus (IgH). Detection of these transcripts correlates with changes in local chromatin structure and accessibility of IgH genes to *trans*-acting factors. Rearrangement of IgH genes proceeds in two steps, first by assembling diversity (D) and joining (J) segments (D–J rearrangements), followed by the joining of variable regions to D–J segments to create mature V–D–J joints. Productively rearranged genes encode heavy chains that assemble the pre-BCR on the plasma membrane, together with Igα and Igβ (encoded by the *mb-1*, *B29* genes), a pair of membrane-spanning polypeptides that are required for display of Ig on the cell surface and for Ig-mediated signal transduction (reviewed in Tamir and Cambier, 1998). Each Ig heavy chain also associates with surrogate light chain polypeptides encoded by the *λ5* and *VpreB* genes. Expression of a functional pre-BCR is required for developmental progression of pre-B cells to subsequent stages, and for initiating rearrangement and expression of Ig light

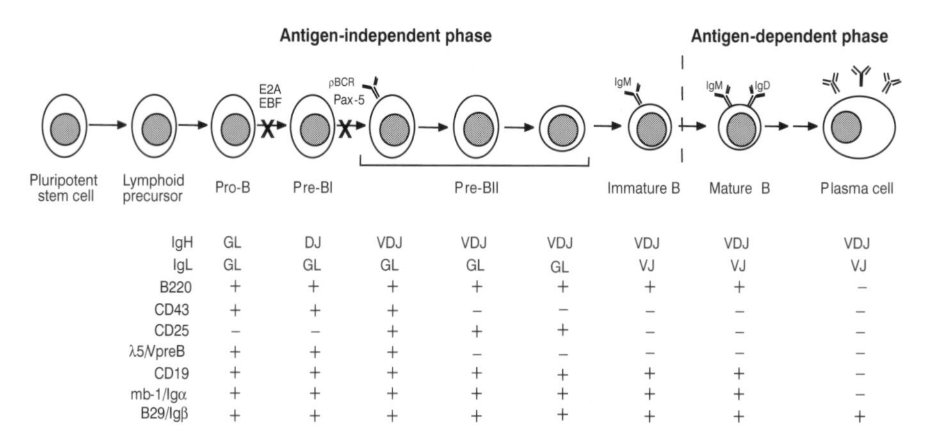

	Pro-B	Pre-BI	Pre-BII			Immature B	Mature B	Plasma cell
IgH	GL	DJ	VDJ	VDJ	VDJ	VDJ	VDJ	VDJ
IgL	GL	GL	GL	GL	GL	VJ	VJ	VJ
B220	+	+	+	+	+	+	+	–
CD43	+	+	+	–	–	–	–	–
CD25	–	–	+	+	+	–	–	–
λ5/VpreB	+	+	+	–	–	–	–	–
CD19	+	+	+	+	+	+	+	–
mb-1/Igα	+	+	+	+	+	+	+	–
B29/Igβ	+	+	+	+	+	+	+	+

Figure 1. Development of normal B lymphocytes. Distinct stages of differentiation are characterized by the status (shown below) of immunoglobulin (Ig) genes and expression of cell surface markers (Rolink *et al.*, 1994). Xs indicate approximately developmental blocks imposed by targeted deletions of *E2A*, *ebf*, or *pax-5* genes in mice (see text). IgH, status of immunoglobulin heavy chain loci. IgL, status of immunoglobulin light chain loci. GL, germline, or unrearranged Ig loci. pBCR, preB cell receptor.

chain genes that will constitute the mature antigen receptor, the BCR. Display of the mature BCR on the plasma membrane results in immature B cells that respond to specific antigens by secreting serum antibodies.

Much of the interest in the regulation of early B cell-specific genes has focused on three transcription factor-encoding genes: *E2A*, *ebf*, and *pax-5*. Although none of these genes is expressed exclusively in B cells, they synergize to activate genes that define the earliest stages of differentiation. Together with biochemical experiments, data generated using mice with targeted deletions of these genes defined their importance for lineage progression and tissue-specific gene expression. Taken together, these data suggest a hierarchy of developmental control.

2.2 Basic helix-loop-helix proteins encoded by the E2A *gene*

Basic helix-loop-helix (bHLH) proteins encoded by the *E2A* gene play important roles for activating lineage-specific gene expression in B lymphocytes. The *E2A* gene products E12 (*Figure 2*) and E47 (referred to collectively as E2A proteins), which are closely related to the muscle-specific factor MyoD, to c-Myc, and to the *daughterless* gene of *D. melanogaster* (Murre *et al.*, 1989), were among the first lineage-specific factors to be studied in differentiated mammalian cells. Early observations of functionally important regulatory regions of Ig genes (e.g., the IgH intronic enhancer) identified binding sites for E2A (and related) proteins termed E boxes (CANNTG), and defined their importance for the control of Ig gene transcription (reviewed in Kadesch, 1992). Early studies detected multiple E boxes in enhancers of the Ig heavy and κ light chain genes. Later studies showed that *E2A*-encoded factors by themselves activate transcription of B lineage-specific genes in non-B cells. Therefore, expression of bHLH proteins may be an initiating event leading to B-cell lineage commitment (Schlissel *et al.*, 1991). An emerging model proposes that collaboration between *E2A* gene

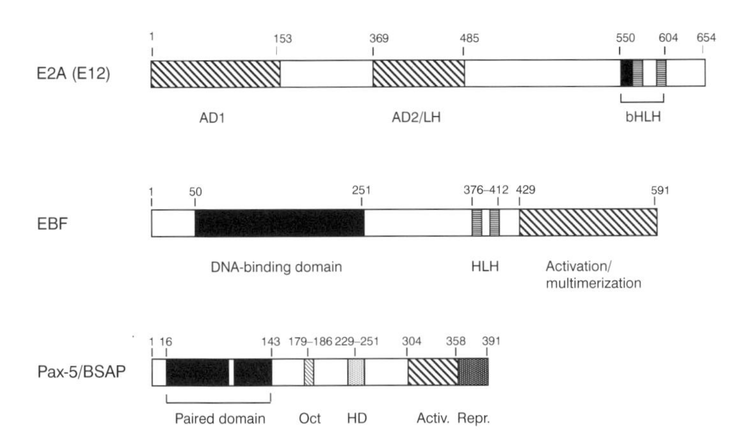

Figure 2. Important transcription factors in early B cells. Schematic diagrams of domains and other sequences of E2A (E12 alternatively spliced form), early B-cell factor (EBF), and Pax-5/BSAP. DNA-binding motifs are shaded black. AD1, activation domain 1. AD2/LH, activation domain 2/loop helix. bHLH, basic helix-loop-helix. HLH, distinct type of helix-loop-helix in EBF. Oct, octapeptide. HD, partial homeodomain.

products and other early B-cell-specific factors including EBF and Pax-5/BSAP defines early patterns of transcribed genes in B cells (see Section 2.5).

The defining characteristic of HLH family proteins is their utilization of a common dimerization motif: each HLH domain includes conserved helix 1 and helix 2 sequences separated by a loop of variable length. Together, two monomers assemble a parallel, four helix bundle that mediates stable homo- or heterodimerization in the absence of DNA. E12 and E47 each result from alternatively spliced transcripts of the *E2A* gene that differ only within a 76 amino acid region of their bHLH domains that determines the range of their partners for dimerization. In this regard, E12:E47 heterodimers and E47:E47 homodimers have each been detected in B cells (E12 also dimerizes with other bHLH proteins in T cells and other tissues).

X-ray crystallographic studies showed that the HLH domain precisely positions residues of a short α-helical basic region for contacts with the major groove of E box DNA (Ellenberger *et al.*, 1994). Contacts between each E47 monomer and the two CA dinucleotides of the E box are similar, however, interactions with the central variable bases of the binding site (5′CACCTG was used to assemble the crystal) are asymmetric; these nonequivalent contacts contribute to sequence discrimination. Side chains within the loop and helix 2 augment DNA binding through contacts with the phosphodiester backbone.

In addition to their DNA-binding/dimerization domains, E12 and E47 each comprise other domains that activate target gene transcription *in vivo* (*Figure 2*). Two transcriptional activation domains, AD1 and AD2/LH, are present in the amino-terminal regions of E2A polypeptides (Aronheim *et al.*, 1993; Quong *et al.*, 1993; Massari *et al.*, 1996). Both are required for activation *in vivo*. Each domain includes conserved loop and α-helical regions (also found in HEB; see Section 3.4.4) that are important for activation. AD1 α-helical domains, including those encoded by *E2A* and the related *E2–2* and *HEB* genes (see Section 3.4), and by an HLH family protein of yeast, Rtg3p, include the sequence LDFS (Massari *et al.*, 1999). Mutation of residues in the LDFS motif of AD1 severely decreased transcriptional activation. Interestingly, the LDFS motif is important for recruiting proteins of the nuclear histone acetyltransferase, or SAGA complex, to DNA. Recruitment of the SAGA complex suggests that bHLH proteins encoded by *E2A* may activate transcription, in part, by directing the reorganization of chromatin through select histone modifications. E47 has also been reported to interact with p300/CBP (Eckner *et al.*, 1996), which catalyzes the acetylation of histones through its interactions with the PCAF complex (Bannister and Kouzarides, 1996; Ogryzko *et al.*, 1996; Yang *et al.*, 1996).

Regulation of target genes by E2A. In addition to *in vitro* DNA binding studies, knockout mice and transfected cells defined roles of both E12 and E47 for B-cell lineage-specific gene expression and commitment. In early experiments, very similar phenotypes were observed in *E2A* knockout mice lacking both E12 and E47 (Bain *et al.*, 1994; Zhuang *et al.*, 1994). High degrees of postnatal lethality were observed with a block of early B lymphopoiesis and a total lack of mature B cells and serum Ig (effects of these gene disruptions on T cells are discussed in Section 3.4.4). B220$^+$CD43$^-$ early and B220$^+$CD43$^-$ late progenitors (see *Figure 1*) were nearly undetectable in fetal livers of these mice. Partial reduction of early progenitor cells was also observed in heterozygous knockout mice, suggesting that E2A levels titrate numbers of B cell progenitors. Ig gene rearrangements were not detected in homozygous knockout mice, which is likely due to the lack of E2A protein interactions with E boxes in their regulatory modules, and by the absence of recombination activating gene (*RAG*)-1 transcripts. As a related observation, the *RAG-1* promoter includes E box sequences that may be a target for regulation by E2A protein (Fuller and Storb, 1997). Transcripts from germline Ig genes ($I\mu$), which are indicative of chromatin remodeling prior to Ig heavy chain gene rearrangements, were also lacking. Transcripts from the $\lambda 5$, *CD19*, and *mb-1(Igα)* genes were also strongly reduced, but *B29(Igβ)* transcripts were not affected. Importantly, transcripts encoding EBF and Pax-5/BSAP (B-cell-specific activator protein), which encode key regulators in early B-cells (see below) were lacking in $E2A^{-/-}$ knockout mice, suggesting a hierarchy of regulatory factors in early B cells (Bain *et al.*, 1994, 1997b).

Knockout mice that express either E12 or E47 have also been generated (Bain *et al.*, 1997b). These mice demonstrated increased viability relative to *E2A* knockout mice that lacked both factors. Targeted deletion of E47-specific exons resulted in a lack of mature B cells and serum Ig, indicating that E12 alone is insufficient for the development of mature B cells. In the presence of normal levels of E12, some $B220^+CD43^+$ progenitors were detected, but these cells did not progress to the $B220^+CD43^-$ stage. Similar to E2A-deficient mice, $E47^{-/-}$ mice lack detectable Ig gene rearrangements, express low levels of transcripts from the $\lambda 5$, *mb-1*, and *CD19* genes, and reduced levels of EBF and Pax-5/BSAP transcripts. E12-specific exons have not been specifically targeted, but introduction of transgenes for expression of E47 and E12 into $E47^{-/-}$ mice resulted in greater numbers of IgM^+ mature B cells versus $E47^{-/-}$ mice reconstituted with the E47 transgene alone. Thus, E12 augments function of E47, but E47 appears to be more critical for B lymphopoiesis.

Ectopic expression of E47 (and to a lesser degree, E12) in fibroblasts strongly induced transcription from endogenous Ig heavy chain intronic enhancers ($I\mu$ transcripts). Transcription of the early lymphocyte-specific terminal deoxynucleotidyl transferase (*TdT*) gene, which adds extra bases to coding joints during recombination of Ig heavy chain genes, was also activated by E47 (Choi *et al.*, 1996). These results showed that *E2A*-encoded factors can activate transcription of quiescent genes in nonlymphoid cells. Moreover, E2A proteins activate Ig gene recombination and transcription in pre-T cells that possess 'inaccessible' Ig gene loci (Schlissel *et al.*, 1991). Even more strikingly, expression of E47/E12, RAG-1, and RAG-2 is sufficient for activating rearrangement of Ig κ light chain genes in embryonic kidney cells, but not Ig λ light chain genes (Massari and Murre, 2000; Romanow *et al.*, 2000). Together, these results demonstrate the role of E2A proteins for activating Ig gene recombination. Recent results have also demonstrated a requirement for these proteins for Ig class switch recombination, which shifts expression from IgM heavy chains to other isotypes (IgG1, IgG2a, IgG2b, IgG3, IgE, or IgA) at later stages of B cell differentiation (Goldfarb *et al.*, 1996; Quong *et al.*, 1999).

Regulation of E2A activity. *E2A* transcripts are detected in a wide range of cell types, but paradoxically, encode factors that regulate gene expression in a tissue-specific fashion. The tissue specificity of these factors is largely due to regulated post-translational modifications that control their dimerization and/or activation functions, and by association with inhibitory subunits termed Id proteins. Two additional levels of control are less well characterized: Notch signaling and competition for E box binding by the zinc finger protein ZEB. The interplay between these various control mechanisms may result in developmental decisions that affect lineage determination (B vs. T vs. natural killer [NK] cells) of progenitor cells.

The activity of E47 is regulated, in part, at the level of dimerization with itself, or with inhibitory proteins. E47:E47 homodimers comprise the B cell-specific factor BCF1 (Shen and Kadesch, 1995). Control of selective dimerization may be due to lineage-specific phosphorylation (Sloan *et al.*, 1996), and/or covalent linkage of monomers through intermolecular disulfide bridges (Benezra, 1994). Functional E2A activity is also regulated by the assembly of heterodimers with other factors that include HLH domains. One group includes ABF-1 and Tal-1, which repress transcription of E-box-containing genes in activated B cells (Massari, 1998) or T cells (Park and Sun, 1998), respectively. Another group of HLH proteins comprises the Id proteins (Id1, Id2, Id3, and Id4), which do not possess basic region α-helices required for interaction with specific DNA (reviewed in Norton *et al.*, 1998). As a result, Id proteins act as dominant–negative antagonists by forming dimers with E2A proteins and preventing their interaction with DNA. Regulation of E2A proteins by multiple Id proteins *in vivo* has been characterized in transfected cells and transgenic mice. Pro-B-like cells express high levels of Id1, and down-regulation of the *Id1* gene occurs prior to the initiation of Ig heavy chain gene recombination (Sun *et al.*, 1991; Wilson *et al.*, 1991). In related experiments, constitutive expression of Id1 in early B cells of transgenic mice resulted in a phenotype much like that of E2A-deficient mice (Sun, 1994). Paradoxically, targeted deletion of *Id1* or *Id2* genes in mice had little effect on their B cell (or T cell) compartments. Deletion of *Id1* genes reduced the neonatal lethality observed with *E2A*-deficient mice, but also promoted the generation of T cell tumors (Yan *et al.*, 1997). Id2-deficient mice showed greatly reduced numbers of NK cells, suggesting that it may play a role in other developmental decisions (Yokota *et al.*, 1999). In light of these data, the actual roles of Id proteins for regulating E2A polypeptides *in vivo* remains to be determined.

Recent studies suggest that Notch signaling contributes to lineage determination in B and T cells (Ordentlich *et al.*, 1998; Osborne and Miele, 1999; see also Section 3.4.6). Expression of activated Notch1 in transgenic mice blocked B lymphopoiesis and increased numbers of immature double-positive T cells in bone marrow, and lymphoid precursors are directed toward a B-cell fate in the absence of Notch1 (Radtke *et al.*, 1999). Activated Notch effectively inhibited co-expressed E47. A model was suggested in which Notch inhibits E47 activity through a cofactor that may be the Notch-interactive partner Deltex. Deltex may block formation of E47 homodimers, while promoting binding of E47 with other partners (e.g., the bHLH protein HEB) that activate the T cell pathway. Interestingly, while E47 transactivation is antagonized by activated Notch, activities of Pax-5/BSAP and EBF are not affected (Pui *et al.*, 1999).

Activity of bHLH proteins at a subset of E boxes may be limited by competition for DNA binding with the 'two handed' zinc finger/homeodomain protein, ZEB (Genetta *et al.*, 1994). ZEB (also known as BZP/AREB6/δEF1) (Franklin *et al.*, 1994; Ikeda and Kawakami, 1995; Sekido *et al.*, 1996) functions may include repression of transcription from E-box-containing genes in the absence of bHLH proteins (see Section 3.4.5 and *Figure 6*). ZEB includes multiple repression domains, including sequences that interact with the co-repressor C-terminal binding protein (Turner and Crossley, 1998; Postigo and Dean, 1999). ZEB blocks p300/CBP histone acetyltransferase activities, and repression by ZEB can be overcome by overexpression of the p300/CBP coactivator. Recently, targeted deletion of genes encoding ZEB in mice resulted in greatly reduced numbers of thymocytes (Brabletz *et al.*, 1999), potentially through biasing of differentiation to the B lineage in the presence of reduced competition between ZEB and E2A proteins for E boxes and/or cofactors.

2.3 Early B cell factor (EBF)

Early B cell factor, or EBF, was first identified as a tissue-specific regulator of the *mb-1* gene promoter (*Figure 3b*; Hagman *et al.*, 1991; Feldhaus *et al.*, 1992). *mb-1* is an early B-cell-specific gene that encodes the Igα transmembrane protein, which is required for docking of Ig on the plasma membrane and for signal transduction by the BCR (Hombach *et al.*, 1988, 1990; Campbell and Cambier, 1990). EBF DNA-binding activity was detected in cell lines representing pro-B, pre-B, and more mature mIg^+ B cells, but not in terminally differentiated plasma cells, T cells, or other cell types. Purification of EBF from pre-B cells and peptide sequencing allowed the cloning of cDNAs encoding a novel 64.5 kDa protein (*Figure 2*), in close agreement with the calculated mass of EBF extracted from cells (Hagman *et al.*, 1993; Travis *et al.*, 1993). EBF exhibits only short stretches of homology with other DNA-binding proteins.

Biochemical analysis identified several novel features in EBF, including a novel DNA-binding domain in its amino terminus and an unusual dimerization

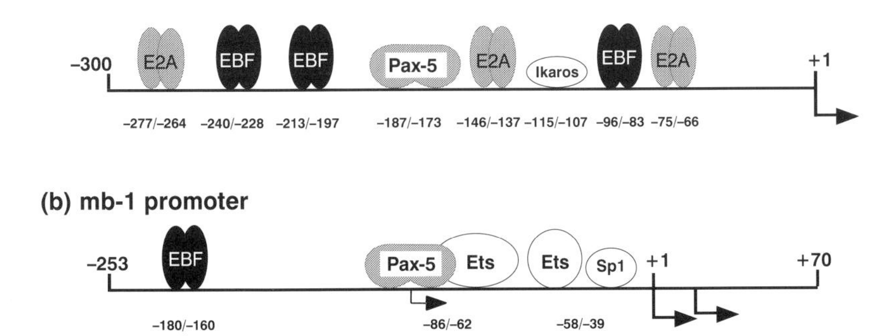

Figure 3. Two early B-cell-specific enhancer/promoters, factor binding sites, and putative regulatory factors. Approximate coordinates (relative to +1) of regulatory factor binding sites are shown below. (a) The murine λ5 enhancer/promoter (Lo *et al.*, 1991; Okabe *et al.*, 1992; Yang *et al.*, 1995; Sigvardsson *et al.*, 1997; Sigvardsson, 2000). (b) The murine mb-1 promoter (Hagman *et al.*, 1991; Travis *et al.*, 1991; Fitzsimmons *et al.*, 1996).

domain (*Figure 2*). Experiments with recombinant polypeptides of different lengths showed that it binds DNA as a homodimer, but can multimerize further with itself to assemble higher order complexes. Serine/threonine-rich sequences in the most carboxy-terminal part of EBF function as a transcriptional activation motif (Hagman *et al.*, 1995).

Detectable binding of DNA by monomers of EBF has been observed, but dimerization is necessary for its high affinity binding to an optimized sequence, AT/ATTCCCNNGGGAA/TT (Travis *et al.*, 1993), or to the consensus site CCC-NNGGG. Three regions of EBF contribute to its multimerization. First, two repeats of a 15 amino acid amphipathic α-helical sequence each exhibit a high degree of homology with helix 2 sequences of HLH domains, including those of Nautilus (the *Drosophila* ortholog of MyoD), and v-Myc (Hagman *et al.*, 1993). The 15 amino acid repeats, which are separated by a linker of seven amino acids, are required for dimerization *in vitro* and for functional activity *in vivo* (Hagman *et al.*, 1993, 1995). Therefore, EBF may be considered an atypical HLH protein, in which dimerization sequences are clearly segregated from its DNA-binding domain. A second region that participates in dimerization is present within the DNA-binding domain, which can mediate its homodimerization in a DNA-dependent fashion in the absence of the α-helical repeats. In addition to the repeats, sequences near the carboxy-terminus contribute to the assembly of higher order multimers.

Mapping of functional sequences in EBF localized its novel DNA-binding domain to approximately 200 amino acids (50–251) within the amino-terminal half of the polypeptide. Within this region, a novel motif, HEIMCSRCCDKKSC, is essential for DNA binding (Hagman *et al.*, 1995). Although significantly different from canonical zinc finger motifs, experiments suggested that this region coordinates metal ions. Mutation of any one of the four underlined amino acids completely ablated the ability of EBF to bind the *mb-1* promoter. Moreover, DNA binding activity of recombinant EBF was dependent upon its denaturation/renaturation in the presence of zinc (or cadmium) ions. The precise functional role of the HCCC metal-binding motif is unknown, but the lack of DNA binding following mutation of the central arginine in the metal-binding motif suggests a direct role for this residue in DNA recognition.

Roles of EBF in vivo. Identification of putative gene targets suggests that EBF coordinately regulates multiple components of the BCR in B cells. In addition to the single site in the *mb-1* promoter (*Figure 3b*; Hagman *et al.*, 1991), multiple, functionally important binding sites for EBF have been detected in promoters of the surrogate light chain genes $\lambda 5$ (*Figure 3a*) and *VpreB1* (Mårtensson and Mårtensson, 1997; Sigvardsson *et al.*, 1997; Persson *et al.*, 1998), B-lymphoid tyrosine kinase gene *blk* (Åkerblad and Sigvardsson, 1999), and *B29* gene (Åkerblad *et al.*, 1999). EBF may also regulate expression of Ig genes themselves, because it can repress the activity of the Ig heavy chain intronic enhancer (Åkerblad *et al.*, 1996). Binding sites for EBF have also been noted in subsets of mouse and human Vκ promoters, which may contribute to their differential activation during early B cell differentiation (Sigvardsson *et al.*, 1996), and in the human CD19 promoter (Gisler *et al.*, 1999).

Targeted deletion of *ebf* (*ebf1*- other EBF family members have been identified; see below) genes blocked B lymphopoiesis at the stage of very early B lymphoid progenitors (*Figure 1*; Lin and Grosschedl, 1995). Like E2A-deficient mice, EBF-deficient mice lack serum Ig and do not possess mature surface Ig^+ B cells. However, $B220^+CD43^+$ cells in bone marrow are present at greater numbers in $ebf^{-/-}$ vs. $E2A^{-/-}$ mice. These cells express germline transcripts ($I\mu$ and $\mu 0$) from Ig heavy chain genes, but exhibit only unrearranged Ig heavy (and light) chain loci. This is due, in part, to the lack of *RAG-1* expression. $ebf^{-/-}$ progenitor B cells also do not express transcripts from the *mb-1*, *B29*, *VpreB*, or $\lambda 5$ genes, reflecting the presence of functional EBF binding sites in their promoters. Transcripts encoding Pax-5/BSAP were not detected in $ebf^{-/-}$ mutant mice, suggesting that, like E2A (see above), EBF is an upstream regulator of *pax-5* gene transcription in B cells. In support of this conclusion, a functional EBF binding site was identified 1121 bp upstream of transcription start sites in the *pax-5* promoter, suggesting that EBF directly regulates Pax-5/BSAP expression (O'Riordan and Grosschedl, 1999).

Because EBF and closely related family members are expressed in other tissues (Wang and Reed, 1993; Garel *et al.*, 1997; Wang *et al.*, 1997), combinatorial interactions between EBF and other DNA-binding proteins are deemed necessary to direct its functions to appropriate gene targets *in vivo*. In this regard, experiments using single and dual knockout mice show that EBF and E2A collaborate to activate expression of early B cell genes (O'Riordan and Grosschedl, 1999), and the effects are quantitative. Thus, B cells of $ebf^{+/-}E2A^{+/-}$ mice attain a more differentiated phenotype than either homozygous knockout, but exhibit defective B lymphopoiesis at very early stages of differentiation, with greatly reduced levels of *RAG-1*, *RAG-2*, $\lambda 5$, and *pax-5*, and reductions of *VpreB*, *mb-1*, *B29*, *CD19*, and *Lymphoid Enhancer Factor-1* (*LEF-1*).

Collaboration between E2A and EBF has also been demonstrated using *in vitro* systems. Ectopic expression of EBF in Ba/F3 cells, which express E2A proteins, activates transcription of endogenous $\lambda 5$ genes (Sigvardsson *et al.*, 1997). More strikingly, co-expression of EBF and 'forced dimers' of E47 strongly synergize to activate $\lambda 5$ transcription in non-lymphoid HeLa cells (Sigvardsson, 2000). Activation of target genes involves assembly of ternary complexes comprising EBF and E47 homodimers bound at adjacent promoter sites (–277/–264 and –240/–228, *Figure 3a*) in the $\lambda 5$ enhancer/promoter (O'Riordan and Grosschedl, 1999; Sigvardsson, 2000), but there is no evidence for heterodimerization between these factors. In another recent study, EBF, but not E2A, activated VJ recombination in embryonic kidney cells that also express RAG-1 and RAG-2 (Romanow *et al.*, 2000). The authors concluded that collaboration between E2A (which can activate heavy chain D-J and κ light chain gene rearrangements; see Section 2.2.1) and EBF results in activation of much of the Ig repertoire in B cells (Pax-5/BSAP also has an important role, see Section 2.4.2). As an important point, expression of EBF together with E47 homodimers activates only a subset of early B cell-specific genes, suggesting that other transcription factors, cell–cell interactions, and/or soluble factors are required for the overall pattern of gene expression in B cells.

Little is understood concerning the regulation of *ebf* gene transcription. However, one recent study demonstrated that it can be induced by ectopically expressed E12 (Kee and Murre, 1998). Moreover, *ebf* transcripts are nearly absent from the bone marrow of E2A-deficient mice and greater numbers of B220⁺CD43⁺ progenitors are present in *ebf*$^{-/-}$ mice versus *E2A*$^{-/-}$ mice (Lin and Grosschedl, 1995; Bain *et al.*, 1997b). Together, these data argue that *E2A*-encoded proteins first act upstream of EBF, potentially by directly activating its transcription in early B cell progenitors.

2.4 Pax-5/B cell-specific activator protein (BSAP)

The *pax-5* gene encodes BSAP (Adams *et al.*, 1992). BSAP was initially identified as a mammalian DNA-binding activity that binds a sequence recognized by TSAP (Barberis *et al.*, 1990), a tissue-specific regulator of histone gene expression in sea urchin tube feet (actually sea urchin Pax-6 (Czerny and Busslinger, 1995). Within the array of hematopoietic lineages, BSAP is only detected in B lymphocytes. Similar to EBF, BSAP activity is detectable in mouse and human cell lines representing pro-B, pre-B, and mIg⁺ B cells, but is absent from terminally differentiated plasma cells. The predominant form of Pax-5/BSAP (sometimes referred to as Pax-5a, the major alternatively spliced form; Zwollo *et al.*, 1997), comprises a 45 kDa protein (*Figure 2*), but post-translational modifications increase its mass to 50 kDa. Pax-5/BSAP binds sites in regulatory regions of many genes expressed in early B cells, including the *CD19* (Kozmik *et al.*, 1992) and *mb-1* genes (Fitzsimmons *et al.*, 1996), and also may function as a regulator of Ig heavy chain isotype switching at later stages of differentiation (reviewed in Hagman *et al.*, 2000).

Pax-5/BSAP binds DNA via its paired domain, a 128 amino acid domain comprised of two subdomains in the amino terminal half of the protein (*Figure 2*). Structural data have not been obtained for Pax-5/BSAP itself, but crystallographic structures are available for two closely related proteins, *D. melanogaster* paired and murine Pax-6 (Xu *et al.*, 1995, 1999). In each of the two subdomains, α-helices are arranged as a helix-turn-helix (HTH) motif to position side chains of helices 2 and 3 (and 5 and 6) for contacts with bases of the major groove and the phosphodiester backbone. Additional contacts with the minor groove are mediated by a β-hairpin motif at the amino terminus of the paired domain, and by the linker region that separates the two subdomains. The bipartite structure of the paired domain provides the basis for the unusual DNA recognition by Pax-5/BSAP, which can be described by the degenerate consensus sequence G/ANGC/ATG/CAA/TGCGT/GG/AA/CC/A (Busslinger and Urbanek, 1995). An unusual feature of Pax-5/BSAP is its recognition of multiple classes of sequences through utilization of two, or only one (the amino terminal subdomain) of its subdomains for binding to DNA (Czerny *et al.*, 1993; Xu *et al.*, 1995, 1999).

Pax-5/BSAP comprises other functionally important sequences, including carboxy-terminal domains that contribute to transcriptional activation or repression in the context of different binding sites (Dörfler and Busslinger, 1996). Deletion of sequences in its carboxy-terminus greatly increased transcriptional activation by the protein. Unlike other Pax proteins, e.g. Pax-4 and Pax-6, Pax-5/BSAP lacks an additional functional homeodomain motif, but does include a remnant of this domain (see below). A highly conserved octapeptide motif that participates in protein–protein interactions is also present in Pax-5/BSAP.

Interaction of Pax-5/BSAP with other proteins. The observation that Pax-5/BSAP recognizes a diverse, or degenerate set of nucleotide sequences raised the hypothesis that interactions with other factors contribute to its nucleotide sequence specificity *in vivo*. This was confirmed by studies of Pax-5/BSAP binding to the *mb-1* promoter (Fitzsimmons *et al.*, 1996). Pax-5/BSAP binds the promoter by itself, but its binding is stabilized through recruitment of proteins of the Ets proto-oncogene family to bind an adjacent site (*Figure 3b*). Interaction with Pax-5/BSAP enables Ets family proteins, including Fli-1, Ets-1, and GABPα, to bind a suboptimal nucleotide sequence (CCGGAG). Without Pax-5/BSAP, binding of Ets proteins to this site is weakened by the underlined G nucleotide, which is an important contact for Pax-5/BSAP. Ets and Pax-5/BSAP interact through specific sequences in their respective DNA-binding domains. For cooperative DNA binding by ETS domains with Pax-5/BSAP, an aspartic acid just following the ETS α3 DNA recognition helix domain is required. This region may directly contact part of the paired domain through a highly conserved glutamine in the β-hairpin motif (Wheat *et al.*, 1999). Binding of the *mb-1* promoter by Pax-5/BSAP and Ets proteins is required for functional promoter activity, but the carboxy-terminal activation domain of Pax-5/BSAP is dispensable for upregulation of *mb-1* gene transcription in *pax-5*$^{+/-}$ pre-B cells (Nutt *et al.*, 1998).

Other proteins may functionally interact with Pax-5/BSAP. Pax-5/BSAP by itself can activate the *blk* gene promoter in transfection assays, but activates more effectively if co-expressed with the runt domain protein CBFα2/PEBP2α/AML1, which binds a nearby site in the promoter (Libermann *et al.*, 1999). Physical interaction between the DNA-binding domains of Pax-5/BSAP with CBFα2/PEBP2α/AML1 in the absence of DNA was demonstrated *in vitro*. Pax-5/BSAP interacts with the conserved core of the TATA-binding protein (TBP) through its partial homeodomain, suggesting that it can interact directly with the basal transcription machinery (Eberhard and Busslinger, 1999). A potentially relevant observation is the lack of a TATA box, but presence of a Pax-5/BSAP binding site, at –30 of the *CD19* promoter (Kozmik *et al.*, 1992). The partial homeodomain can also interact with the regulatory 'pocket' and carboxy-terminal sequences of the retinoblastoma protein (pRb). This intriguing observation indicates that Pax-5/BSAP activity may be controlled in a cell cycle-specific fashion, but additional studies are needed to explore this hypothesis.

Another recent study suggests that corepressors of the Groucho family contribute to context-dependent repression by Pax-5/BSAP (Eberhard *et al.*, 2000). The Groucho family member Grg4 interacts with the transactivation domain of Pax-5/BSAP *in vivo*. Interactions with full length Grg4 also require the conserved octapeptide motif of Pax-5/BSAP. Increased phosphorylation of Grg4 (or the related Groucho family members Grg1 and 3a) was observed following its interaction with Pax-5/BSAP. This interesting observation suggests that interactions with Pax-5/BSAP may expose Grg4 to modification by a specific kinase, or alternatively, Pax-5/BSAP may recruit a kinase to the complex. Importantly, coexpression of Grg4 represses transcriptional activation by Pax-5/BSAP in transfected cells.

Functional roles of Pax-5/BSAP in early B cells. The large number of potential target genes for Pax-5/BSAP in early B cells suggested that this factor plays an important role for B lymphopoiesis. It was, therefore, not surprising that mice with targeted disruptions of *pax-5* genes lack serum Ig and cannot mount a humoral immune response (Urbánek *et al.*, 1994). Similar to E2A- or EBF-deficient mice, Pax-5-deficient mice exhibit an early block of B-cell differentiation, but B220^{+}CD43^{+} cells differentiate further than do B cells of the other two knockout mice (*Figure 1*). Interestingly, though present in bone marrow, early B-cell progenitors are absent from fetal livers of these mice (Nutt *et al.*, 1997). B cells of Pax-5-deficient mice partially recombine their Ig heavy chain genes to complete D–J rearrangements, but show a greatly reduced ability to recombine V region segments necessary for the production of Ig heavy chains (Nutt *et al.*, 1997). These cells also show greatly reduced expression of the *mb-1*, *LEF-1*, *CD19*, and *N-myc* genes (Nutt *et al.*, 1998). Interestingly, levels of *PD-1* transcripts, which encode an immunoregulatory protein in lymphocytes, are increased in the absence of Pax-5/BSAP. Reduced expression of other putative target genes, including *λ5*, *VpreB1*, and *blk*, was not detected (Bain *et al.*, 1994).

Pax-5-deficient pre-B cells can be maintained in culture in the presence of interleukin (IL)-7 and stromal cells (Nutt *et al.*, 1998, 1999). Restoration of Pax-5/BSAP expression using recombinant retroviruses partially reactivated *CD19* expression, and completely reactivated expression of *mb-1*, *LEF-1*, and *N-myc*. Provocatively, restoration of *CD19* and *N-myc* expression required full length Pax-5/BSAP, while restoration of *mb-1* and *LEF-1* expression only required its DNA-binding domain. These data suggest that *CD19* transcription requires transcriptional activation domains in the carboxy terminus of Pax-5/BSAP, while the activation of *mb-1* may be a function of protein–protein interactions between the paired domain and other proteins, e.g., with Ets proteins (Fitzsimmons *et al.*, 1996; Wheat *et al.*, 1999).

Reconstitution of Pax-5/BSAP expression restored, in part, developmental progression of Pax-5-deficient pre-B cells as evidenced by V-D-J recombination of IgH genes and expression of IgM and CD40 (Nutt *et al.*, 1999). Surprisingly, in the absence of Pax-5/BSAP, it was found that pre-B cells promiscuously adopt other hematopoietic cell fates upon exhausting their supply of IL-7. In the absence of IL-7, the cells readily adopt a promyeloid phenotype when grown in the presence of the cytokine M-CSF. Further differentiation to dendritic cells, macrophages, or osteoclasts proceeds in the presence of GM-CSF, M-CSF, or Trance factor respectively. If IL-7 is replaced with IL-3, IL-6, stem cell factor, and G-CSF, granulocytes appear in the culture dish. Cells with NK cell markers appear if IL-7 is replaced by IL-2. Most strikingly, Pax-5-deficient pre-B cells mediate long term reconstitution of T-cell development in *RAG*$^{-/-}$ immunodeficient mice (Rolink *et al.*, 1999). Attainment of non-B-cell fates is accompanied by the loss of transcripts from B-lineage-specific genes, including *ebf*. Together, these results assert that despite expression of both E2A and EBF, the maintenance of B lineage characteristics requires expression of functional Pax-5/BSAP.

Pax-5/BSAP appears to have at least two major roles for B cell commitment: promoting B lymphocyte differentiation through regulation of B cell-specific genes, and blocking differentiation to attain other hematopoietic cell fates. Pax-5/BSAP regulates both activation and repression of B-lineage-specific genes, including targets for combinatorial regulation with E2A and EBF. Second, Pax-5-deficient pre-B cells express cytokine receptors that are not expressed on normal

B-cell progenitors, including the M-CSF receptor (Nutt *et al.*, 1999). Reintroduction of Pax-5/BSAP shuts off expression of these receptors, suggesting that it normally represses 'promiscuous' gene transcription. In this manner, Pax-5/BSAP limits developmental outcomes by blocking the ability of cells to respond to cytokines that promote differentiation to non-B-cell fates.

2.5 A model for B cell lineage commitment

At least three regulatory factors, E2A, EBF, and Pax-5/BSAP, function together to promote B lymphopoiesis *in vivo*. An early event leading to B-lineage commitment is the expression of functional bHLH proteins encoded by *E2A*. E2A expression may be required for expression of EBF, and these two factors synergistically activate early B-cell-specific genes, e.g. λ5. E2A and EBF collaborate to activate Ig gene recombination by activating RAG gene expression, and by promoting D–J and light chain gene recombination. E2A, and possibly EBF, activates *pax-5* transcription in early B cells. In turn, Pax-5/BSAP is necessary for activating expression of B lineage-specific genes including *mb-1* and for activating V to D–J recombination by an as yet undetermined mechanism. Pax-5/BSAP also represses transcription of *PD-1* and multiple cytokine receptor genes that are expressed in its absence. Following the induction of these factors, progenitor cells can progress to later stages of differentiation, while other developmental options are prevented. The nature of the B-lineage commitment signal(s) is unknown. One possible contributing mechanism is assembly of the functional pre-BCR, which signals a block to further Ig heavy chain gene rearrangements (allelic exclusion) and stimulates progression to later stages of differentiation. However, B cells of RAG-1- and RAG-2-deficient mice completely lack Ig gene rearrangements and do not receive pre-BCR-mediated signals. They do not show the developmental plasticity observed in B cells that lack Pax-5/BSAP. Therefore, Pax-5/BSAP likely promotes B lineage commitment and progression in multiple ways. An important feature may be its ability to repress expression of cytokine receptors that can influence fate decisions. Another issue is whether early B-cell progenitors (prior to expression of the preBCR) are irreversibly committed to B-lineage progression, or whether maintenance of Pax-5/BSAP is required to continuously block promiscuous differentiation of these cells. Clearly, more research is required to further elucidate these important regulatory mechanisms.

3. Developmental decisions in T cells: regulation of *CD4* transcription

3.1 T cell development

T lymphocytes exhibit multiple functions as regulators of the immune response, as both effectors and long term memory cells. Because of the requirements for antigen recognition and the correlation of function with MHC specificity, the T-cell developmental process is extremely complex. Two major types of mature T cells play different roles in the immune response: cytotoxic T_C cells kill virally-infected cells, and helper T_H cells secrete soluble protein factors that regulate the immune response (Fowlkes and Pardoll, 1989). Both T_H and T_C cells recognize antigens via their cell-surface T-cell antigen receptors (TCR), but in general, the two cell types recognize antigens bound to different MHC proteins on antigen-presenting cells (APCs). Most T_H cells recognize foreign antigen bound only to MHC class II molecules; this specificity is augmented by interactions of MHC class II with CD4 co-receptor molecules expressed exclusively on T_H cells. CD4 contributes to antigen recognition by binding to non-polymorphic regions of the MHC class II proteins, which increases the avidity of the T cell for APCs and sends stimulatory signals to the T cell via the tyrosine kinase *lck* (Perlmutter, 1993). Conversely, most T_C cells recognize antigens bound to MHC class I molecules and express the CD8 co-receptor.

CD4 and CD8 each play important roles for MHC class recognition that contribute to positive and negative selection during early T-cell development. Therefore, it is not surprising that they each exhibit complex patterns of expression in thymocytes. T_H and T_C cells derive from common precursors, achieving their mature functional capabilities and antigenic specificity during a multistep developmental process that occurs primarily in the thymus (*Figure 4*). Expression, or lack of expression of CD4 and CD8 is an important component of each stage of this process. T-cell precursors that first arrive in the thymus do not express cell-surface CD4 or CD8 and are referred to as $CD4^-CD8^-$, or double-negative (DN) thymocytes. Subsequently, these cells express low levels of CD8 alone, followed by increasing levels of CD4, CD8, and TCR molecules. These double-positive (DP or $CD4^+CD8^+$) cells represent the bulk of thymocytes and are the first population of cells to undergo receptor-mediated selection processes that are critical for generating the appropriate repertoire of antigen-specific, non-self-reactive T cells. The vast majority of thymocytes die at this stage, as most either lack a functional TCR, or express TCRs that are either autoreactive or are incapable of recognizing the MHC (Bevan *et al.*, 1994; Fink and Bevan, 1995; Guidos, 1996). The survivors downregulate either CD4 or CD8, leading to mature $CD4^+CD8^-$ T_H or $CD4^-CD8^+$ T_C populations that leave the thymus as functional single positive (SP) T cells.

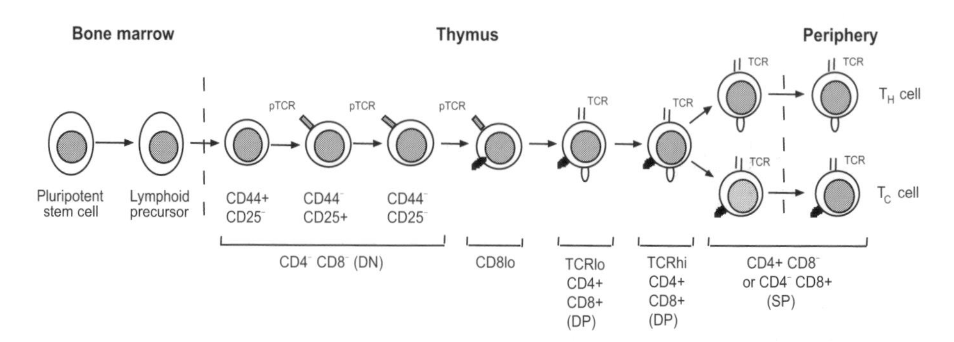

Figure 4. T-cell development in the thymus. Bone marrow, thymus, and peripheral environments are indicated above. Distinct stages of differentiation are characterized by expression of the pre-TCR (pTCR) or TCR complexes, cell surface markers, and CD4 and/or CD8 expression (indicated by white or black ovals, respectively).

3.2 *Control of* CD4 *as a model system*

The control of CD4 expression provides an ideal model system to study molecular mechanisms that drive T-cell development (Sarafova and Siu, 1999a). CD4 transcription is induced or repressed throughout thymopoiesis in response to developmental cues. In addition, the final downregulation of CD4 expression, which is central to the development of $CD8^+$ SP T_C cells from $CD4^+CD8^+$ thymocytes, comprises a critical lineage decision. Thus, transcription factors that control CD4 expression are also likely to mediate proper T-cell development and selection. This hypothesis predicts that these factors will be responsive to signals transmitted during the selection process. In this section, we will review the transcription factors that control *CD4* gene expression, and how they respond to development cues to mediate both CD4 expression and T cell development. Clearly, mechanisms that control expression of CD4 may also be involved in the regulation of CD8; CD8 regulation is discussed in detail elsewhere (Ellmeier *et al.*, 1999).

3.3 *Control modules of the* CD4 *gene*

The genomic locus encompassing the human *CD4* gene has been sequenced (Ansari-Lari *et al.*, 1996). Work by several groups has identified regulatory modules that act at different stages of development to selectively induce or repress *CD4* gene transcription (*Figure* 5): 1) the promoter; 2) the thymocyte enhancer; 3) the mature enhancer; and 4) the silencer. Interestingly, the modules are dispersed throughout the locus; whether they function only on the *CD4* promoter and not on nearby genes remains unclear. Our discussion will focus on the promoter, enhancers, and silencer, however, it should be noted that other studies have suggested the existence of additional regulatory sequences (Ellmeier *et al.*, 1999).

The CD4 *promoter.* The *CD4* promoter is a tissue-specific, TATA-less promoter that functions effectively only with the addition of an enhancer; it is capable of functioning by itself in activated $CD4^+$ T_H cells, but not in other cell types (Siu *et al.*, 1992; Nakayama *et al.*, 1993; Salmon *et al.*, 1993; Hanna *et al.*, 1994; Duncan *et al.*, 1995; Zhao-Emonet *et al.*, 1998; Sarafova and Siu, 1999b). Transcription begins at a sequence that resembles an 'initiator', a CA dinucleotide surrounded by pyrimidines. Deletion studies indicated that the minimal promoter required for full function is a 101 bp fragment immediately 5′ to the site of initiation (Siu *et al.*, 1992). The promoter contains four functional binding sites that are important for promoter function (*Figure 5a*); mutation of any of these sites significantly decreases promoter activity (Duncan *et al.*, 1995). Three of the sites are recognized by previously characterized transcription factors. The P1 site (–76) is recognized by c-Myb, which is important for hematopoiesis (Lipsick, 1996) and for T-cell development (Siu *et al.*, 1992; Nakayama *et al.*, 1993). In addition, the Myc-Associated Zinc finger (MAZ) factor and Elf-1, a member of the ETS family of transcription factors, bind to promoter P2 (–60) and P4 (–1) sites, respectively (Duncan *et al.*, 1995; Sarafova and Siu, 1999b). The factor that binds the P3 site (–31) is unknown. Although identification of these factors solved a considerable part of this regulatory puzzle, the mechanism by

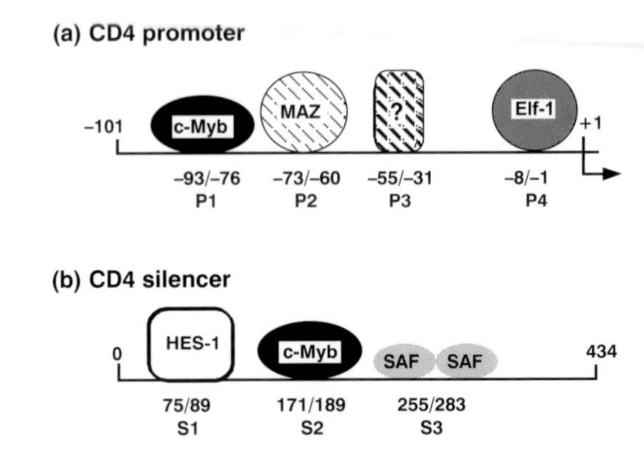

Figure 5. CD4 transcriptional control modules and factor binding sites. For each regulatory region, functional sites are indicated with previously published nomenclature; nucleotide positions indicated for each site were defined by functional mutations. The mature and thymocyte enhancers have not been analyzed in detail and are not shown. (a) The CD4 promoter (101 bp). (b) The murine CD4 intronic silencer (434 bp). Sites shown include S1, S2, and S3 sites (Duncan *et al.*, 1996). S3 comprises two binding sites for SAF.

which the *CD4* promoter is induced in a differentiation stage- and subclass-specific manner remains unclear.

The CD4 *locus enhancers.* Although the mature enhancer (13 kb upstream from the promoter) was originally characterized as a T-cell-specific enhancer using transient transfection approaches (Sawada and Littman, 1991), transgenic studies in multiple laboratories demonstrated that it is not functional in immature thymocytes (Duncan *et al.*, 1996; Salmon *et al.*, 1996; Adlam *et al.*, 1997; Uematsu *et al.*, 1997). Interestingly, it functions only in mature SP T cells that have successfully survived the positive selection process; for this reason we refer to it as the mature enhancer. Factors that bind to the mature enhancer are induced by thymic selection and thus may be targets of a signaling process that drives T cell development at mature stages of differentiation. Multiple factor-binding sites in the mature enhancer include a core region comprising two consensus E-box-like recognition sites separated by one base pair (AACAGGTGTCAGCTGGC; Sawada and Littman, 1993). Transient transfection analyses demonstrated that only the first of the E boxes is critical for mature enhancer function. This site is recognized by a heterodimer of E12 and the related bHLH protein HEB (see *Figure 6*). Induction of *CD4* mature enhancer function was one of the first roles identified for bHLH factors for regulating T-cell development (Section 3.4.3). Although this study suggested that the E12/HEB binding site is important for mature enhancer function, recent experiments in transgenic mice suggest that other sites are more important (Adlam and Siu, in preparation). Therefore, the importance of each putative site within the mature enhancer needs to be rigorously tested *in vivo*.

Early DP thymocytes utilize a separate enhancer to induce *CD4* transcription (Adlam *et al.*, 1997). The thymocyte enhancer is located 70 kb 3′ to the *CD4* promoter in the first intron of the constitutively-expressed IsoT gene (Adlam and Siu, 1999). Little is understood concerning the functional specificity of the thymocyte

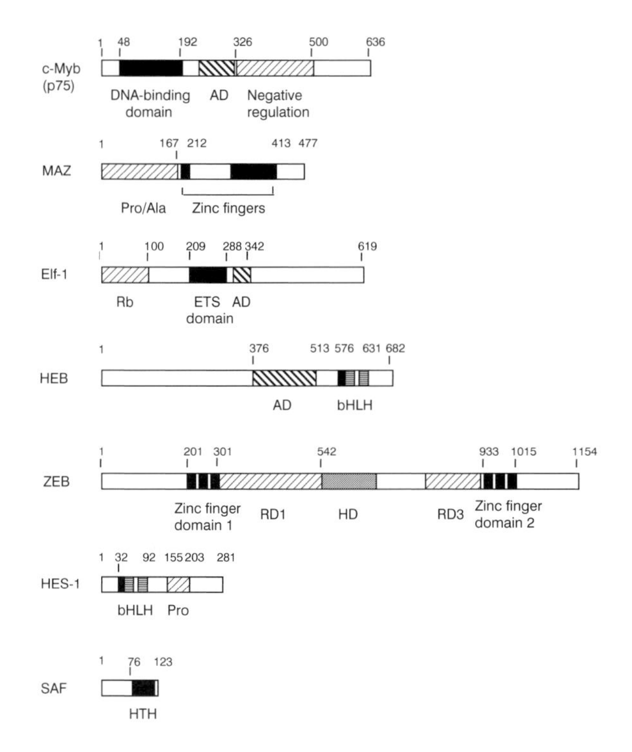

Figure 6. Regulators of CD4 transcription. Schematic diagrams of domains and other sequences of the p75 isoform of c-Myb (Oh and Reddy, 1999), MAZ (Bossone *et al.*, 1992), Elf-1 (Bassuk and Leiden, 1997), ZEB (Genetta *et al.*, 1994; Postigo and Dean, 1999), HES-1 (Sasai *et al.*, 1992), and SAF (Kim and Siu, 1999). DNA-binding motifs are shaded black. AD, activation domain. Pro/Ala, alternating proline- and alanine-rich regions. RD domains shown in ZEB demonstrated cell type-specific repression; the intervening region 2 did not show repression activity, but includes a conserved homeodomain (HD) (Postigo and Dean, 1999). Polyalanine tracts in MAZ are similar to repression domains of other factors, e.g., *D. melanogaster* Krüppel. Rb, retinoblastoma protein-interactive domain. RD, repression domain. HD, homeodomain. BHLH, basic helix-loop-helix domain. Pro, proline-rich domain. HTH, helix-turn-helix, or homeodomain-like DNA-binding domain of SAF.

enhancer except that it is essential for the induction of expression of a *CD4* minigene gene in DP thymocytes.

A region identified 24 kb upstream of the *CD4* promoter may include an additional (distal) enhancer (Wurster *et al.*, 1994). However, inclusion of this module did not result in appropriate expression of a *CD4* minigene in transgenic mice (Siu *et al.*, 1992). The distal enhancer may be more important for regulating the upstream *LAG-3* gene.

The CD4 *silencer.* To date, the *CD4* silencer is the only extensively characterized mammalian transcriptional silencer. Similar to the yeast mating-type silencer, the *CD4* silencer functions in a position- and orientation-independent manner and can decrease activities of heterologous promoters and enhancers (Sawada *et al.*, 1994; Siu *et al.*, 1994). The silencer is the critical module that represses *CD4* gene transcription during T-cell development. First, repression of *CD4* transcription in DN thymocytes is mediated by the *CD4* silencer. During the later transition to the DP stage, *CD4* silencer function ceases and expression is maintained by the thymocyte enhancer. The subsequent differentiation of DP to *CD4*$^{+}$ SP T_H cells is concurrent with activation of mature enhancer function after TCR-mediated selection, thus maintaining *CD4* gene transcription (Adlam *et al.*, 1997). Alternatively, CD8 SP T_C cells exhibit repression of mature enhancer activity by the silencer, and thus, lack *CD4* gene transcription. Therefore, the factors that mediate silencer function are directly linked to the processes that drive these important developmental transitions.

Three factor binding sites identified in the silencer are required for its function (Duncan *et al.*, 1996). These sites are recognized by HES-1, a target of the mammalian Notch pathway (Kim and Siu, 1998); c-Myb (Allen *et al.*, submitted); and SAF, a novel homeodomain-like transcription factor (Kim and Siu, 1999). SAF is especially interesting in that its subcellular compartmentalization appears to regulate *CD4* silencer function (see Section 3.4.7).

3.4 Transcription factors that control CD4 expression and T-cell development

Factors that mediate the control of *CD4* gene transcription are also critical for other aspects of T lymphocyte development. The most important of these factors are described below and in *Figure 6*.

c-Myb. c-Myb (*Figure 6*) has multiple critical roles during T cell development, because it binds to positive and negative regulatory elements in the *CD4* locus and within critical transcriptional control modules of many other T-cell-specific genes (Lipsick, 1996). c-Myb positively activates transcription by binding nucleotide sequences between –93 and –76 in the murine *CD4* promoter (Siu *et al.*, 1992; Nakayama *et al.*, 1993). Interestingly, c-Myb also contributes to the negative regulation of *CD4* by binding its silencer. How c-Myb exerts these apparently contradictory effects at different control modules is unknown.

Analysis of chimeric mice generated by introducing homozygous *c-Myb*$^{-/-}$ ES cells into *RAG-1*$^{-/-}$ blastocysts showed that the lack of c-Myb results in failure to develop mature macrophages and lymphocytes (Allen *et al.*, 1999). Using a panel of early stem cell and T-lineage markers, a block to further T-cell development was detected in an early CD4^{-}CD8^{-} DN population. Interestingly, the phenotype of the *c-Myb*$^{-/-}$ thymocyte population in these chimeric mice is very similar to that of multipotent stem cells that have just arrived in the thymus from the bone marrow. This population is capable of differentiating into T or B lymphocytes, NK cells, and dendritic cells and is believed to represent the last stage of stem cell development before commitment to the T-cell lineage (Wu *et al.*, 1991). These data suggest that c-Myb is required for the commitment of stem cells to the T-cell lineage. c-Myb may also play important roles at later stages of T-cell development. However, the failure of *c-Myb*$^{-/-}$ progenitors to mature past the early DN thymocyte stage obscures this phenotype in *RAG-1*$^{-/-}$ *c-Myb*$^{-/-}$ chimeric mice. Further genetic experiments with mice bearing conditionally deleting *c-Myb* genes are necessary to address this issue directly.

MAZ. The Myc-Associated Zinc finger factor (MAZ; *Figure 6*) was originally described as binding to the c-Myc promoter at a critical functional site essential for the control of transcriptional initiation and elongation (Bossone *et al.*, 1992). To date, however, the murine *CD4* promoter is the only T-cell-specific transcriptional control module recognized by MAZ (Duncan *et al.*, 1995). Although MAZ binds to a critical *CD4* promoter site, exogenously expressed MAZ is unable to transactivate the *CD4* promoter in transient transfection experiments. This observation supports the hypothesis that MAZ does not act directly as a transcriptional activator, but instead activates transcription by inducing a bend in the DNA helix to facilitate interactions between other promoter-binding factors (in the manner reported for LEF-1 binding to the TCRα enhancer; Giese *et al.*, 1995). Recent experiments have demonstrated that only one of the multiple MAZ consensus recognition sites in the *CD4* promoter is required for its function (Sarafova and Siu, 2000). Alteration of the spacing distance between the functional MAZ site and other sites renders the promoter nonfunctional, but the addition of a new MAZ site with proper phasing and spacing relative to other *CD4* promoter sites restores promoter function. These data indicate that the proper positioning of MAZ with respect to the other promoter-binding factors is critical for full promoter function, consistent with the hypothesis that MAZ plays an 'architectural' role in *CD4* promoter function.

Elf-1. The murine *CD4* promoter includes a consensus binding site for Ets family proteins, CAGGAAG, near the site of transcription initiation (*Figure 5a*). The Ets family comprises a large group of proteins with distinctive DNA-binding domains that feature a conserved winged HTH motif. One member of this family, Elf-1 (*Figure 6*), has been implicated as an important factor in early T-cell development (Bassuk *et al.*, 1998). Elf-1 is expressed at high levels in the thymus, binds the P4 site *in vitro*, and can activate transcription through this functionally important site (Sarafova and Siu, 1999b). The introduction of a mutation that favors DNA binding by Elf-1, and not Ets-1 (and other Ets proteins) did not affect *CD4* promoter function, suggesting that Elf-1 (or a factor with similar DNA-binding specificity) is indeed a regulator of *CD4* transcription.

The observation that Elf-1 is a putative regulator of *CD4* gene expression is especially interesting in light of the roles of Ets transcription factors in lymphocyte development, and by observations that Ets transcription factors are targets of the Ras/mitogen-activated protein kinase (MAPK) pathway (Wasylyk *et al.*, 1998). By analogy, Elf-1 (or another Ets family member) may integrate CD4 regulation into the Ras/MAPK pathway, which is important for thymic selection (Alberola-Ila *et al.*, 1995; Michie *et al.*, 1999; Sugawara *et al.*, 1999). One attractive hypothesis is that signaling from the Ras/MAPK pathway activates Elf-1 function, thus influencing the expression of CD4. Along these lines, it is interesting to note that overexpression of Elf-1 does not transactivate the *CD4* promoter in $CD4^-$ T cells, but does in $CD4^+$ T cells, indicating that the potential for Elf-1 function is different in T cells with different developmental phenotypes (Sarafova and Siu, 1999b). Further studies, including analysis of Elf-1-deficient mice, are necessary before a link between Elf-1, the Ras/MAPK pathway and *CD4* promoter function can be established with certainty.

E12/HEB. As described above (Section 2.2), the *E2A* gene encodes transcription factors of the bHLH family, including the alternatively spliced factor E12 (*Figure 2*). A second bHLH gene encodes Hela E-box binding protein, or HEB (*Figure 6*; also termed ME1). HEB is expressed in skeletal muscle and in B and T lymphocytes, and was identified as a partner for assembling heterodimers with E2A-encoded proteins (Hu *et al.*, 1992; Bain *et al.*, 1993). In E2A-deficient mice, detection of functional DNA-binding complexes that include HEB is greatly reduced (Rivera *et al.*, 2000).

The roles of bHLH factors in T-cell development have only recently been studied; indeed, the demonstration that an E12/HEB heterodimer binds to a critical functional site of the *CD4* mature enhancer was the first indication that such a role existed (Sawada and Littman, 1993). Targeted disruption of the *HEB* and the *E2A* genes blocks the maturation of early thymocytes into $CD4^+CD8^+$ DP cells (Bain *et al.*, 1997a). Heterozygous $E2A^{+/-}HEB^{+/-}$ mice also develop the DN to DP developmental block, indicating that this phenotype may not be the result of the loss of a specific bHLH protein, but rather the overall decrease of available bHLH proteins (Zhuang *et al.*, 1996). Similarly, overexpression of Id2, which functions as a dominant negative partner by heterodimerizing with these factors, also blocks early thymocyte development at this stage (Morrow *et al.*, 1999). In the absence of E47 expression, increased production of mature $CD4^+$ and $CD8^+$ cells was detected, suggesting increased positive selection (Bain *et al.*, 1999). Taken together, these genetic studies indicate that the bHLH proteins and Id proteins play an important role in T-cell development, but the details remain to be determined.

ZEB. In addition to roles in B cells (see Section 2.2.2), the zinc finger protein ZEB (*Figure 6*) has also been implicated as a regulator of CD4 expression (Brabletz *et al.*, 1999). ZEB binds the 5′-most E box of the mature (proximal) enhancer, which is also a target for DNA binding by HEB/E12. A model has been proposed for the reduction of enhancer activity by ZEB in $CD4^+$ SP T cells, but not in immature DP thymocytes. Interestingly, sequences in the amino-terminus of ZEB act to repress activator proteins bound at adjacent sites (Postigo and Dean, 1999). When their binding sites are linked to a reporter gene, ZEB can repress activation functions of c-Myb, Ets proteins, and at least one bHLH protein, TFE3, in T cells.

HES-1. The Notch signaling pathway plays a critical role in developmental cell-fate decisions in many different systems (Greenwald, 1998). In the mammalian immune system, the Notch pathway has been shown to influence developmental outcomes during early T-cell differentiation. Overexpression of an activated form of murine Notch1 biases T-cell development towards CD8 SP T cells, with implications for the control of CD4 gene transcription (Robey *et al.*, 1996). More recently, experiments have led to the proposal that Notch signaling is more important for promoting the survival of early $CD4^-CD8^-$ DN thymocytes and does not play a direct role in T cell selection processes (Deftos *et al.*, 1998; Radtke *et al.*, 1999; Tomita *et al.*, 1999). However, these data do not rule out an additional role for the Notch pathway during later stages of T cell development.

In *D. melanogaster*, genetic studies have determined that the *enhancer of split*, or *E(spl)*, locus encodes a nuclear effector of the Notch pathway that represses specific batteries of developmentally important genes. Using a variety of different biochemical and molecular techniques, it was determined that a murine ortholog of E(spl)m8 protein, Hairy/Enhancer of Split factor-1 (HES-1; *Figure 6*) binds a functional site in the *CD4* silencer (*Figure 5b*) to mediate silencer function (Kim and Siu, 1998). These observations suggest that HES-1 is a mediator of *CD4* silencing in response to Notch pathway signaling in differentiating thymocytes. However, HES-1 expression and E12/HEB DNA-binding activity are detected in both $CD4^+$ and $CD8^+$ mature T cells, indicating that the mechanism of silencer functional specificity is not likely to be mediated by the differential expression of HES-1 alone. Thus, although these data indicate that signaling from the Notch pathway may contribute to *CD4* silencer function, it is still unclear whether Notch signaling mediates the specificity of silencer function during the T-cell lineage decision process.

Silencer-associated factor (SAF). The critical control modules that mediate the specificity of *CD4* transcription during T-cell development have been identified, but until recently none of the factors that interact with these modules exhibited biochemical properties specific to T-cell subclasses. This requirement has now been met by a recently cloned, novel transcription factor termed silencer associated factor (SAF), which binds a critical functional site of the *CD4* silencer and localizes to different subcellular compartments in $CD4^+$ versus $CD8^+$ cells (Kim and Siu, 1999).

Using a yeast one-hybrid approach, SAF was cloned via its binding to the functional S3 site of the silencer (*Figure 5b*). SAF is a 14 kDa protein (*Figure 6*) that binds to each of two consensus CTGTGC recognition sites in the S3 region. SAF comprises an HTH DNA-binding motif similar to that of the homeodomain class of proteins. In addition to the DNA binding domain, SAF has a glutamine- and glutamic acid-rich domain reminiscent of the Sp1 class of protein–protein interaction domains. Deletion of this region inhibits the ability of SAF to alter transcription from an SAF-dependent promoter but does not affect its stability or DNA-binding activity (Kim and Sui, unpublished data). There are no obvious consensus nuclear localization signals; however, SAF contains a consensus nuclear export signal.

SAF protein is detected in T cells at all stages of differentiation. However, in immunofluorescence studies, specific antisera detected SAF in the cytoplasm of $CD4^+$ SP and $CD4^+CD8^+$ DP cells most intensely. Cytoplasmic staining correlates with the expression of CD4 in these two T-cell developmental subclasses. In contrast, SAF localizes to the nucleus in $CD4^-CD8^-$ DN and $CD8^-$ SP T cells. Therefore, SAF localization to the nucleus correlates with the silencing of CD4 transcription, suggesting that the developmental stage-specific nuclear transport of SAF is critical for mediating *CD4* silencer function.

How is SAF transport controlled in developing DP T cells? Although SAF does not have a consensus nuclear localization signal, its small size might permit it to migrate passively through nuclear pores. Post-translational modification(s) of SAF may control its subcellular localization, similar to other transcriptional regulators in T cells, e.g. NF-AT. Dephosphorylation of NF-AT initiates translocation of its cytoplasmic component to the nucleus (Rao *et al.*, 1997). Alternatively, SAF may bind another factor that sequesters it in the cytoplasm, similar to the mechanism used to regulate nuclear localization of NF-κB (reviewed in Karin, 1999). As SAF is a homeodomain-like protein, the *D. melanogaster* factor Extradenticle (EXD) is a potentially relevant example. EXD mediates cell fates during embryonic development, and its compartmentalization is mediated by binding to another homeodomain protein, Homothorax, which translocates the heterodimer to the nucleus (Mann and Chan, 1996; Rieckhof *et al.*, 1997; Pai *et al.*, 1998). As development progresses, the nuclear export factor crm1 binds to EXD and transports it from the nucleus (Abu-Shar *et al.*, 1999). Interestingly, SAF contains a consensus crm1 binding site; therefore, an ortholog of crm1 may mediate SAF transport during T-cell development. Experiments are currently being conducted to address these issues.

T-cell lineage commitment and the control of CD4 *gene expression.* In addition to Notch, other signaling pathways (e.g. the Ras/MAPK pathway) are known to be important for T-cell development and selection. However, it is unknown how activation of these pathways in DP thymocytes results in selective expression of CD4 versus CD8 in SP T cells. One difficulty is that molecular targets of these pathways have not been identified. Because the downregulation of CD4 expression is the defining marker for positive selection of CD8 SP T cells from DP precursors, it is likely that factors that mediate the repression of CD4 transcription during this process are linked to the other important signals that mediate thymopoiesis. In this regard, the differential subcellular compartmentalization of SAF in T cells of different developmental phenotypes suggests a paradigm for the control of CD4 gene expression and T-cell development.

4. Conclusions

Considerable progress has been made toward identifying nuclear proteins that control early events in B- and T-lineage development. As described in this review, early developmental decisions in both B and T cells are a function of lineage-specific transcription factors that collaborate to activate or repress specific genes at discrete stages of development. Clearly, these factors are likely targets for additional regulation, e.g., by post-translational modifications that further modify their activities *in vivo*. Perhaps the greatest challenge in the next few years is that of making sense of the extensive signal transduction pathways that impinge on factors that are themselves regulated at multiple levels. The availability of tools for these studies, including knockout mice, *in vivo* systems for genetic reconstitution of transcription factor function, and more conventional biochemical methods, provides exciting opportunities for understanding the genesis of these important cells. Determining the nature of these mechanisms is essential for understanding how the immune system is generated, how regulation of lymphocyte functions can fail, and for designing interventions for autoimmune disease and other types of immune dysfunction.

Acknowledgments

The authors wish to thank Dr. Lawrence Abraham for his critical comments and helpful discussions. J. Hagman wishes to thank Daniel Fitzsimmons and Boyd Jacobson for assistance with figures.

References

Abu-Shar, M., Ryoo, H.D. and Mann, R.S. (1999) Control of the nuclear localization of Extradenticle by competing nuclear import and export signals. *Genes Dev.* **13**: 935–945.

Adams, B., Dorfler, P., Aguzzi, A., Kozmik, Z., Urbanek, P., Maurer-Fogy, I. and Busslinger, M. (1992) Pax-5 encodes the transcription factor BSAP and is expressed in B lymphocytes, the developing CNS, and adult testis. *Genes Dev.* **6**: 1589–1607.

Adlam, M. and Siu, G. (2000) A thymocyte enhancer in the CD4 locus. *In preparation.*

Adlam, M., Duncan, D.D., Ng, D.K. and Siu, G. (1997) Positive selection induces CD4 promoter and enhancer function. *Int. Immunol.* **9**: 877–887.

Åkerblad, P. and Sigvardsson, M. (1999) Early B cell factor is an activator of the B lymphoid kinase promoter in early B cell development. *J. Immunol.* **163**: 5453–5461.

Åkerblad, P., Sigvardsson, M. and Leanderson, T. (1996) Early B-cell factor (EBF) down-regulates immunoglobulin heavy chain intron enhancer function in a plasmacytoma cell line. *Scan. J. Immunol.* **44**: 89–92.

Åkerblad, P., Rosberg, M., Leanderson, T. and Sigvardsson, M. (1999) The B29 (immunoglobulin β-chain) gene is a genetic target for early B cell factor. *Mol. Cell. Biol.* **19**: 392–401.

Alberola-Ila, J., Forbush, K.A., Seger, R., Krebs, E.G. and Perlmutter, R.M. (1995) Selective requirement for MAP kinase activation in thymocyte differentiation. *Nature* **373**: 620–623.

Allen, R.D., 3rd, Bender, T.P. and Siu, G. (1999) c-Myb is essential for early T cell development. *Genes Dev.* **13**: 1073–1078.

Allen, R.D., Sarafova, S.D. and Siu, G. (Submitted). Positive and negative roles for c-Myb in the regulation of CD4 gene expression.

Ansari-Lari, M.A., Muzny, D.M., Lu, J., Lu, F., Lilley, C.E., Spanos, S., Malley, T. and Gibbs, R.A. (1996) A gene-rich cluster between the CD4 and triosephosphate isomerase genes at human chromosome 12p13. *Genome Res.* **6**: 314–326.

Aronheim, A., Shiran, R., Rosen, A. and Walker, M.D. (1993) The E2A gene product contains two separable and functionally distinct activation domains. *Proc. Natl Acad. Sci. USA* **90**: 8063–8067.

Bain, G., Gruenwald, S. and Murre, C. (1993) E2A and E2–2 are subunits of B-cell-specific E2-box DNA-binding proteins. *Mol. Cell. Biol.* **13**: 3522–3529.

Bain, G., Maandag, E.C., Izon, D.J., Amsen, D., Kruisbeek, A.M., Weintraub, B.C., Krop, I., Schlissel, M.S., Feeney, A.J., van Roon, M., van der Valk, M., te Reile, H.P.J., Berns, A. and Murre, C. (1994) E2A proteins are required for proper B cell development and initiation of immunoglobulin gene rearrangements. *Cell* **79**: 885–892.

Bain, G., Engel, I., Maandag, E.C., te Reile, H.P.J., Voland, J.R., Sharp, L.L., Chun, J., Huey, B., Pinkel, D. and Murre, C. (1997a) E2A deficiency leads to abnormalities in αβ T-cell development and to rapid development of T-cell lymphomas. *Mol. Cell. Biol.* **17**: 4782–4791.

Bain, G., Maandag, E.C., te Riele, H.P., Feeney, A.J., Sheehy, A., Schlissel, M., Shinton, S.A., Hardy, R.R. and Murre, C. (1997b) Both E12 and E47 allow commitment to the B cell lineage. *Immunity* **6**: 145–154.

Bain, G., Quong, M.W., Soloff, R.S., Hedrick, S.M. and Murre, C. (1999) Thymocyte maturation is regulated by the activity of the helix-loop-helix protein, E47. *J. Exp. Med.* **190**: 1605–1616.

Bannister, A.J. and Kouzarides, T. (1996) The CBP co-activator is a histone acetyltransferase. *Nature* **384**: 641–643.

Barberis, A., Widenhorn, K., Vitelli, L. and Busslinger, M. (1990) A novel B-cell lineage-specific transcription factor present at early but not late states of differentiation. *Genes Dev.* **22**: 37–43.

Bassuk, A.G. and Leiden, J.M. (1997) The role of the Ets transcription factors in the development and function of the mammalian immune system. *Adv. Immunol.* **64**: 65–104.

Bassuk, A.G., Barton, K.P., Anandappa, R.T., Lu, M.M. and Leiden, J.M. (1998) Expression pattern of the Ets-related transcription factor Elf-1. *Mol. Med.* **4**: 392–401.

Benezra, R. (1994). An intermolecular disulfide bond stabilizes E2A homodimers and is required for DNA binding at physiological temperatures. *Cell* **79**: 1057–1067.

Bevan, M.J., Hogquist, K.A. and Jameson, S.C. (1994) Selecting the T cell receptor repertoire. *Science* **264**: 796–797.

Bossone, S.A., Asselin, C., Patel, A.J. and Marcu, K.B. (1992) MAZ, a zinc finger, binds to a c-*myc* and C2 gene sequences regulating transcriptional initiation and termination. *Proc. Natl Acad. Sci. USA* **89**: 7452–7456.

Brabletz, T., Jung, A., Hlubeck, F., Lohberg, C., Meiler, J., Suchy, U. and Kirchner, T. (1999) Negative regulation of CD4 expression in T cells by the transcriptional repressor ZEB. *Int. Immunol.* **11**: 1701–1708.

Busslinger, M. and Urbanek, P. (1995) The role of BSAP (Pax-5) in B-cell development. *Curr. Opin. Genet. Dev.* **5**: 595–601.

Campbell, K.S. and Cambier, J.C. (1990) B lymphocyte antigen receptors (mIg) are non-covalently associated with a disulfide linked, inducibly phosphorylated glycoprotein complex. *EMBO J.* **9**: 441–448.

Choi, J.K., Shen, C.-P., Radomska, H.S., Eckhardt, L.A. and Kadesch, T. (1996) E47 activates the Ig-heavy chain and Tdt loci in non-B cells. *EMBO J.* **15**: 5014–5021.

Czerny, T. and Busslinger, M. (1995) DNA-binding and transactivation properties of Pax-6: three amino acids in the paired domain are responsible for the different sequence recogntion of Pax-6 and BSAP (Pax-5). *Mol. Cell. Biol.* **15**: 2858–2871.

Czerny, T., Schaffner, G. and Busslinger, M. (1993) DNA sequence recognition by Pax proteins: bipartite structure of the paired domain and its binding site. *Genes Dev.* **7**: 2048–2061.

Deftos, M.L., He, Y.W., Ojala, E.W. and Bevan, M.J. (1998) Correlating notch signaling with thymocyte maturation. *Immunity* **9**: 777–786.

Dörfler, P. and Busslinger, M. (1996) C-terminal activating and inhibitory domains determine the transactivation potential of BSAP (Pax-5), Pax-2, and Pax-8. *EMBO J.* **15**: 1971–1982.

Duncan, D.D., Stupakoff, A., Hedrick, S.M., Marcu, K.B. and Siu, G. (1995) A Myc-Associated Zinc-finger protein (MAZ) binding site is one of four important functional regions in the CD4 promoter. *Mol. Cell. Biol.* **15**: 3179–3186.

Duncan, D.D., Adlam, M. and Siu, G. (1996) Asymmetric redundancy in CD4 silencer function. *Immunity* **4**: 301–311.

Eberhard, D. and Busslinger, M. (1999) The partial homeodomain of the transcription factor Pax-5(BSAP) is an interaction motif for the retinoblastoma and TATA-binding proteins. *Cancer Res.* **59(7 Suppl)**: 1716s–1724s.

Eberhard, D., Jiménez, G., Heavey, B. and Busslinger, M. (2000) Transcriptional repression by Pax5 (BSAP) through interaction with corepressors of the Groucho family. *EMBO J.* **19**: 2292–2303.

Eckner, R., Yao, T.-P., Oldread, E. and Livingston, D.M. (1996) Interaction and functional collaboration of p300/CBP and bHLH proteins in muscle and B-cell differentiation. *Genes Dev.* **10**: 2478–2490.

Ellenberger, T., Fass, D., Arnaud, M. and Harrison, S.C. (1994) Crystal structure of transcription factor E47: E box recognition by a basic region helix-loop-helix dimer. *Genes Dev.* **8**: 970–980.

Ellmeier, W., Sawada, S. and Littman, D.R. (1999) The regulation of CD4 and CD8 coreceptor gene expression during T cell development. *Ann. Rev. Immunol.* **17**: 523–554.

Feldhaus, A., Mbangkollo, D., Arvin, K., Klug, C. and Singh, H. (1992) BlyF, a novel cell-type- and stage-specific regulator of the B-lymphocyte gene *mb*-1. *Mol. Cell. Biol.* **12**: 1126–1133.

Fink, P.J. and Bevan, M.J. (1995) Positive selection of thymocytes. *Adv. Immunol.* **59**: 99–133.

Fitzsimmons, D., Hodsdon, W., Wheat, W., Maira, S.-M., Wasylyk, B. and Hagman, J. (1996) Pax-5 (BSAP) recruits Ets proto-oncogene family proteins to form functional ternary complexes on a B-cell-specific promoter. *Genes Dev.* **10**: 2198–2211.

Fowlkes, B.J. and Pardoll, D. (1989) Molecular and cellular events of T cell development. *Adv. Immunol.* **44**: 207–264.

Franklin, A.J., Jetton, T.L., Shelton, K.D. and Magnuson, M.A. (1994) BZP, a novel serum-responsive zinc finger protein that inhibits transcription. *Mol. Cell. Biol.* **14**: 6773–6788.

Fuller, K. and Storb, U. (1997) Identification and characterization of the murine Ragl promoter. *Mol. Immunol.* **34**: 939–954.

Garel, S., Marin, F., Mattei, M.G., Vesque, C., Vincent, A. and Charnay, P. (1997) Family of Ebf/Olf-1-related genes potentially involved in neuronal differentiation and regional specification in the central nervous system. *Dev. Dyn.* **210**: 191–205.

Genetta, T., Ruezinsky, D. and Kadesch, T. (1994) Displacement of an E-box-binding repressor by basic helix-loop-helix proteins: implications for B-cell specificity of the immunoglobulin heavy-chain enhancer. *Mol. Cell. Biol.* **14**: 6153–63.

Giese, K., Kingsley, C., Kirshner, J.R. and Grosschedl, R. (1995) Assembly and function of a TCRα enhancer complex is dependent on LEF-1-induced DNA bending and multiple protein-protein interactions. *Genes Dev.* **9**: 995–1008.

Gisler, R., Åkerblad, P. and Sigvardsson, M. (1999) A human early B-cell factor-like protein participates in the regulation of the human CD19 promoter. *Mol. Immunol.* **36**: 1067–1077.

Goldfarb, A.N., Flores, J.P. and Lewandowska, K. (1996) Involvement of the E2A basic helix-loop-helix protein in immunoglobulin heavy chain class switching. *Mol. Immunol.* **33**: 947–956.

Greenwald, I. (1998) LIN-12/Notch signaling: lessons from worms to flies. *Genes Dev.* **12**: 1751–1762.

Guidos, C.J. (1996) Positive selection of CD4+ and CD8+ T cells. *Curr. Opin. Immunol.* **8**: 225–232.

Hagman, J., Travis, A. and Grosschedl, R. (1991) A novel lineage-specific nuclear factor regulates mb-1 gene transcription at the early stages of B cell differentiation. *EMBO J.* **10**: 3409–3417.

Hagman, J., Belanger, C., Travis, A., Turck, C.W. and Grosschedl, R. (1993) Cloning and functional characterization of early B-cell factor, a regulator of lymphocyte-specific gene expression. *Genes Dev.* **7**: 760–73.

Hagman, J., Gutch, M.J., Lin, H. and Grosschedl, R. (1995) EBF contains a novel zinc coordination motif and multiple dimerization and transcriptional activation domains. *EMBO J.* **14**: 2907–2916.

Hagman, J., Wheat, W., Fitzsimmons, D., Hodsdon, W., Negri, J. and Dizon, F. (2000) Pax-5/BSAP: regulator of specific gene expression and differentiation in B lymphocytes. *Curr. Top. Micro. Immunol.* **245**: 169–194.

Hanna, Z., Simard, C., LaPerriere, A. and Jolicoeur, P. (1994) Specific expression of the human CD4 gene in mature $CD4^{+}CD8^{-}$ and immature $CD4^{+}CD8^{+}$ T cells and in macrophages of transgenic mice. *Mol. Cell. Biol.* **14**: 1084–1094.

Hombach, J., Leclercq, L., Radbruch, A., Rajewsky, K. and Reth, M. (1988) A novel 34-kd protein co-isolated with the IgM molecule in surface IgM-expressing cells. *EMBO J.* **7**: 3451–3456.

Hombach, J., Tsubata, T., Leclerq, L., Stappert, H. and Reth, M. (1990) Molecular components of the B-cell antigen receptor complex of the IgM class. *Nature* **343**: 760–762.

Hu, J.S., Olson, E.N. and Kingston, R.E. (1992) HEB, a helix-loop-helix protein related to E2A and ITF2 that can modulate the DNA-binding ability of myogenic regulatory factors. *Mol. Cell. Biol.* **12**: 1031–1042.

Ikeda, K. and Kawakami, K. (1995) DNA binding through distinct domains of zinc-finger-homeodomain protein AREB6 has different effects on gene transcription. *Eur. J. Biochem.* **233**: 73–82.

Kadesch, T. (1992) Helix-loop-helix proteins in the regulation of immunoglobulin gene transcription. *Immunol. Today* **13**: 31–36.

Karin, M. (1999) How NF-κB is activated: the role of the IκB kinase (IKK) complex. *Oncogene* **18**: 6867–6874.

Kee, B.L. and Murre, C. (1998) Induction of early B cell factor (EBF) and multiple B lineage genes by the basic helix-loop-helix transcription factor E12. *J. Exp. Med.* **188**: 699–713.

Kim, H.K. and Siu, G. (1998) The Notch pathway intermediate HES-1 silences CD4 gene expression. *Mol. Cell. Biol.* **18**: 7166–7175.

Kim, W.W.S. and Siu, G. (1999) Subclass-specific nuclear localization of a novel CD4 silencer-binding factor. *in press.*

Kozmik, Z., Wang, S., Dorfler, P., Adams, B. and Busslinger, M. (1992) The promoter of the CD19 gene is a target for the B-cell-specific transcription factor BSAP. *Mol. Cell. Biol.* **12**: 2662–2672.

Libermann, T.A., Pan, Z., Akbarali, Y., Hetherington, C.J., Boltax, J., Yergeau, D.A. and Zhang, D.E. (1999) AML1 (CBFα2) cooperates with B cell-specific activating protein (BSAP/PAX5) in activation of the B cell-specific BLK gene promoter. *J. Biol. Chem.* **274**: 24671–24676.

Lin, H. and Grosschedl, R. (1995) Failure of B-cell differentiation in mice lacking the transcription factor EBF. *Nature* **376**: 263–7.

Lipsick, J.S. (1996) One billion years of Myb. *Oncogene* **13**: 223–235.

Lo, K., Landau, N.R. and Smale, S.T. (1991) LyF-1, a transcriptional regulator that interacts with a novel class of promoters for lymphocyte-specific genes. *Mol. Cell. Biol.* **11**: 5229–5243.

Mann, R.S. and Chan, S.K. (1996) Extra specificity from extradenticle: the partnership between HOX and PBX/EXD homeodomain proteins. *Trends Genet* **12**: 258–262.

Mårtensson, A. and Mårtensson, I.L. (1997) Early B cell factor binds to a site critical for lambda5 core enhancer activity. *Eur. J. Immunol.* **27**: 315–320.

Massari, M.E. and Murre, C. (2000) Helix-loop-helix proteins: regulators of transcription in eucaryotic organisms. *Mol. Cell. Biol.* **20**: 429–440.

Massari, M., Jennings, P. and Murre, C. (1996) The AD1 transactivation domain of E2A contains a highly conserved helix which is required for its activity in both *Saccharomyces cerevisiae* and mammalian cells. *Mol. Cell. Biol.* **16**: 121–129.

Massari, M.E., Grant, P.A., Pray-Grant, M.G., Berger, S.L., Workman, J.L. and Murre, C. (1999) A conserved motif present in a class of helix-loop-helix proteins activates transcription by direct recruitment of the SAGA complex. *Mol. Cell* **4**: 63–73.

Michie, A.M., Trop, S., Wiest, D.L. and Zúniga-Pflücker, J.C. (1999) Extracellular signal-regulated kinase (ERK) activation by the pre-T cell receptor in developing thymocytes *in vivo*. *J. Exp. Med.* **190**: 1647–1655.

Morrow, M.A., Mayer, E.W., Perez, C.A., Adlam, M. and Siu, G. (1999) Overexpression of the helix-loop-helix protein Id2 blocks T cell development at multiple stages. *Submitted.*

Murre, C., McCaw, P.S. and Baltimore, D. (1989) A new DNA binding and dimerization motif in immunoglobulin enhancer binding- *daughterless*, *MyoD*, and *myc* proteins. *Cell* **56**: 777–783.

Nakayama, K., Yamamoto, R., Ishii, S. and Nakauchi, H. (1993) Binding of c-Myb to the core sequence of the CD4 promoter. *Int. Immunol.* 5: 817–824.

Norton, J.D., Deed, R.W., Craggs, G. and Sablitsky, F. (1998) Id helix-loop-helix proteins in cell growth and differentiation. *Trends Cell. Biol.* **8**: 58–65.

Nutt, S.L., Urbánek, P., Rolink, A. and Busslinger, M. (1997) Essential functions of Pax5 (BSAP) in pro-B cell development: difference between fetal and adult B lymphopoiesis and reduced *V-to-DJ* redcombination at the *IgH* locus. *Genes Develop.* **11**: 476–491.

Nutt, S.L., Morrison, A.M., Dörfler, P., Rolink, A. and Busslinger, M. (1998) Identification of BSAP (Pax-5) target genes in early B-cell development by loss- and gain-of-function experiments. *EMBO J.* **17**: 2319–2333.

Nutt, S.L., Heavy, B., Rolink, A. and Busslinger, M. (1999) Commitment to the B-lymphoid lineage depends on the transcription factor Pax5. *Nature* **401**: 556–562.

Ogryzko, V.V., Schiltz, R.L., Russanova, V., Howard, B.H. and Nakatani, Y. (1996) The transcriptional coactivators p300 and CBP are histone acetyltransferases. *Cell* **87**: 953–959.

Oh, I.-H. and Reddy, E.P. (1999) The *myb* family in cell growth, differentiation, and apoptosis. *Oncogene* **18**: 3017–3033.

Okabe, T., Watanabe, T. and Kudo, A. (1992) A pre-B and B cell-specific DNA-binding protein, EBB-1 which binds to the promoter of the *VpreB1* gene. *EMBO J.* **12**: 2753–2772.

Ordentlich, P., Lin, A., Shen, C.-P., Blaumueller, C., Matsuno, K., Artvanis-Tsakonas, S. and Kadesch, T. (1998) Notch inhibition of E47 supports the existence of a novel signaling pathway. *Mol. Cell. Biol.* **18**: 2230–2239.

O'Riordan, M. and Grosschedl, R. (1999) Coordinate regulation of B cell differentiation by the transcription factors EBF and E2A. *Immunity* **11**: 21–31.

Osborne, B. and Miele, L. (1999) Notch and the immune system. *Immunity* **11**: 653–663.

Pai, C.-Y., Kuo, T.-S., Jaw, T.J., Kurant, E., Chen C.-T., Bessarab, D.A., Salzberg, A. and Sun, Y.H. (1998) The Homothorax homeoprotein activates the nuclear localization of another homeoprotein, Extradenticle, and suppresses eye development in Drosophila. *Genes Dev.* **12**: 435–446.

Park, S.T. and Sun, X.-H. (1998) The Tal1 oncoprotein inhibits E47-mediated transcription. Mechanism of inhibition. *J. Biol. Chem.* **273**: 7030–7037.

Perlmutter, R.M. (1993) *In vivo* dissection of lymphocyte signaling pathways. *Clin. Immunol. Immunopath.* **67**: S44–S49.

Persson, C., Martensson, A. and Martensson, I.L. (1998) Identification of a tissue- and differentiation stage-specific enhancer of the VpreB1 gene. *Eur. J. Immunol.* **28**: 787–798.

Postigo, A.A. and Dean, D.C. (1999) ZEB represses transcription through interaction with the corepressor CtBP. *Proc. Natl Acad. Sci. USA* **96**: 6683–6688.

Pui, J.C., Allman, D., Xu, L., DeRocco, S., Karnell, F.G., Bakkour, S., Lee, J.Y., Kadesch, T., Hardy, R.R., Aster, J.C. and Pear, W.S. (1999) Notch1 expression in early lymphopoiesis influences B versus T lineage determination. *Immunity* **11**: 299–308.

Quong, M.W., Massari, M.E., Zwart, R. and Murre, C. (1993) A new transcriptional-activation motif restricted to a class of helix-loop-helix proteins is functionally conserved in both yeast and mammalian cells. *Mol. Cell. Biol.* **13**: 792–800.

Quong, M.W., Harris, D.P., Swain, S.L. and Murre, C. (1999) E2A activity is induced during B-cell activation to promote immunoglobulin class switch recombination. *EMBO J.* **18**: 6307–6318.

Radtke, F., Wilson, A., Stark, G., Bauer, M., van Meerwijk, J., Robson MacDonald, H. and Aguet, M. (1999) Deficient T cell fate specification in mice with an induced inactivation of *Notch1*. *Immunity* **10**: 547–558.

Rao, A., Luo, C. and Hogan, P.G. (1997) Transcription factors of the NFAT family: regulation and function. *Ann. Rev. Immunol.* **15**: 707–747.

Rieckhof, G., Casares, F., Ryoo, H.D., Abu-Shaar, M. and Mann, R.S. (1997) Nuclear translocation of Extradenticle requires homothorax, which encodes an Extradenticle-related homeodomain protein. *Cell* **91**: 171–183.

Rivera, R.R., Johns, C.P., Quan, J., Johnson, R.S. and Murre, C. (2000) Thymocyte selection is regulated by the helix-loop-helix inhibitor protein, Id3. *Immunity* **12**: 17–26.

Robey, E., Chang, D., Itano, A., Cado, D., Alexander, H., Lans, D., Weinmaster, G. and Salmon, P. (1996) An activated from of Notch influences the choice between CD4 and CD8 T cell lineages. *Cell* **87**: 483–492.

Rolink, A., Grawunder, U., Winkler, T.H., Karasuyama, H. and Melchers, F. (1994) IL-2 receptor α chain (CD25, TAC) expresion defines a crucial stage in pre-B cell development. *Int. Immunol.* **6**: 1257–1264.

Rolink, A.G., Nutt, S.L., Melchers, F. and Busslinger, M. (1999) Long-term *in vivo* reconstitution of T-cell development by Pax5-deficient B-cell progenitors. *Nature* **401**: 603–606.

Romanow, W.J., Langerak, A.W., Goebel, P., Wolvers-Tettero, I.L.M., van Dongen, J.J.M., Feeney, A.J. and Murre, C. (2000) E2A and EBF act in synergy with the V(D)J recombinase to generate a diverse immunoglobulin repertoire in nonlymphoid cells. *Mol. Cell* **5**: 343–353.

Salmon, P., Giovane, A., Wasylyk, B. and Klatzmann, D. (1993) Characterization of the human CD4 gene promoter: Transcription from the CD4 gene core promoter is tissue-specific and is activated by Ets proteins. *Proc. Natl. Acad. Sci. USA* **90**: 7739–7743.

Salmon, P., Boyer, O., Lores, P., Jami, J. and Klatzmann, D. (1996) Characterization of an intronless CD4 minigene expressed in mature CD4 and CD8 T cells, but not expressed in immature thymocytes. *J. Immunol.* **156**: 1873–1879.

Sarafova, S.D. and Siu, G. (1999a) Control of CD4 gene expression: connecting signals to outcomes in T cell development. *Braz. J. Med. Biol. Res.* **32**: 785–803.

Sarafova, S.D. and Siu, G. (1999b) A potential role for Elf-1 in CD4 promoter function. *J. Biol. Chem.* **274**: 16126–16134.

Sarafova, S.D. and Siu, G. (2000) Precise arrangement of factor-binding sites is required for murine CD4 promotor function. *Nucleic Acids Res.* **28**: 2664–2671.

Sasai, Y., Kageyama, R., Tagawa, Y., Shigemoto, R. and Nakanishi, S. (1992) Two mammalian helix-loop-helix factors structurally related to Drosophila hairy and Enhancer of split. *Genes Dev.* **6**: 2620–2634.

Sawada, S. and Littman, D.R. (1991) Identification and characterization of a T-cell-specific enhancer adjacent to the murine CD4 gene. *Mol. Cell. Biol.* **11**: 5506–5515.

Sawada, S. and Littman, D.R. (1993) A heterodimer of HEB and an E12-related protein interacts with the CD4 enhancer and regulates its activity in T-cell lines. *Mol. Cell. Biol.* **13**: 5620–5628.

Sawada, S., Scarborough, J.D., Killeen, N. and Littman, D.R. (1994) A lineage-specific transcriptional silencer regulates CD4 gene expression during T lymphocyte development. *Cell* **77**: 917–929.

Schlissel, M., Voronova, A. and Baltimore, D. (1991) Helix-loop-helix transcription factor E47 activates germ-line Ig heavy-chain gene transcription and rearrangement in a pre-T-cell line. *Genes Dev.* **5**: 1367–1376.

Sekido, R., Takagi, T., Okanami, M., Moribe, H., Yamamura, M., Higashi, Y. and Kondoh, H. (1996) Organization of the gene encoding transcriptional repressor δEF1 and cross-species conservation of its domains. *Gene* **173**: 227–232.

Shen, C.-P. and Kadesch, T. (1995) B-cell-specific DNA binding by an E47 homodimer. *Mol. Cell. Biol.* **15**: 4518–4524.

Sigvardsson, M. (2000) Overlapping expression of Early B cell Factor and basic helix-loop-helix proteins as a mechanism to dictate B lineage specific activity of the λ5 promoter. *Mol. Cell. Biol.* **20**: 3640–3654.

Sigvardsson, M., Akerblad, P. and Leanderson, T. (1996) Early B cell factor interacts with a subset of kappa promoters. *J. Immunol.* **156**: 3788–3796.

Sigvardsson, M., O'Riordan, M. and Grosschedl, R. (1997) EBF and E47 collaborate to induce expression of the endogenous immunoglobulin surrogate light chain genes. *Immunity* **7**: 25–36.

Siu, G., Wurster, A.L., Lipsick, J.S. and Hedrick, S.M. (1992) Expression of the CD4 gene requires a Myb transcription factor. *Mol. Cell. Biol* **12**: 1592–1604.

Siu, G., Wurster, A.L., Duncan, D.D., Soliman, T.M. and Hedrick, S.M. (1994) A transcriptional silencer controls the developmental expression of the CD4 gene. *EMBO J.* **13**: 3570–3579.

Sloan, S.R., Shen, C.-P., McCarrick-Walmsley, R. and Kadeschm, T. (1996) Phosphorylation of E47 as a potential determinant of B-cell-specific activity. *Mol. Cell. Biol.* **16**: 6900–6908.

Sugawara, T., Moriguchi, T., Nishida, E. and Takahama, Y. (1999) Differential roles of ERK and p38 MAP kinase pathways in positive and negative selection of T lymphocytes. *Immunity* **9**: 565–574.

Sun, X.-H. (1994) Constitutive expression of the *Id1* gene impairs mouse B cell development. *Cell* **79**: 893–900.

Sun, X.-H., Copeland, N.G., Jenkins, N.A. and Baltimore, D. (1991) Id proteins Idl and Id2 selectively inhibit DNA binding by one class of helix-loop-helix proteins. *Mol. Cell. Biol.* **11**: 5603–5611.

Tamir, I. and Cambier, J.C. (1998) Antigen receptor signaling: integration of protein tyrosine kinase functions. *Oncogene* **17**: 1353–1364.

Tomita, K., Hattori, M., Nakamura, E., Nakanishi, S., Minato, N. and Kageyama, R. (1999) The bHLH gene Hesl is essential for expansion of early T cell precursors. *Genes Dev.* **13**: 1203–1210.

Travis, A., Hagman, J. and Grosschedl, R. (1991) Heterogeneously initiated transcription from the pre-B- and B-cell-specific *mb*-1 promoter: analysis of the requirement for upstream factor-binding sites and initiation site sequences. *Mol. Cell. Biol.* **11**: 5756–5755.

Travis, A., Hagman, J., Hwang, L. and Grosschedl, R. (1993) Purification of early-B-cell factor and characterization of its DNA- binding specificity. *Mol. Cell. Biol.* **13**: 3392–3400.

Turner, J. and Crossley, M. (1998) Cloning and characterization of mCtBP2, a corepressor that associates with basic Kruppel-like factor and other mammalian transcriptional regulators. *EMBO J.* **17**: 5129–5140.

Uematsu, Y., Donda, A. and De Libero, G. (1997) Thymocytes control the CD4 gene differently from mature T lymphocytes. *Int. Immunol.* **9**: 179–187.

Urbánek, P., Wang, Z.-Q., Fetka, I., Wagner, E.F. and Busslinger, M. (1994) Complete block of early B cell differentiation and altered patterning of the posterior midbrain in mice lacking Pax5/BSAP. *Cell* **79**: 901–912.

Wang, M.M. and Reed, R.R. (1993) Molecular cloning of the olfactory neuronal transcription factor OLF-1 by genetic selection in yeast. *Nature* **364**: 121–126.

Wang, S.S., Tsai, R.Y.L. and Reed, R.R. (1997) The characterization of the Olf-1/EBF-like HLH transcription factor family: implications in olfactory gene regulation and neuronal development. *J. Neurosci.* **17**: 4149–4158.

Wasylyk, B., Hagman, J. and Gutierrez-Hartmann, A. (1998) Ets transcription factors: nuclear effectors of the Ras-MAP-kinase signaling pathway. *Trends Bioc. Sci.* **23**: 213–216.

Wheat, W., Fitzsimmons, D., Lennox, H., Krautkramer, S.R., Gentile, L.N., McIntosh, L.P. and Hagman, J. (1999) The highly conserved β-hairpin of the paired DNA-binding domain is required for the assembly of Pax: Ets ternary complexes. *Mol. Cell. Biol.* **19**: 2231–2241.

Wilson, R.B., Kiledjian, M., Shen, C.-P., Benezra, R., Zwollo, P., Dymecki, S.M., Desiderio, S.V. and Kadesch, T. (1991) Repression of immunoglobulin enhancers by the helix-loop-helix protein Id: implications for B-lymphoid-cell development. *Mol. Cell. Biol.* **11**: 6185–6191.

Wu, L., Antica, M., Johnson, G.R., Scollay, R. and Shortman, K. (1991) Developmental potential of the earliest precursor cells from the adult mouse thymus. *J. Exp. Med.* **174**: 1617–1627.

Wurster, A.L., Siu, G., Leiden, J. and Hedrick, S.M. (1994) Elf-1 binds to a critical element in a second CD4 enhancer. *Mol. Cell. Biol.* **14**: 6452–6463.

Xu, W., Rould, M.A., Jun, S., Desplan, C. and Pabo, C.O. (1995) Crystal structure of a paired domain-DNA complex at 2.5 – resolution reveals structural basis for Pax developmental mutations. *Cell* **80**: 639–650.

Xu, H.E., Rould, M.A., Xu, W., Epstein, J.A., Maas, R.L. and Pabo, C.O. (1999) Crystal structure of the human Pax6 paired domain-DNA complex reveals specific roles for the linker region and carboxy-terminal subdomain in DNA binding. *Genes Dev.* **13**: 1263–1275.

Yan, W., Young, A.Z., Soares, V.C., Kelley, R., Benezra, R. and Zhuang, Y. (1997) High incidence of T-cell tumors in E2A-null mice and E2A/Id1 double knockout mice. *Mol. Cell. Biol.* **17**: 7317–7327.

Yang, J., Glozak, M.A. and Blomberg, B.B. (1995) Identification and localization of a developmental stage-specific promoter activity from the murine λ5 gene. *J. Immunol.* **155**: 2498–2514.

Yang, X.-J., Ogryzko, V.V., Nishikawa, J., Howard, B.H. and Nakatani, Y. (1996) A p300/CBP-associated factor that competes with the adenoviral oncoprotein E1A. *Nature* **382**: 319–324.

Yokota, Y., Mansouri, A., Mori, S., Sugawara, S., Adachi, S., Nishikawa, S. and Gruss, P. (1999) Development of peripheral lymphoid organs and natural killer cells depends on the helix-loop-helix inhibitor Id2. *Nature* **397**: 702–706.

Zhao-Emonet, J.C., Boyer, O., Cohen, J.L. and Klatzmann, D. (1998) Deletional and mutational analyses of the human CD4 promoter: characterization of a minimal tissue-specific promoter. *Biochim. Biophys. Acta* **1442**: 109–119.

Zhuang, Y., Soriano, P. and Weintraub, H. (1994) The helix-loop-helix gene E2A is required for B cell formation. *Cell* **79**: 875–884.

Zhuang, Y., Cheng, P. and Weintraub, H. (1996) B-lymphocyte development is regulated by the combined dosage of three basic helix-loop-helix genes, E2A, E2–2, and HEB. *Mol. Cell. Biol.* **16**: 2898–2905.

Zwollo, P., Arrieta, H., Ede, K., Molinder, K., Desiderio, S. and Pollock, R. (1997) The *Pax-5* gene is alternatively spliced during B-cell development. *J. Biol. Chem.* **272**: 10160–10168.

8

Transcriptional integration of hormone and metabolic signals by nuclear receptors

Arndt Benecke, Claudine Gaudon and Hinrich Gronemeyer

1. Introduction

Multicellular organisms require specific intercellular communication to properly organize the complex body plan during embryogenesis and maintain the physiological properties and functions during the entire life span. While growth factors, neurotransmitters and peptide hormones bind to membrane receptors thereby inducing the activity of intracellular kinase cascades or the JAK-STAT/Smad signaling pathways, other small hydrophobic signaling molecules such as steroid hormones, certain vitamins and metabolic intermediates enter the target cells and bind to cognate members of a large family of nuclear receptors. Nuclear receptors are of major importance for intercellular signaling in the animal, as they converge different intra- and extracellular signals on the regulation of genetic programs. They are transcription factors that, in addition to regulating cognate gene programs:

(i) respond directly through physical association with a large variety of hormonal and metabolic signals;
(ii) integrate diverse signaling pathways as they themselves are targets of post-translational modifications;
(iii) regulate the activities of other major signaling cascades (commonly referred to as 'signal transduction crosstalk').

The genetic programs that they establish or modify affect virtually all aspects of the life of multicellular organism, covering such diverse aspects as embryogenesis, homeostasis, reproduction, cell growth or death. Their gene regulatory power and selectivity has prompted intense research on these key factors, which is now starting to decipher the complex network of molecular events that account for their transcription regulatory capacity. The study of these molecular processes also sheds light on general

Transcription Factors, edited by J. Locker.

mechanisms of transcription regulation, and it will be a future challenge to uncover the molecular rules that define spatial and temporal control of gene expression.

2. The family and its ligands

To date, 65 different nuclear receptors have been identified throughout the animal kingdom ranking from nematodes to man (*Figure 1*). They constitute a family of transcription factors that shares a modular structure of five to six conserved domains encoding specific functions (*Figure 2*) (Gronemeyer and Laudet, 1995). The most prominent distinction that they have from other transcription factors is their capacity to specifically bind small hydrophobic molecules. These ligands constitute regulatory signals, which, after binding, change the transcriptional activity of the corresponding nuclear receptor. For some time a distinction was made between classical nuclear receptors with known ligands and so-called 'orphan' receptors, that is, receptors without a known ligand. However, work done in recent years has identified ligands for many of these orphan receptors, making this distinction rather superficial (Giguere, 1999; Kliewer *et al.*, 1999). Moreover, the classification of nuclear receptors into six to seven phylogenetic subfamilies with groups that comprise both orphan and non-orphan receptors further dismisses such a discrimination (Nuclear Receptors Nomenclature Committee, 1999). The classification of nuclear receptors is done by virtue of the homology of the DNA binding domain (region C in *Figure 2*) and the ligand binding domain (region E), which have the highest evolutionary conservation among family members. A phylogenetic tree for the nuclear receptor family is shown in *Figure 1*. *Table 1* lists all known nuclear receptors with their acronyms according to a unified nomenclature system (Nuclear Receptors Nomenclature Committee, 1999). It has been proposed that nuclear receptors have evolved from an ancestral orphan 'receptor' through early diversification, and only later acquired ligand binding ability (Escriva *et al.*, 1997). Nonetheless, as long as the identification of ligands for previous orphan receptors continues, it cannot be formally excluded that all nuclear receptors may have cognate ligands. On the other hand, the concept that the nuclear receptor family has evolved from an ancestral orphan receptor and ligand binding has been acquired during evolution has found broad recognition. It is possible that a number of orphan receptors function exclusively as constitutive repressors or activators of transcription. Rev-erb is an example of such a constitutive repressor which recruits corepressors but lacks a functional activation function AF2 (see Section 4.4). For historical reasons, research focused initially on the classical steroid hormone receptors. Presently, however, there is significant emphasis both from basic scientists and industrial pharmacologists on the systematic screening for agonists and antagonists that bind with high affinity to the ligand-binding domain of several orphan receptors. Interestingly, some recently identified ligands are metabolic intermediates. It appears therefore, that in certain systems, like the metabolism of cholesterols or the fatty-acid β-oxidation pathways, the control of build-up, breakdown and storage of metabolically active substances is regulated at the level of gene expression, and that in many cases this 'intracrine' signaling is brought about by nuclear receptors (*Figure 3*). Furthermore, gene knock-out experiments suggest that metabolic intermediates, such as the ligands for SF1 (NR5A1) or peroxisome proliferator-activated receptor-γ (PPARγ) (NR1C3), may have regulatory function

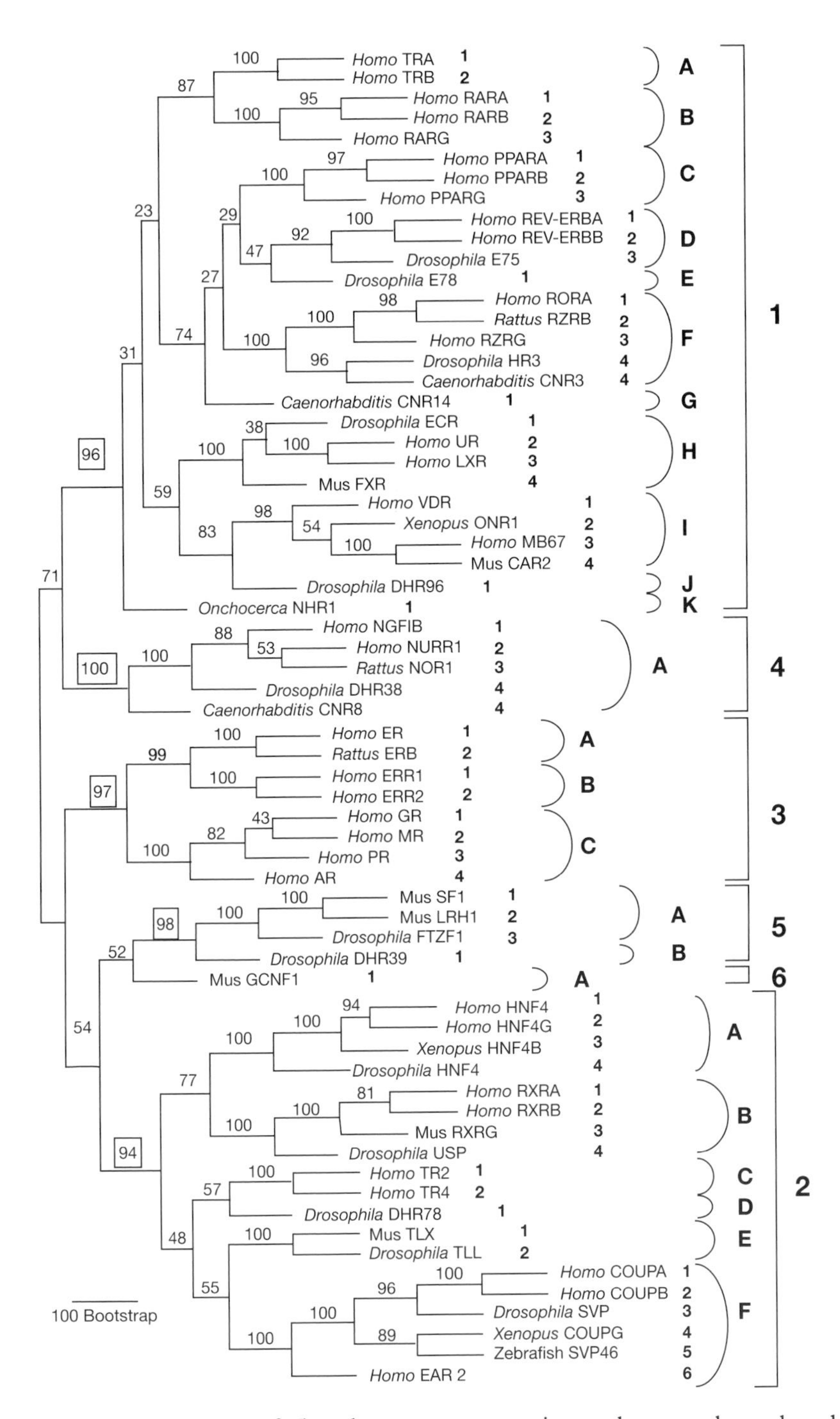

Figure 1. Phylogenetic tree of 65 nuclear receptor genes in vertebrates, arthropods and nematodes. Redrawn from *Cell* 1999; 97: 161–163, A unified nomenclature system for the nuclear receptor superfamily, with permission from Cell Press. (For a detailed description see Nuclear Receptors Nomenclature Committee, 1999 and the regular updates at *http://www.ens-lyon.fr/LBMC/LAUDET/nomenc.html*.)

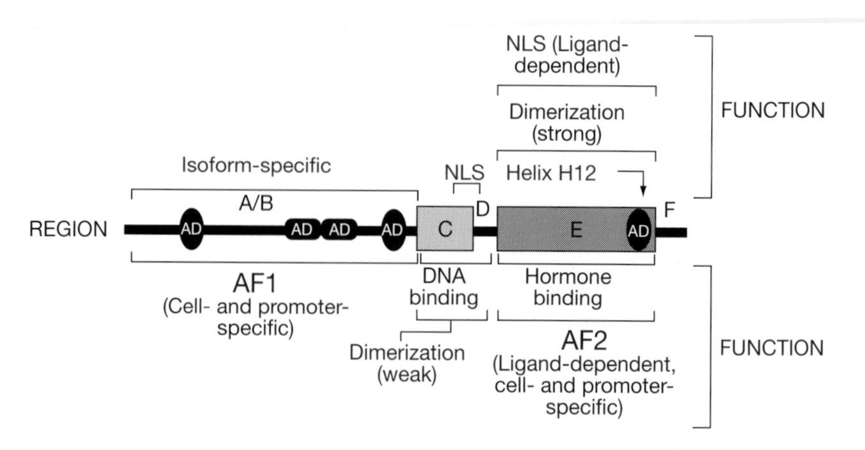

Figure 2. Schematic illustration of the structural and functional organization of nuclear receptors. The evolutionary conserved regions C and E are indicated as boxes and a black bar represents the divergent regions A/B, D, and F. Note that region F may be absent in some receptors. Domain functions are depicted below and above the scheme; most of these are derived from structure–function studies of steroid, thyroid and retinoid receptors. Two transcription activation functions (AFs) have been described in several nuclear receptors, a constitutively active (if taken out of the context of the receptor) AF1 in region A/B and a ligand-inducible AF2 in region E. Within these activation functions, autonomous transactivation domains (ADs) have been defined in the estrogen (ER) and progesterone receptor (PR) N-terminal regions; they may even be specific to particular receptor isoforms (such as the isoform B of PR). In the case of the estrogen, retinoid and thyroid hormone receptors a weak autonomous activation domain (AF2 AD) has been detected at the *C*-terminal end of the ligand-binding domain E.

Table 1. A proposed nomenclature for nuclear receptors

Subfamilies and group	Genes	Trivial names	Accession numbers
1A	NR1A1	TRα, c-erbA-1, THRA	M24748
	NR1A2	TRβ, c-erbA-2, THRB	X04707
1B	NR1B1	RARα	X06538
	NR1B2	RARβ, HAP	Y00291
	NR1B3	RARγ, RARD	M57707
1C	NR1C1	PPARα	L02932
	NR1C2	PPARβ, NUC1, PPARδ, FAAR	L07592
	NR1C3	PPARγ	L40904
1D	NR1D1	REVERBα, EAR1, EAR1A	M24898
	NR1D2	REVERBβ, EAR1β, BD73, RVR, HZF2	L31785
	NR1D3	E75	X51548
1E	NR1E1	E78, DR-78	U01087
1F	NR1F1	RORα, RZRα	U04897
	NR1F2	RORβ, RZRβ Y08639	
	NR1F3	RORγ, TOR	U16997
	NR1F4	HR3, DHR3, MHR3, GHR3,	M90806
		CNR3, CHR3	U13075
1G	NR1G1	CNR14	U13074
1H	NR1H1	ECR	M74078

	NR1H2	UR, OR-1, NER1, RIP15, LXRβ	U07132
	NR1H3	RLD1, LXR, LXRα	U22662
	NR1H4	FXR, RIP14, HRR1	U09416
1I	NR1I1	VDR	J03258
	NR1I2	ONR1, PXR, SXR, BXR	X75163
	NR1I3	MB67, CAR1	Z30425
	NR1I4	CAR2, CARβ	AF00932
1J	NR1J1	DHR96	U36792
1K	NR1K1	NHR1	U19360
2A	NR2A1	HNF4	X76930
	NR2A2	HNF4G	Z49826
	NR2A3	HNF4B	Z49827
	NR2A4	DHNF4, HNF4D	U70874
2B	NR2B1	RXRα	X52773
	NR2B2	RXRβ, H-2RIIBP, RCoR-1	M84820
	NR2B3	RXRγ	X66225
	NR2B4	USP, Ultraspiracle, 2C1, CF1	X52591
2C	NR2C1	TR2, TR2–11	M29960
	NR2C2	TR4, TAK1	L27586
2D	NR2D1	DHR78	U36791
2E	NR2E1	TLL, TLX, XTLL	S72373
	NR2E2	TLL, Tailless	M34639
2F	NR2F1	COUP-TFI, COUPTFA, EAR3, SVP44	X12795
	NR2F2	COUP-TFII, COUPTFB, ARP1, SVP40	M64497
	NR2F3	SVP, COUP-TF	M28863
	NR2F4	COUP-TFIII, COUPTFG,	X63092
	NR2F5	SVP46	X70300
	NR2F6	EAR2	X12794
3A	NR3A1	ERα	X03635
	NR3A2	ERβ	U57439
3B	NR3B1	ERR1, ERRα	X51416
	NR3B2	ERR2, ERRβ	X51417
3C	NR3C1	GR	X03225
	NR3C2	MR	M16801
	NR3C3	PR	M15716
	NR3C4	AR	M20132
4A	NR4A1	NGFIB, TR3, N10, NUR77, NAK1	L13740
	NR4A2	NURR1, NOT, RNR1, HZF-3, TINOR	X75918
	NR4A3	NOR1, MINOR	D38530
	NR4A4	DHR38, NGFIB	U36762
		CNR8, C48D5	U13076
5A	NR5A1	SF1, ELP, FTZ-F1, AD4BP	D88155
	NR5A2	LRH1, xFF1rA, xFF1rB, FFLR, PHR, FTF	U93553
	NR5A3	FTZ-F1	M63711
5B	NR5B1	DHR39, FTZF1B	L06423
6A	NR6A1	GCNF1, RTR	U14666
0A	NR0A1	KNI, Knirps	X13331
	NR0A2	KNRL, Knirps related	X14153
	NR0A3	EGON, Embryonic gonad, EAGLE	X16631
	NR0A4	ODR7	U16708
	NR0A5	Trithorax	M31617
0B	NR0B1	DAX1, AHCH	S74720
	NR0B2	SHP	L76571

Note: Subfamilies and Groups are defined as referred in the text. The groups contains highly related genes with often paralogous relationship in vertebrates (e.g. RARα, RARβ and RARγ). The term isoform is reserved for different gene products originating from the same gene due to alternative promoter usage or splicing, or alternative initiation of translation.

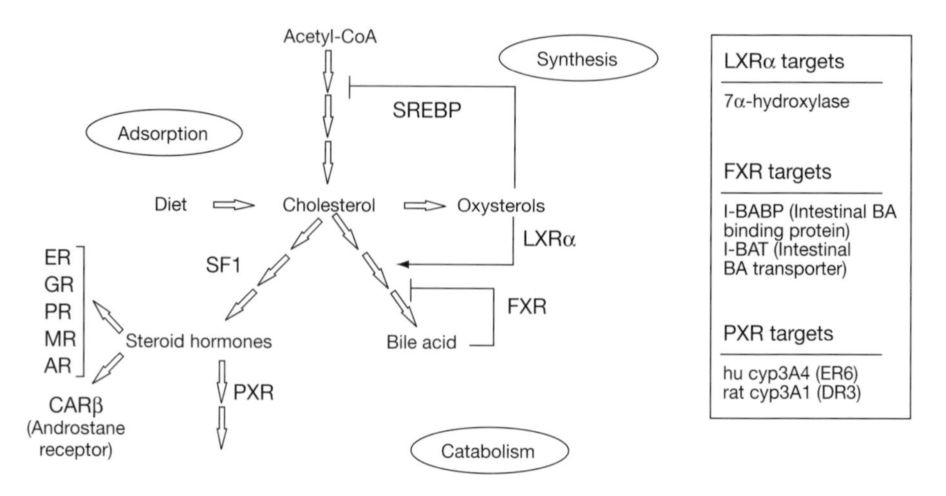

Figure 3. Cholesterol metabolism: an example of intracrine regulation by nuclear receptors for which ligands were discovered only recently. Three members of the nuclear receptor family regulate cholesterol metabolism, oxysterol receptor (LXRα), chendoxycholate receptor (FXR) and SF1. The C21 steroid receptors (PXR), which includes the progesterone receptor, regulates steroid catabolism. Sterol regulatory element binding protein (SREBP) is a bHLHZ factor that also regulate the pathway. Some known transcriptional targets of these receptors are depicted in the box. For details and original references see the recent review by Repa and Mangelsdorf (1999).

in specifying organ development (Luo *et al.*, 1994; Barak *et al.*, 1999). Prominent metabolic ligands are bile and fatty acids, eicosanoids and oxysterols. The group of steroid hormones encompasses estrogens, progestins, mineralocorticoids, glucocorticoids, androgens, and ecdysterones. Examples for vitamins are vitamin D or the vitamin A derivative, retinoic acid. Thus, nuclear receptors function in endocrine (steroid hormone receptors), auto/paracrine (retinoid receptors) and intracrine (metabolic receptors such as LXR, SF1, FXR, PXR, BXR, SXR, PPAR, CARβ) signaling pathways. Certainly, with the identification of more ligands for orphan receptors, new surprises are likely to be encountered.

3. Targets and genetics

3.1 Target gene recognition

Through their DNA-binding domain, nuclear receptors are able to specifically bind to regulatory *cis*-acting DNA sequence elements in the promoters/enhancers of target genes. In keeping with the high conservation of the DNA-binding domain, these response elements have a similar structure; the canonical core recognition sequence is PuGGTCA (*Figure 4*) (Gronemeyer and Laudet, 1995). Nuclear receptors bind to DNA as monomers or, more frequently, as homo- and

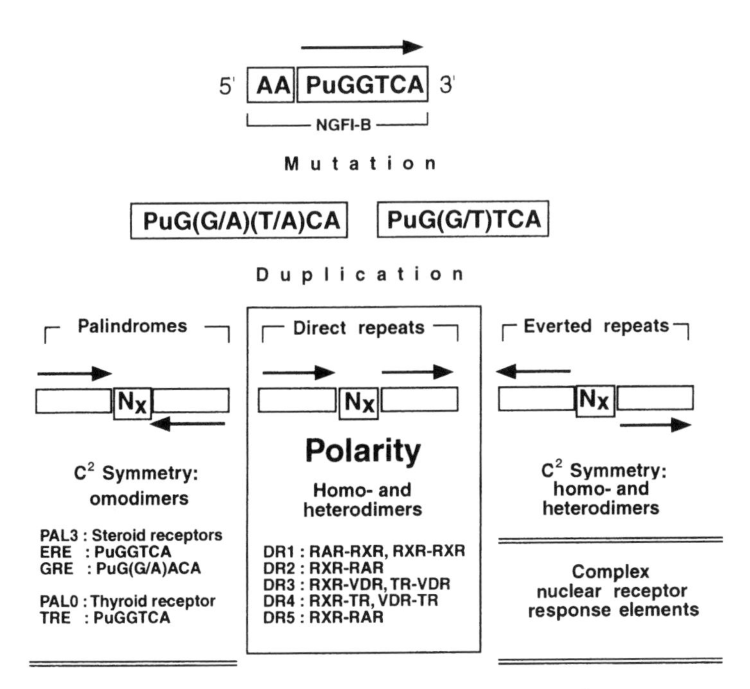

Figure 4. Response elements of nuclear receptors. The canonical core recognition sequence is 5′-PuGGTCA (arrow indicates the 5′ to 3′ direction) which, together with two 5′ As, is a response element of the orphan receptor, NGFI-B. Duplication of the core sequence generates symmetrical palindromes (PALs) and everted repeats, and polar direct repeats (DRs), with x base-pairs separating the two half-sites. PAL3 is an estrogen response element (ERE), while Pal0 corresponds to a thyroid hormone response element (TRE). A single mutation at position 4 of the core sequence from T to A leads to PAL3 response elements recognized by the glucocorticoid receptor (GREs) as well as androgen, progesterone and mineralocorticoid receptors. Note that most response elements are far from ideal; often one of the half-sites contains one or more mutations. Whereas PALs bind homodimers, DRs can bind homo- or heterodimers with the specificities given in the dark box (polarity of the receptors on their cognate DRs: left, 5′; right, 3′). Note that the 3′-positioned receptor makes minor grove DNA contacts in the spacer. Some everted repeats are response elements for homo or heterodimers of the thyroid or retinoid receptors. Response elements are known that are comprised of complex arrangements of the core motif.

heterodimers. In heterodimers it is mainly, if not exclusively, RXR that acts as a promiscuous heterodimerization partner. Some nuclear receptors can bind to DNA response elements as monomers or heterodimers. For example, NGFI-B (NR4A1) can either recognize the (extended) hexameric core sequence (termed 'NBRE') as a monomer or binds to DR5 elements as a NGFI-B/RXR heterodimer (Forman *et al.*, 1995; Perlmann and Jansson, 1995). As the canonical core sequence occurs statistically every 256 base-pairs, nuclear receptors that bind as monomers recognize additional nucleotides 5′ to the core sequence to increase efficiency and selectivity of DNA binding (Meinke and Sigler, 1999). To accommodate dimers, the canonical core is duplicated and arranged with different polarities and spacings between both half-sites of the response element, thus forming direct, inverted (palindromic) or everted repeats (*Figure 4*). In general, but not always, different homo- and heterodimers have different response element specificities. The spacing (usually 0–8 nucleotides) and polarity therefore discriminates largely between different nuclear receptor dimers. Polarity, as it is used here, reflects the anisotropic nature of direct repeat elements that lack the rotational symmetry of inverted repeats. Thus, heterodimers bound to such an element exhibit an inherent polarity. Although not yet formally proven, such a polarity (5′-NR1-NR2–3′, versus 5′-NR2-NR1–3′) may effect promoter activity, as the cooperativity or synergy with other transcription factors may depend on the relative orientation of the complex. The polarities of several anisotropic complexes have been determined. As an example, on DR2 and DR5 elements retinoic acid receptors (RAR) and RXRs bind in the 5′-RXR-RAR-3′ orientation whereas this orientation is most likely reversed on DR1 elements to 5′-RAR-RXR-3′ (Zechel *et al.*, 1994). The polarity originates from the alternate use of dimerization interfaces and is also affected by steric constraints (compare Kurokawa *et al.*, 1994; Rastinejad *et al.*, 1995; Zechel *et al.*, 1994). Note also that some receptors have a preference for 5′ flanking sequences of the core recognition motif (Kurokawa *et al.*, 1993; Meinke and Sigler, 1999; Rastinejad *et al.*, 1995; Zhao *et al.*, 1998) and that some receptors prefer PuGTTCA over PuGGACA.

The structural basis of the selectivity of DNA recognition by nuclear receptors is now well understood (see also Chapter 4). This is due to a wealth of crystal structure data, in particular on the complexes between the DNA-binding domains of the glucocorticoid and estrogen receptors with their cognate palindromic response elements (Luisi *et al.*, 1991; Schwabe *et al.*, 1993); on the RXR heterodimer with a direct repeat (DR4) element (Rastinejad *et al.*, 1995); on the Rev-erbα homodimer with a DR2 element (Zhao *et al.*, 1998); and on the NGFI-B monomer on an 'extended half-site' element (Meinke and Sigler, 1999). The latter demonstrated the additional selectivity of DNA recognition due to minor groove contacts 5′ of the core motif. Moreover, structural principles that allow the interaction between non-optimal response elements and nuclear receptors also have been elucidated (Gewirth and Sigler, 1995; Schwabe *et al.*, 1995). For a more detailed discussion of the DNA binding of nuclear receptors see the reviews by Gronemeyer and Laudet, 1995 and Schwabe, 1997.

3.2 Genetics of nuclear receptors

Genetic programs consist typically of hundreds of genes that are expressed in a spatially and temporally controlled fashion. Nuclear receptors act as master 'switches' that initiate specific genetic programs which lead, for example, to cell differentiation, proliferation or apoptosis, or regulate homeostasis. In the context of other programs these genetic activities support or initiate complex physiological phenomena, such as reproduction and organ function. Once activated by the cognate ligand, nuclear receptors regulate the primary and secondary target gene expressions that make up the corresponding physiological event. Throughout the life cycle of a multicellular organism the coordinate interplay between programs defines cell fates in the different tissues and organs, and finally the entire body is at the foundation of the organisms' development and subsistence. This is fully supported by the analysis of mice bearing mutations or deletions of one or several receptors (a searchable mouse knockout and mutation database can be found at http://www.biomednet.com/db/mkmd). Several nuclear receptor knockout animals, in particular compound knockout animals, die *in utero* or around birth, displaying severe malformations of organs that render them unviable (Beato *et al.*, 1995; Kastner *et al.*, 1995; 1997a; Mascrez *et al.*, 1998). Others are viable under laboratory conditions but display a reduced life span and are often unfertile (e.g., Luo *et al.*, 1994; Parker *et al.*, 1996). These knockout animal models have been of great help in deciphering the physiology of nuclear receptor action. Often they provided initial or additional evidence for new, yet undiscovered, functions exerted by the receptor, and thus stimulated further research on previously unknown signaling pathways. One example is the involvement of RARβ (NR1B3) and RXRγ (NR2B3) in long-term memory potentiation and depression (Chiang *et al.*, 1998). Furthermore, knockouts have also provided insight into the distinct modes of transcriptional regulation by nuclear receptors. An example is the mouse *NR3C1* gene encoding for the glucocorticoid receptor (GR). $GR^{-/-}$ mice die at birth due to respiratory failure (Cole *et al.*, 1995). However, replacement of the glucocorticoid receptor gene by a glucocorticoid receptor mutant, GrA(458T) (Heck *et al.*, 1994), which impairs binding to consensus glucocorticoid receptor-response elements, generated mice (termed GR dim/dim, with 'dim' indicating DNA-binding domain dimerization deficiency) that were fully viable and exhibited only minor phenotypic alteration (Reichardt *et al.*, 1998). These results demonstrated that the indirect transcriptional effects ('signaling crosstalk') of the glucocorticoid receptor are as crucial for glucocorticoid receptor-mediated signaling as the direct regulation of cognate target genes (Reichardt et *al.*, 1998, Tronche *et al.*, 1998). Transrepression of activating protein AP1 is one of the main indirect effects of the glucocorticoid receptor, a mechanism that is discussed further in Section 6.1.

An important result of studies with nuclear receptor gene-deletion models has been the discovery of redundancy and adaptivity among family members of the same group. In this respect, the interpretation of vitamin A signaling, which is of remarkable complexity and displays a high degree of apparent RAR redundancy, may serve as an example. Vitamin A derivatives are bound by two different groups of nuclear receptors. RARα, RARβ and RARγ (NR1B1, NR1B2, NR1B3) bind both all-*trans*-retinoic acid and the isomer 9-*cis*-retinoic acid, whereas the RXRα, RXRβ and RXRγ (NR2B1, NR2B2, NR2B3) bind exclusively 9-*cis*-retinoic acid. RXRs are heterodimerization partners for a great number of nuclear receptors and they also heterodimerize with RARs to form active signaling molecules. All six retinoid receptor genes give rise to at least two different isoforms through alternative splicing or differential promoter usage. Considering the necessity of heterodimerization between one member of each group, theoretically, 36 different combinations can be formed from the existing pool of genes. As indicated through single and compound knock-outs, the removal of a single gene results in rather restricted phenotypes, while the removal of multiple isotypes or of RAR and RXR group members results in animals displaying the full vitamin A deficiency syndrome (Kastner *et al.*, 1995, 1997a, b; Mark *et al.*, 1999; Smith *et al.*, 1998; Sucov *et al.*, 1994). This indicates functional redundancy between different retinoid receptor genes *in vivo*. However, molecular and biochemical analyses have demonstrated significant functional differences among group members. For example, cell differentiation assays have revealed that one receptor is able to induce a certain differentiation event, while another is not, even though it is equally expressed (e.g., Benoit *et al.*, 1999; Chen *et al.*, 1996). These seemingly contradictory findings reveal another important feature that contributes to the complexity of nuclear receptor signaling *in vivo* — adaptivity. It is assumed, though it has not yet been convincingly demonstrated, that the organism (in this case the developing mouse) overcomes the lack of a given receptor gene by increasing the activities of the remaining genes to cope with this deficiency. Thus, the organism compensates for most of the deleterious effects caused by absent or weak phenotypes at least under laboratory conditions. This phenomenon assures that the organism can survive under standard conditions (e.g. the spontaneous somatic mutation of a given receptor gene will not necessarily lead to the loss of the affected cell), but may fail to deal with more extreme physiological conditions. This phenomenon is therefore distinct from true functional redundancy. It is possible that redundancy and adaptivity allow mutations to accumulate that would otherwise be lethal. Such 'buffering' of mutations may result in a more rapid evolution of phenotypes based on the accumulation of multiple and/or more complex gene alterations.

4. Modular structure and function

4.1 Region A/B harbors activation function 1 (AF1)

As schematized in *Figure 2,* nuclear receptors are composed of five to six regions (A to F) (Krust *et al.*, 1986) that have modular character. Distinct A and B regions were originally characterized in receptors with long N-terminal regions adjacent to the DNA-binding domain, but for most receptors, this is considered as a combined A/B region. This region harbors one (or more) autonomous transcriptional activation function domains (AF1), which when linked to a heterologous DNA-binding domain can activate transcription in a constitutive manner. Note, however, that in the context of the full-length receptor AF1 is silent in the absence of an agonist and with certain antagonists. When comparing nuclear receptors from different subfamilies and groups, the A/B region displays the weakest evolutionary conservation, and the distinction between A and B regions is not always evident. A/B regions differ significantly in their length, ranging from 23 (vitamin D receptor, NR1I1) to 550 (androgen, NR3C4, mineralocorticoid, NR3C2, and glucocorticoid receptors, NR3C1) amino acids. No 3D-structure of a nuclear receptor A/B region has been solved up to now, and structure prediction is not straightforward. A/B regions are subject to alternative splicing and differential promoter usage, with the majority of known nuclear receptor isoforms differing in their N-terminal region. Through alternative splicing, as in the case of RAR and TR, and differential promoter usage, which generates the isoforms A and B of the progesterone receptor, the absence or presence of different activation functions found in the AB regions can be regulated (Gronemeyer and Laudet, 1995). Moreover, the N-terminus of nuclear receptors has reportedly been found to be a subject of post-translational events such as phosphorylation (Adam-Stitah *et al.*, 1999; Hammer *et al.*, 1999; Shao and Lazar, 1999; Taneja *et al.*, 1997; Tremblay *et al.*, 1999). The role of phosphorylation of the AB domains for the transactivation potential of AF1, as well as for synergy and cooperativity with AF2, which is located in the E domain of the receptor, is currently being intensively studied. Finally, the activation function(s) AF1 display cell and promoter specificity (Berry *et al.*, 1990; Bocquel *et al.*, 1989), the origin of which is still elusive but may be related to the cell-specific action and/or expression of AF1 coactivators. These issues will be discussed in more detail later.

4.2 The DNA-binding domain encompasses region C

The highly conserved region C encodes the DNA-binding domain of nuclear receptors and confers sequence-specific DNA recognition. This domain has been extensively investigated, especially in view of the previously discussed complex features of response element recognition and dimerization properties (Gronemeyer and Laudet, 1995). Several X-ray and NMR data sets are available for different nuclear receptor C domains in their DNA complexed and uncomplexed forms. The domain is mainly composed of four zinc-finger motifs, the N-terminal motif, Cys-X_2-Cys-X_{13}-Cys-X_2-Cys (CI), and the C-terminal motif, Cys-X_5-Cys-X_9-Cys-X_2-Cys (CII). In each motif, two cysteine residues chelate one Zn^{2+} ion. Within the C domain several sequence elements (termed P, D, T and A-boxes) have been characterized that define or contribute to response-element specificity and the dimerization interface, and contacts with residues outside the core recognition sequence. *Figure* 5 displays a schematic illustration of the RXR DNA-binding domain, with an illustration of the different Zn^{2+} fingers and the various boxes.

4.3 Region D, a hinge with compartmentalization functions

The D region of nuclear receptors is less conserved than the surrounding regions of C and E and can significantly vary in length. This domain appears to correspond to a 'hinge' between the highly structured C and E domains. It might allow the DNA and ligand-binding domains to adopt several different conformations without creating steric hindrance problems. Note in this respect that the C and E regions contribute separate dimerization interfaces that accommodate different heterodimerization partners on very different response elements. Region D contains a nuclear localization signal, or at least some elements of a functional nuclear localization signal (reviewed in Gronemeyer and Laudet 1995). The intracellular localization of nuclear receptors is a result of a dynamic equilibrium between nuclear–cytoplasmic and cytoplasmic–nuclear shuttling (Guiochon-Mantel *et al.*, 1994). At equilibrium the large majority of receptors are nuclear, although there is some controversy in the case of corticoid receptors (glucocorticoid and mineralocorticoid receptors), which have been reported to reside at cytoplasmic locations in the absence of their cognate ligands and translocate to the nucleus in a ligand-induced fashion (Baumann *et al.*, 1999).

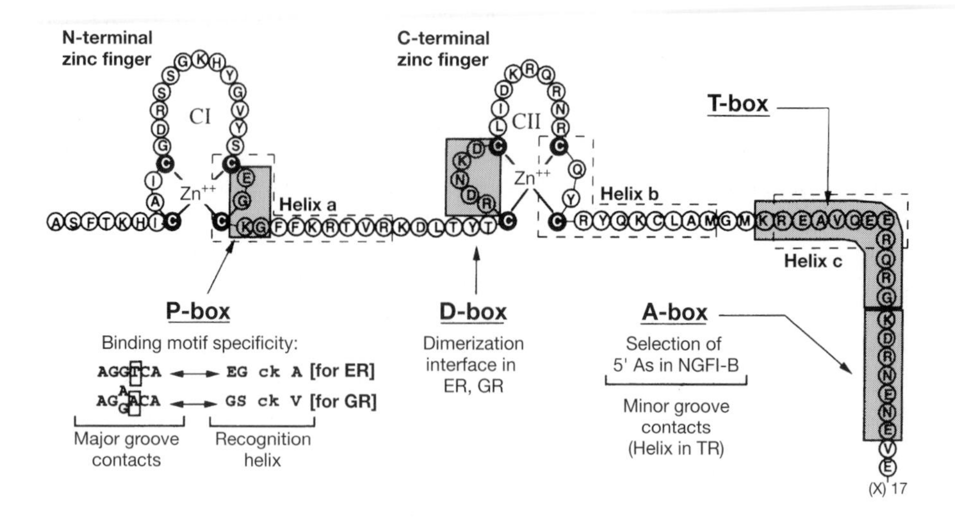

Figure 5. Schematic illustration of the retinoid X receptor (RXR) DNA-binding domain. Boxes illustrate regions involved in response element selection. The P-box is part of the DNA recognition helix (helix a; framed by broken lines) and swapping of the EGA residues of the ER P-box with the corresponding GSV residues of glucocorticoid receptor switches estrogen/glucocorticoid response element (ERE/GRE) recognition as is outlined below the illustration. The D-box is responsible for PAL3/PAL0 selection by estrogen or thyroid hormone receptors (ER and TR respectively) and contributes to the homodimerization interfaces of ER and GR DNA-binding domains. The T-box region forms a helix (helix c) and corresponds to a dimerization surface in RXR homodimers. The A-box was originally defined as a NGFIB segment, which selects the minor-groove binding at the 5′-end of the site, as in the NGFI-B response element (NBRE). In the RXR:TR DNA-binding domain co-crystal on DR4, the TR A-box forms a helix that interacts with the spacer DNA in the minor groove (Rastinejad *et al.*, 1995). In addition, the crystal structure of the NGFI-B DNA-binding domain:NBRE complex has been solved recently (Meinke and Sigler, 1999). CI and CII are the two zinc-finger motifs. PRF, pre-finger region. For further details see Gronemeyer and Laudet (1995).

4.4 Region E encompasses the ligand-binding domain and activation function 2 (AF-2)

The hallmark of a nuclear receptor is its ligand-binding domain in the E region. This domain is highly structured, and encodes a wealth of distinct functions, most of which operate in a ligand-dependent manner. The ligand-binding domain harbors the ligand-dependent activation function AF2, a major dimerization interface, and often a repression function (Chambon, 1996; Gronemeyer and Laudet, 1995; Moras and Gronemeyer, 1998; Tsai and O'Malley, 1994). Detailed molecular insights into the structure–function relation of signal integration by nuclear receptors have been gained by the elucidation of the crystal structures of the E region alone or in presence of agonists, antagonists and coregulator peptides. The first 3D-structures reported for nuclear receptor ligand-binding domains were those of the unliganded RXRα (NR2B1), the all-*trans*-retinoic acid-bound RARγ (NR1B3), and the agonist-bound thyroid receptor TRβ (NR1A2)

(Bourguet *et al.*, 1995; Renaud *et al.*, 1995; Wagner *et al.*, 1995). Unliganded receptors are frequently referred to as apo-receptor forms, while the liganded receptor corresponds to the holo-form. These structures demonstrated that apo and holo ligand-binding domains have a common fold. Moreover, a structure-based sequence alignment of all known nuclear receptor primary amino-acid sequences strongly supported a common fold for all nuclear receptor ligand-binding domains (Wurtz *et al.*, 1996). This hypothesis was fully confirmed when the crystal structures of the ligand-binding domains of the holo-estrogen receptor (NR3A1), progesterone (NR3C3) and the holo-PPAR (NR1C3) were solved (Brzozowski *et al.*, 1997; Nolte *et al.*, 1998; Uppenberg *et al.*, 1998; Williams and Sigler, 1998; Xu *et al.*, 1999a). The general fold of the ligand-binding domain consists of a three-layered α-helical sandwich. Further structural features are one β-hairpin and connecting loops of variable lengths. The helices have been designated H1 to H12, starting with the most N-terminal H1, and form a hydrophobic cavity that accommodates the hydrophobic ligands. For an example of a detailed analysis of ligand–ligand-binding domain interactions in the case of the RAR and the structural basis of isoform specificity see Gehin *et al.* (1999) and Klaholz *et al.* (1998). The different crystal structures of apo and holoforms of receptor ligand-binding domains, as well as extensive mutagenesis, demonstrated that the E domain undergoes a major conformational change upon ligand binding (*Figure 6*). The structural transition upon ligand binding has been described as a 'mouse-trap' mechanism (Moras and Gronemeyer, 1998; Renaud *et al.*, 1995): pushed by the ligand the helix H11 is repositioned in the continuity of helix H10, and the concomitant swinging of helix H12 unleashes the Ω-loop, which flips over underneath helix H6 carrying along the N-terminal part of helix H3. In its final position helix H12 seals the 'lid' in the ligand-binding pocket and further stabilizes ligand binding by contributing to the hydrophobic environment, and, in some cases by making additional contacts with the ligand itself. The repositioning of helix H12 back onto the core of the ligand-binding domain is apparent from the structures of the apo-RXRα and holo-RARγ ligand-binding domains shown in *Figure 6*. The critical implication in transcription activation of helix H12, also referred to as the activating domain of the AF2 function (the features of this AF2 activating domain are conserved in all transcriptionally active receptors) was previously deduced from multiple mutagenesis studies (Chambon, 1996; Gronemeyer and Laudet, 1995). Moreover, helix H12 not only coincided with residues critically required for transcriptional activation, but also those critical for coactivator recruitment (coactivators and their interaction with the ligand-binding domains are discussed later). The structural data reveal that helix H12, when folded back onto the core of the ligand-binding domain, forms a hydrophobic cleft together with other surface-exposed amino-acids that accommodates the 'NR box' of coactivators (Heery *et al.*, 1997; Le Douarin *et al.*, 1996). Crystal structure data have revealed the details of the interaction of the LXXLL NR-box motif with the accommodating surface on the ligand-binding domain (Darimont *et al.*, 1998; Nolte *et al.*, 1998; Shiau *et al.*, 1998). As an example the interaction of the transcription intermediary factor-2 (TIF2)/glucocorticoid receptor interacting protein-1 (GRIP1) NR-box 2 peptide with the agonist (diethylstilbestrol, DES)-bound estrogen receptor ligand-binding domain is illustrated in *Figure 7* (left).

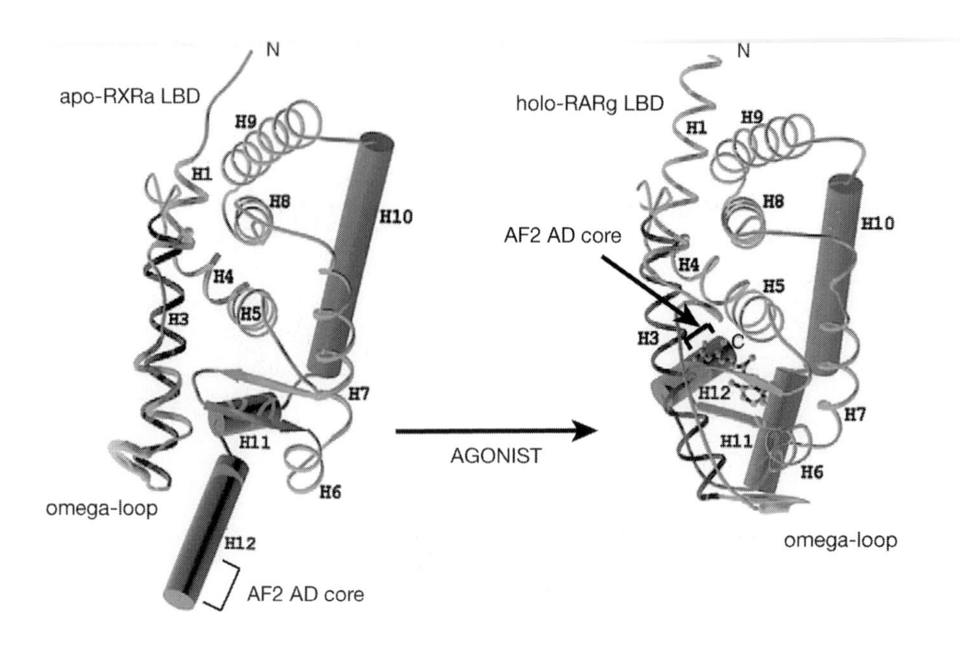

Figure 6. A comparison of the crystal structures of the retinoid X receptor (RXRα) apo-ligand binding domain (apo-LBD) and the retinoic acid receptor (RARγ) holo-LBD reveals the ligand-induced conformational change that generates the transcriptionally active form of a nuclear receptor. In this model, ligand-binding induces a structural transition that triggers a mousetrap-like mechanism: in continuity with H10, H11 is pushed by the ligand and repositioned. The concomitant swinging of helix H12 unleashes the Ω-loop, which flips underneath H6 carrying along the *N*-terminal part of H3. In its final position, H12 seals the ligand-binding cavity and further stabilizes ligand binding by contributing to the hydrophobic pocket. Note that helix H12, which encompasses the core of the AF2 activation domain (AF2 AD core), is now positioned in an entirely different environment of the LBD where it is able to interact with the LxxLL NR-box motif of *bona fide* coactivators (Darimont *et al.*, 1998; Moras and Gronemeyer, 1998; Nolte *et al.*, 1998; Shiau *et al.*, 1998).

Note that helix H12 is crucial in stabilizing this interaction (Shiau *et al.*, 1998). The corresponding crystal structure of the (tamoxifen) antagonist-bound estrogen receptor ligand-binding domain was particularly illuminating, as it demonstrated that this type of antagonist induces a positioning of H12 that is uncompatible with the binding of coactivator NR boxes; indeed H12 binds to the identical cleft that the NR box binds to (compare the two structures in *Figure 7* (Bourguet *et al.*, 2000; Shiau *et al.*, 1998). Hence, the information generated by the synthesis or secretion of a nuclear-receptor ligand is converted, first into a molecular recognition (ligand–receptor) process, and subsequently into an allosteric event that allows secondary recognition of a 'downstream' mediator which is more closely connected to the cellular machineries involved in transcription activation.

Some nuclear receptors can also function as DNA-bound repressors of transcription. This phenomenon occurs in the absence of agonists, may be enhanced by certain antagonists, and is attributable to the recruitment of corepressors. Corepressor

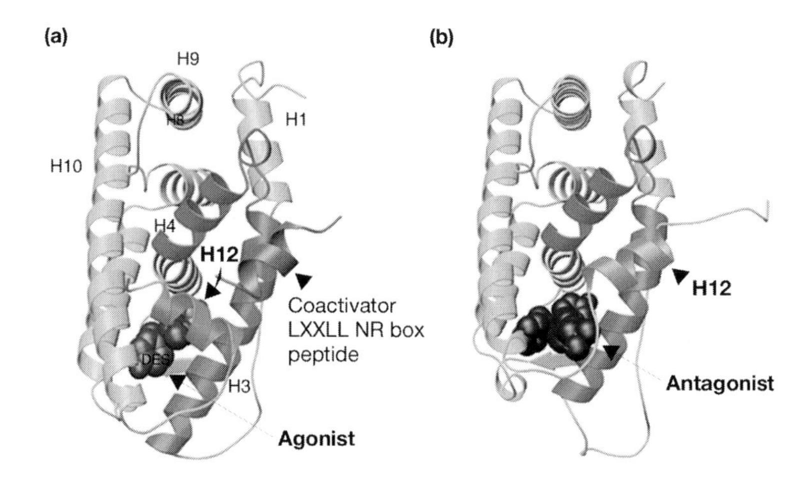

Figure 7. Mechanism of estrogen receptor antagonist action. (a) Comparison of the ligand-binding domain/NR box peptide complex in the presence of the agonist, diethylstilbestrol or (b) the antagonist tamoxifen. Note that helix H12 in the antagonist structure occupies the identical position as the NR-box peptide in the agonist ligand-binding domain. The same position of H12 was observed in the crystal structure of the retinoic acid receptor-α ligand-binding domain/BMS614 antagonist complex (Bourguet *et al*., 2000) indicating a general mechanism of antagonism. In the retinoic acid receptor case, steric hindrance problems are responsible for the 'antagonistic' positioning of helix H12. Reproduced from Shiau, A.K. *et al.* (1998) Structural basis of estrogen receptor/coactivator recognition and the antagonism of this interaction by tamoxifen. *Cell* 95: 927–937, with permission from Cell Press. The structures presented here are derived from IDS 3ERD and 3ERT in the Brookhaven Protein DataBank (PDB).

binding also occurs on the surface of the non-liganded ligand-binding domain and is similar to coactivator binding in that it is also mediated by a short signature motif, the so called 'CoRNR box', which binds to a surface topologically related to that involved in coactivator interaction (Hu and Lazar, 1999). Upon ligand binding, the conformational *trans* configuration of the α-helical sandwich displaces the corepressors by shifting the equilibrium of the E domain from the apo to the holo form.

Some nuclear receptors form only homodimers (e.g., steroid hormone receptors) while others can form both homodimers (e.g. RXR, TR) and heterodimers with the promiscuous heterodimerization partner, RXR (e.g. RAR:RXR, TR:RXR, VDR:RXR, PPAR:RXR). Although structural studies have highlighted the dimerization function of the second zinc finger in region C, the ligand-binding domain is actually the major domain that contributes to dimerization. The recent crystallization of the RAR:RXR ligand-binding domain heterodimer, and the comparison with homodimer interfaces, has provided detailed information on the structural elements governing homo and heterodimerization (Bourguet *et al*., 2000). Based on studies *in vitro* it has also been reported that ligand binding can affect the dimerization properties of nuclear receptors. For the thyroid hormone and vitamin D receptors, which can both form homodimers and RXR heterodimers, ligand binding appears to favor heterodimerization (Cheskis and Freedman, 1994; Collingwood *et al*., 1997; Kakizawa *et al*., 1997).

4.5 The C-terminal region F

Some receptors possess C-terminal of the ligand-binding domain, a region F which displays little evolutionary conservation. There are no clues to its function and it is unclear why it is present in some, but not all, receptors. Recent literature indicates that the F region might play a role in coactivator recruitment to the E domain and in determining the specificity of the ligand-binding domain coactivator interface (Peters and Khan, 1999; Sladek *et al.*, 1999). It is clear that this domain is poorly conserved, and we anticipate learning that the F domain fine-tunes the molecular events associated with the transcriptional properties of the E domain or the entire receptor (Montano *et al.*, 1995; Nichols *et al.*, 1997).

As for A/B domains, the E and F domains are also targets of post-translational modifications, which add another level of complexity to nuclear receptor signal integration and is covered later.

5. Molecular mechanisms of transcriptional regulation

The expression of a given gene can be regulated at several different levels (transcription, translation or RNA processing, post-translation). Transcription itself has multiple levels at which regulation might occur. As other chapters have discussed, the transcriptional activity of a gene can be controlled epigenetically via methylation, at the level of its chromatin structure, and at the level of the assembly and activity of the initiating and elongating polymerase complexes. The reported effects of nuclear receptors on transcription are so far restricted to the initiation of transcription (Chapters 1 and 2). Below we will summarize our current knowledge about the molecular mechanisms/interactions through which nuclear receptors can positively or negatively regulate the expression of targate genes. For further details and references, see the recent review by Glass and Rosenfeld (2000).

5.1 Chromatin modifying nuclear receptor coregulators

So-called squelching experiments (Gill and Ptashne, 1988) paved the way to predict the existence of factors that would transmit the signal generated by the holo-receptor to the transcription machineries. Squelching occurs if an excess of receptor molecules, most unbound to the DNA, inhibits the activity of the same ('autosquelching') or a different receptor ('heterosquelching') (Bocquel *et al.*, 1989; Meyer *et al.*, 1989; Shemshedini *et al.*, 1992; Tasset *et al.*, 1990). True squelching requires an agonist and an intact AF2 domain. Such squelching was thought to result from sequestering of so-called transcription intermediary factors (TIFs) that mediate the action of the activation or repression functions of nuclear receptors and are the limiting constituents of the machineries required for transcription initiation. This concept predicted the existence of TIFs that are shared between, and are critically involved in, the action of different receptors. Indeed, the subsequent cloning and characterization of TIFs, also known as coactivators and corepressors (Chapter 2), has fully justified this concept.

According to the squelching data, *bona fide* coactivators were expected to fulfill the following criteria:

(i) interact directly with nuclear receptor ligand-binding domains in an agonist- and activation function-dependent manner;
(ii) enhance nuclear receptor-dependent reporter-gene activity when transiently expressed in mammalian cells;
(iii) activate transcription autonomously when tethered to DNA via a heterologous DNA-binding domain;
(iv) relieve squelching.

The development of yeast 'two hybrid' and direct cDNA expression library-screening approaches has identified a great number of coactivators, corepressors and other coregulators that are believed to transmit the nuclear receptor signal to its molecular targets. A summary of these coregulators, their characteristics and their putative mode of action is given in *Table 2* (and have been reviewed by Chen and Li, 1998; Collingwood *et al.*, 1999; Glass and Rosenfeld, 2000; McKenna *et al.*, 1999; Moras and Gronemeyer, 1998; Torchia *et al.*, 1998; Xu *et al.*, 1999b). We will now discuss the general emerging features of the different classes of coregulators.

The cloning of coregulators (*Table 2*) was followed by the definition of the coactivator signature NR box (LXXLL, where X is any amino acid) embedded in a short α-helical peptide (Heery *et al.*, 1997; Le Douarin *et al.*, 1996). These NR boxes are necessary and sufficient for ligand-dependent direct interaction with the transcriptional activation function AF2 in the nuclear receptor ligand-binding domains. This domain contains a hydrophobic cleft surrounded by 'charge clamps'. Upon agonist binding, helix H12 becomes repositioned on the surface of the ligand-binding domain. The repositioned helix H12, together with surrounding amino acids, accommodates the short amphipathic NR box helix found in the nuclear receptor-interacting domain of coactivators (Darimont *et al.*, 1998; Nolte *et al.*, 1998; Shiau *et al.*, 1998). Some coactivators contain multiple LXXLL motifs (up to nine in RIP140), most of which appear to be functional in terms of nuclear receptor binding *in vitro*. Whether this multiplicity reflects redundancy or confers some specificity to the interface is unresolved. Indeed different coactivators, even highly related coactivators from the TIF2/SRC1/RAC3 family, display some degree of nuclear receptor selectivity (Darimont *et al.*, 1998; Ding *et al.*, 1998; Hong *et al.*, 1999; McInerney *et al.*, 1998). Residues adjacent to the core LXXLL motif are thought to make contacts with the charged residues (the charge clamp) at both extremities of the hydrophobic cleft within the holo-ligand-binding domain in order to generate some receptor selectivity. However, such selectivity has yet to be confirmed by studies *in vivo*. In this respect the analysis of different coactivator knockout mice will be particularly instructive. Preliminary studies from our laboratory support the concept that *bona fide* coactivators are not fully redundant *in vivo*.

Recently, a second contact site for coactivators has been identified in nuclear receptors. Proteins from the TIF2 family are able to interact specifically with the A/B domains of estrogen and androgen receptor (Alen *et al.*, 1999; Berrevoets *et al.*, 1998; Bevan *et al.*, 1999; He *et al.*, 1999; Ma *et al.*, 1999). These interactions result in a stimulation of the transcriptional activity originating from AF1.

Table 2. Nuclear receptor-interacting factors

Human factors	Synonyms	NID motif	Accession number	Comments	Reference
1. Nuclear receptor-binding subunits of coactivator complexes					
Bona fide nuclear receptor coactivators					
Src1a	hERAP160 mNCoA1	Yes	U90661 NM_010881	First identified coactivator; binds nuclear receptor and CBP/p300; displays acetyltransferase activity and contacts basal transcription factors	Cavaillès *et al.*, 1994 Onate *et al.*, 1995
Src1e		Yes	U19179	Isoform of Src1a; lacks the C-terminal NR box of Src1a	
TIF2	hNCoA2 mGRIP1 Src3	Yes	X97674 U39060	Member of the same coactivator family as Src1	Voegel *et al.*, 1996 Hong *et al.*, 1997
ACTR	hAib1 hRac3 hTRAM1 mpCip	Yes	U59302 AF000581	Member of the same coactivator family as Src1/TIF2	Anzick *et al.*, 1997 Li *et al.*, 1997 Takeshita *et al.*, 1997 Torchia *et al.*, 1996
Cointegrators					
CBP	mCBP	Yes	U47741 sw:P45481	Transcriptional cointegrator for several signaling pathways in addition to those involving nuclear receptors (CREB, STATs, API, etc.); interacts with pCAF, Src1, TIF2 and pCIP; displays acetyltransferase activity	Torchia *et al.*, 1996 Kamei *et al.*, 1996
p300		Yes	U01877	Functional homologue of CBP; p300 and CBP functions are not entirely redundant	
pCAF			U57317	Similar to hGcn5; interacts with CBP/p300; possesses intrinsic acetyltransferase activity; present in a 2 MDa complex	Blanco *et al.*, 1998
Others					
SRA	mSRA		AF092038 AF092039	An RNA molecule, identified by using the *N*-terminal domain of PR as a bait; the *SRA* gene does not encode a protein; SRA is proposed to be a part of a complex containing Src1	Lanz *et al.*, 1999

Nuclear receptor-binding subunits of corepressor complexes

Bona fide nuclear receptor corepressors					
N-CoR	hNcoR1 mN-CoR Rip13 (partial)	Yes	NM_006311 U35312	Interacts with, and corepresses, TR, RAR, COUP-TFI and Rev-erb	Horlein *et al.*, 1995 Zamir *et al.*, 1996
SMRT	hSMRT hNCoR2 Trac2 (partial)	Yes	U37146 NM_006312	Corepresses TR and RAR and has sequence similarity with N-CoR	Chen and Evans, 1995
Sun-CoR	mSun-CoR		AF031426	Corepresses TR and Rev-erb	Zamir *et al.*, 1997
Others					
TRUP	Surf3		M36072	Interacts with the hinge region and the *N*-terminal portion of the ligand-binding domain of TR in a hormone-independent manner; exerts its inhibiting activity by interfering with DNA binding of the receptor to this element	Burris *et al.*, 1995
Tsg101				Interacts with AF1 of GR and represses transcriptional activity of GR in mammalian and in yeast cells	Hittelman *et al.*, 1999
REA	Bap37	Unclear		Isolated by interaction with the dominant negative ER receptor (L540Q); interacts also with the anti-estrogen-liganded ER; represses transcriptional activity of ER but not of PR, RAR or VP16; contains an LXXLL NR box motif but its function in interaction has not been demonstrated	Montano *et al.*, 1999
ALIEN	hALIEN dALIEN		AF120268 U57758	Shows expected properties as a corepressor of nuclear receptors in *Drosophila*; apparently plays the same role in mammalian nuclear receptors	Dressel *et al.*, 1999

continued overleaf

Table 2 *continued*

Human factors	Synonyms	NID motif	Accession number	Comments	Reference
Nuclear receptor-binding subunits of the SRB/Mediator containing complex (SMCC)					
Trap220	hDrip205 hDrip230 hTrip2 (partial)	Yes	AF055994	Included in the SMCC complex that interacts with liganded TR and VDR; enhances *in vitro* transcription by these receptors	Hittelman *et al.,* 1999; Ito *et al.*, 1999; Rachez *et al.*, 1999; Rachez *et al.*, 1998;
	mPBP		AF000294	mPBP was originally isolated by interaction with PPAR	Yuan *et al.*, 1998; Zhu *et al.*, 1997
Trap100	hDrip100	Unclear	AF055995	Drip100 does not interact directly with, VDR in a ligand-dependent manner; overexpression leads to coactivation first and then to squelching; nuclear receptor-interacting domain (NID) does not promote direct interaction with nuclear receptor	Zhang and Fondell, 1999 Rachez *et al.*, 1998
Trap170	hDrip150 hRgr1 hExlm1		AF135802	Drip150 interacts with, the coactivates AF1 of GR; may work in association with Drip205 (interacts with and coactivates GR AF2)	Hittelman *et al.*, 1999
Nuclear receptor-binding subunits of the SWI/SNF complex					
hSnf2a hSnf2b	hBrm Brg1		X72889	Co-immunoprecipitates with GR; both hBrm and Brg1 interact with ER in a ligand-dependent manner in yeast	Chiba *et al.*, 1994; Ichinose *et al.*, 1997; Muchardt and Yaniv, 1993
Factors interacting with nuclear receptors					
In a ligand and/or AF2-dependent manner					
Rip140	hErap140 mRip140	Yes	sw:P48542 AF053062	Interacts with, and coactivates, ER but corepresses TR2 orphan receptor and retinoid receptors	Cavaillès *et al.*, 1995; Halachmi *et al.*, 1994
Ara70	hEle1a hRfg hEle1b	Unclear	L49399	Reported to coactivate AR in DU145 cells; coactivates PPARγ activity in a ligand-dependent manner; possesses one LXXLL motif but this motif is not implicated in the interaction; isoform of hEle1a	Gao *et al.*, 1999; Yeh and Chang, 1996

Ara55			AF116343	Interacts with AR through LIM domain	Fujimoto *et al*., 1999
Ara54			AF049330	Another AR putative coactivator that contains a RING finger domain	Kang *et al*., 1999
Trip 1	mSug1	Yes	L38810 Z54219	Trip1 interacts with RXR, mSUG1 with different members of the nuclear receptor superfamily; Trip1 is a component of the 26S proteasome	Lee *et al*., 1995; vom Baur *et al*., 1996
Asc1			AF168418	Interacts with basal transcription factors (TBP and TFIIA), transcription integrators (Src1 and CBP/p300) and nuclear receptors. Although the interaction domain with nuclear receptor seems to be the hinge region, AF2 integrity is required	Kim *et al*., 1999
Asc2	Alb3	Unclear	AF177388	Interacts with RAR, TR, ERa and GR and binds also TFIIA, TBP, Src1 and CBP/p300; micro-injection of anti Asc2 antibody abrogated the ligand-dependent transactivation by RAR; is amplified in human cancers; contains two LXXLL motifs but implication in the interaction has not been demonstrated	Lee *et al*., 1999
p120		Unclear	AF016270	Interacts with TR ligand-binding domain; possesses a LSELL motif in the nuclear receptor interacting region	Monden *et al*., 1999
Tif1α	mTif1a	Yes	AF009353 S78221	Interacts with several nuclear receptors; belongs to the RBCC family of proteins; interacts also with KRAB domain-containing proteins, which are involved in determining heterochromatin structures	Le Douarin *et al*., 1995; vom Baur *et al*., 1996
Nrif3	EnL and EnS	Yes		Claimed to be a specific coactivator for TR and interacts through a variant of the NID motif LXXIL	Li *et al*., 1999
Trip230			AAD09135	Binds to, and coactivates, TR; binds also to pRb	Chang *et al*., 1997
L7	SPA			Interacts with, and coactivates, RU486-bound PR and enhances its partial agonist activity	Jackson *et al*., 1997
Tip60		Unclear	U74667	Interaction with AR ligand-binding domain enhanced by the ligand; able to coactivate PR and ER in transfections; contains a LXXLL motif but its function has not been demonstrated	Brady *et al*., 1999

continued overleaf

Table 2 *continued*

Human factors	Synonyms	NID motif	Accession number	Comments	Reference
Rap250		Yes	AF128458	Ligand-dependent or ligand-enhanced interaction with nuclear receptors; interaction involves the only LXXLL NR box; intrinsic glutamine-rich activation domain; widely expressed in reproductive organs	Caira *et al.*, 2000
Fhl2	DRAL	Unclear	NM_001450	Human four and a half LIM domain protein 2 (Fhl2); claimed to act as tissue-specific coactivator of the androgen receptor; nuclear protein; expressed in myocardium and prostate epithelium; contains autonomous transactivation function; coactivates AR but not other nuclear receptors	Muller *et al.*, 2000
In a ligand-independent manner					
Mbf1	hMbf1a hMbf1b		AB002282 AB002283	Example of a coactivator conserved throughout evolution; interaction implicates the DNA-binding domain and increases DNA binding	Kabe *et al.*, 1999
Ara160	TMF		NM_007114	Interacts with the AB domain of AR; however, the ligand enhances interaction suggesting that Ara160 may work in association with Ara70 to coactivate AR	Hsiao and Change, 1999
ORCA	hORCA		U46751	Factor that binds COUP-TFII *in vitro* and allows COUP-TFII to function as a transcriptional activator in mammalian cells; identical to a recently described ligand of the tyrosine kinase-signaling molecule p56(lck), suggesting that ORCA mediates crosstalk between mitogenic and nuclear receptor signal transduction pathways	Marcus *et al.*, 1996
Pgc1	mPgc1	Unclear	ÀF049330	First report of an inducible coactivator for a nuclear receptor; originally described for PPARγ but also interacts with other nuclear receptors; interaction is ligand independent and implicates the DNA-binding and the hinge region of the receptor; NID motif is not required for PPAR binding	Puigserver *et al.*, 1998
Pgc2	mPgc2		AF017433	Isolated using the AF1 region of PPARγ as bait, binds to, and increases, the transcriptional activity of PPARγ but does not interact with other PPARs or most other nuclear receptors	Castillo *et al.*, 1999

SNURF	rSNURF		AF022081	Isolated by two hybrid interaction using AR DNA-binding domain as bait. Interaction is ligand dependent but SNURF is able to stimulate both basal and ligand induced transcription	Moilanen *et al*., 1998
NCoA-62			AF045184	Fished out with VDR as bait; interacts with the ligand-binding domain even in the absence of the ligand and coactivates vitamin D3, retinoic acid, estrogen, and glucocorticoid-mediated gene expression in mammalian cells	Baudino *et al*., 1998
Rap46	hBag1 hBag1L		Z35491	Originally identified as a protein able to interact with several nuclear receptors including GR, ER and TR; the ligand was not strictly required for interaction; subsequent cloning of a longer form, Bag1L that co-immunoprecipitates and coactivates AR in PC3 transfections; Rap46 was described as a negative regulator of GR and RAR	Froesch *et al*., 1998; Kullmann *et al*., 1998; Liu *et al*., 1998; Zeiner *et al*., 1997
Mip224	TBP7		NM_006503	Isolated using the two hybrid system and the orphan receptor MB67 as a bait; in transfection experiments Mip224 inhibits the transcriptional activation by MB67; Mip224 is a component of the 26S proteasome	Choi *et al*., 1996
Pias1		Unclear	AF167160	Binds the DNA-binding domain + ligand-binding domain of AR in a ligand-dependent manner in yeast; *in vitro*, binds the isolated DNA-binding domain; coactivates AR and PR ligand-dependent transactivation but represses PR in mammalian cells; first cloned as an inhibitor of Stat1; contains an LXXLL motif but its role in interaction has not been demonstrated	Tan *et al*., 2000
Arip3	hPiasxy rArip3	Unclear	 AF044058	Binds the DNA-binding domain of AR and modulates AR-dependent transcription; belongs to the PIAS family of proteins; contains an LXXLL motif but its role in interaction has not been demonstrated	Moilanen *et al*., 1999
p68		Unclear	AF015812	Interacts with the A/B domain of hERα; phosphorylation of hERαSer(118) potentiates the interation with p68; enhances the AF1 activity; contains an LXXLL motif but its role in interaction has not been demonstrated	Endoh *et al*., 1999

continued overleaf

Table 2 *continued*

Human factors	Synonyms	NID motif	Accession number	Comments	Reference
With both liganded and non-liganded receptors					
Nsd1	mNSD1	Yes	AF064553	Interacts through NID-L with unliganded RAR and TR and through NID+L with liganded RAR, TR, RXR and ER; has properties of both coactivator and corepressor	Huang *et al.*, 1998
Zac1b	Zac1		NM_009538	Variant of Zac1, a putative transcriptional activator involved in apoptosis and cell-cycle regulation; isolated as a protein that binds to the C-terminal region of the coactivator Tif2/Grip1; interacts with CBP, p300 and nuclear receptors in yeast 2 hybrid and *in vitro* experiments; contains autonomous transactivation domain; acts as coactivator but in some cell lines, and with certain promoters, Zac1b acts as repressor of nuclear receptor action	Huang and Stallcup, 2000
Other factors that interact directly or indirectly with nuclear receptors					
Stat3	mStat3		U30709	Associates with ligand-bound GR to form a transactivating/signaling complex that can function through either an IL6-responsive element or a glucocorticoid-responsive element	Zhang *et al.*, 1997
Stat5	mStat5		U48730 U21110	GR can act as a transcriptional coactivator for Stat5 and enhance Stat5-dependent transcription	Stocklin *et al.*, 1996
E6-AP	mE6-AP	Unclear	AF016708 U96636	Interacts with, and coactivates, the transcriptional activity of PR in a ligand-dependent manner; however, in the Angelman syndrome, the phenotype results from the defect in the ubiquitin-proteosome-mediated degradation of E6-AP	Nawaz *et al.*, 1999
Cyclin D1		Unclear	M64349	Direct physical binding of Cyclin D1 to the ligand-binding domain increases binding of the receptor to its element and either activates transcription in the absence of estrogen or enhances transcription in its presence	Zwijsen *et al.*, 1997
Hmg1	mHmg1		X12597 U00431	Coregulatory proteins that increase the DNA-binding and transcriptional activity of steroid nuclear receptors	Boonyaratanakornkit *et al.*, 1998

Hmg2	mHmg2		X62534 U00431	cf Hmg1	Boonyaratanakornkit *et al.*, 1998
Bcl3		Unclear	U05681	Able to interact with RXR both in a ligand-independent manner, with a fragment comprising the ABC domain, and in a ligand-dependent manner, with the ligand-binding domain; interacts with the general transcription factors TFIIB, TBP and TFIIA, but not with TFIIEα in the GST pull-down assays and enhances the 9-*cis* RA-2 induced transactivation of RXR; possesses an LXXLL motif but mutation in the motif doesn't abolish interaction	Na *et al.*, 1998
Vpr		Yes	U71182	Interacts directly with the GR and general transcription factors, acting as a coactivator via a LXXLL signature motif	Kino *et al.*, 1999
TLS			AF071213	Interaction with the DNA-binding domain of RXR, ER, TR and GR	Powers *et al.*, 1998
dUTPase			U62891	PPARα-interacting protein. Interacts *in vitro* with all three isoforms of mouse PPAR, but not with RXR and TR; this interaction seems to inhibit PPAR/RXR resulting in an inhibition of PPAR activity in a ligand-independent manner	Chu *et al.*, 1996
Smad3	mSmad3		AB004930 AF016189.1t	Specific coactivator for ligand-induced transactivation of VDR by forming a complex with a member of the steroid receptor coactivator-1 protein family	Yanagisawa *et al.*, 1999
E1A				Interacts directly with TR via the DNA-binding domain and carboxy-terminal part of the ligand-binding domain in a ligand-independent manner	Wahlstrom *et al.*, 1999
hnRNP U	Grip120 mhnRNP U		NM_004501 AF073992	Scaffold attachment region and RNA-binding protein that co-immunoprecipitates with GR	Eggert *et al.*, 1997
Tif1b	KAP 1		NM_005762	Enhances the transcription of the *agp* gene by GR in a ligand and GRE dependent manner, even if the interaction is ligand independent	Chang *et al.*, 1998
TDG			NM_003211	Implicated in the reparation of TG mismatch; interacts through a region containing α-helix1 of the ligand-binding domain with RAR and RXR	Um *et al.*, 1998
Tsc2			NM_000548	Cloned by interaction with RXR in yeast; coactivates PPARγ and VDR in mammalian cells but no clear demonstration of a direct interaction	Henry *et al.*, 1998

Moreover, it appears that simultaneous interaction of coactivators with both the AF1 and the AF2 of a nuclear receptor accounts for the synergy between the two transcriptional activation functions. It is thus possible that AF1 and AF2 may not be, in all cases, independent from each other in the context of the entire receptor. Rather the two are separate parts of a single coactivator-recruiting surface. Hence, coactivator specificity will have to be reanalyzed, particularly for receptors with proven AF1 activities, after taking account of the effects of both the AF1 and AF2-binding surface(s). It should also be noted that for some nuclear receptors the N and C-terminal domains interact with each other even in the absence of coactivators (Alen *et al.*, 1999; He *et al.*, 1999; Tetel *et al.*, 1999). In keeping with the great sequence divergence between the different A/B domains of nuclear receptors, for AF1 recognition, the coactivators do not invoke a single short interaction motif similar to the NR box. However, a 'Q-rich' region (Voegel *et al.*, 1996; 1998) appears to be critical for AF1 interaction. The structural features of the A/B domains that are recognized by the coactivator have not yet been defined. Notably, in the case of the estrogen receptor ERβ, specific phosphorylation of the A/B domain by MAPK pathways results in enhanced recruitment of the coactivator SRC1 (Tremblay *et al.*, 1999). In addition to the *bona fide* coactivators, other putative coactivator proteins have been reported to interact with AF1 and contribute to its activity (*Table 2*). An interesting candidate is p68 (Endoh *et al.*, 1999), which interacts with the A/B domain of hERα; phosphorylation of hERαSer118 potentiates the interaction with p68.

The identification of specific nuclear receptor coactivators has prompted the question of how they function on a molecular level in transcription. To this end several observations have been made. It is now generally accepted that nuclear receptor coactivators possess or directly recruit enzymatic activities, and that they form large coactivator complexes. CBP, p300, P/CAF, SRC1, P/CIP and GCN5 are all histone acetyltransferases (Chen, 2000; Glass and Rosenfeld, 2000; Kuo and Allis, 1998). They can acetylate specific residues in the *N*-terminal tails of different histones, a process that is believed to play an important role in the opening of chromatin during transcriptional activation (Chapter 3) as well as in non-histone targets (Bayle and Crabtree, 1997; Chen *et al.*, 1999a; Gu *et al.*, 1997; Imhof *et al.*, 1997). Other coactivators, such as TIF2 and RIP140, do not have direct activity, but probably recruit histone acetyl transferases. Specifically, the activation domain AD1 of TIF2 has been demonstrated to function by recruiting the cointegrator CBP (Voegel *et al.*, 1998), which apparently in turn acetylates TIF2 (Chen *et al.*, 1999a). Besides histone acetyl transferase other enzymatic activities have been attributed to nuclear receptor coactivator complexes. Thus TIF2 proteins can interact functionally with a protein methyltransferase via their activation domain AD2 (Chen *et al.*, 1999b). Although unproven, it is believed that the methyltransferase activity changes the activity of the basal transcription machinery, and it will be of great interest to identify its specific targets.

In conclusion, *bona fide* coactivators, i.e. members of the TIF2/SRC1/RAC3 family, together with the CBP/p300 cointegrators (see *Table 2*) function by rendering the chromatin environment of a nuclear receptor-target gene favorable for transcription. This open chromatin environment is the result of intrinsic or recruited histone acetyltransferase activity. The histone acelyltransferase activities of different coactivators/coactivator complexes targets (i) the N-termini of histones, which have reduced DNA-binding activity upon acetylation, (ii) certain basal transcription factors and (iii) at least some *bona fide* coactivators themselves. The chromatin modification step represents the first of two independent steps in transcription activation by nuclear receptors (see *Figure 8*).

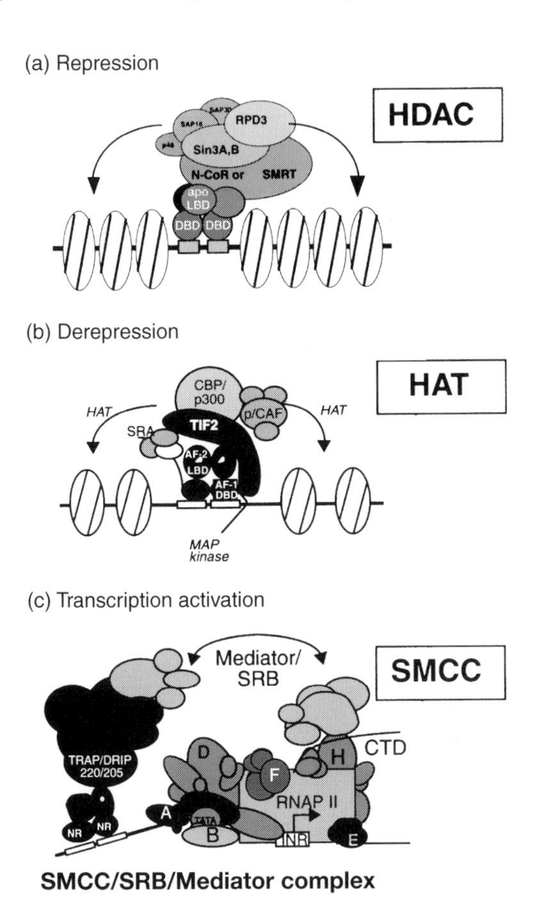

Figure 8. Three-step mechanism of nuclear receptor action. (a) Some nuclear receptors act as silencers of target gene transcription in the absence of a ligand (or in the presence of certain antagonists). This repression step is due to the recruitment by the apo-nuclear receptor of a corepressor complex that exerts histone deacetylase (HDAC) activity. Note that for repression to occur by this mechanism the receptor has to be able to interact with its target gene promoter in the absence of ligand. (b) Ligand binding dissociates this complex and recruits the coactivator complex that displays histone acetyltransferase (HAT) activity. The subsequent chromatin decondensation (derepression) is believed to be necessary, but not sufficient, for target gene activation. (c) At the third step, the HAT complex dissociates, possibly due to acetylation of the coactivator, which decreases its ability to interact with the receptor, and the SMCC/DRIP/TRAP complex is assembled through its TRAP220/DRIP205 subunit. The SMCC complex is able to establish contacts with the basal transcription machinery, resulting in transcription initiation. Modified from Freedman, L.P. (1999) Increasing the complexity of coactivation in nuclear receptor signaling. *Cell* 97: 5–8, with permission from Cell Press.

5.2 Nuclear receptor corepressors and HDACs

The second class of nuclear receptor coregulators comprises the corepressors (*Table 2*). Early on, it had been learned that some nuclear receptors do actively repress transcription when in the apo form. This phenomenon had been particularly well established for retinoic acid and thyroid hormone receptors (Burcin *et al.*, 1994). Soon after the identification of ligand-recruited coactivators, similar approaches identified proteins that recognize the ligand-free nuclear receptor (Chen and Evans, 1995; Hörlein *et al.*, 1995; Zamir *et al.*, 1997) of which nuclear receptor co-repressor (NCoR) and silencing mediator for retinoid and thyroid hormone receptors (SMRT) are the best characterized. Both contain a conserved corepressor nuclear receptor (CoRNR) box motif that interacts with a surface on the ligand-binding domain. This surface is topologically very similar to that recognized by coactivator LXXLL motifs, except that it does not include helix H12 (Hu *et al.*, 1999; Nagy *et al.*, 1999; Perissi *et al.*, 1999). Corepressors are believed to reside in, or recruit, high-molecular-weight complexes that oppose the activity of coactivator complexes. While coactivator complexes acetylate histones thereby weakening the interaction of the *N*-terminal histone tails with the nucleosomal DNA, corepressors recruit histone deacetylase activities that reverse this process (*Figure 8a*). Deacetylated histones are associated with silent regions of the genome, and it is generally accepted that histone acetylation and deacetylation shuffle nucleosomal targets between condensed and relaxed chromatin configurations, the latter being requisite for transcriptional activation. It is unresolved whether all nuclear receptors, especially steroid receptors, are capable of active repression. However, recent studies have shown that estrogen, glucocorticoid and progesterone receptors all bind to corepressors in presence of certain antagonists (Jackson *et al.*, 1997; Wagner *et al.*, 1998; Zhang *et al.*, 1998a).

An intriguing example of a nuclear receptor coregulator is Nsd1 (Huang *et al.*, 1998). This factor has been reported to contain both transcriptional activation and repression domains. Nsd1 is a STET family protein and contains coactivator-binding LXXLL motifs. It will be of interest to further elucidate the molecular mechanisms by which this factor is able to switch from corepressor to coactivator upon ligand binding to the associated nuclear receptor.

Given this high number of different coregulators for nuclear receptors, two principal questions emerge: what defines coactivator selectivity and how is the assembly of different coactivator complexes with different intrinsic transcription activities regulated? Future research will have to address such questions, especially in view of potential therapeutic applications (see Section 7).

5.3 Recruitment of the RNA polymerase II holoenzyme – the second step of coactivation

The initial chromatin-modifying step carried out by nuclear receptor coactivators (that has been described already) is followed be the actual recruitment of the pol II holoenzyme and initiation of transcription (*Figure 8c*) (Freedman, 1999). The mechanism of recruitment of the pol II holoenzyme by nuclear receptors was clarified by the identification and cloning of the mammalian Mediator complex, which was found to consist of thyroid hormone and vitamin D-receptor coactivators (Fondell *et al.*, 1996b, Ito, *et al.*, 1999; Rachez *et al.*, 1998). The mammalian mediator comes in several 'flavors'. It was identified as the so-called SMCC (SRB and Mediator protein-containing complex) (Gu *et al.*, 1999), the thyroid hormone receptor-associated protein (TRAP) complex (Fondell *et al.*, 1996b), and the vitamin D-receptor-interacting protein (DRIP) complex (Rachez *et al.*, 1998). Furthermore, common subunits are shared with PC2, the so-called positive coactivator-2 (Kretzschmar *et al.*, 1994), as well as the activator-recruited cofactor (ARC) (Näär *et al.*, 1999), cofactor required for Sp1 (CRSP) (Ryu and Tjian, 1999) and negative regulator of activated transcription (NAT) (Sun *et al.*, 1998) complexes. Hereafter, all of these related complexes are referred to as 'SMCC'. It is a large multi-subunit protein complex that contains several homologues of the yeast Mediator complex (Rgr1, Med6, Med7, Srb7, Srb10, Srb11, Nut2, Soh1) (Gu *et al.*. 1999; Ito *et al.*, 1999) as well as additional proteins of unknown function. As expected for a Mediator complex, SMCC associates with pol II to form pol II holoenzyme (Ito *et al.*, 1999; R.G. Roeder, personal communication). The SMCC complex is also able to interact functionally with different transcription factors such as p53 and VP16 (Ito *et al.*, 1999). Furthermore, due to its identification as thyroid hormone and vitamin D receptor-interacting complex, it is believed to function as a nuclear receptor coactivator. This notion is supported by the demonstration that SMCC can enhance thyroid hormone and vitamin D transcription activation in *in vitro* transcription systems (Fondell *et al.*, 1996b, 1999, Rachez *et al.*, 1998; 1999). The SMCC subunit responsible for interaction with the agonist-bound ligand-binding domain of nuclear receptors has been identified as TRAP220/DRIP205, which contains a functional LXXLL NR box motif (Yang and Freedman, 1999; Yuan *et al.*, 1998). Interestingly, another subunit of the SMCC complex, DRIP150, interacts with the N-terminal AF1 region of the glucocorticoid receptor (Hittelmann *et al.*, 1999). The current working hypothesis is that once the chromatin environment at target gene promoters has been decondensed by coactivator complexes containing members of the TIF2 and CBP families, the nuclear receptor recruits pol II holoenzyme via its association with the TRAP220/DRIP205 subunit of the SMCC. This switch between coactivators and the SMCC complex might be regulated by acetylation of the coactivators within the histone acetyltransferase complex, which results in their dissociation from the nuclear receptor and thus allows the recruitment of factors such as SMCC via the LXXLL motif of the TRAP220/DRIP205 subunit (Chen *et al.*, 1999a). The recruitment of the pol II holoenzyme might also be enhanced by interactions of nuclear receptors with components of the SWI/SNF complex (Gusin and Wolffe, 1999), which is also part of the pol II holoenzyme (*Table 2*).

5.4 Interaction of nuclear receptors with components of the basal transcription machinery

The first hypotheses about possible mechanisms of transcription activation by some nuclear receptors were based on *in vitro* studies demonstrating direct interactions with components of the basal machinery. Amongst those are TBP and TAFs (for a detailed list and short description see *Table 3*). These interactions were reported to promote the transcriptional activity of nuclear receptors, at least in transient transfection studies. Whether active nuclear receptors recruit both the holoenzyme and the TAF complex(es) independently during the second activation step remains to be established. $TF_{II}H$ has been found to phosphorylate certain nuclear receptors at AF1 (Rochette-Egly *et al.*, 1997). The fact that some nuclear receptor coactivators interact in a phosphorylation-dependent manner with AF1s suggests an important relationship (Hammer *et al.*, 1999; Tremblay *et al.*, 1999). However, these issues require further investigation.

Table 3. Factors of the basal machinery reported to interact, directly or indirectly, with nuclear receptors

Basal factor	References
TBP	
RXR interaction requires ligand and intact AF2	Schulman *et al.*, 1995
ER interaction does not require ligand or intact AF2	Sadovsky *et al.*, 1995
TR interaction could be involved in repression	Fondell *et al.*, 1996a
TFIIB	
COUP-TFI, ER and PR COUP/TFI interaction is direct; ER interaction seems to involve AF2 function	Ing *et al.*, 1992
TR interaction seems to promote either AF1 activation or AF2 repression	Baniahmad *et al.*, 1993 Fondell *et al.*, 1996
VDR is cell specific; depending on the cell type, in the presence of ligand, promotes either activation of transcription (P19 cells) or repression (3T3 cells); 2 hybrid interaction and *in vitro* techniques have been used to show this interaction; however, the requirement for ligand in this interaction is controversial	Blanco *et al.*, 1995 MacDonald *et al.*, 1995 Masuyama *et al.*, 1997
TFIID	
$hTAF_{II}30$:ER interaction requires the DEF region of the receptor; ligand independent; an antibody directed against $hTAF_{II}30$ inhibits ER transactivation	Jacq *et al.*, 1994
$dTAF_{II}110$:PR interaction – implicates the DNA-binding domain of the receptor and is ligand independent	Schwerk *et al.*, 1995
$dTAF_{II}110$:RXR interacts in the 2 hybrid system in a AF2 and ligand-dependent manner but not *in vitro*	Schulman *et al.*, 1995
$dTAF_{II}110$:TR interacts both *in vitro* and in the 2 hybrid system; the ligand inhibits *in vitro* but has no effect in the two hybrid system	Petty *et al.*, 1996
$hTAF_{II}28$ stimulates AF2 function in RXR, ER, VDR and RAR in transfection but no direct interaction has been shown	May *et al.*, 1996
$hTAF_{II}135$ stimulates AF2 function in RAR, VDR and TR but no direct interaction has been shown	Mengus *et al.*, 1997
$hTAF_{II}55$ interacts with α-helices H3 to H5 of VDR and TR in a ligand-independent manner	Lavigne *et al.*, 1996
TFIIH	
RAR binds TFIIH and free CAK *in vitro*; Cdk7 phosphorylation of Ser77, located in the AB region enhances transcriptional activation	Rochette-Egly *et al.*, 1997

The allosteric changes induced by nuclear receptor-ligand binding are of fundamental importance. These allosteric changes define protein interaction and recruitment cascades which assemble complexes that (i) modify chromatin condensation through complex-associated enzymatic activities and (ii) recruit, allow, or enhance the formation of transcription initiation complexes at target gene promoters.

6. Platforms for multiple signal integration

It has become increasingly well documented in the past few years that nuclear receptor action is not confined to the positive and negative regulation of the expression of cognate target genes. Indeed these receptors, and most likely their 'downstream' mediators, are targets of other signaling pathways and reciprocally, can modify the activity of such pathways. The best known example of such signal transduction 'crosstalk' is the mutual repression of nuclear receptor and AP1 (c-Fos/c-Jun) activities. A distinct type of crosstalk is the modification of nuclear receptor AF activity by phosphorylation, e.g., by the mitogen-activated protein (MAP) kinase pathway. The existence of signal transduction crosstalk is likely to provide a mechanism for integrating nuclear receptor action with the functional state of the cell in which it is expressed. The importance of crosstalk in 'real life' was impressively demonstrated by the observation that glucocorticoid receptor null mice die at birth (Heck *et al.*, 1994), whereas mice harboring a glucocorticoid receptor mutant (GR dim/dim) that can still crosstalk with AP1 but not activate glucocorticoid receptor-target genes are viable (Reichardt *et al.*, 1998).

6.1 AP1 nuclear receptor crosstalk

AP1/glucocorticoid receptor transrepression is mutual and requires a specific but uncharacterized state of the receptor that can be induced by both agonists and certain, but not all, antagonists (Göttlicher *et al.*, 1998; Pfahl, 1993; Resche-Rigon and Gronemeyer, 1998). It is important to point out that the nuclear receptor–AP1 crosstalk does not *per se* imply negative regulation of transcription; several reports show that under certain conditions this crosstalk can lead to positive transcriptional effects (Bubulya *et al.*, 1996; Pearce *et al.*, 1998; Shemshedini *et al.*, 1991).

AP1 transrepression has been studied intensively because of the possibility of exploiting this mechanism pharmacologically for anti-inflammatory treatment. Moreover, AP1 is an immediate–early gene product that has been implicated in the proliferative responses of cells, and AP1 tranrepression could potentially provide a means of inhibiting tumor cell proliferation. AP1 transrepression is found to be dependent on the promoter structure and dimer composition of the AP1 transcription factor (Pearce *et al.*, 1998). Furthermore, AP1 nuclear receptor crosstalk is also cell type-dependent (Shemshedini *et al.*, 1991), which probably reflects the involvement of other cellular factors with varying expression as mediators of this mutual transrepression event.

Controversial insight into this phenomenon has been gained through studies with CBP. This coactivator, which functions for both AP1 and nuclear receptors, can overcome transrepression when overexpressed in cells with limiting endogenous levels of this protein (Fronsdal *et al.*, 1998; Kamei *et al.*, 1996). Other studies however, have shown that this mechanism of sequestration of CBP can only partially account for transrepression. In particular, some synthetic ligands, even antagonists, can induce AP1 transrepression without coactivator recruitment (Chen *et al.*, 1995). A second mechanism that has been proposed is based on the observation that estrogen receptors are capable of downregulating the activity of Jun-kinase, which leads to reduced AP1 activity (Caelles *et al.*, 1997). Thirdly, nuclear receptor-mediated effects on the dimerization of the AP1 subunits have been observed (Zhou *et al.*, 1999). A fourth mechanism proposed to be involved in AP1 nuclear receptor crosstalk is direct physical contact between both factors when they are bound to so-called 'composite elements' on a promoter (Pearce and Yamamoto, 1993; Pearce *et al.*, 1998). Composite elements are thought to recruit both AP1 proteins and nuclear receptors, which brings them into close physical contact. Depending on the nature of such response elements and on the activity of the participating proteins, steric effects might lead to differential regulation. It is currently not clear whether this last mechanism applies only to promoters that carry composite elements or whether it has a more general significance. Further analysis is required to understand the relative contributions of these four mechanisms to receptor/AP1 crosstalk.

An intriguing addition to the AP1 signaling crosstalk theme was the recent discovery that the two estrogen receptors, ERα and ERβ, differ in their ability to modulate transcription driven from AP1 sites in response to synthetic estrogens, notably raloxifene (Paech *et al.*, 1997; Webb *et al.*, 1995; 1999). The molecular explanations for such different activities in transcriptional interference will provide new ideas about how synthetic ligands might discriminate between, and hence differentially affect, the various functions associated with nuclear receptors.

6.2 NF-κB and nuclear receptor crosstalk

Another good example of transcription factor crosstalk is the mutual interference between glucocorticoid receptors and NF-κB proteins. This pathway is again highly significant, since it may also contribute to the anti-inflammatory as well as the osteoporotic action of glucocorticoids. Crosstalk may not be the only anti-NF-κB activity of glucocorticoids as it also stimulates production of the NF-κB inhibitory molecule IκB, which in turn can sequester active NF-κB in the cell nucleus (Auphan *et al.*, 1995; Scheinmann *et al.*, 1995). However, studies with mutant receptors and 'dissociated' glucocorticoids (glucocorticoids that are able to transrepress AP1 activity but have little or no activation potential for cognate GR target genes) show that IκB-independent repression of NF-κB activity is incompatible with a simple IκB-mediated mechanism (Heck *et al.*, 1997). The underlying molecular events are still elusive, but may be related to those discussed earlier on AP1 crosstalk. As for AP1 (Chen *et al.*, 1995), the squelching of limiting amounts of CBP, which also coactivates NF-κB, is unlikely to be involved because glucocorticoid receptor ligands that dissociate transactivation from transrepression still induce transcriptional interference with NF-κB signaling (Heck *et al.*, 1997).

Several other transcription factors are also transregulated by nuclear receptors. These include Oct 2A, RelA (another NF-κB family member), Stat5 and Spil/PU.1 (Göttlicher *et al.*, 1998).

6.3 Post-translational modification by phosphorylation

Phosphorylation is increasingly being recognized as a signaling crosstalk mechanism that affects most, if not all, nuclear receptors. The major phosphorylation targets are the A/B, E and F regions. Phosphorylation has been reported to occur via the MAPK, protein kinase A and C (PKA and PKC) pathways, as well as by the Cdk7 subunit of the general transcription factor TFIIH and the CyclinA/CDK2 complex (Adams *et al.*, 1997; Camp *et al.*, 1999; Chen *et al.*, 1999c; Hu *et al.*, 1996; Joel *et al.*, 1995; Kato *et al.*, 1995; Rochette-Egly *et al.*, 1997; Rogatsky *et al.*, 1999; Shao *et al.*, 1998). The functional consequences of receptor phosphorylation have been defined in a few cases. For example, phosphorylation of estrogen receptors in their A/B regions by MAP kinases enhances the transactivation potential of the associated activation function AF1 (e.g. Kato *et al.*, 1995). Conversely, MAP kinase-dependent phosphorylation of PPARγ inhibited its transcriptional activity (Hu *et al.*, 1996). It is tempting to speculate that phosphorylation of the A/B region of nuclear receptors can, positively or negatively, modulate coregulator association. Indeed, it has been recently reported that phosphorylation by MAP kinase of ERβ region A/B led to a constitutive recruitment of SRC1 (Tremblay *et al.*, 1999). Notably, enhanced receptor-SRC1 binding was also induced *in vivo* in cells treated with epidermal growth factor or expressing activated Ras (Tremblay *et al.*, 1999). On the other hand, it can not be concluded that phosphorylation is critically required for AF1 activity. Thus RAR, phosphorylated by $TF_{II}H$, show moderate enhancement of transcriptional activity. However, reintroduction of RAR mutated at the corresponding serine residues into RAR knock-out F9 cell lines still rescues some, but not all, of the observed phenotypes (Taneja *et al.*, 1997). It is tempting to speculate that the phosphorylation fine-tunes the activity of AF1 but is dispensable for basal activity. Both AF1 activity (Berry *et al.*, 1990) and the phosphorylation of serine residues in the A/B domains are cell and promoter-context dependent, so the full significance of AF1 phosphorylation warrants further investigation.

Other sites of phosphorylation are found in the E and F domains of nuclear receptors. The functional consequences of region E/F phosphorylation is not understood. Preliminary data suggest roles in coregulator specificity (Hong *et al.*, 1998) and protein degradation (Zhang *et al.*, 1998b). One remarkable phosphorylation event has been described for the estrogen receptor ERα. This receptor is phosphorylated by a protein kinase A-dependent pathway on Ser537. This serine is located in the helix H3 of the ligand-binding domain. Once targeted, the phosphoserine induces a conformational change that renders the ERα constitutively active (Yudt *et al.*, 1999). Alternatively, this phosphoserine might, together with other residues in the vicinity, constitute a novel coactivator-binding surface. This effect is highly significant since it bypasses the requirement for a specific ligand to activate the receptor.

7. Deregulation in disease and novel therapeutic targets

Given the major impact of nuclear receptor signaling on animal physiology it is no surprise to find aberrant nuclear receptor function at the basis of multiple pathologies. Indeed, synthetic agonists and antagonists are in clinical use for endocrine cancer therapies and hormone replacement therapies in osteoporosis. Thiazolidinediones that were known as insulin 'sensitizers' in the treatment of non-insulin-dependent diabetes have now been recognized as PPARγ agonists. Furthermore, the recent establishment of a link between human type 2 diabetes and PPARγ mutation has proven that PPARγ malfunction can lead to severe insulin resistance, diabetes mellitus and hypertension (Barroso *et al.*, 1999). Moreover, the impact of the contraceptive pill on social life is evident, as is that of RU486 for pregnancy termination. But this is only the tip of the iceberg. Novel nuclear receptor-type drugs are expected to be developed based on our increasing knowledge of the structural and molecular details of nuclear receptor and ligand function, and on the elucidation of the signaling pathways involved in (patho) physiological events.

7.1 Novel perspectives for therapy

In addition to the well-established endocrine therapies of breast and prostate cancers with estrogen and androgen receptor antagonists, respectively, and on the more recent differentiation therapy of acute promyelocytic leukemia by retinoids (Chapter 12), novel synthetic nuclear receptor ligands are of considerable interest for the therapy and prevention of other types of cancers (Lotan, 1996; Hong and Sporn, 1997). The development of novel nuclear receptor ligands has been facilitated by recent pharmacological and chemical developments, such as (i) combinatorial chemistry, computer-assisted ligand docking based on ligand-binding domain crystal structure and ultra-high throughput screening with nuclear receptor-based reporter systems, (ii) the ability to dissociate nuclear receptor-associated functions, such as transactivation and crosstalk, from other signaling pathways, and (iii) the ability to generate receptor and receptor isotype-selective ligands.

It is tempting to speculate that coactivators are not entirely promiscuous in their choice of nuclear receptors. For example only AIB1/RAC3 is found to be amplified in breast cancer cells while the expression level of the other two family members remains constant, which reflects estrogen receptor specificity (Anzick *et al.*, 1997). Furthermore, in some types of acute myeloid leukemia a chromosomal translocation specifically fuses a monocytic zinc-finger protein of unknown function (MOZ) to the *C*-terminus of TIF2 and not to any of the other two TIF2 family members, again reflecting a bias towards one specific coactivator (Carapeti *et al.*, 1998; Liang *et al.*, 1998). Interestingly, the observation has been made that overexpression of coactivators of the TIF2 family can lead, in some systems, to ligand-independent activity under certain conditions (Voegel *et al.*, 1998). This suggests that transcriptional mediators may possibly be involved in the origin and/or progression of neoplastic diseases and may become novel pharmacological targets. Indeed, mutations of the CBP gene have been implicated in the mental retardation and malformations of the Rubinstein-Taybi syndrome, and alterations of the p300 gene were found associated with gastric and colorectal carcinomas (Giles *et al.*, 1998). Of interest are also studies in which the oncogenic capacity of E1A was compromised by coexpression of p300 (Chakravarti *et al.*, 1999). Based on these various data it is tempting to speculate that alterations in the cellular abundance of coregulators, or altered substrate specificity of coactivator-associated acetylation or deacetylation functions, may lead to pathological states.

Synthetic ligands of nuclear receptors are classified as agonist and antagonist with respect to a particular receptor-associated function (Resche-Rigon and Gronemeyer, 1998). This discrimination is not always obvious, since a particular ligand might antagonize some activities while functioning as agonist for other activities. The ERα antagonists hydroxytamoxifen and ICI164.384 exemplify this ambiguity. Hydroxytamoxifen antagonizes the activation function AF2 but it acts as an agonist for AF1, whereas ICI164.384 antagonizes both AF1 and AF2 (Berry *et al.*, 1990). Similarly certain retinoid receptor antagonist are agonists for AP1 repression (Chen *et al.*, 1995). It is therefore important to consider, where possible, the molecular basis of the anticipated action of a nuclear receptor-based drug to increase efficacy and limit side-effects. If the molecular mechanism is unknown, it may be wise to use screening paradigms that will consider the multiple dimensions of receptor activities.

Another twist to the classification of synthetic nuclear receptor ligands results from the availability of isotype-specific ligands. These compounds affect one isotype of a nuclear receptor but not another. The results obtained with such isotype selective retinoids, together with gene ablation studies and the known interspecies conservation of receptor isoforms, have established that each of the three RAR genes has a distinctive spectrum of functions (Kastner *et al.*, 1995). Given the pharmacological potential of retinoids, the development of isotype-specific ligands has attracted much attention. Today a wealth of synthetic retinoids exist that display either isotype specificity or act as mixed agonists/antagonists for the three RARs (Gehin *et al.*, 1999). Some of these retinoids were found to display cell-specific effects and their pharmacological potential is currently under investigation. RXR isotype-specific ligands are also being developed, which is of particular interest since RXR is a promiscuous heterodimerization partner in a number of signaling pathways. A recent report suggests that RXR ligands may stimulate insulin action in non-insulin-dependent-diabetes (Mukherjee *et al.*, 1997) through a PPARγ/RXR heterodimer that is responsive to thiazolidinediones. It is thus conceivable that pathway-specific RXR ligands can be generated.

The recent gain in understanding of nuclear receptor coregulator function at the molecular level sets the ground for new strategies of pharmacological interference within nuclear receptor-signaling pathways. The nuclear receptor-coactivator and nuclear receptor-corepressor interfaces are composed of precisely defined features. The coactivator's amphipatic α-helical chain containing the LXXLL NR box motif, the corepressor's CoRNR box motif, and the nuclear receptor's hydrophobic cleft plus charge clamp on the surface of the ligand-binding domain (Darimont *et al.*, 1998; Feng *et al.*, 1998; Hu and Lazar, 1999; Nolte *et al.*, 1998; Shiau *et al.*, 1998), are all potential targets for disruption by interaction with small synthetic molecules. From current structural information, it is feasible to screen combinatorial peptides containing the core LXXLL motif for high-affinity binding to the hydrophobic groove on the ligand-binding domain. A phage display approach has already provided a library of nuclear receptor-interacting peptides (Chang *et al.*, 1999). These peptides have been found to be active (in transfected cells) and could be used in probing surfaces of nuclear receptors that have been differentially generated in the presence of agonists and antagonists (Norris *et al.*, 1999; Paige *et al.*, 1999).

8. Conclusions

Nuclear receptors are ligand-regulated transcription factors that have evolved from an ancestral orphan receptor into a highly diverse family present throughout the entire animal kingdom. They encompass receptors for the steroid and non-steroid hormones, vitamins, corticoids and metabolic intermediates. These receptors signal through endocrine, paracrine, autocrine and intracrine modes of action to regulate multiple aspects of animal physiology, including homeostasis, development and reproduction. They regulate target genes through direct binding as mono, homo or heterodimers at cognate response elements. They also indirectly modulate other gene expression programs ('signal transduction crosstalk'). Through the coordinated expression of genetic programs, nuclear receptors contribute to cell fate-determining processes thereby shaping and sustaining the organism. All of these processes result from one fundamental interaction, the signal binding of the ligand, which induces a major allosteric change in the ligand-binding domain. This allosteric change is transformed into cascades of protein–protein recognitions, which lead to the establishment of coregulator/cointegrator complexes that are able to recruit and/or activate the basal transcriptional machinery at target gene promoters.

For some nuclear receptors, direct transcriptional repression in the absence of ligand or the presence of certain antagonists is mediated by corepressor complexes that are associated with the unliganded receptor. These corepressors condense the chromatin environment at the promoter region through histone deacetylation. Corepressors interact by virtue of their CoRNR boxes with non-liganded nuclear receptors. Upon ligand binding, the allosteric change in the ligand-binding domain induces corepressor dissociation, and coactivator complexes are recruited. *Bona fide* coactivators recognize the active nuclear receptor ligand-binding domain (AF2) via conserved LXXLL NR boxes and often also recognize the *N*-terminal activation function AF1. The NR boxes of coactivators and the CoRNR boxes of corepressors bind to topologically similar sites in the ligand-binding domain, but the surfaces of these sites are entirely distinct due to the agonist-induced conformational changes. In particular the H12 helix of the apo-receptor is required for coactivator binding but is incompatible with corepressor binding. Coactivator complexes reverse the repression by acetylated chromatin and allow access by the basal transcription machinery. In a subsequent step the mammalian SMCC mediator is recruited to the nuclear receptor and apparently stabilizes the formation of the pre-initiation complex at target gene promoters. SMCC recruitment might also be regulated by acetylation, with subsequent dissociation of TIF2 family members, thus allowing SMCC-receptor association.

Despite their direct actions on the chromatin environment and the transcription machinery, nuclear receptors also regulate transcription by positive and negative interference with other signaling pathways. Different mechanisms for such transcription factor crosstalk have been described, but none have been fully accepted or explain all aspects of the particular crosstalk. The activity of nuclear receptors is also regulated by phosphorylation that may serve to fine-tune the signaling and establish a link to other signaling pathways. Finally, the promoter context of target genes, and the temporal order of incoming signals on a particular promoter probably adjust the transcriptional potential of nuclear receptors to particular situations. Taken together, nuclear receptors serve as platforms to coordinate cognate signals with those emanating from other signaling pathways thereby integrating the nuclear receptor signal into the functional context of cellular state and activity.

Nuclear receptors and their coregulators have been implicated in several diseases. Their role as key regulatory molecules in a wide variety of signaling pathways qualifies them as novel pharmacological targets. The ongoing improvement of synthetic nuclear receptor ligands with altered specificity is likely to improve therapy and reduce side-effects.

Future research on nuclear receptors still has to answer important questions. What are the constituents of the genetic programs that are governed by a given nuclear receptor? How are the nuclear receptor signals matched and complemented with other signaling cascades? What are the precise molecular events leading to the variety of transcriptional effects exerted by nuclear receptors? Once these questions have been addressed adequately, specific interference into these immensely complex systems might lead to the successful control and reprogramming of an organism's physiology and pathology. Understanding of nuclear receptor-controlled transcription will shed light on the general and signaling pathway-selective control of all gene expression. In this respect the use of gene arrays, together with the information derived from the genome sequencing, will have enormous impact.

9. Acknowledgments

Thanks to William Bourguet for his help in generating the crystal structure figures. Arnott Benecke was an EC Marie Curie fellow. This work was supported by funds from the Institut National de la Santé et de la Recherche Médicale, the Centre National de la Recherche Scientifique, the Ligue Nationale contre le Cancer, the Hôpital Universitaire de Strasbourg, and Bristol-Myers-Squibb.

When constructing this review we were faced with an enormous amount of original literature. In order to keep the list of references to a reasonable size we have referred to recent comprehensive reviews. We apologize to all our colleagues whose original contributions have not been cited directly.

References

Adam-Stitah, S., Penna, L., Chambon, P. and Rochette-Egly, C. (1999) Hyperphosphorylation of the retinoid X receptor alpha by activated c-Jun NH_2-terminal kinases. *J. Biol. Chem.* **274**: 18932–18941.

Adams, M., Reginato, M.J., Shao, D., Lazar, M.A. and Chatterjee, V.K. (1997) Transcriptional activation by peroxisome proliferator-activated receptor gamma is inhibited by phosphorylation at a consensus mitogen-activated protein kinase site. *J. Biol. Chem.* **272**: 5128–5132.

Alen, P., Claessens, F., Verhoeven, G., Rombauts, W. and Peeters, B. (1999) The androgen receptor amino-terminal domain plays a key role in p160 coactivator-stimulated gene transcription. *Mol. Cell. Biol.* **19**: 6085–6097.

Anzick, S.L., Kononen, J., Walker, R.L. *et al.* (1997) AIB1, a steroid receptor coactivator amplified in breast and ovarian cancer. *Science* **277**: 965–968.

Auphan, N., DiDonato, J.A., Rosette, C., Helmberg, A. and Karin, M. (1995) Immunosuppression by glucocorticoids: inhibition of NF-kappa B activity through induction of I kappa B synthesis. *Science* **270**: 286–290.

Baniahmad, A., Ha, I., Reinberg, D., Tsai, S., Tsai, M.J. and O'Malley, B.W. (1993) Interaction of human thyroid hormone receptor beta with transcription factor TFIIB may mediate target gene derepression and activation by thyroid hormone. *Proc. Natl Acad. Sci. USA* **90**: 8832–8836.

Barak, Y., Nelson, M.C., Ong, E.S. *et al.* (1999) PPAR gamma is required for placental, cardiac, and adipose tissue development. *Mol. Cell* **4**: 585–595.

Barroso, I., Gurnell, M., Crowley, V.E. *et al.* (1999) Dominant negative mutations in human PPARgamma associated with severe insulin resistance, diabetes mellitus and hypertension. *Nature* **402**: 880–883.

Baudino, T.A., Kraichely, D.M., Jefcoat, S.C., Jr., Winchester, S.K., Partridge, N.C. and MacDonald, P.N. (1998) Isolation and characterization of a novel coactivator protein, NCoA-62, involved in vitamin D-mediated transcription. *J. Biol. Chem.* **273**: 16434–16441.

Baumann, C.T., Lim, C.S. and Hager, G.L. (1999) Intracellular localization and trafficking of steroid receptors. *Cell. Biochem. Biophys.* **31**: 119–127.

Bayle, J.H. and Crabtree, G.R. (1997) Protein acetylation: more than chromatin modification to regulate transcription. *Chem. Biol.* **4**: 885–888.

Beato, M., Herrlich, P. and Schutz, G. (1995) Steroid hormone receptors: many actors in search of a plot. *Cell* **83**: 851–857.

Benoit, G., Altucci, L., Flexor, M. *et al.* (1999) RAR-independent RXR signaling induces t(15;17) leukemia cell maturation. *EMBO J.* **18**: 7011–7018.

Berrevoets, C.A., Doesburg, P., Steketee, K., Trapman, J. and Brinkmann, A.O. (1998) Functional interactions of the AF2 activation domain core region of the human androgen receptor with the amino-terminal domain and with the transcriptional coactivator TIF2. *Mol. Endocrinol.* **12**: 1172–1183.

Berry, M., Metzger, D. and Chambon, P. (1990) Role of the two activating domains of the oestrogen receptor in the cell-type and promoter-context dependent agonistic activity of the anti-oestrogen 4-hydroxytamoxifen. *EMBO J.* **9**: 2811–2818.

Bevan, C.L., Hoare, S., Claessens, F., Heery, D.M. and Parker, M.G. (1999) The AF1 and AF2 domains of the androgen receptor interact with distinct regions of SRC1. *Mol. Cell Biol.* **19**: 8383–8392.

Blanco, J.C., Minucci, S., Lu, J. *et al.* (1998) The histone acetylase PCAF is a nuclear receptor coactivator. *Genes Dev.* **12**: 1638–1651.

Blanco, J.C., Wang, I.M., Tsai, S.Y. *et al.* (1995) Transcription factor TFIIB and the vitamin D receptor cooperatively activate ligand-dependent transcription. *Proc. Natl Acad. Sci. USA* **92**: 1535–1539.

Bocquel, M.T., Kumar, V., Stricker, C., Chambon, P. and Gronemeyer, H. (1989) The contribution of the *C*-terminal regions of steroid receptors to activation of transcription is both receptor and cell-specific. *Nucl. Acids Res.* **17**: 2581–2595.

Boonyaratanakornkit, V., Melvin, V., Prendergast, P. *et al.* (1998) High-mobility group chromatin proteins 1 and 2 functionally interact with steroid hormone receptors to enhance their DNA binding *in vitro* and transcriptional activity in mammalian cells. *Mol. Cell Biol.* **18**: 4471–4487.

Bourguet, W., Vivat, V., Wurtz, J.-M., Chambon, P., Gronemeyer, H. and Moras, D. (2000) Crystal structure of a heterodimeric complex of RAR and RXR ligand-binding domains. *Mol. Cell* **5**: 289–329.

Bourguet, W., Ruff, M., Chambon, P., Gronemeyer, H. and Moras, D. (1995) Crystal structure of the ligand-binding domain of the human nuclear receptor RXR-alpha. *Nature* **375**: 377–382.

Brady, M.E., Ozanne, D.M., Gaughan, L. *et al.* (1999) Tip60 is a nuclear hormone receptor coactivator. *J. Biol. Chem.* **274**: 17599–17604.

Brzozowski, A.M., Pike, A.C., Dauter, Z. *et al.* (1997) Molecular basis of agonism and antagonism in the oestrogen receptor. *Nature* **389**: 753–758.

Bubulya, A., Wise, S.C., Shen, X.Q., Burmeister, L.A. and Shemshedini, L. (1996) c-Jun can mediate androgen receptor-induced transactivation. *J. Biol. Chem.* **271**: 24583–24589.

Burcin, M., Kohne, A.C., Runge, D., Steiner, C. and Renkawitz, R. (1994) Factors influencing nuclear receptors in transcriptional repression. *Semin. Cancer Biol.* **5**: 337–346.

Burris, T.P., Nawaz, Z., Tsai, M.J. and O'Malley, B.W. (1995) A nuclear hormone receptor-associated protein that inhibits transactivation by the thyroid hormone and retinoic acid receptors. *Proc. Natl Acad. Sci. USA* **92**: 9525–9529.

Caelles, C., Gonzalez-Sancho, J.M. and Munoz, A. (1997) Nuclear hormone receptor antagonism with AP1 by inhibition of the JNK pathway. *Genes Dev.* **11**: 3351–3364.

Caira, F., Antonson, P., Pelto-Huikko, M., Treuter, E. and Gustafsson, J.A. (2000) Cloning and characterization of RAP250, a novel nuclear receptor coactivator. *J. Biol. Chem.* **275**: 5308–5317.

Camp, H.S., Tafuri, S.R. and Leff, T. (1999) c-Jun *N*-terminal kinase phosphorylates peroxisome proliferator-activated receptor-gamma1 and negatively regulates its transcriptional activity. *Endocrinology* **140**: 392–397.

Carapeti, M., Aguiar, R.C.T., Goldman, J.M. and Cross, N.C.P. (1998) A novel fusion between MOZ and the nuclear receptor coactivator TIF2 in acute myeloid leukemia. *Blood* **91**: 3127–3133.

Castillo, G., Brun, R.P., Rosenfield, J.K. *et al.* (1999) An adipogenic cofactor bound by the differentiation domain of PPARgamma. *EMBO J.* **18**: 3676–3687.

Cavaillès, V., Dauvois, S., L'Horset, F. *et al.* (1995) Nuclear factor RIP140 stimulates transcriptional activation by the estrogen receptor. *EMBO J.* **14**: 3741–3751.

Cavaillès, V., Dauvois, S., Danielian, P.S. and Parker, M.G. (1994) Interaction of proteins with transcriptionally active estrogen receptors. *Proc. Natl Acad. Sci. USA* **91**: 10009–10013.

Chakravarti, D., Ogryzko, V., Kao, H.Y., Nash, A., Chen, H., Nakatani, Y. and Evans, R.M. (1999) A viral mechanism for inhibition of p300 and PCAF acetyltransferase activity. *Cell* **96**: 393–403.

Chambon, P. (1996) A decade of molecular biology of retinoid acid receptors. *FASEB J* **10**: 940–954.

Chang, K.H., Chen, Y., Chen, T.T. *et al.* (1997) A thyroid hormone receptor coactivator negatively regulated by the retinoblastoma protein. *Proc. Natl Acad. Sci. USA* **94**: 9040–9045.

Chang, C.J., Chen, Y.L. and Lee, S.C. (1998) Coactivator TIF1beta interacts with transcription factor C/EBPβ and glucocorticoid receptor to induce alpha1-acid glycoprotein gene expression. *Mol. Cell. Biol.* **18**: 5880–5887.

Chang, C.Y., Norris, J.D., Gron, H. *et al.* (1999) Dissection of the LXXLL nuclear receptor-coactivator interaction motif using combinatorial peptide libraries: discovery of peptide antagonists of estrogen receptors alpha and beta. *Mol. Cell. Biol.* **19**: 8226–8239.

Chen, H., Lin, R.J., Xie, W., Wilpitz, D. and Evans, R.M. (1999a) Regulation of hormone induced histone hyperacetylation and gene activation via acetylation of an acetylase. *Cell* **98**: 675–686.

Chen, J.D. (2000) Steroid/nuclear receptor coactivators. *Vitam. Horm.* **58**: 391–448.

Chen, J.D., Ma, H., Hong, H. *et al.* (1999b) Regulation of transcription by a protein methyltransferase. *Science* **284**: 2174–2177.

Chen, J.D., Pace, P.E., Coombes, R.C. and Ali, S. (1999c) Phosphorylation of human estrogen receptor alpha by protein kinase A regulates dimerization. *Mol. Cell Biol.* **19**: 1002–1015.

Chen J.D. and Li, H. (1998) Coactivation and corepression in transcriptional regulation by steroid/nuclear hormone receptors. *Crit. Rev. Eukaryot. Gene Expr.* **8**: 169–190.

Chen, J.D. and Evans, R.M. (1995) A transcriptional co-repressor that interacts with nuclear hormone receptors. *Nature* **377**: 454–457.

Chen, J.Y., Clifford, J., Zusi, C. *et al.* (1996) Two distinct actions of retinoid-receptor ligands. *Nature* **382**: 819–822.

Chen, J.Y., Penco, S., Ostrowski, J. *et al.* (1995) RAR-specific agonist/antagonists which dissociate transactivation and AP1 transrepression inhibit anchorage-independent cell proliferation. *EMBO J.* **14**: 1187–1197.

Cheskis, B. and Freedman, L.P. (1994) Ligand modulates the conversion of DNA-bound vitamin D3 receptor (VDR) homodimers into VDR-retinoid X receptor heterodimers. *Mol. Cell Biol.* **14**: 3329–3338.

Chiang, M.Y., Misner, D., Kempermann, G. *et al.* (1998) An essential role for retinoid receptors RARb and RXRg in long-term potentiation and depression. *Neuron* **21**: 1353–1361.

Chiba, H., Muramatsu, M., Nomoto, A. and Kato, H. (1994) Two human homologues of *Saccharomyces cerevisiae* SWI2/SNF2 and *Drosophila* brahma are transcriptional coactivators cooperating with the estrogen receptor and the retinoic acid receptor. *Nucl. Acids Res.* **22**: 1815–1820.

Choi, H.S., Seol, W. and Moore, D.D. (1996) A component of the 26S proteasome binds on orphan member of the nuclear hormone receptor superfamily. *J. Steroid Biochem. Mol. Biol.* **56**: 23–30.
Chu, R., Lin, Y., Rao, M.S. and Reddy, J.K. (1996) Cloning and identification of rat deoxyuridine triphosphatase as an inhibitor of peroxisome proliferator-activated receptor alpha. *J. Biol. Chem.* **271**: 27670–27676.
Cole, T.J., Blendy, J.A., Monaghan, A.P. *et al.* (1995) Targeted disruption of the glucocorticoid receptor gene blocks adrenergic chromaffin cell development and severely retards lung maturation. *Genes Dev.* **9**: 1608–1621.
Collingwood, T.N., Urnov, F.D. and Wolffe, A.P. (1999) Nuclear receptors: coactivators, corepressors and chromatin remodeling in the control of transcription. *J. Mol. Endocrinol.* **23**: 255–275.
Collingwood, T.N., Butler, A., Tone, Y., Clifton-Bligh, R.J., Parker, M.G. and Chatterjee, V.K. (1997) Thyroid hormone-mediated enhancement of heterodimer formation between thyroid hormone receptor beta and retinoid X receptor. *J. Biol. Chem.* **272**: 13060–13065.
Darimont, B.D., Wagner, R.L., Apriletti, J.W. *et al.* (1998) Structure and specificity of nuclear receptor-coactivator interactions. *Genes Dev.* **12**: 3343–3356.
Ding, X.F., Anderson, C.M., Ma, H., Hong, H. *et al.* (1998). Nuclear receptor-binding sites of coactivators glucocorticoid receptor interacting protein 1 (GRIP1) and steroid receptor coactivator 1 (SRC-1): multiple motifs with different binding specificities. *Mol. Endocrinol.* **12**: 302–313.
Dressel, U., Thormeyer, D., Altincicek, B. *et al.* (1999) Alien, a highly conserved protein with characteristics of a corepressor for members of the nuclear hormone receptor superfamily. *Mol. Cell. Biol.* **19**: 3383–3394.
Eggert, M., Michel, J., Schneider, S. *et al.* (1997) The glucocorticoid receptor is associated with the RNA-binding nuclear matrix protein hnRNP U. *J. Biol. Chem.* **272**: 28471–28478.
Endoh, H., Maruyama, K., Masuhiro, Y *et al.* (1999) Purification and identification of p68 RNA helicase acting as a transcriptional coactivator specific for the activation function 1 of human estrogen receptor alpha. *Mol. Cell Biol.* **19**: 5363–5372.
Escriva, H., Safi, R., Hanni, C. *et al.* (1997) Ligand binding was acquired during evolution of nuclear receptors. *Proc. Natl Acad. Sci. USA* **94**: 6803–6808.
Feng, W.F., Ribeiro, R.C.J., Wagner, R.L. *et al.* (1998) Hormone dependent coactivator binding to a hydrophobic cleft on nuclear receptors. *Science* **280**: 1747–1749.
Fondell, J.D., Guermah, M., Malik, S. and Roeder, R.G. (1999) Thyroid hormone receptor-associated proteins and the general positive cofactor mediate thyroid hormone receptor function in the absence of TATA box-binding protein-associated factors of TFIID. *Proc. Natl Acad. Sci. USA* **96**: 1959–1964.
Fondell, J.D., Brunel, F., Hisatake, K. and Roeder, R.G. (1996a) Unliganded thyroid hormone receptor alpha can target TATA-Binding protein for transcriptional repression. *Mol. Cell Biol.* **16**: 281–287.
Fondell, J.D., Ge, H. and Roeder, R.G. (1996b) Ligand induction of a transcriptionally active thyroid hormone receptor coactivator complex. *Proc. Natl Acad. Sci.* **93**: 8329–8333.
Forman, B.M., Umesono, K., Chen, J. and Evans, R.M. (1995) Unique response pathways are established by allosteric interactions among nuclear hormone receptors. *Cell* **81**(4): 541–550.
Freedman, L.P. (1999) Increasing the complexity of coactivation in nuclear receptor signaling. *Cell* **97**: 5–8.
Froesch, B.A., Takayama, S. and Reed, J.C. (1998) BAG-1L protein enhances androgen receptor function. *J. Biol. Chem.* **273**: 11660–11666.
Fronsdal, K., Engedal, N., Slagsvold, T. and Saatcioglu, F. (1998) CREB binding protein is a coactivator for the androgen receptor mediated cross-talk with AP1. *J.Biol. Chem.* **273**: 31853–31859.
Fujimoto, N., Yeh, S., Kang, H.Y. *et al.* (1999) Cloning and characterization of androgen receptor coactivator, ARA55, in human prostate. *J. Biol. Chem.* **274**: 8316–8321.
Gao, T., Brantley, K., Bolu, E. and McPhaul, M.J. (1999) RFG (ARA70, ELE1) interacts with the human androgen receptor in a ligand-dependent fashion, but functions only weakly as a coactivator in cotransfection assays. *Mol. Endocrinol.* **13**: 1645–1656.
Gehin, M., Vivat, V., Wurtz, J.M. *et al.* (1999) Stuctural basis for engineering of retinoic acid receptor isotype-selective agonists and antagonists. *Chem. Biol.* **6**: 519–529.
Gewirth, D.T. and Sigler, P.B. (1995) The basis for half-site specificity explored through a non-cognate steroid receptor-DNA complex. *Nat. Struct. Biol.* **2**: 386–394.

Giguere, V. (1999) Orphan nuclear receptors: from gene to function. *Endocrinol. Rev.* **20**: 689–725.

Giles, R.H., Peters, D.J.M. and Breuning, M.H. (1998) Conjunction dysfunction: CBP/p300 in human disease. *Trends Genet.* **14**: 178–183.

Gill, G. and Ptashne, M. (1988) Negative effect of the transcriptional activator GAL4. *Nature* **334**(6184): 721–724.

Glass, C.K. and Rosenfeld, M.G. (2000) The coregulator exchange in transcriptional functions of nuclear receptors. *Genes Dev.* **14**: 121–141.

Göttlicher, M., Heck, S. and Herrlich, P. (1998) Transcriptional cross-talk, the second mode of steroid hormone receptor action. *J. Mol. Med.* **76**: 480–489.

Gronemeyer, H. and Laudet, V. (1995) Protein Profile 2(11). *Transcription factors 3: nuclear receptors*, 1173–1308.

Gu, W., Malik, S., Ito, M. *et al.* (1999) A novel human SRB/MED-containing cofactor complex, SMCC, involved in transcription regulation. *Mol. Cell* **3**: 97–108.

Gu, W. and Roeder, R.G. (1997). Activation of p53 sequence-specific DNA binding by acetylation of the p53 *C*-terminal domain. *Cell* **90**: 595–606.

Guiochon-Mantel, A., Delabre, K., Lescop, P. and Milgrom, E. (1994) Nuclear localization signals also mediate the outward movement of proteins from the nucleus. *Proc. Natl Acad. Sci. USA* **91**: 7179–7183.

Gushin, D. and Wolffe, A.P. (1999) Transcriptional control: SWItched on mobility. *Curr. Biol.* **9**: 742–746.

Halachmi, S., Marden, E., Martin, G., MacKay, H., Abbondanza, C. and Brown, M. (1994) Estrogen receptor-associated proteins: possible mediators of hormone-induced transcription. *Science* **264**: 1455–1458.

Hammer, G.D., Krylova, I., Zhang, Y. *et al.* (1999) Phosphorylation of the nuclear receptor SF-1 modulates cofactor recruitment: integration of hormone signaling in reproduction and stress. *Mol. Cell.* **3**: 521–526.

He, B., Kemppainen, J.A., Voegel, J.J., Gronemeyer, H. and Wilson, E.M. (1999) Activation function 2 in the human androgen receptor ligand binding domain mediates interdomain communication with the NH(2)-terminal domain. *J. Biol. Chem.* **274**: 37219–37225.

Heck, S., Kullmann, M., Gast, A. *et al.* (1994) A distinct modulating domain in glucocorticoid receptor monomers in the repression of activity of the transcription factor AP-1. *EMBO J.* **13**: 4087–4095.

Heck, S., Bender, K., Kullmann, M., Gottlicher, M., Herrlich, P. and Cato A.C. (1997) I kappaB alpha-independent downregulation of NF-κB activity by glucocorticoid receptor. *EMBO J.* **16**: 4698–4707.

Heery, D.M., Kalkhoven, E., Hoare, S. and Parker, M.G. (1997) A signature motif in transcriptional co-activators mediates binding to nuclear receptors. *Nature* **387**: 733–736.

Henry, K.W., Yuan, X., Koszewski, N.J., Onda, H., Kwiatkowski, D.J. and Noonan, D.J. (1998) Tuberous sclerosis gene 2 product modulates transcription mediated by steroid hormone receptor family members. *J. Biol. Chem.* **273**: 20535–20539.

Hittelman, A.B., Burakov, D., Iniguez-Lluhi, J.A., Freedman, L.P. and Garabedian, M.J. (1999) Differential regulation of glucocorticoid receptor transcriptional activation via AF-1-associated proteins. *EMBO J.* **18**: 5380–5388.

Hong, W.K. and Sporn, M.B. (1997) Recent advances in chemoprevention of cancer. *Science* **278**: 1073–1077.

Hong, S.H., Darimont, B.D., Ma, H., Yang, L., Yamamoto, K.R. and Stallcup, M.R. (1999) An additional region of coactivator GRIP1 required for interaction with the hormone-binding domains of a subset of nuclear receptors. *J. Biol. Chem.* **274**: 3496–3502.

Hong, S.H., Wong, C.W. and Privalski, M.L. (1998) Signaling by tyrosine kinases negatively regulates the interaction between transcription factors and SMRT (silencing mediator of retinoic acid and thyroid hormone receptor) corepressor. *Mol. Endocrinol.* **12**: 1161–1171.

Hong, S.H., Kohli, K., Garabedian, M.J. and Stallcup, M.R. (1997) GRIP1, a transcriptional coactivator for the AF-2 transactivation domain of steroid, thyroid, retinoid, and vitamin D receptors. *Mol. Cell Biol.* **17**: 2735–2744.

Hörlein, A.J., Näär, A.M., Heinzel, T. *et al.* (1995) Ligand-independent repression by the thyroid hormone receptor mediated by a nuclear receptor co-repressor [see comments]. *Nature* **377**: 397–404.

Hsiao, P.W. and Chang, C. (1999) Isolation and characterization of ARA160 as the first androgen receptor N-terminal-associated coactivator in human prostate cells. *J. Biol. Chem.* **274**: 22373–22379.

Hu, E., Kim, J.B., Sarraf, P. and Spiegelman, B.M. (1996) Inhibition of adipogenesis through MAP kinase-mediated phosphorylation of PPARγ. *Science* **274**: 2100–2103.

Hu, X. and Lazar, M.A. (1999) The CoRNR motif controls the recruitment of corepressors by nuclear hormone receptors. *Nature* **402**: 93–96.

Huang, N., vom Baur, E., Garnier, J.M. *et al.* (1998) Two distinct nuclear receptor interaction domains in NSD1, a novel SET protein that exhibits characteristics of both corepressors and coactivators. *EMBO J.* **17**: 3398–3412.

Huang S.M. and Stallcup, M.R. (2000) Mouse zac1, a transcriptional coactivator and repressor for nuclear receptors. *Mol. Cell. Biol.* **20**: 1855–1867.

Ichinose, H., Garnier, J.M., Chambon, P. and Losson, R. (1997) Ligand-dependent interaction between the estrogen receptor and the human homologues of SWI2/SNF2. *Gene* **188**: 95–100.

Imhof, A., Yang, X.J., Ogryzko, V.V., Nakatani, Y., Wolffe, A.P. and Ge, H. (1997) Acetylation of general transcription factors by histone acetyltransferases. *Curr. Biol.* **7**: 689–692.

Ing, N.H., Beekman, J.M., Tsai, S.Y., Tsai, M.J. and O'Malley, B.W. (1992) Members of the steroid hormone receptor superfamily interact with TFIIB (S300-II). *J. Biol. Chem.* **267**: 17617–17623.

Ito, M., Yuan, C.X., Malik, S. *et al.* (1999) Identity between TRAP and SMCC complexes indicates novel pathways for the function of nuclear receptors and diverse mammalian activators. *Mol. Cell.* **3**: 361–370.

Jackson, T.A., Richer, J.K., Bain, D.L. *et al.* (1997) The partial agonist activity of antagonist-occupied steroid receptors is controlled by a novel hinge domain-binding coactivator L7/SPA and the corepressors N-CoR or SMRT. *Mol. Endocrinol.* **11**: 693–705.

Jacq, X., Brou, C., Lutz, Y., Davidson, I., Chambon, P. and Tora, L. (1994) Human TAFII30 is present in a distinct TFIID complex and is required for transcriptional activation by the estrogen receptor. *Cell* **79**: 107–117.

Joel, P.B., Traish, A.M. and Lannigan, D.A. (1995) Estradiol and phorbol ester cause phosphorylation of serine 118 in the human estrogen receptor. *Mol. Endocrinol.* **9**: 1041–1052.

Kabe, Y., Goto, M., Shima, D. *et al.* (1999) The role of human MBF1 as a transcriptional coactivator. *J. Biol. Chem.* **274**: 34196–34202.

Kakizawa, T., Miyamoto, T., Kaneko, A., Yajima, H., Ichikawa, K. and Hashizume, K. (1997) Ligand-dependent heterodimerization of thyroid hormone receptor and retinoid X receptor. *J. Biol. Chem.* **272**: 23799–23804.

Kamei, Y., Xu, L., Heinzel, T *et al.* (1996) A CBP integrator complex mediates transcriptional activation and AP-1 inhibition by nuclear receptors. *Cell* **85**: 403–414.

Kang, H.Y., Yeh, S., Fujimoto, N. and Chang, C. (1999) Cloning and characterization of human prostate coactivator ARA54, a novel protein that associates with the androgen receptor. *J. Biol. Chem.* **274**: 8570–8576.

Kastner, P., Mark, M. and Chambon, P. (1995) Nonsteroid nuclear receptors: what are genetic studies telling us about their role in real life? *Cell* **83**: 859–869.

Kastner, P., Mark, M., Ghyselinck, N. *et al.* (1997a) Genetic evidence that the retinoid signal is transduced by heterodimeric RXR/RAR functional units during mouse development. *Development* **124**: 313–326.

Kastner, P., Messaddeq, N., Mark, M. *et al.* (1997b) Vitamin A deficiency and mutations of RXRalpha, RXRbeta and RARalpha lead to early differentiation of embryonic ventricular cardiomyocytes. *Development* **124**: 4749–4758.

Kato, S., Endoh, H., Masuhiro, Y. *et al.* (1995) Activation of the estrogen receptor through phosphorylation by mitogen-activated protein kinase. *Science* **270**: 1491–1494.

Kim, H.J., Yi, J.Y., Sung, H.S., Moore, D.D., Jhun, B.H., Lee, Y.C. and Lee, J.W. (1999) Activating signal cointegrator 1, a novel transcription coactivator of nuclear receptors, and its cytosolic localization under conditions of serum deprivation. *Mol. Cell. Biol.* **19**: 6323–6332.

Kino, T., Gragerov, A., Kopp, J.B., Stauber, R.H., Pavlakis, G.N. and Chrousos, G.P. (1999) The HIV-1 virion-associated protein vpr is a coactivator of the human glucocorticoid receptor. *J. Exp. Med.* **189**: 51–62.

Klaholz, B.P., Renaud, J.P., Mitschler, A. *et al.* (1998) Conformational adaptation of agonists to the human nuclear receptor RARγ. *Nat. Struct. Biol.* **5**: 199–202.

Kliewer, S.A., Lehmann, J.M. and Willson, T.M. (1999) Orphan nuclear receptors: shifting the endocrinology into reverse. *Science* **284**: 757–760.

Kretzschmar, M., Stelzer, G., Roeder, R.G. and Meisterernst, M. (1994) RNA polymerase II cofactor PC2 faciliates activation of transcription by GAL4-AH *in vitro*. *Mol. Cell. Biol.* **14**: 3927–3937.

Krust, A., Green, S., Argos, P *et al.* (1986) The chicken oestrogen receptor sequence: homology with v-erbA and the human oestrogen and glucocorticoid receptors. *EMBO J.* 5: 891–897.

Kullmann, M., Schneikert, J., Moll, J. *et al.* (1998) RAP46 is a negative regulator of glucocorticoid receptor action and hormone-induced apoptosis. *J. Biol. Chem.* **273**: 14620–14625.

Kuo, M.-H. and Allis, C.D. (1998) Roles of histone acetyltransferases and deacetylases in gene regulation. *BioEssays* **20**: 615–626.

Kurokawa, R., Kalafus, D., Ogliastro, M.H. *et al.* (1998) Differential use of CREB binding protein coactivator complexes. *Science* **279**: 700–703.

Kurokawa, R., DiRenzo, J., Boehm, M. *et al.* (1994) Regulation of retinoid signalling by receptor polarity and allosteric control of ligand binding. *Nature* **371**: 528–531.

Kurokawa, R., Yu, V.C., Naar, A. *et al.* (1993) Differential orientations of the DNA-binding domain and carboxy-terminal dimerization interface regulate binding site selection by nuclear receptor heterodimers. *Genes Dev.* **7**: 1423–1435.

Lanz, R.B., McKenna, N.J., Onate, S.A. *et al.* (1999) A steroid receptor coactivator, SRA, functions as an RNA and is present in an SRC-1 complex. *Cell* **97**: 17–27.

Lavigne, A.C., Mengus, G., May, M. *et al.* (1996) Multiple interactions between hTAF(II)55 and other TFIID subunits – Requirements for the formation of stable ternary complexes between hTAF(II)55 and the TATA-binding protein. *J. Biol. Chem.* **271**: 19774–19780.

Le Douarin, B., Nielsen, A.L., Garnier, J.M. *et al.* (1996) A possible involvement of TIF1α and TIF1β in the epigenetic control of transcription by nuclear receptors. *EMBO J.* **15**: 6701–6715.

Le Douarin, B., Zechel, C., Garnier, J.M. *et al.* (1995) The *N*-terminal part of TIF1, a putative mediator of the ligand-dependent activation function (AF-2) of nuclear receptors, is fused to B-raf in the oncogenic protein T18. *EMBO J.* **14**: 2020–2033.

Lee, J.W., Ryan, F., Swaffield, J.C., Johnston, S.A. and Moore, D.D. (1995) Interaction of thyroid-hormone receptor with a conserved transcriptional mediator. *Nature* **374**: 91–94.

Lee, S.K., Anzick, S.L., Choi, J.E. *et al.* (1999) A nuclear factor, ASC-2, as a cancer-amplified transcriptional coactivator essential for ligand-dependent transactivation by nuclear receptors *in vivo*. *J. Biol. Chem.* **274**: 34283–34293.

Li, H., Gomes, P.J. and Chen, J.D. (1997) RAC3, a steroid/nuclear receptor-associated coactivator that is related to SRC-1 and TIF2. *Proc. Natl Acad. Sci. USA* **94**: 8479–8484.

Li, D., Desai-Yajnik, V., Lo, E., Schapira, M., Abagyan, R. and Samuels, H.H. (1999) NRIF3 is a novel coactivator mediating functional specificity of nuclear hormone receptors. *Mol. Cell. Biol.* **19**: 7191–7202.

Liang, J., Prouty, L., Williams, B.J., Dayton, M.A. and Blanchard, K.L. (1998) Acute mixed lineage leukemia with an inv(8)(p11q13) resulting in fusion of the genes for MOZ and TIF2. *Blood* **92**: 2118–2122.

Liu, R., Takayama, S., Zheng, Y. *et al.* (1998) Interaction of BAG-1 with retinoic acid receptor and its inhibition of retinoic acid-induced apoptosis in cancer cells. *J. Biol. Chem.* **273**: 16985–16992.

Lotan, R. (1996) Retinoids in cancer chemoprevention. *FASEB J.* **10**: 1031–1039.

Luisi, B.F., Xu, W.X., Otwinowski, Z., Freedman, L.P., Yamamoto, K.R. and Sigler, P.B. (1991) Crystallographic analysis of the interaction of the glucocorticoid receptor with DNA. *Nature* **352**: 497–505.

Luo, X., Ikeda, Y. and Parker, K.L. (1994) A cell-specific nuclear receptor is essential for adrenal and gonadal development and sexual differentiation. *Cell* **77**: 481–490.

Ma, H., Hong, H., Huang, S.M. *et al.* (1999) Multiple signal input and output domains of the 160-kilodalton nuclear receptor coactivator proteins. *Mol. Cell. Biol.* **19**: 6164–6173.

MacDonald, P.N., Sherman, D.R., Dowd, D.R., Jefcoat, S.C., Jr. and De Lisle, R.K. (1995) The vitamin D receptor interacts with general transcription factor IIB. *J. Biol. Chem.* **270**: 4748–4752.

Marcus, S.L., Winrow, C.J., Capone, J.P. and Rachubinski, R.A. (1996) A p56(lck) ligand serves as a coactivator of an orphan nuclear hormone receptor. *J. Biol. Chem.* **271**: 27197–27200.

Mark, M., Ghyselinck, N.B., Wendling, O. *et al.* (1999) A genetic dissection of the retinoid signalling pathway in the mouse. *Proc. Nutr. Soc.* **58**: 609–613.

Mascrez, B., Mark, M., Dierich, A., Ghyselinck, N.B., Kastner, P. and Chambon, P. (1998) The RXRalpha ligand-dependent activation function 2 (AF-2) is important for mouse development. *Development* **125**: 4691–4707.

Masuyama, H., Jefcoat, S.C. and MacDonald, P.N. (1997) The N-terminal domain of transcription factor IIB is required for direct interaction with the vitamin D receptor and participates in vitamin D-mediated transcription. *Mol. Endocrinol.* **11**: 218–228.

May, M., Mengus, G., Lavigne, A.C., Chambon, P. and Davidson, I. (1996) Human TAF(II)28 promotes transcriptional stimulation by activation function 2 of the retinoic X receptors. *EMBO J.* **15**: 3093–3104.

McInerney, E.M., Rose, D.W., Flynn, S.E. *et al.* (1998) Determinants of coactivator LXXLL motif specificity in nuclear receptor transcriptional activation. *Genes Dev.* **12**: 3357–3368.

McKenna, N.J., Lanz, R.B. and O'Malley, B.W. (1999b) Nuclear receptor coregulators: cellular and molecular biology. *Endocrinol. Rev.* **20**: 321–344.

Meinke, G. and Sigler, P.B. (1999) DNA-binding mechanism of the monomeric orphan nuclear receptor NGF1-B. *Nat. Struct. Biol.* **6**: 471–477.

Mengus, G., May, M., Carre, L., Chambon, P. and Davidson, I. (1997) Human TAF(II)135 potentiates transcriptional activation of AF-2s of the retinoic acid, vitamine D3, and thyroid hormone receptors in mammalian cells. *Genes Dev.* **11**: 1381–1395.

Meyer, M.E., Gronemeyer, H., Turcotte, B., Bocquel, M.T., Tasset, D. and Chambon, P. (1989) Steroid hormone receptors compete for factors that mediate their enhancer function. *Cell* **57**: 433–442.

Moilanen, A.M., Karvonen, U., Poukka, H. *et al.* (1999) A testis-specific androgen receptor coregulator that belongs to a novel family of nuclear proteins. *J. Biol. Chem.* **274**: 3700–3704.

Moilanen, A.M., Poukka, H., Karvonen, U., Hakli, M., Janne, O.A. and Palvimo, J.J. (1998) Identification of a novel RING finger protein as a coregulator in steroid receptor-mediated gene transcription. *Mol. Cell. Biol.* **18**: 5128–5139.

Monden, T., Kishi, M., Hosoya, T. *et al.* (1999) p120 acts as a specific coactivator for 9-*cis*-retinoic acid receptor (RXR) on peroxisome proliferator-activated receptor-gamma/RXR heterodimers. *Mol. Endocrinol.* **13**: 1695–1703.

Montano, M.M., Ekena, K., Delage-Mourroux, R., Chang, W., Martini, P. and Katzenellenbogen, B.S. (1999) An estrogen receptor-selective coregulator that potentiates the effectiveness of antiestrogens and represses the activity of estrogens. *Proc. Natl Acad. Sci. USA* **96**: 6947–6952.

Montano, M.M., Muller, V., Trobaugh, A. and Katzenellenbogen, B.S. (1995) The carboxy-terminal F domain of the human estrogen receptor: role in the transcriptional activity of the receptor and the effectiveness of antiestrogens as estrogen antagonists. *Mol. Endocrinol.* **9**: 814–825.

Moras, D. and Gronemeyer, H. (1998) The nuclear receptor ligand-binding domain: structure and function. *Curr. Opin. Cell. Biol.* **10**: 384–391.

Muchardt, C. and Yaniv, M. (1993) A human homologue of *Saccharomyces cerevisiae* SNF2/SWI2 and *Drosophila brm* genes potentiates transcriptional activation by the glucocorticoid receptor. *EMBO J.* **12**: 4279–4290.

Mukherjee, R., Davies, P.J., Crombie, D.L. *et al.* (1997) Sensitization of diabetic and obese mice to insulin by retinoid X receptor agonists. *Nature* **386**: 407–410.

Muller, J.M., Isele, U., Metzger, E. *et al.* (2000) FHL2, a novel tissue-specific coactivator of the androgen receptor. *EMBO J.* **19**: 359–369.

Na, S.Y., Choi, H.S., Kim, J.W., Na, D.S. and Lee, J.W. (1998) Bcl3, an IkappaB protein, as a novel transcription coactivator of the retinoid X receptor. *J. Biol. Chem.* **273**: 30933–30938.

Näär, A.M., Beaurang, P.A., Zhou, S., Abraham, S., Solomon, W. and Tjian, R. (1999) Composite coactivator ARC mediates chromatin-directed transcriptional activation. *Nature* **398**: 828–832.

Nagy, L., Kao, H.Y., Love, J.D. *et al.* (1999) Mechanism of corepressor binding and release from nuclear hormone receptors. *Genes Dev.* **13**: 3209–3216.

Nawaz, Z., Lonard, D.M., Smith, C.L. *et al.* (1999) The Angelman syndrome-associated protein, E6-AP, is a coactivator for the nuclear hormone receptor superfamily. *Mol. Cell. Biol.* **19**: 1182–1189.

Nolte, R.T., Wisely, G.B., Westin, S. *et al.* (1998) Ligand binding and co-activator assembly of the peroxisome proliferator-activated receptor-gamma. *Nature* **395**: 137–143.

Norris, J.D., Paige, L.A., Christensen, D.J. *et al.* (1999) Peptide antagonists of the human estrogen receptor. *Science* **285**: 744–746.

Nuclear Receptors Nomenclature Committee (1999) A unified nomenclature system for the nuclear receptor superfamily. *Cell* **97**: 161–163.

Onate, S.A., Tsai, S.Y., Tsai, M.J. and O'Malley, B.W. (1995) Sequence and characterization of a coactivator for the steroid hormone receptor superfamily. *Science* **270**: 1354–1357.

Paech, K., Webb, P., Kuiper, G.G. *et al.* (1997) Differential ligand activation of estrogen receptors ERalpha and ERbeta at AP1 sites. *Science* **277**: 1508–1510.

Paige, L.A., Christensen, D.J., Gron, H. *et al.* (1999) Estrogen receptor (ER) modulators each induce distinct conformational changes in ER alpha and ER beta. *Proc. Natl Acad. Sci. USA* **96**: 3999–4004.

Parker, K.L., Ikeda, Y. and Luo, X. (1996) The roles of steroidogenic factor-1 in reproductive function. *Steroids* **61**: 161–165.

Pearce, D., Matsui, W., Miner, J.N. and Yamamoto, K.R. (1998) Glucocorticoid receptor transcriptional activity determined by spacing of receptor and nonreceptor DNA sites. *J. Biol. Chem.* **273**: 30081–30085.

Pearce, D. and Yamamoto, K.R. (1993) Mineralcorticoid and glucocorticoid receptor activities distinguished by nonreceptor factors at a composite response element. *Science* **259**: 1161–1165.

Perissi, V., Staszewski, L.M., McInerney, E.M. *et al.* (1999) Molecular determinants of nuclear receptor-corepressor interaction. *Genes Dev.* **13**: 3198–3208.

Perlmann, T. and Jansson, L. (1995) A novel pathway for vitamin A signaling mediated by RXR heterodimerization with NGFI-B and NURR1. *Genes Dev.* **9**: 769–782.

Peters, G.A. and Khan, S.A. (1999) Estrogen receptor domains E and F: role in dimerization and interaction with coactivator RIP-140. *Mol. Endocrinol.* **13**: 286–296.

Petty, K.J., Krimkevich, Y.I. and Thomas, D. (1996) A TATA binding protein-associated factor functions as a coactivator for thyroid hormone receptors. *Mol. Endocrinol.* **10**: 1632–1645.

Pfahl, M. (1993) Nuclear receptor/AP-1 interaction. *Endocr. Rev.* **14**: 651–658.

Powers, C.A., Mathur, M., Raaka, B.M., Ron, D. and Samuels, H.H. (1998) TLS (translocated-in-liposarcoma) is a high-affinity interactor for steroid, thyroid hormone, and retinoid receptors. *Mol. Endocrinol.* **12**: 4–18.

Puigserver, P., Wu, Z., Park, C.W., Graves, R., Wright, M. and Spiegelman, B.M. (1998) A cold-inducible coactivator of nuclear receptors linked to adaptive thermogenesis. *Cell* **92**: 829–839.

Rachez, C., Lemon, B.D., Suldan, Z. *et al.* (1999) Ligand-dependent transcription activation by nuclear receptors requires the DRIP complex. *Nature* **398**: 824–828.

Rachez, C., Suldan, Z., Ward, J. *et al.* (1998) A novel protein complex that interacts with the vitamin D3 receptor in a ligand-dependent manner and enhances VDR transactivation in a cellfree system. *Genes Dev.* **12**: 1787–1800.

Rastinejad, F., Perlmann, T., Evans, R.M. and Sigler, P.B. (1995) Structural determinants of nuclear receptor assembly on DNA direct repeats. *Nature* **375**: 203–211.

Reichardt, H.M., Kaestner, K.H., Tuckermann, J. *et al.* (1998) DNA binding of the glucocorticoid receptor is not essential for survival. *Cell* **93**: 531–541.

Renaud, J.P., Rochel, N., Ruff, M. *et al.* (1995) Crystal structure of the RAR-gamma ligand-binding domain bound to all-*trans* retinoic acid. *Nature.* **378**: 681–689.

Repa, J.J. and Mangelsdorf, D.J. (1999) Nuclear receptor regulation of cholesterol and bile acid metabolism. *Curr. Opin. Biotechnol.* **10**: 557–563.

Resche-Rigon, M. and Gronemeyer, H. (1998) Therapeutic potential of selective modulators of nuclear receptor action. *Curr. Opin. Chem. Biol.* **2**: 501–507.

Rochette-Egly, C., Adam, S., Rossignol, M., Egly, J. M. and Chambon, P. (1997) Stimulation of RAR alpha activation function AF-1 through binding to the general transcription factor TFIIH and phosphorylation by CDK7. *Cell* **90**: 97–107.

Rogatsky, I., Trowbridge, J.M. and Garabedian, M.J. (1999) Potentiation of human estrogen receptor alpha transcriptional activation through phosphorylation of serines 104 and 106 by the cyclin A-CDK2 complex. *J. Biol. Chem.* **274**:(32) 22296–22302.

Ryu, S. and Tjian, R. (1999) Purification of transcription cofactor complex CRSP. *Proc. Natl Acad. Sci.* **96**: 7137–7142.

Sadovsky, Y., Webb, P., Lopez, G. *et al.* (1995) Transcriptional activators differ in their responses to overexpression of TATA-box-binding protein. *Mol. Cell. Biol.* **15**: 1554–1563.

Scheinman, R.I., Gualberto, A., Jewell, C.M., Cidlowski, J.A. and Baldwin, A.S. Jr. (1995) Characterization of mechanisms involved in transrepression of NF-kappa B by activated glucocorticoid receptors. *Mol. Cell. Biol.* **15**: 943–953.

Schulman, I.G., Chakravarti, D., Juguilon, H., Romo, A. and Evans, R.M. (1995) Interactions between the retinoid X receptor and a conserved region of the TATA-binding protein mediate hormone-dependent transactivation. *Proc. Natl Acad. Sci. USA* **92**: 8288–8292.
Schwabe, J.W. (1997) The role of water in protein–DNA interactions. *Curr. Opin. Struct. Biol.* **7**: 126–134.
Schwabe, J.W., Chapman, L. and Rhodes, D. (1995) The oestrogen receptor recognizes an imperfectly palindromic response element through an alternative side-chain conformation. *Structure* **3**: 201–213.
Schwabe, J.W., Chapman, L., Finch, J.T. and Rhodes, D. (1993) The crystal structure of the estrogen receptor DNA-binding domain bound to DNA: how receptors discriminate between their response elements. *Cell* **75**: 567–578.
Schwerk, C., Klotzbucher, M., Sachs, M., Ulber, V. and Klein-Hitpass, L. (1995) Identification of a transactivation function in the progesterone receptor that interacts with the TAFII110 subunit of the TFIID complex. *J. Biol. Chem.* **270**: 21331–21338.
Shao, D. and Lazar, M.A. (1999) Modulating nuclear receptor function: may the phos be with you. *J. Clin. Invest.* **103**: 1617–1618.
Shao, D., Rangwala, S.M., Bailey, S.T., Krakow, S.L., Reginato, M.J. and Lazar, M.A. (1998) Interdomain communication regulating ligand binding by PPAR-gamma. *Nature* **396**: 377–380.
Shemshedini, L., Ji, J.W., Brou, C., Chambon, P. and Gronemeyer, H. (1992) In vitro activity of the transcription activation functions of the progesterone receptor. Evidence for intermediary factors. *J. Biol. Chem.* **267**: 1834–1839.
Shemshedini, L., Knauthe, R., Sassone-Corsi, P., Pornon, A. and Gronemeyer, H. (1991) Cell-specific inhibitory and stimulatory effects of Fos and Jun on transcription activation by nuclear receptors. *EMBO J.* **10**: 3839–3849.
Shiau, A.K., Barstad, D., Loria, P.M. *et al.* (1998) The structural basis of estrogen receptor coactivator recognition and the antagonism of this interaction by tamoxifen. *Cell* **95**: 927–937.
Sladek, F.M., Ruse, M.D. Jr., Nepomuceno, L., Huang, S.M. and Stallcup, M.R. (1999) Modulation of transcriptional activation and coactivator interaction by a splicing variation in the F domain of nuclear receptor hepatocyte nuclear factor 4alpha1. *Mol. Cell. Biol.* **19**: 6509–6522.
Smith, S.M., Dickman, E.D., Power, S.C. and Lancman, J. (1998) Retinoids and their receptors in vertebrate embryogenesis. *J. Nutr.* **128**: 467–470.
Stocklin, E., Wissler, M., Gouilleux, F. and Groner, B. (1996) Functional interactions between Stat5 and the glucocorticoid receptor. *Nature* **383**: 726–728.
Sucov, H.M., Dyson, E., Gumeringer, C.L., Price, J., Chien, K.R. and Evans, R.M. (1994) RXR alpha mutant mice establish a genetic basis for vitamin A signaling in heart morphogenesis. *Genes Dev.* **8**: 1007–1018.
Sun, X., Zhang, Y., Cho, H. *et al.* (1998) NAT, a human complex containing Srb polypeptides that functions as a negative regulator of activated transcription. *Mol. Cell.* **2**: 213–222.
Takeshita, A., Cardona, G.R., Koibuchi, N., Suen, C.S. and Chin, W.W. (1997) TRAM-1, a novel 160-kDa thyroid hormone receptor activator molecule, exhibits distinct properties from steroid receptor coactivator-1. *J. Biol. Chem.* **272**: 27629–27634.
Tan, J., Hall, S.H., Hamil, K.G. *et al.* (2000) Protein inhibitor of activated STAT-1 (signal transducer and activator of transcription-1) is a nuclear receptor coregulator expressed in human testis. *Mol. Endocrinol.* **14**: 14–26.
Taneja, R., Rochette-Egly, C., Plassat, J.L., Penna, L., Gaub, M.P. and Chambon, P. (1997) Phosphorylation of activation function AF1 and AF2 of RARa and RARg is indispensable for differentiation of F9 cells upon retinoic acid and cAMP treatment. *EMBO J.* **16**: 6452–6465.
Tasset, D., Tora, L., Fromental, C., Scheer, E. and Chambon, P. (1990) Distinct classes of transcriptional activating domains function by different mechanisms. *Cell* **62**: 1177–1187.
Tetel, M.J., Giangrande, P.H., Leonhardt, S.A., McDonnell, D.P. and Edwards D.P. (1999) Hormone-dependent interaction between the amino- and carboxyl-terminal domains of progesterone receptor *in vitro* and *in vivo*. *Mol. Endocrinol.* **13**: 910–924.
Torchia, J., Rose, D.W., Inostroza, J. *et al.* (1996) The transcriptional co-activator p/CIP binds CBP and mediates nuclear-receptor function. *Nature* **387**: 677–684.
Torchia, J., Glass, C. and Rosenfeld, M.G. (1998) Co-activators and co-repressors in the integration of transcriptional responses. *Curr. Opin. Cell. Biol.* **10**: 373–383.
Tremblay, A., Tremblay, G.B., Labrie, F. and Giguere, V. (1999) Ligand-independent recruitment of SRC-1 to estrogen receptor beta through phosphorylation of activation function AF-1. *Mol. Cell.* **3**: 513–519.

Tronche, F., Kellendonk, C., Reichardt, H.M. and Schutz, G. (1998) Genetic dissection of glucocorticoid receptor function in mice. *Curr. Opin. Genet. Dev.* **8**: 532–538.
Tsai, M.J. and O'Malley, B.W. (1994) Molecular mechanisms of action of steroid/thyroid receptor superfamily members. *Annu. Rev. Biochem.* **63**: 451–486.
Um, S., Harbers, M., Benecke, A., Pierrat, B., Losson, R. and Chambon, P. (1998) Retinoic acid receptors interact physically and functionally with the T:G mismatch-specific thymine-DNA glycosylase. *J. Biol. Chem.* **273**: 20728–20736.
Uppenberg, J., Svensson, C., Jaki, M., Bertilsson, G., Jendeberg, L. and Berkenstam, A. (1998) Crystal structure of the ligand binding domain of the human nuclear receptor PPARgamma. *J. Biol. Chem.* **273**: 31108–31112.
Utley, R.T., Ikeda, K., Grant, P.A. *et al.* (1998) Transcriptional activators direct histone acetyltransferase complexes to nucleosomes. *Nature* **394**: 498–502.
Voegel, J.J., Heine, M.J., Tini, M., Vivat, V., Chambon, P. and Gronemeyer, H. (1998) The coactivator TIF2 contains three nuclear receptor-binding motifs and mediates transactivation through CBP binding-dependent and -independent pathways. *EMBO J.* **17**: 507–519.
Voegel, J.J., Heine, M.J.S., Zechel, C., Chambon, P. and Gronemeyer, H. (1996) TIF2, a 160 kDa transcriptional mediator for the ligand-dependent activation function AF-2 of nuclear receptors. *EMBO J.* **15**: 3667–3675.
vom Baur, E., Harbers, M., Um, S.J., Benecke, A., Chambon, P. and Losson, R. (1998) The yeast ADA complex mediates the ligand-dependent activation function AF-2 of retinoic X and estrogen receptors. *Genes Dev.* **12**: 1278–1289.
vom Baur, E., Zechel, C., Heery, D. *et al.* (1996) Differential ligand-dependent interactions between the AF-2 activating domain of nuclear receptors and the putative transcriptional intermediary factors mSUG1 and TIF1. *Embo J.* **15**: 110–124.
Wagner, B.L., Norris, J.D., Knotts, T.A., Weigel, N.L. and McDonnell, D.P. (1998) The nuclear corepressors NCoR and SMRT are key regulators of both ligand- and 8-bromo-cyclic AMP-dependent transcriptional activity of the human progesterone receptor. *Mol. Cell. Biol.* **18**: 1369–1378.
Wagner, R.L., Apriletti, J.W., McGrath, M.E., West, B.L., Baxter, J.D. and Fletterick, R.J. (1995) A structural role for hormone in the thyroid hormone receptor. *Nature* **378**: 690–697.
Wahlstrom, G.M., Vennstrom, B. and Bolin, M.B. (1999) The adenovirus E1A protein is a potent coactivator for thyroid hormone receptors. *Mol. Endocrinol.* **13**: 1119–1129.
Webb, P., Nguyen, P., Valentine, C. *et al.* (1999) The estrogen receptor enhances AP-1 activity by two distinct mechanisms with different requirements for receptor transactivation functions. *Mol. Endocrinol.* **13**: 1672–1685.
Webb, P., Lopez, G.N., Uht, R.M. and Kushner, P.J. (1995) Tamoxifen activation of the estrogen receptor/AP-1 pathway: potential origin for the cell-specific estrogen-like effects of antiestrogens. *Mol. Endocrinol.* **9**: 443–456.
Williams, S.P. and Sigler, P.B. (1998) Atomic structure of progesterone complexed with its receptor. *Nature* **393**: 392–396.
Wurtz, J.M., Bourguet, W., Renaud, J.P. *et al.* (1996) A canonical structure for the ligand-binding domain of nuclear receptors. *Nat. Struct. Biol.* **3**: 87–94.
Xu, H.E., Lambert, M.H., Montana, V.G. *et al.* (1999a) Molecular recognition of fatty acids by peroxisome proliferator-activated receptors. *Mol. Cell.* **3**: 397–403.
Xu, L., Glass, C.K. and Rosenfeld, M.G. (1999b) Coactivator and corepressor complexes in nuclear receptor function. *Curr. Opin. Genet. Dev.* **9**: 140–147.
Yanagisawa, J., Yanagi, Y., Masuhiro, Y. *et al.* (1999) Convergence of transforming growth factor-beta and vitamin D signaling pathways on SMAD transcriptional coactivators. *Science* **283**: 1317–1321.
Yang, W. and Freedman, L.P. (1999) 20-Epi analogues of 1,25-dihydroxyvitamin D3 are highly potent inducers of DRIP coactivator complex binding to the vitamin D3 receptor. *J. Biol. Chem.* **274**: 16838–16845.
Yeh, S. and Chang, C. (1996) Cloning and characterization of a specific coactivator, ARA70, for the androgen receptor in human prostate cells. *Proc. Natl Acad. Sci. USA* **93**: 5517–5521.
Yuan, C.X., Ito, M., Fondell, J.D., Fu, Z.Y. and Roeder, R.G. (1998) The TRAP220 component of a thyroid hormone receptor-associated protein (TRAP) coactivator complex interacts directly with nuclear receptors in a ligand-dependent fashion. *Proc. Natl Acad. Sci. USA* **95**: 7939–7944.

Yudt, M.R., Vorojeikina, D., Zhong, L. *et al.* (1999) Function of estrogen receptor tyrosine 537 in hormone binding, DNA binding, and transactivation. *Biochemistry* **38**: 14146–14156.

Zamir, I., Dawson, J., Lavinsky, R.M., Glass, C.K., Rosenfeld, M.G. and Lazar, M.A. (1997) Cloning and characterization of a corepressor and potential component of the nuclear hormone receptor repression complex. *Proc. Natl Acad. Sci. USA* **94**: 14400–14405.

Zamir, I., Harding, H.P., Atkins, G.B. *et al.* (1996) A nuclear hormone receptor corepressor mediates transcriptional silencing by receptors with distinct repression domains. *Mol. Cell. Biol.* **16**: 5458–5465.

Zechel, C., Shen, X.Q., Chen, J.Y., Chen, Z.P., Chambon, P. and Gronemeyer, H. (1994) The dimerization interfaces formed between the DNA binding domains of RXR, RAR and TR determine the binding specificity and polarity of the full-length receptors to direct repeats. *EMBO J.* **13**: 1425–1433.

Zeiner, M., Gebauer, M. and Gehring, U. (1997) Mammalian protein RAP46: an interaction partner and modulator of 70 kDa heat shock proteins. *EMBO J.* **16**: 5483–5490.

Zhang, Z., Jones, S., Hagood, J.S., Fuentes, N.L. and Fuller, G.M. (1997) STAT3 acts as a co-activator of glucocorticoid receptor signaling. *J. Biol. Chem.* **272**: 30607–30610.

Zhang, X., Jeyakumar, M., Petukhov, S. and Bagchi, M.K. (1998a) A nuclear receptor corepressor modulates transcriptional activity of antagonist-occupied steroid hormone receptor. *Mol. Endocrinol.* **12**: 513–524.

Zhang, J. and Fondell, J.D. (1999) Identification of mouse TRAP100: a transcriptional coregulatory factor for thyroid hormone and vitamin D receptors. *Mol. Endocrinol.* **13**: 1130–1140.

Zhang, J., Guenther, M.G., Carthew, R.W. and Lazar, M.A. (1998b) Proteasomal regulation of nuclear receptor corepressor mediated repression. *Genes Dev.* **12**: 1775–1780.

Zhao, Q., Khorasanizadeh, S., Miyoshi, Y., Lazar, M.A. and Rastinejad, F. (1998) Structural elements of an orphan nuclear receptor–DNA complex. *Mol. Cell.* **1**: 849–861.

Zhou, X.F., Shen, X.Q. and Shemshedini, L. (1999) Ligand-activated retinoic acid receptor inhibits AP-1 transactivation by disrupting c-Jun/c-Fos dimerization. *Mol. Endocrinol.* **13**: 276–285.

Zhu, Y., Qi, C., Jain, S., Rao, M.S. and Reddy, J.K. (1997) Isolation and characterization of PBP, a protein that interacts with peroxisome proliferator-activated receptor. *J. Biol. Chem.* **272**: 25500–25506.

Zwijsen, R.M.L., Wientjens, E., Klompmaker, R., vanderSman, J., Bernards, R. and Michalides, R.J.A.M. (1997) CDK-independent activation of estrogen receptor by cyclin D1. *Cell* **88**: 405–415.

9

Developmental control by Hox transcriptional regulators and their cofactors

Thomas Lufkin

1. Introduction

The establishment of distinct cell types within mammalian embryos likely shares many similarities to development in *Drosophila*, whereby specific boundaries between different groups of cells are achieved through overlapping gradients of signaling molecules that in turn activate or repress particular signal transduction pathways. These signals consequently affect the activity (both positive and negative) of transcriptional regulators (Frasch, 1999; Skeath, 1999; Wolpert, 1996). Thus initially broad domains of cell-signaling molecules (or even transcription factors) in the developing embryo can be transformed with time and cell divisions into specific subregions of gene expression and distinct cell types. Specificity results from the interplay between modest activation threshold requirements and overlapping gradients of competing regulators. In the developing mammalian embryo, recent progress in cell-fate mapping studies has provided a clearer understanding of which cells give rise to the different tissues and organs (Lawson and Hage, 1994; Lawson and Pederson, 1992). Coincident with this understanding, *in situ* analysis of gene expression in embryos, combined with modern molecular genetics approaches, have demonstrated the specific role of developmental transcription factors in directing cellular differentiation, tissue formation and body-plan formation during embryogenesis.

A large number of developmentally important transcription factors have been identified that contain a homeodomain (*Figure 1*), which is a HTH type of DNA-binding motif of 60–63 amino acids that makes contact with the major and minor grooves of its DNA target recognition site (Chapter 4). The homeodomain derives its name from the segment of DNA that encodes it, referred to as the homeobox. The amino-acid sequence of the homeodomain, and the presence of

Transcription Factors, edited by J. Locker.

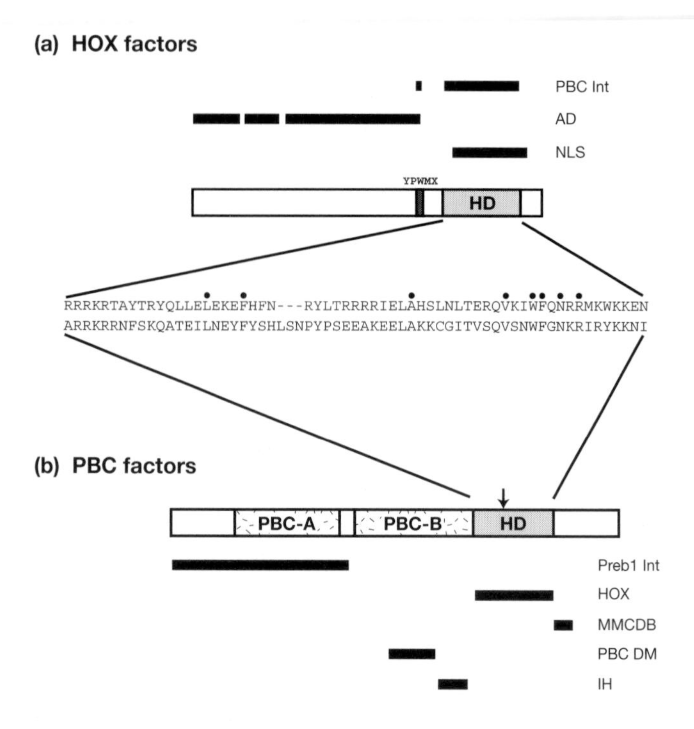

Figure 1. Functional motifs of a Hox and a PBC protein. The diagrams are a compilation from different studies on various members of each family. Sizing and positioning of the homeodomain (HD) is based on Hoxa5 (a) and Pbx1 (b) from mouse, which were chosen as representative proteins for each family. The position of the YPMWX pentapeptide motif is indicated, and the downward arrow in b represents the position of the three-amino-acid insertion (LSN) in the PBC homeodomain relative to HOX. Consensus sequences for HOX and PBC homeodomains are shown. The least variant residues are indicated with a circle. AD, activation domains; HOX, HOX dimerization domain; HD, homeodomain; IH, inhibitory helix; MMCDB, maximal monomer and cooperative DNA-binding domain; NLS, nuclear localization signal; PBC-A and PBC-B, two highly conserved domains found in PBC proteins across all species; PBC Int, PBC interaction domain; PBC DM, Pbx dimerization domain; Prep1 Int, Prep1 interaction domain.

characteristic accessory domains, has allowed the classification of homeodomain-containing proteins (homeoproteins) into over 20 separate families (e.g. *Dlx, Msx, Nkx, Hmx, PBC, Hox,*) with usually between 3–10 members per family in vertebrates. In the exceptional case of the *Hox* family, which includes 39 members in mammals, this number is considerably larger (Burglin, 1994; Gehring *et al.*, 1994). Most of the homeoprotein families have been shown genetically to affect some aspect of development ranging from cell-type specification to global patterning of the embryo.

This chapter will focus on the Hox proteins, a critically important and evolutionary conserved family of transcriptional regulators that control the embryonic development of the animal body plan. The *Hox* gene family directs morphological development and tissue differentiation along all the major and minor axes of the developing organism. The genes encoding this family of 60 amino acid homeodomain-containing proteins (*Figure 1*) are clustered together into four independent complexes (referred to as A, B, C and D) that are located on different chromosomes in mammals (*Figure 2*). These complexes are also linked on the chromosome to the genes for the distantly related Dlx homeoproteins. The homeobox-containing genes of the *Hox* complexes have been shown, via genetics, to encode relative positional information within the developing organism, meaning that they are able to instruct cells to produce morphologically appropriate structures according to the cells location within the developing organism. *Hox* genes control the morphogenesis of the different body parts, such as the vertebrae and limbs, directing one vertebra to differ slightly in shape and size from the adjacent ones. Body-plan development is controlled by the properties of the Hox proteins and their cofactors, and by the enhancers and promoters that regulate expression of the *Hox* gene complexes.

The protein sequences of the Hox homeodomain, the chromosomal arrangement of the *Hox* genes and their patterns of expression during embryogenesis have all been highly conserved during evolution of the animal kingdom (*Figure 2*). In addition, there is an intriguing colinearity between the order of mammalian *Hox* genes on the chromosome, their temporal and spatial expression, and responsiveness to retinoic acid, a molecule with morphogen-like characteristics (Lufkin, 1997). In vertebrates, the expression of *Hox* genes along the anteroposterior axis occurs in a staggered overlapping fashion. The expression of 3′ located *Hox* genes extends caudally from the anterior hindbrain along the entire length of the embryo to its caudal tip. Only genes located at the most 3′ end of a *Hox* complex are expressed in the rostral hindbrain, whereas more 5′ positioned genes are expressed with increasingly more caudally-restricted (staggered) boundaries of expression, such that in posterior parts of the embryo, there are increasing numbers of *Hox* genes expressed in any given cell type. Hence one mechanism by which *Hox* genes might function in anteroposterior patterning is through the precise 'combination' of Hox proteins expressed in a particular cell. Thus, each cell along the embryonic axis would express its phenotype based upon the repertoire of *Hox* genes active within. This model is supported by gain-of-function experiments where posteriorly expressed *Hox* genes were ectopically expressed in anterior regions, resulting in their transformation to posterior body structures (Krumlauf, 1994). However, an alternate model for *Hox* gene function was indicated by gene substitution (knock-in) experiments in mice (Greer *et al*., 2000). These experiments indicated that the total'concentration' of Hox proteins in a given cell could play just as important a role as the combination of different Hox proteins. Clearly the results from these and similar studies indicate that continued investigations into the area of the mechanism of *Hox* gene function are required. Forthcoming models will probably integrate the effects of both the combination and concentration of Hox proteins.

Although closely related, properties vary among the different Hox proteins. All can bind to DNA *in vitro* as monomers to a core consensus-containing TAAT. The similarity with which different Hox proteins can bind this consensus and the widespread occurrence of this site in the genome have raised questions about how HOX proteins achieve specificity of function. One solution to the problem of specificity is provided by the interaction of Hox proteins with another class of atypical homeodomain proteins known as the TALE (three-amino acids-loops-extension). The TALE proteins contain a 63-amino acid homeodomain, which is distinguished by three additional amino acid residues inserted between helixes 1 and 2 of the homeodomain (*Figure 1*). The best-characterized members of the TALE family include the PBC proteins, Extradentical (Exd) from *Drosophila*, and its homologs pre B-cell leukemia transcription factor (Pbx)1, Pbx2 and Pbx3 from mammals. The interaction of HOX and PBC increases both the affinity and specificity for DNA binding (see Section 4).

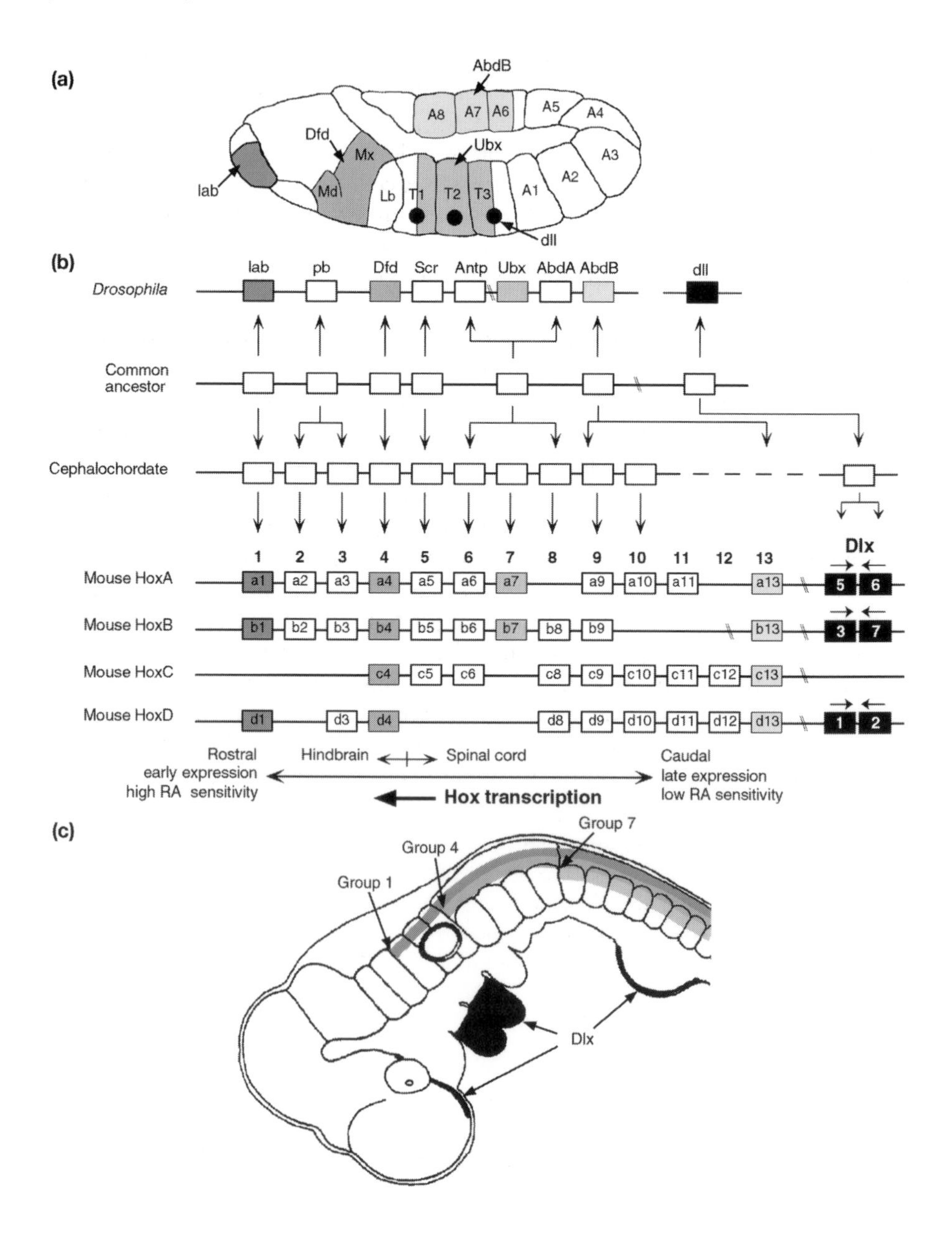

2. Summary of early embryonic development of the mammalian body plan

Several days following fertilization, the mammalian embryo is composed of two cell types: trophectoderm cells, which will give rise to primarily supportive structures (placenta); and the inner cell mass, which will give rise primarily to the embryo proper. Around the time of uterine implantation, the inner cell mass differentiates into a single epithelial layer of definitive ectoderm cells and another layer of primitive endoderm cells. At this point, at the molecular level of gene expression analysis, the embryo has already begun to establish its anteroposterior (head–tail) polarity (Gardner, 1999). The definitive ectoderm (henceforth referred to as ectoderm) is the source of cells that will give rise to all the organs of the developing fetus. The first morphological/histological appearance of an anteroposterior polarity in the amniote embryo is defined by the appearance of the primitive streak, a region of the ectoderm where cells in a process of delamination and migration, move into a central furrow and come to lay between the overlying ectoderm, and the underlying endoderm. During this process (termed gastrulation), this new sheet of intermediate cells is transformed into mesoderm, which eventually gives rise to tissues such as muscle and the skeleton. The primitive streak originates at the posterior pole of the embryo and extends anteriorly at the same time as the embryo continues to grow in all dimensions. The anterior tip of the primitive streak is termed 'Hensen's node' in the chick or the dorsal lip of the blastopore in the frog, and delimits the most anterior portion of the embryo through which cells can invaginate to form mesoderm. The node extends along the anteroposterior axis more than half the length of the embryo, but does not make it to the anterior pole, before it stops and begins to retreat posteriorly along the anterior–posterior axis. As the node retreats towards the posterior end of the embryo, an extremely important narrow rod of mesodermal cells termed the 'notochord' is laid down in its wake. During the process of gastrulation, the ectoderm delaminates a group of cells that will come to directly underlay it as a sheet of epithelium (the definitive endoderm), which replaces the primitive endoderm. It is the definitive endoderm that is the eventual source of cells for the gut and certain internal organs (Tam and Behringer, 1997). The field of ectodermal cells both overlying and anterior to the notochord is termed the neural plate, and will

Figure 2. Chromosomal arrangement and expression of evolutionarily-conserved *Hox*/*Dlx* complexes. (a) Diagram of an elongated germband-stage *Drosophila* embryo with anterior to the left. (b) Diagrams of the *Hox*/*Dlx* complexes of *Drosophila*, common ancestor, cephalochordate (amphioxus), and mouse. (c) Expression patterns of mammalian *Dlx5* along with *Hox* Paralogue groups 1, 4, 7 and 13. The expression of Paralogue group 13 does not appear in the mammalian embryo as shown, as its expression is too caudally restricted. The collinear arrangement of the *Hox* genes in the complex is indicated relative to spatial expression (rostral versus caudal), timing of expression (early versus late) and retinoic acid (RA) sensitivity (high versus low). The division between *Hox* genes that have anterior borders of expression in the hindbrain or in the spinal cord is shown. The transcriptional orientation of *Dlx* genes is convergent (indicated with an arrow above each mammalian gene). The transcriptional orientation of *Hox* genes is identical (indicated with a bold arrow below the complexes).

eventually give rise to the nervous system. Ectoderm that is adjacent to the neural plate and contiguous with it gives rise to surface ectoderm (presumptive skin) and in anterior regions of the embryo, to important cranial sensory placodes. The demarcation within the ectoderm to form either neural plate or adjacent surface ectoderm is established through the positive actions of the cell–cell signaling molecules, Shh, and the bone morphogenetic proteins (primarily bmp4) and the negative influences of certain extracellular blocking molecules e.g. Chordin, Noggin, Follistatin, Cerberus and Gremlin (Briscoe and Ericson, 1999; Streit and Stern, 1999).

3. Hox protein function: regulation of transcription

In vitro studies have shown the homeodomain to be the principal portion of the Hox protein that contacts DNA. The homeodomain can bind to DNA as a monomer, although this is probably not the manner in which it functions *in vivo* (see Section 4). The DNA target consensus site for Hox protein interaction has been examined in a number of *in vitro* and *in vivo* systems and the most common degenerate consensus-binding site includes the sequence TNAT(G/T)(G/A) with a core consensus sequence of TAAT (Biggin and McGinnis, 1997; Vigano *et al.*, 1998). The absence of a more restricted DNA-binding site consensus for the different Hox paralog groups has raised the question of how *Hox* genes obtain specificity of function in the developing embryo and suggest a role for Hox cofactors in modulating the DNA-binding specificity of Hox proteins. Moreover, the Hox-effector domains may either positively or negatively influence the basal transcriptional apparatus (Gross and Gruss, 1994; Vigano *et al.*, 1998; Zhao *et al.*, 1996). The analysis of the Hoxa5 protein in tissue culture and transgenic embryos revealed that the *N*-terminal region of Hoxa5 plays a role in transcriptional activity and homeotic transformation potential, but has no role in determining functional specificity (Zhao *et al.*, 1996). The homeodomain was shown to be essential for nuclear targeting and overall protein function, and the conserved YPWMX motif, located *N*-terminal to the homeodomain, was shown to be essential for biological specificity yet did not contribute to the potential for Hoxa5 transcriptional activation. The YPWMX motif has subsequently been shown to be an important site of protein–protein interaction with the Hox cofactors Pbx1-Pbx3/Exd, proteins collectively referred to as PBC (Mann and Chan, 1996). Dissection of the transcription activation/repression function of Hoxa5, Hoxb1, Hoxb3, Hoxd8 and Hoxd9 proteins was examined in cell culture transfection experiments (Vigano *et al.*, 1998; Zappavigna *et al.*, 1994) and in embryos (Zhao *et al.*, 1996). These studies determined that the Hox activation domains could be identified in different regions of the protein depending upon the type of assay employed and the manner in which the Hox protein interacted with DNA (target site selection). Transcriptional activation domains were identified by deletion analysis of each protein and assayed by the activation/repression of a specific target either as a Hox monomer binding to the TAAT-binding site, as a Hox/PBC dimer on the TGAT(T/G)NAT-binding site, as a protein fusion with the GAL4 DNA-binding domain on its consensus recognition element, or finally as a transactivator of the *forkhead* gene, a downstream target of Hoxa5 (Zhao *et al*., 1996). When tested in these different contexts, each protein revealed different amino acids *N*-terminal to the homeodomain which were important for transcriptional activation. The *C*-terminal portion also contained a potent activation domain, but this was revealed only in the context of a GAL4 homodimer or PBC heterodimer, and not in the context of the Hox monomer (Vigano *et al*., 1998). The conclusions are that Hox proteins can perform multiple transcriptional functions that are influenced by the specific cofactor or component of the basal transcriptional machinery with which the Hox protein interacts. Function is also likely to be influenced by the DNA sequence of the target recognition site.

An additional layer of complexity can be added when one considers the role of phosphorylation in the modulation of Hox protein function. Phosphorylation and dephosphorylation have been shown to be important in the control of many DNA-binding transcription factors, and recent evidence suggests that the serine/threonine kinase, casein kinase II modulates the activity of Hox proteins during embryonic development (Jaffe *et al*., 1997). Mutation of a wildtype Hox protein at its casein kinase II target sites to prevent phosphorylation led to increased homeotic activity in embryos. In contrast, when the casein kinase II target sites were mutated to represent a constitutively phosphorylated form of the Hox protein, its *in vivo* activity was dramatically reduced and coincided with its reduced ability to bind to DNA as a Hox/PBC heterodimer (Jaffe *et al*., 1997). The reduction of Hox protein function following phosphorylation suggests an alternate mechanism for repressing Hox protein activity.

4. Homeodomain cofactors: TALE proteins

In lower organisms, homeodomain proteins have been shown to heterodimerize with one another based upon the interaction of sequences adjacent to the homeodomain that permit cooperative binding of the heterodimers to DNA (Vershon and Johnson, 1993). TALE proteins are a superclass of atypical homeodomain-containing proteins that are distinguished by three additional amino acid residues located between helixes 1 and 2 of the homeodomain. Four classes of TALE proteins have been identified: PBC, Meis, TG interacting factor (TGIF) and Iroquois (Burglin, 1997; 1998). The three amino acid insertions result in an enlarged four-helix DNA-binding domain. The PBC family is comprised of three vertebrate *Pbx* genes (*Pbx1, Pbx2* and *Pbx3*) and one *Drosophila* gene, *Exd*, as well as orthologous genes from other organisms. In vertebrates, Hox proteins derived from Paralogue Groups 1–8 (*Figure 2*) have a conserved pentapeptide (YPWMX) positioned generally about 5–20 amino acids *N*-terminal to the homeodomain. In early reports, this conserved pentapeptide was shown to be important for binding the TALE homeodomain protein family, PBC and for increasing both the affinity and specificity of heterodimer–DNA interaction and biological function (Chang *et al*., 1995; DiRocco *et al*., 1997; Green *et al*., 1998; Johnson *et al*., 1995; Knoepfler and Kamps, 1995; Lu *et al*., 1995; Mann and Chan, 1996; Maconochie *et al*., 1997; Neuteboom and Murre, 1997; Neuteboom *et al*., 1995; Peltenburg and Murre, 1996; Phelan and Featherstone, 1997; Phelan *et al*., 1995; Pöpperl *et al*., 1995;

Shanmugam *et al.*, 1997; Vandijk *et al.*, 1995). Further work has shown that members of Hox Paralogue group 9 and 10 can also form heterodimers with PBC proteins, although they lack the conserved pentapeptide (Shen *et al.*, 1997). Groups 11–13, which also lack the conserved pentapeptide, fail to show any PBC cooperativity.

The minimal PBC homeodomain is capable of binding to DNA cooperatively with a HOX partner; however strongest cooperativity is found with PBC proteins containing the conserved *C*-terminal tail, which spans 16 amino acids immediately following the homeodomain and is highly conserved between mammals and flies (Green *et al.*, 1998; Lu and Kamps, 1996). PBX and Hox proteins can interact with one another in the absence of DNA binding (Sanchez *et al.*, 1997) although whether or not this forms a functional complex is unclear. A bipartite 10 base-pair Hox/PBC consensus DNA-binding sequence has been determined — TGATNNAT(G/T)(G/A) — with TGATNN representing the PBC half-site and NNAT(G/T)(G/A) representing the Hox half-site (Mann and Chan, 1996). The central two nucleotides (NN) make contact with the Hox N terminal arm, and can influence which Hox partner forms with PBC. The nucleotides GG favor the binding of Paralogue group 1 proteins, and the nucleotides TA favor the binding of Paralogue group 4 proteins, preference that appears to be evolutionarily conserved (Chan *et al.*, 1997).

The crystal structure of a Hox/PBX/DNA ternary complex (*Figure 3*) revealed that Hox and PBX bind to overlapping binding sites positioned on opposite faces of the DNA. The conserved Hox YPWMX pentapeptide motif inserts into a hydrophobic pocket present in PBX, and mediates the interaction between the two proteins (Jabet *et al.*, 1999; Passner *et al.*, 1999; Piper *et al.*, 1999). Within the heterodimer, the conserved *C*-terminal tail residues of PBX enlarges the DNA-binding domain of PBX to include a fourth helix positioned immediately *C*-terminal to the homeodomain. The additional helix of PBX is present in an unfolded state when PBX is not in contact with DNA, and folding is activated by PBX/DNA interaction, although these residues do not make direct contact with DNA. The folding of the PBX fourth helix appears to be independent of PBX interaction with Hox prior to DNA binding. This is in contrast to the third helix that is folded in the absence of DNA. A resulting model for the sequence of events in Hox/PBX/DNA interaction puts the PBX/DNA interaction first, which then opens and stabilizes the PBX pocket for interaction with the Hox YPWMX motif that would subsequently bind to form the ternary Hox/DNA/PBX complex (Jabet *et al.*, 1999).

5. Other HOX and PBC cofactors: PREP, Hth and Meis

Pbx regulating protein 1 (Prep1) is another member of the TALE class of homeodomain proteins and is most closely related to Meis and Homothorax (Hth). Members of the TALE family have been shown to form a strong heterodimeric complex with PBX proteins *in vitro* (Abu-Shaar *et al.*, 1999; Berthelsen *et al.*, 1998a, 1998b; Bischof *et al.*, 1998; Calvo *et al.*, 1999; Ferretti *et al.*, 1999; Goudet *et al.*, 1999; Shen *et al.*, 1999; Swift *et al.*, 1998). Prep1 was first isolated as a subunit of

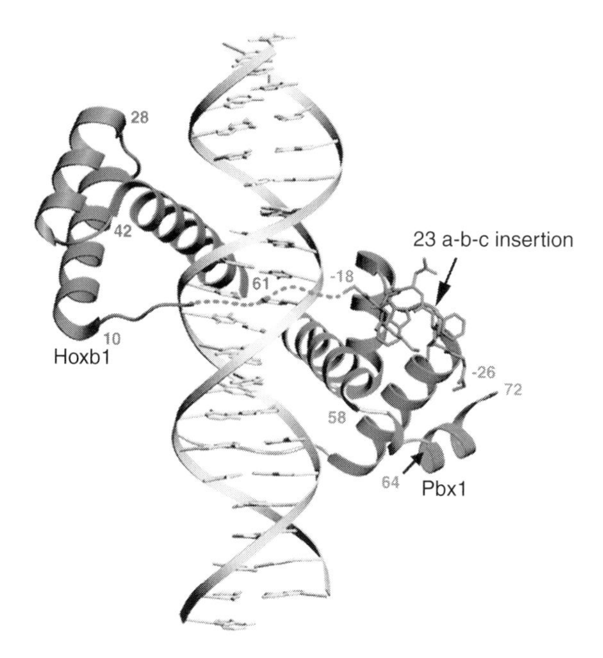

Figure 3. The Hoxb1/Pbx1/DNA crystal structure. Ribbon diagram of the Pbx1 and Hoxb1 proteins bound to DNA (Reproduced from Piper, D.E. *et al.* (1999) Structure of a HoxB1–Pbx1 heterodimer bound to DNA: role of the hexapeptide and a fourth homeodomain helix in complex formation. *Cell* 96: 587–597, with permission from Cell Press). The 20 disordered residues between the Hoxb1 N-terminal pentapeptide (YPWMK) and the homeodomain are indicated with a dotted line. Homeodomain numbering (residues 1–60) is shown for each protein. The three-residue insertion in the Pbx1 homeodomain is numbered 23a–c. Residues N-terminal to the Hoxb1 homeodomain are illustrated using negative numbers, the more negative numbers indicating the more N-terminal residues (Jabet *et al.*, 1999; Piper *et al.*, 1999).

interacting proteins forming the Uef3 transcription factor complex (Berthelsen *et al.*, 1998) which also contains PBX. In this complex, Prep1 functions to modify the affinity and DNA-binding specificity of PBX. Unlike the Hox proteins, Prep1 actively forms a strong and stable heterodimer with PBX in the absence of DNA, and, furthermore, it does not contain the conserved YPWMX pentapeptide motif necessary for Hox/PBC interaction. Heterodimerization of Prep1 and PBX results in a high-binding affinity for the TGACAG DNA sequence, yet neither protein binds to this sequence alone (Berthelsen *et al.*, 1998a, 1998b). In addition, *in vitro* cotranslation of both proteins is necessary for efficient protein–protein interaction and DNA-binding, simple mixing of the individually translated proteins does not result in significant complex formation. The interaction of Prep1 with PBX is independent of the homeodomains of the two proteins, although these are necessary for DNA binding. Instead, the *N*-terminal region of both proteins modulates the Prep1/PBX interaction. In the case of PBX, this interaction domain resides in the first 140 amino acids that include the conserved PBC-A domain, but does not affect the conserved PBC-B domain. For Prep1, the interaction domain is contained within amino acids 58–137 and includes the two conserved Meis homology regions (Berthelsen *et al.*, 1998a, 1998b). During embryonic development, Prepl/PBX heterodimers are present as early as E7.5, a time period where the most anterior HOX genes begin to exert their biological function (Ferretti *et al.*, 1999). The expression of Prepl/PBX is predominately nuclear and is present throughout gestation. Prepl expression is also seen in the hindbrain in a region that co-expresses Hoxbl, which can also interact with Prepl *in vitro*. The *Hoxbl* gene is itself regulated by Hox proteins, and this regulation is mediated by a ternary complex of one Hox and two TALE class factors.

Cotransfection of Prepl, PBX and Hoxbl with a Hoxbl autoregulatory element in COS cells showed that the reporter activity was not affected by either Prepl, PBX or Hoxbl alone (Berthelsen *et al.*, 1998a, 1998b). Cotransfection of PBX+Hoxbl gave a 4–5-fold increase in expression, whereas neither Prepl+PBX nor Prepl+Hoxbl had any effect. However, cotransfection of Prepl+PBX+Hoxbl resulted in a 9-fold stimulation. Analysis of mutant forms of each protein showed that the homeodomain of Prepl was dispensable, but the Prepl *N*-terminal heterodimerization domain was absolutely required. Deletion of the corresponding Pbx heterodimerization domain also prevented full stimulation. Electrophoretic mobility shift assay showed that Prepl/Pbx/Hoxbl forms a ternary complex on the Hoxbl autoregulatory element, which is dependent upon the Prepl/PBX interaction domain, but not the Prepl homeodomain. Thus the Hoxbl regulatory element can be bound by (Prepl/PBX) (Hoxbl/PBX) dimers or a Prepl/Hoxbl/ PBX trimer (Berthelsen *et al.*, 1998a, 1998b).

6. Subcellular localization

Drosophila Homothorax is another member of the TALE gene family, most closely related to Meis. In the fly, Hth protein is necessary for the nuclear localization of the PBC protein, Exd. In turn, Exd has a positive role in stabilizing the Hth protein and ultimately establishing Hth-expressing and non-Hth expressing domains in the developing embryo (Abu-Shaar and Mann, 1998).The control of the subcellular localization of TALE proteins is tightly regulated during embryonic development (Abu-Shaar *et al.*, 1999; Berthelsen *et al.*, 1999; Gonzalez-Crespo *et al.*, 1998). Hth and Prepl are primarily located in the cytoplasm and require association with PBC proteins (either Pbxl or Exd) for localization to the nucleus where they function. The PBC proteins contain at least two independent antagonizing domains that specify nuclear localization and nuclear export, respectively, and a third domain that interacts with Hth or Prepl. In the absence of combined PBC/(Hth or Prepl) interaction, the PBC nuclear export signal dominates and the uncomplexed PBC proteins are found primarily in the cytoplasm. The export signal probably interacts with chromosomal region maintenance (Crm)l/exportinl. In contrast, Hth or Prepl appear to interact directly with PBC and repress the nuclear export activity, probably through a conformational change. The nuclear localization signals of the combined PBC/(Hth or Prepl) complex predominates resulting in the sequestration of both proteins in the nucleus (Abu-Shaar *et al.*, 1999; Berthelsen *et al.*, 1999). Meisl is also highly similar in amino acid sequence to Hth and Prepl and can substitute for Hth in the regulation of Exd nuclear import and export (Rieckhof *et al.*, 1997).

7. 'Coselective binding' model for Hox functional specificity versus the 'widespread binding/activity regulation' model

A major question from the onset of Hox protein analysis was how to rectify the discrepancy between the relatively promiscuous binding of Hox monomers to DNA with the high specificity of *Hox* gene function in the developing organism. Similar to what is known about the function of other transcriptional regulators, it was believed that Hox proteins achieved functional specificity through the specific binding of discrete target recognition elements. The discovery that HOX cofactors increased the selective binding of Hox/cofactor heterodimers to DNA has bolstered the notion that Hox proteins function to activate or repress a discreet number of downstream target (effector) genes. This has been termed the 'coselective binding' model of *Hox* gene function (*Figure 4*) (Biggin and McGinnis, 1997; Li and McGinnis, 1999; Li *et al.*, 1999; Pinsonneault *et al.*, 1997). The alternate 'widespread binding/activity regulation' model suggests that the role of cofactors is not so much to increase the selectivity of a certain Hox-cofactor heterodimer for a particular DNA sequence, but to increase or decrease the activity of the Hox protein as a transcription factor (Biggin and McGinnis, 1997; Li and McGinnis, 1999; Li *et al.*, 1999; Pinsonneault *et al.*, 1997). Recent evidence for this second model was obtained from the analysis of a Hox protein and a Hox/VP16 hybrid protein (Li and McGinnis, 1999; Li *et al.*, 1999). These studies demonstrated that the Hox protein could interact with a DNA-binding site in the developing embryo in the absence of the PBC cofactor, but was unable to activate the target gene. In the presence of the PBC cofactor the Hox protein and the Hox/VP16 protein could strongly activate target gene expression, but with no increase in DNA-binding activity (*Figure 5*). This suggested that a principal function of the PBC cofactor could be to activate Hox protein transcriptional activity, possibly through the release of repression that is mediated by the homeodomain (Li and McGinnis, 1999; Li *et al.*, 1999).

8. Transcriptional regulation of *Hox* gene expression in the developing embryo

Some aspects of *Hox* gene regulation may have been conserved throughout evolution, yet there are important differences among species. The contribution of cellular location, timing, and cell lineage appear to differ substantially from one organism to the next in establishing correct spatiotemporal *Hox* gene expression (Kenyon, 1994; Krumlauf, 1994; Lawrence and Morata, 1994; Lufkin, 1997). In *Drosophila*, some of the regulators of *Hox* genes have been identified, and the products of the *gap* genes such as the Zinc-finger proteins Hunchback and Krüppel are thought to directly activate the transcription of homeotic genes in broad domains of the embryo along the anteroposterior axis (Irish *et al.*, 1989).

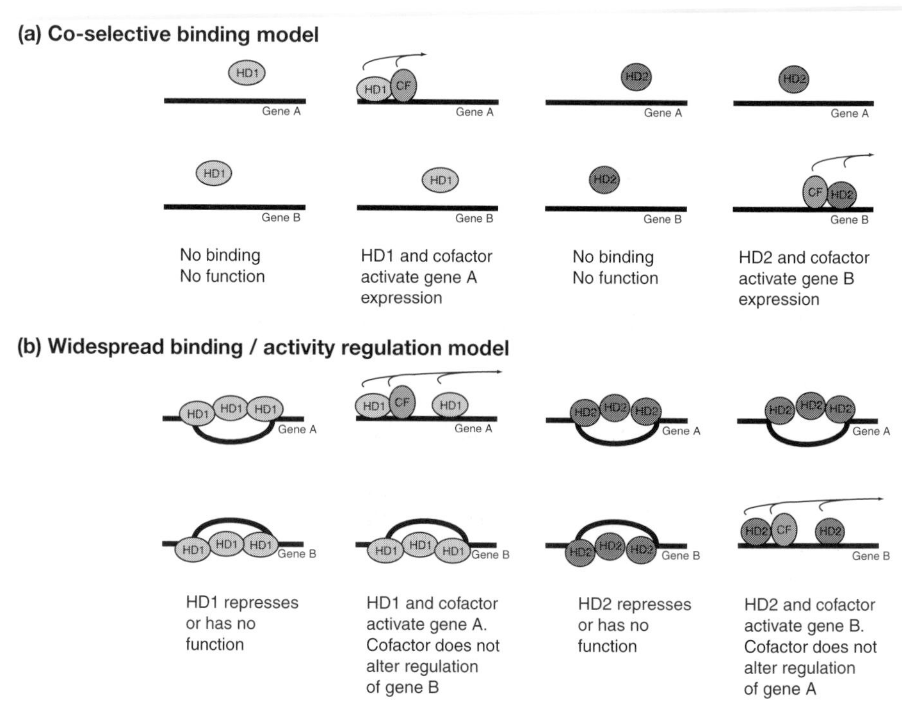

Figure 4. Two different models by which cofactors (CF) might affect the function of two HOX homeodomain proteins (HD1 and HD2). From left to right are shown potential outcomes in cells that express only HD1 (left), in cells expressing HD1 plus a cofactor (center, left), in cells expressing only HD2 (center, right) and in cells expressing HD2 plus a cofactor (right). In either case, to achieve the particular regulatory outcomes shown, the resulting heterodimers would have to have different DNA-binding specificities. In the examples shown, the cofactors are required for activation, but similar models can also be imagined for cofactors that repress. In (b), HD1 and HD2 are shown to bind cooperatively to multiple DNA sites in the absence of the cofactor. It has been assumed that interaction with the cofactor masks one of the interaction surfaces on the HOX proteins. This may reduce binding of HOX proteins at some lower affinity sites and change the overall structure of the promoter region to an active configuration. Figure reproduced with permission from Biggin and McGinnis (1997).

These expression domains become further refined through the action of the *pair-rule* gene products, such as the homeodomain protein Ftz (Muller and Bienz, 1992). In the mouse, the regulation of *Hoxb2* by the Zinc-finger protein Krox20 (Maconochie *et al.*, 1997), may parallel the function of *gap* genes in *Drosophila* homeotic gene regulation. Other similarities between the two systems include important cross-regulatory interactions among the *Hox* genes themselves. Moreover, following their initial activation or repression, the level of activity of *Hox* appears to be maintained during subsequent cell divisions by chromatin-associated proteins, some of which are conserved between *Drosophila* and mouse (Strutt *et al.*, 1997). Retinoic acid acts as a regulator of *Hox* gene expression in mouse and other vertebrates, whereas in *Drosophila*, there is no evidence for the

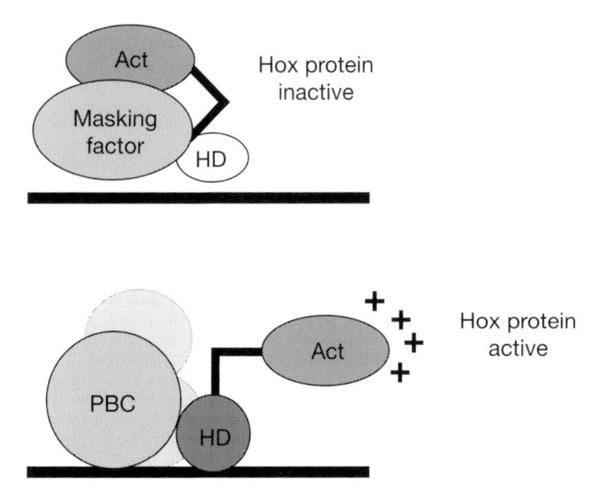

Figure 5. Potential mechanism for widespread binding/activity regulation of the Hox proteins. Hox binds via the homeodomain (HD) to the DNA target site of a variety of downstream genes; however, on certain of these sites, no transcriptional activation is achieved as the function of the Hox activation domain (Act) is suppressed by interaction with the homeodomain. The suppression may be mediated by a masking factor(s). Activation domain function is released when PBC and other cofactors bind and interact with the Hox homeodomain. Figure reproduced with permission from Li *et al.*, 1999.

existence of retinoic acid. Moreover, *ultraspiracle* does not appear to be involved in the regulation of *Drosophila Hox* genes. *ultraspiracle* is the *Drosophila* homologue of the RXR, which is the dimerization partner of the RARs in vertebrates (Henrich *et al.*, 1990; Oro *et al.*, 1990). In recent years, transgenic mice have demonstrated that *Hox* gene DNA regulatory elements can recapitulate certain aspects of the pattern of endogenous gene expression. However, exact reproduction of the full expression pattern is uncommon, even when large genomic regions spanning several *Hox* genes are analyzed. This suggests that important regulatory regions may be located at large distances from the genes they control, or that their specific chromosomal context may be important for proper regulation. Four groups of transcriptional regulators have been shown to directly regulate *Hox* gene expression in the mammalian embryos: RAR, *Krox20*, members of the *PBC* family, and the *Hox* genes themselves (Lufkin, 1997). To illustrate the complexity of *Hox* gene regulation, and the interspersion of positive and negative regulatory elements between *Hox* genes, a few representative examples are considered here.

8.1 Hoxa1 *and* Hoxa2

A genomic DNA segment spanning from 6.5 kilobases 5′ to 3.7 kilobases 3′ of the *Hoxa1* coding sequence, which therefore includes *Hoxa2* and the entire *Hoxa1* gene and genomic sequence, was examined for its ability to direct *lacZ* expression in transgenic embryos (Thompson *et al.*, 1998). The *lacZ* gene was inserted in-frame in exon I of *Hoxa1* resulting in a 44-base-pair deletion of *Hoxa1* sequences. In parallel, a nearly identical transgene was generated, which differed only by an eight base-pair mutation in a conserved element previously designated CE2 and located several kilobases 3′ to *Hoxa1* (Langston *et al.*, 1997). The wildtype construct gave *lacZ* expression at E8.0–9.5 in the neuroepithelium, adjacent mesenchyme and somites, with expression extending from the caudal tip of the embryo to the rhombomere 3/4 border, and included an ectopic *Hoxa2*-like expression domain in rhombomere 2 (Thompson *et al.*, 1998). The CE2-mutant transgene showed a similar pattern of expression in the neural tube, but there was a complete lack of expression in somites and adjacent mesenchyme. By E10.5 both transgene constructs showed a decreased activity that was similar to the endogenous *Hoxa1* gene. When challenged with retinoic acid, both transgenes were capable of showing an anteriorization of their expression, the effect of which decreased with embryonic age, again similar to the endogenous *Hoxa1* gene. Thus separate transcription regions regulate *Hoxa1* expression in neural and non-neural tissues and the CE2 element appears independent of retinoic acid regulation.

In a parallel study, 15 kilobases of genomic DNA surrounding the *Hoxa1* gene was tested for its ability to direct reproducible *lacZ* expression from the *Hoxa1* promoter. This DNA segment spanned 7 kilobases of *Hoxa1* 5′ flanking sequences (extending into the *Hoxa2* gene), the entire *Hoxa1* coding sequences, and 5 kilobases of 3′ flanking sequences (Frasch *et al.*, 1995). This large transgene was able to direct reproducible expression that faithfully recapitulated the correct onset of the endogenous *Hoxa1* expression pattern, but the transgene was not efficiently downregulated, and *lacZ* expression persisted up to 48 hours longer than the endogenous *Hoxa1* gene. This may be due, in part, to an increased stability of the *lacZ* mRNA or the beta-galactosidase protein, since only a histochemical analysis was performed. Deletion analysis identified three independent fragments that could function as enhancers by directing expression in either orientation from a heterologous promoter. Enhancer I is a 1.2-kilobase fragment that overlaps the two exons and single intron of the neighboring *Hoxa2* gene. This enhancer can drive expression in rhombomere 4 (r4), which is a critical expression domain for both *Hoxa1* and *Hoxa2* as shown by the respective null phenotypes obtained from the genetic targeting of these two loci (Gendron-Maguire *et al.*, 1993; Lufkin *et al.*, 1991; Mark *et al.*, 1993). Enhancer II is a 2.5-kilobase fragment overlapping the second exon and 3′ flanking sequences of the *Hoxa2*, which is capable of directing expression to r2 (a domain unique to the *Hoxa2* gene) and to the most caudal somites. Examination of the activity of this murine enhancer in *Drosophila* embryos, revealed a discrete expression domain in the developing head region which overlapped precisely with the *Drosophila Hoxa2* homologue, *proboscipedia*, indicating that not only have the homeodomains of *Hoxa2* and *proboscipedia* been conserved during evolution, but so have certain aspects of the *cis*-acting elements and upstream regulators of these two genes. Enhancer III, a 0.5 kilobase fragment located 2 kilobases 3′ to the *Hoxa1* gene, was shown to be the most important enhancer for directing expression from the *Hoxa1* promoter since its deletion from the larger 15 kilobase construct resulted in a dramatic loss of ~95% of the *lacZ* positive cells. When placed upstream of a heterologous promoter, Enhancer III directs expression to a limited domain that includes the notochord, floor plate, gut endoderm and dorsal portions of the rostral spinal cord. Sequence analysis of Enhancer III revealed a perfect consensus retinoic acid response element (RARE) which had been previously isolated by DNAse I hypersensitivity mapping of chromatin (Langston and Gudas, 1992). The contribution of this element to *Hoxa1* regulation was examined with point mutations in just the RARE within the context of the large 15-kilobase transgene. This resulted in a loss of expression within the neural tube caudal to the r4/5 boundary, indicating that retinoic acid controls *Hoxa1* expression throughout a significant portion of its expression domain, and more importantly, in the region where the *Hoxa1* protein exerts its unique biological function (r4–r6). Additional work (Nonchev *et al.*, 1995) has revealed that late-phase upregulation of *Hoxa2* in r3 and r5 is mediated by Krox20.

8.2 Hoxa4

Transgenic analysis of the regulatory region surrounding the *Hoxa4* gene was performed with a 5-kilobase fragment spanning 4 kilobases of 5′ flanking sequences and 1 kilobase of the *Hoxa4* structural gene fused to *lacZ*. This genomic DNA segment was sufficient to essentially recapitulate the entire expression pattern of the endogenous *Hoxa4* gene in embryos and adult tissues (Behringer *et al.*, 1993; Packer *et al.*, 1998). Transgene expression was initiated at E8.0 in mesoderm and by E8.5 in the neural tube. At E9.5, neural tube expression extended along its entire length, with an anterior border just caudal to the hindbrain. Expression was observed at later stages in the central nervous system (with primarily a dorsal restriction) spinal ganglia, somites, and in the mesenchyme of several internal organs. Deletion of the 5′ flanking sequence showed no expression in transgenic embryos or adult mice, indicating that an important *Hoxa4* regulatory element was located between 2 and 4 kilobases 5′ to the transcription start site (Behringer *et al.*, 1993). When mice carrying the largest transgene were crossed to *Hoxa4* null mice, a decrease or loss of *lacZ* expression was observed in the *Hoxa4* null background relative to the *Hoxa4* heterozygous background. Specifically, expression was downregulated in the gut and mesonephros in E10.5 embryos, and additionally in the peripheral nerves and spinal ganglia at E12.5 (Packer *et al.*, 1998). These results suggest that the *Hoxa4* gene is autoregulated similarly to its *Drosophila* homologue *Deformed* (Bergson and McGinnis, 1990; Lou *et al.*, 1995). Three potential retinoic acid response elements were identified in the *Hoxa4* 5′ flanking region, and treatment of the wildtype transgene resulted in an increase, in *lacZ* expression. However, the same stimulation was not observed in the *Hoxa4* null background which indicated that *Hoxa4* autoregulation is necessary for maintenance of retinoic acid-induced expression. In a separate study, the intron of the *Hoxa4* gene was examined for activity in *Drosophila* embryos. This intron contains a cluster of three homeodomain-binding sites that have been conserved between mammals and flies and are referred to as the HB1 element (Haerry and Gehring, 1997). The HB1 element also functions as an enhancer and responds to several homeobox genes in a manner similar to the autoregulation of *Deformed*. In addition, point mutation within its binding sites abolished HB1 activity.

8.3 Hoxb1

Hoxb1 is the paralogue of *Hoxa1* and thus might potentially have regulatory mechanisms similar to *Hoxa1*. Endogenous *Hoxb1* expression appears initially expressed in the neural tube from a rostral boundary at the presumptive hindbrain rhombomere r3/4 border, and extends caudally along the length of the neural tube. With time, *Hoxb1* becomes restricted to r4 as expression fades in a rostrocaudal gradient in the neural tube. Both positive and negative enhancer elements appear to control the restriction of *Hoxb1* to r4 (Studer *et al.*, 1994). A 0.3-kilobase rhombomere r3–5-enhancer located immediately 5′ to the *Hoxb1* gene is sufficient to weakly drive early and late expression in r3 and r5 but strongly drives expression in r4. It is a region larger than the normal *Hoxb1* late-expression domain that is limited only to r4. An adjacent 0.7-kilobase negative-regulatory element is necessary to restrict the activity of this enhancer to r4, by specifically inhibiting enhancer activity in rhombomeres r3 and r5 but not r4. The ability of this negative regulatory element to inhibit activity in rhombomeres r3 and r5 is not unique to the *Hoxb1* r3, r4 and r5 enhancer as it can also block the activity of a r3+r5-specific enhancer from the *Hoxb2* gene. The latter normally functions to upregulate *Hoxb2* in rhombomeres r3 and r5 (but not r4) following the initial homogeneous activation of *Hoxb2* in the hindbrain (Studer *et al.*, 1994). The r3+r5 negative regulatory element is functionally conserved in chicks, and DNA-sequence analysis of this element identified a conserved RARE in both mouse and chick enhancers. The RARE is of the DR2 class (Chambon, 1994), hence it is similar to the RARE found 3′ to the *Hoxb1* gene (Marshall *et al.*, 1994), but different from the DR5 RARE found 3′ to the *Hoxa1* gene (Frasch *et al.*, 1995). Point mutation of the RARE was sufficient to abolish the negative regulatory activity of this element, indicating that the RARE mediates an essential part of the inhibiting activity of this negative regulatory element. This was surprising, because RAREs in general, and the *Hoxa1* and the *Hoxb1* 3′ RAREs in particular, function as activating elements (Frasch, 1995; Marshall *et al.*, 1994).

Two independent enhancers were identified 3′ to the *Hoxb1* gene that are critical for establishing the early phase of *Hoxb1* expression (Marshall *et al.*, 1994). The first of these enhancers, a 4-kilobase fragment lying approximately 1 kilobase 3′ to the *Hoxb1* gene is necessary for expression of *Hoxb1* in the somites, node and lateral plate mesoderm. The second enhancer was identified as an 0.8 kilobase fragment essential for expression in the neural tube and primitive streak. This second enhancer was responsive to retinoic acid, mimicking the early response of *Hoxb1* to retinoic acid. This involved both an anteriorization of the *Hoxb1* rostral border of expression as well as an overall increase in the levels of *Hoxb1* expression (Marshall *et al.*, 1992; 1994). RARE is functionally conserved between mouse, pufferfish and chick, which all contain a DR2-type RARE (Chambon, 1994) that can functionally interact with RAR/RXR heterodimers in an electrophoretic mobility shift assay. Mutation of RARE resulted in a loss of expression in the neural tube that did not affect expression in the somitic mesoderm or notochord, which is similar to the results obtained from mutation of the *Hoxa1* RARE (Frasch, 1995). Hence RARE was essential for expression of *Hoxb1* within the developing central nervous system. It is unclear if the activation of *Hoxb1* following retinoic acid exposure is driven solely by RAR interaction with the *Hoxb1* RARE, or if it is also mediated in part by *Hoxa1* overexpression following retinoic acid exposure. The latter is suggested by the observation that ectopic expression of *Hoxa1* is capable of activating the endogenous *Hoxb1* gene (Zhang *et al.*, 1994).

9. Conclusion

Hox proteins determine the fate of cells along the principal axes of the developing animal embryo. They function as transcriptional regulators that control the expression of large numbers of target genes and act as both activators and repressors. *Hox* genes themselves are targets for auto and cross-regulatory interactions that are mediated by highly conserved enhancer elements located within the *Hox* complex. The interdigitation of closely linked *Hox* genes with enhancers that work across several genes may have contributed to the maintenance of *Hox* gene complexes during evolution. However, the correct recapitulation of certain *Hox* gene expression patterns from transgenes integrated into non-*Hox* chromosomes argues against a strict dependence for positioning within a complex or a unique higher-order chromatin configuration of the *Hox* chromosome. At early stages of vertebrate development, the conservation of paralog group order, staggered temporal and spatial expression, and collinear sensitivity to retinoic acid may all be linked to the movement of cells through the node, which is a proposed site of retinoic acid synthesis. The observation of homeotic mutations in mice lacking RARs supports this idea.

The correct function of different *Hox* genes appears to depend on their ability to regulate different groups of target genes. They also regulate a highly overlapping set of target genes differentially. When binding alone, Hox proteins have poor specificity and a low affinity for DNA. However, in conjunction with cofactors of the homeodomain TALE class, both the specificity and affinity increase, resulting in increased occupancy by the Hox/TALE complex. Furthermore, genetic and molecular analysis has shown that the association of Hox and TALE proteins is necessary for the *in vivo* function of both groups of proteins. However, with regard to specificity of function, it is currently unclear whether it is the combination of Hox proteins, the total concentration of Hox proteins, or an integration of both that is most important for *Hox* gene function within a given cell. Additionally, given the widespread binding observed for certain Hox proteins in the presence and absence of TALE cofactors, the role of these cofactors in modulating the activity (both positive and negative) of the Hox/TALE complex, rather than the occupancy of binding sites, needs to be further explored.

References

Abu-Shaar, M., Ryoo, H.D. and Mann, R.S. (1999) Control of the nuclear localization of *Extradenticle* by competing nuclear import and export signals. *Genes Dev.* **13**: 935–945.

Abu-Shaar, M. and Mann, R.S. (1998) Generation of multiple antagonistic domains along the proximodistal axis during *Drosophila* leg development. *Development* **125**: 3821–3830.

Behringer, R.R., Crotty, D.A., Tennyson, V.M., Brinster, R.L., Palmiter, R.D. and Wolgemuth, D.J. (1993) Sequences 5′ of the homeobox of the *Hox-1.4* gene direct tissue-specific expression of *lacZ* during mouse development. *Development* **117**: 823–833.

Bergson, C. and McGinnis, W. (1990) An autoregulatory enhancer element of the *Drosophila* homeotic gene *Deformed*. *EMBO* J. **9**: 4287–4297.

Berthelsen, J., Kilstrup-Nielsen, C., Blasi, F., Mavilio, F. and Zappavigna, V. (1999) The subcellular localization of PBX1 and EXD proteins depends on nuclear import and export signals and is modulated by association with PREP1 and HTH. *Genes Dev.* **13**: 946–953.

Berthelsen, J., Zappavigna, V., Ferretti, E., Mavilio, F. and Blasi, F. (1998a) The novel homeoprotein Prep1 modulates Pbx–Hox protein cooperativity. EMBO J. **17**: 1434–1445.

Berthelsen, J., Zappavigna, V., Mavilio, F. and Blasi, F. (1998b) Prep1, a novel functional partner of Pbx proteins. *EMBO J.* **17**: 1423–1433.

Biggin, M.D. and McGinnis, W. (1997) Regulation of segmentation and segmental identity by *Drosophila* homeoproteins: the role of DNA binding in functional activity and specificity. *Development* **124**: 4425–4433.

Bischof, L.J., Kagawa, N., Moskow, J.J. *et al.* (1998) Members of the meis1 and pbx homeodomain protein families cooperatively bind a cAMP-responsive sequence (CRS1) from bovine CYP17. *J. Biol. Chem.* **273**: 7941–7948.

Briscoe, J. and Ericson, J. (1999) The specification of neuronal identity by graded Sonic Hedgehog signalling. *Semin. Cell Dev. Biol.* **10**: 353–362.

Burglin, T.R. (1998) The PBC domain contains a MEINOX domain: coevolution of Hox and TALE homeobox genes? *Dev. Genes Evol.* **208**: 113–116.

Burglin, T.R. (1997) Analysis of TALE superclass homeobox genes (MEIS, PBC, KNOX, Iroquois, TGIF) reveals a novel domain conserved between plants and animals. *Nucl. Acids Res.* **25**: 4173–4180.

Burglin, T. (1994) A comprehensive classification of homeobox genes. In: *Guidebook to the Homeobox Genes* (ed D. Duboule). Oxford University Press: Oxford. pp. 25–72.

Calvo, K.R., Knoepfler, P., McGrath, S. and Kamps, M.P. (1999) An inhibitory switch derepressed by pbx, hox, and Meis/Prep1 partners regulates DNA-binding by pbx1 and E2a–pbx1 and is dispensable for myeloid immortalization by E2a-pbx1. *Oncogene* **18**: 8033–8043.

Chambon, P. (1994) The retinoid signalling pathway: molecular and genetic analysis. *Semin. Cell Biol.* 5: 115–125.

Chan, S.K., Ryoo, H.D., Gould, A., Krumlauf, R. and Mann, R.S. (1997) Switching the *in vivo* specificity of a minimal Hox-responsive element. *Development* **124**: 2007–2014.

Chang, C.P., Shen, W.F., Rozenfeld, S., Lawrence, H.J., Largman, C. and Cleary, M.L. (1995) Pbx proteins display hexapeptide-dependent cooperative DNA binding with a subset of Hox proteins. *Genes Dev.* **9**: 663–674.

DiRocco, G., Mavilio, F. and Zappavigna, V. (1997) Functional dissection of a transcriptionally active, target-specific Hox–Pbx complex. *EMBO J.* **16**: 3644–3654.

Ferretti, E., Schulz, H., Talarico, D., Blasi, F. and Berthelsen, J. (1999) The PBX-regulating protein PREP1 is present in different PBX-complexed forms in mouse. *Mech. Dev.* **83**: 53–64.

Frasch, M. (1995) Induction of visceral and cardiac mesoderm by ectodermal Dpp in the early *Drosophila embryo. Nature* **374**: 464–467.

Frasch, M. (1999) Intersecting signalling and transcriptional pathways in *Drosophila* heart specification. *Semin. Cell Dev. Biol.* **10**: 61–71.

Frasch, M., Chen, X. and Lufkin, T. (1995) Evolutionary-conserved enhancers direct region-specific expression of the murine Hoxa-1 and Hoxa-2 loci in both mice and *Drosophila*. *Development* **121**: 957–974.

Gardner, R.L. (1999) Polarity in early mammalian development. *Curr. Opin. Genet. Dev.* **9**: 417–421.

Gehring, W.J., Affolter, M. and Burglin, T. (1994) Homeodomain proteins. *Annu. Rev. Biochem.* **63**: 487–526.

Gendron-Maguire, M., Mallo, M., Zhang, M. and Gridley, T. (1993) *Hoxa-2* mutant mice exhibit homeotic transformations of skeletal elements derived from cranial neural crest. *Cell* **75**: 1317–1331.

Gonzalez-Crespo, S., Abu-Shaar, M., Torres, M., Martinez, A.C., Mann, R.S. and Morata, G. (1998) Antagonism between extradenticle function and Hedgehog signalling in the developing limb. *Nature* **394**: 196–200.

Goudet, G., Delhalle, S., Biemar, F., Martial, J.A. and Peers, B. (1999) Functional and cooperative interactions between the homeodomain PDX1, Pbx, and Prep1 factors on the somatostatin promoter. *J. Biol. Chem.* **274**: 4067–4073.

Green, N.C., Rambaldi, I., Teakles, J. and Featherstone, M.S. (1998) A conserved *C*-terminal domain in PBX increases DNA binding by the PBX homeodomain and is not a primary site of contact for the YPWM motif of HOXA1. *J. Biol. Chem.* **273**: 13273–13279.

Greer, J.M., Puetz, J., Thomas, K.R. and Capecchi, M.R. (2000) Maintenance of functional equivalence during paralogous *Hox* gene evolution [see comments]. *Nature* **403**: 661–665.

Gross, M.K. and Gruss, P. (1994) Functional analysis of mouse *Hoxa-7* in *Saccharomyces cerevisiae*: sequences outside the homeodomain base contact zone influence binding and activation. *Mol. Cell. Biol.* **14**: 238–254.

Haerry, T.E. and Gehring, W.J. (1997) A conserved cluster of homeodomain binding sites in the mouse *Hoxa-4* intron functions in *Drosophila* embryos as an enhancer that is directly regulated by *Ultrabithorax. Dev. Biol.* **186**: 1–15.

Henrich, V.C., Sliter, T.J., Lubahn, D.B., MacIntyre, A. and Gilbert, L.I. (1990) A steroid/thyroid hormone receptor superfamily member in *Drosophila* melanogaster that shares extensive sequence similarity with a mammalian homologue. *Nucl. Acids Res.* **18**: 4143–4148.

Irish, V.F., Martinez-Arias, A. and Akam, M. (1989) Spatial regulation of the *Antennapedia* and *Ultrabithorax* homeotic genes during *Drosophila* early development. *EMBO J.* **8**: 1527–1537.

Jabet, C., Gitti, R., Summers, M.F. and Wolberger, C. (1999) NMR studies of the pbx1 TALE homeodomain protein free in solution and bound to DNA: proposal for a mechanism of HoxB1–Pbx1–DNA complex assembly. *J. Mol. Biol.* **291**: 521–530.

Jaffe, L., Ryoo, H.D. and Mann, R.S. (1997) A role for phosphorylation by casein kinase II in modulating *Antennapedia* activity in *Drosophila. Genes Dev.* **11**: 1327–1340.

Johnson, F.B., Parker, E. and Krasnow, M.A. (1995) Extradenticle protein is a selective cofactor for the *Drosophila* homeotics: role of the homeodomain and YPWM amino acid motif in the interaction. *Proc. Natl Acad. Sci. USA* **92**: 739–743.

Kenyon, C. (1994) If birds can fly, why can't we? Homeotic genes and evolution. *Cell* **78**: 175–180.

Knoepfler, P.S. and Kamps, M.P. (1995) The pentapeptide motif of hox proteins is required for cooperative DNA binding with pbx1, physically contacts pbx1 and enhances DNA binding by pbx1. *Mol. Cell. Biol.* **15**: 5811–5819.

Krumlauf, R. (1994) *Hox* genes in vertebrate development. *Cell* **78**: 191–201.

Langston, A.W., Thompson, J.R. and Gudas, L.J. (1997) Retinoic acid-responsive enhancers located 3′ of the Hox A and Hox B homeobox gene clusters. Funct. analysis. *J. Biol. Chem.* **272**: 2167–2175.

Langston, A.W. and Gudas, L.J. (1992) Identification of a retinoic acid responsive enhancer 3′ of the murine homeobox gene *Hox-1.6. Mech. Dev.* **38**: 217–228.

Lawrence, P.A. and Morata, G. (1994) Homeobox genes: their function in *Drosophila* segmentation and pattern formation. *Cell* **78**: 181–189.

Lawson, K.A. and Hage, W.J. (1994) Clonal analysis of the origin of primordial germ cells in the mouse. In: *Germline Development*, eds. J. Marsh and J. Goode. John Wiley & Sons, Chichester, pp. 68–84.

Lawson, K.A. and Pederson, R.A. (1992) Clonal analysis of cell fate during gastrulation and early neurulation in the mouse. In: *Postimplantation Development in the Mouse*, eds. D. J. Chadwick and J. Marsh. John Wiley & Sons, Chichester, pp. 3–21.

Li, X. and McGinnis, W. (1999) Activity regulation of Hox proteins, a mechanism for altering functional specificity in development and evolution. *Proc. Natl Acad. Sci. USA* **96**: 6802–6807.

Li, X., Murre, C. and McGinnis, W. (1999. Activity regulation of a Hox protein and a role for the homeodomain in inhibiting transcriptional activation. EMBO J. **18**: 198–211.

Lou, L., Bergson, C. and McGinnis, W. (1995) Deformed expression in the *Drosophila* central nervous system is controlled by an autoactivated intronic enhancer. *Nucl. Acids Res.* **23**: 3481–3487.

Lu, Q. and Kamps, M.P. (1996) Selective repression of transcriptional activators by pbx1 does not require the homeodomain. *Proc. Natl Acad. Sci. USA* **93**: 470–474.

Lu, Q., Knoepfler, P.S., Scheele, J., Wright, D.D. and Kamps, M.P. (1995) Both Pbx1 and E2A-Pbx1 bind the DNA motif ATCAATCAA cooperatively with the products of multiple murine *Hox* genes, some of which are themselves oncogenes. *Mol. Cell. Biol.* **15**: 3786–3795.

Lufkin, T. (1997) Transcriptional regulation of mammalian *Hox* genes during embryogenesis. *Crit. Rev. Eukary. Gene Express.* **7**: 193–213.

Lufkin, T., Dierich, A., LeMeur, M., Mark, M. and Chambon, P. (1991) Disruption of the *Hox-1.6* homeobox gene results in defects in a region corresponding to its rostral domain of expression. *Cell* **66**: 1105–1119.

Maconochie, M.K., Nonchev, S., Studer, M. *et al.* (1997) Cross-regulation in the mouse *HoxB* complex: the expression of *Hoxb2* in rhombomere 4 is regulated by *Hoxb1. Genes Dev* **11**: 1885–1895.

Mann, R.S. and Chan, S.K. (1996) Extra specificity from *extradenticle*: the partnership between HOX and PBX/EXD homeodomain proteins. *Trends Genet* **12**: 258–262.

Mark, M., Lufkin, T., Vonesch, J. *et al.* (1993) Two rhombomeres are altered in *Hoxa-1* mutant mice. *Development* **119**: 319–338.

Marshall, H., Studer, M., Pöpperl, H. ***et al.*** (1994) A conserved retinoic acid response element required for early expression of the homeobox gene *Hoxb-1*. *Nature* **370**: 567–571.

Marshall, H., Nonchev, S., Sham, M.H., Muchamore, I., Lumsden, A. and Krumlauff, R. (1992) Retinoic acid alters hindbrain *Hox* code and induces transformation of rhombomeres 2/3 into a 4/5 identity. *Nature* **360**: 737–741.

Muller, J. and Bienz, M. (1992) Sharp anterior boundary of homeotic gene expression conferred by the fushi tarazu protein. *EMBO J.* **11**: 3653–3661.

Neuteboom, S.T. and Murre, C. (1997) Pbx raises the DNA binding specificity but not the selectivity of antennapedia Hox proteins. *Mol. Cell. Biol.* **17**: 4696–4706.

Neuteboom, S.T.C., Peltenburg, L.T.C., Vandijk, M.A. and Murre, C. (1995) The hexapeptide LFPWMR in *hoxb-8* is required for cooperative DNA binding with pbxl and pbx2 proteins. *Proc. Natl Acad. Sci. USA* **92**: 9166–9170.

Nonchev, S., Vesque, C., Maconochie, M. *et al.* (1995) Segmental expression of *Hoxa-2* in the hindbrain is directly regulated by *Krox-20*. *Development* **122**: 543–554.

Oro, A.E., McKeown, M. and Evans, R.M. (1990) Relationship between the product of the *Drosophila ultraspiracle* locus and the vertebrate retinoid X receptor. *Nature* **347**: 298–301.

Packer, A.I., Crotty, D.A., Elwell, V.A. and Wolgemuth, D.J. (1998) Expression of the murine *Hoxa4* gene requires both autoregulation and a conserved retinoic acid response element. *Development* **125**: 1991–1998.

Passner, J.M., Ryoo, H.D., Shen, L., Mann, R.S. and Aggarwal, A. K. (1999) Structure of a DNA-bound Ultrabithorax–Extradenticle homeodomain complex. *Nature* **397**: 714–719.

Peltenburg, L.T.C. and Murre, C. (1996) Engrailed and hox homeodomain proteins contain a related pbx interaction motif that recognizes a common structure present in pbx. *EMBO J.* **15**: 3385–3393.

Phelan, M.L. and Featherstone, M.S. (1997) Distinct HOX *N*-terminal arm residues are responsible for specificity of DNA recognition by HOX monomers and HOX–PBX heterodimers. *J. Biol. Chem.* **272**: 8635–8643.

Phelan, M.L., Rambaldi, I. and Featherstone, M.S. (1995) Cooperative interactions between HOX and PBX proteins mediated by a conserved peptide motif. *Mol. Cell. Biol.* **15**: 3989–3997.

Pinsonneault, J., Florence, B., Vaessin, H. and McGinnis, W. (1997) A model for extradenticle function as a switch that changes HOX proteins from repressors to activators. *EMBO J.* **16**: 2032–2042.

Piper, D.E., Batchelor, A.H., Chang, C.P., Cleary, M. L. and Wolberger, C. (1999) Structure of a HoxB1–Pbx1 heterodimer bound to DNA: role of the hexapeptide and a fourth homeodomain helix in complex formation. *Cell* **96**: 587–597.

Pöpperl, H., Bienz, M., Studer, M. *et al.* (1995) Segmental expression of *Hoxb-1* is controlled by a highly conserved autoregulatory loop dependent upon *exd/pbx*. *Cell* **81**: 1031–1042.

Rieckhof, G.E., Casares, F., Ryoo, H.D., AbuShaar, M. and Mann, R.S. (1997) Nuclear translocation of extradenticle requires homothorax, which encodes an extradenticle-related homeodomain protein. *Cell* **91**: 171–183.

Sanchez, M., Jennings, P.A. and Murre, C. (1997) Conformational changes induced in Hoxb-8/Pbx-1 heterodimers in solution and upon interaction with specific DNA. *Mol. Cell. Biol.* **17**: 5369–5376.

Shanmugam, K., Featherstone, M.S. and Saragovi, H. U. (1997) Residues flanking the HOX YPWM motif contribute to cooperative interactions with PBX. *J. Biol. Chem.* **272**: 19081–19087.

Shen, W.F., Rozenfeld, S., Kwong, A., Kömüves, L.G., Lawrence, H.J. and Largman, C. (1999) HOXA9 forms triple complexes with PBX2 and MEIS1 in myeloid cells. *Mol. Cell. Biol.* **19**: 3051–3061.

Shen, W.F., Rozenfeld, S., Lawrence, H.J. and Largman, C. (1997) The Abd-B-like Hox homeodomain proteins can be subdivided by the ability to form complexes with Pbxla on a novel DNA target. *J. Biol. Chem.* **272**: 8198–8206.

Skeath, J.B. (1999) At the nexus between pattern formation and cell-type specification: the generation of individual neuroblast fates in the *Drosophila* embryonic central nervous system. *Bioessays* **21**: 922–931.

Streit, A. and Stern, C.D. (1999) Neural induction. A bird's eye view. *Trends Genet* **15**: 20–24.

Strutt, H., Cavalli, G. and Paro, R. (1997) Co-localization of Polycomb protein and GAGA factor on regulatory elements responsible for the maintenance of homeotic gene expression. *EMBO J.* **16**: 3621–3632.

Studer, M., Pöpperl, H., Marshall, H., Kuroiwa, A. and Krumlauf, R. (1994) Role of a conserved retinoic acid response element in rhombomere restriction of *Hoxb-1*. *Science* **265**: 1728–1732.

Swift, G.H., Liu, Y., Rose, S.D. (1998) An endocrine–exocrine switch in the activity of the pancreatic homeodomain protein PDX1 through formation of a trimeric complex with PBX1b and MRG1 (MEIS2). *Mol. Cell. Biol.* **18**: 5109–5120.

Tam, P.P. and Behringer, R.R. (1997) Mouse gastrulation: the formation of a mammalian body plan. *Mech. Dev.* **68**: 3–25.

Thompson, J.R., Chen, S.W., Ho, L., Langston, A.W. and Gudas, L.J. (1998) An evolutionary conserved element is essential for somite and adjacent mesenchymal expression of the *Hoxa1* gene. *Dev. Dyn.* **211**: 97–108.

Vandijk, M.A., Peltenburg, L.T.C. and Murre, C. (1995) *Hox* gene products modulate the DNA binding activity of Pbx1 and Pbx2. *Mech. Dev.* **52**: 99–108.

Vershon, A.K. and Johnson, A. D. (1993) A short, disordered protein region mediates interactions between the homeodomain of the yeast alpha 2 protein and the MCM1 protein. *Cell* **72**: 105–112.

Vigano, M.A., Di Rocco, G., Zappavigna, V. and Mavilio, F. (1998) Definition of the transcriptional activation domains of three human HOX proteins depends on the DNA-binding context. *Mol. Cell. Biol.* **18**: 6201–6212.

Wolpert, L. (1996) One hundred years of positional information. *Trends Genet.* **12**: 359–364.

Zappavigna, V., Sartori, D. and Mavilio, F. (1994) Specificity of HOX protein function depends on DNA–protein and protein–protein interactions, both mediated by the homeo domain. *Genes Dev.* **8**: 732–744.

Zhang, M.B., Kim, H.J., Marshall, H. *et al.* (1994) Ectopic *Hoxa-1* induces rhombomere transformation in mouse hindbrain. *Development* **120**: 2431–2442.

Zhao, J.J.G., Lazzarini, R.A. and Pick, L. (1996) Functional dissection of the mouse *hox-a5* gene. *EMBO J.* **15**: 1313–1322.

10

Tissue-specific regulation by transcription factors

Joseph Locker

1. Introduction

Animal tissues are made up of cells with distinctive properties that result from selective gene expression. Each cell type expresses a specific set of protein products and has a characteristic pattern of growth and cell cycle regulation. To mediate this selective gene expression, each cell type expresses a distinctive mixture of transcription factors. The mixture consists of general 'housekeeping' transcription factors, which are common to most cells, and specialized 'tissue-enriched' transcription factors. Multiple factors bind to each gene and combine their activities, efficiently integrating multiple signals into a single transcription unit (Carey, 1998). Genes that define a cell phenotype, e.g., serum albumin in the liver, are regulated by a combination of housekeeping and tissue-enriched factors. The latter generate the phenotype of the cell, but are not sufficient to activate transcription without the housekeeping factors.

Regulatory modules (i.e., promoters and enhancers) always contain multiple factor-binding sites even though *in vitro* analysis has suggested that most factors have equivalent or redundant function (Wang and Gralla, 1991), for example, to recruit the basal transcription complex or stabilize its binding to the promoter (see Chapters 1 and 2). While these mechanisms are fundamental, they provide a limited view of tissue-specific regulation. Recent studies have fully defined two regulatory modules, the β-interferon and T-cell receptor α gene enhanceosomes. In these regulatory modules, every site is essential and contributes a specific function (Carey, 1998; Kim and Maniatis, 1997). A central concept of this chapter is that all tissue-specific promoters and enhancers are like these defined enhanceosomes, i.e., each *cis* element binds factors that contribute either distinct cooperative binding or regulation to the overall regulatory module. At this level, current knowledge is limited. Few promoters and enhancers have been completely defined, and the specific mechanisms of action of most transcription factors are incompletely understood. Nevertheless, focusing on the distinct mechanisms of individual factors will lead to

Transcription Factors, edited by J. Locker.

a more complete understanding of tissue-specific regulation by explaining how the factors work together and by highlighting the missing contributions that remain to be defined.

To understand tissue-specific regulation by transcription factors, one should consider two additional levels of control. First, regulation at single binding sites is complex. Though a single factor generally acts through a characteristic *cis*-DNA-binding site, a single *cis*-site is generally capable of binding a number of different factors. Since these factors may have different regulatory effects, each individual *cis*-site links a gene to a program of changing regulation. Such dynamic regulation of factors allows cell phenotype to be coordinated with proliferation, diurnal regulation, homeostatic signals, or tissue injury. The second control level reflects development, via the history of transcription factor expression in precursors of the differentiated cell. The transcription factors appear sequentially during development, leading to gradual and sequential activation of the genes that define the mature phenotype.

Because liver is easily studied and synthesizes massive amounts of protein as its primary function, more transcriptional regulatory modules have been studied in liver than in any other tissue (*Table 1*). For this reason, this chapter concentrates on the liver, while additional aspects of tissue-specific regulation are presented in Chapters 7 and 11. Since the direct review of liver factors only begins to address the more fundamental question of how a tissue-specific phenotype is established by transcription factors, liver expression will be compared with that of two other tissues also derived from the foregut, namely the intestine and pancreas. Both tissues have a developmental sequence and transcription-factor mixture similar to those of liver, but express mostly different genes. This comparison highlights a fundamental problem: the known differences in the transcription factors of those tissues are not sufficient to explain why they express different genes.

2. Regulation of hepatocyte expression

2.1 Hepatocyte gene-regulatory modules

Some of the best-characterized liver gene regulatory modules are summarized in *Table 10.1*. Serum protein genes are exemplified by albumin, α-fetoprotein, transthyretin, and apolipoprotein B. The high-level transcription of these serum protein genes is generally constitutive and regulated by combinations of enhancers and promoters. Hepatitis B is a unique virus that is liver-specific by virtue of the fact that it has similar enhancers and promoters, which also mediate high-level transcription. Liver also synthesizes many metabolic regulatory enzymes, which may be either constitutive or inducible (Lemaigre and Rousseau, 1994). Glucagon, insulin and other hormones regulate phosphoenolpyruvate carboxykinase, while tyrosine aminotransferase is an important model of glucocorticoid gene regulation. L-pyruvate kinase is constitutive, and provides an interesting model of transcriptional regulation by typical liver-enriched transcription factors because the gene is also expressed in the small intestine and pancreas. *Table 1* also summarizes the current knowledge about genes for three liver-enriched transcription factors, hepatocyte nuclear factors HNF1α, HNF4α, and CCAAT/enhancer-binding protein-α (C/EBPα), but these have received more limited study.

The promoters and enhancers listed in *Table 1* contain binding sites for four classes of factors: (i) the families of liver-enriched factors, especially HNF1, HNF3, HNF4, HNF6, and C/EBP (*Figure 1*); (ii) constitutively active housekeeping factors; (iii) factors that mediate specific inducible responses; (iv) specialized developmental factors. Except for the presence of basal promoter elements, it is hard to distinguish promoters from enhancers, although it appears that HNF3 and C/EBP sites are critical for almost all enhancers, while proximal HNF1 sites are particularly common in promoters. Such distinctions might be clearer were it not for two aspects of regulatory module architecture. First, an enhancer may be adjacent to a promoter, causing its activities to be confused with promoter-specific functions.

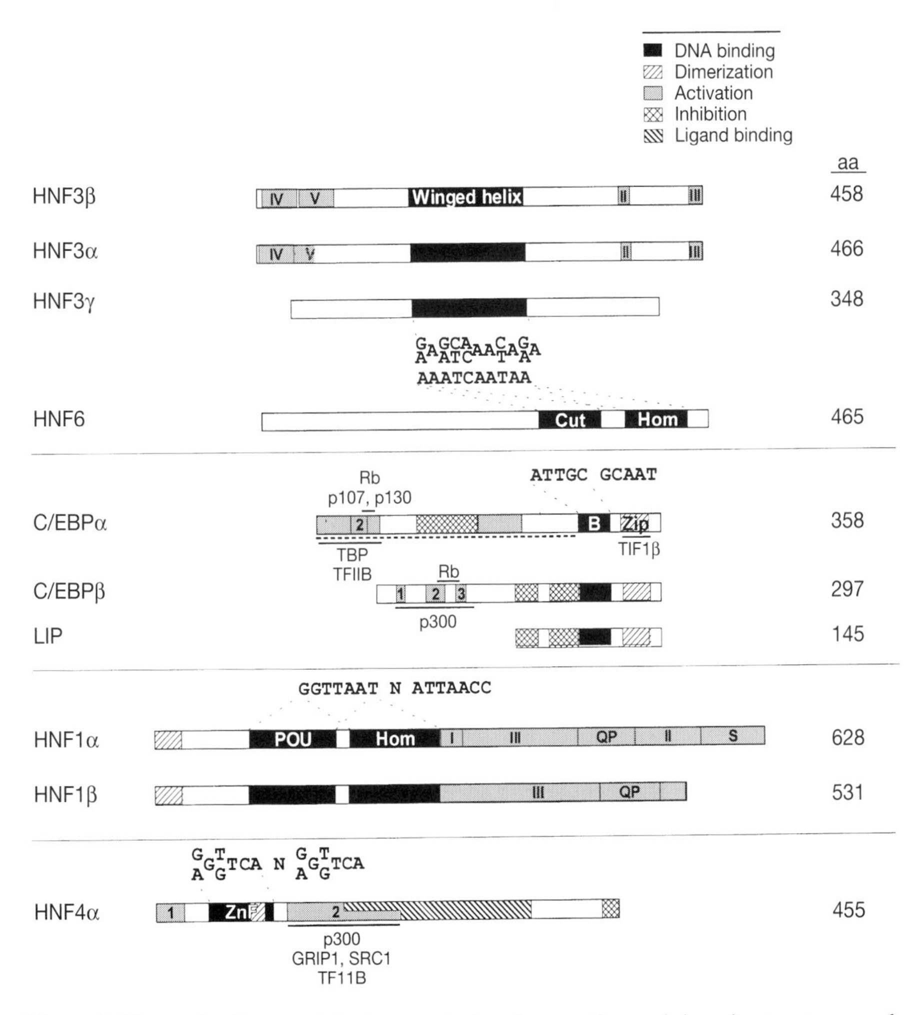

Figure 1. The major liver-enriched transcription factors. General domain structures and the labeled activation domains are described in the text. For factors that bind as dimers, the relationship of a monomer to a half-site motif is shown with the full binding site.

Table 10.1. Selected liver gene regulatory modules

Gene	Type	Regulation	Representative modules	Transcription factors			References
				Liver-enriched	General	Specialized (regulator)	
Albumin	Serum protein	Expressed only in liver; high levels in fetus and adult, decreased in liver regeneration; enhancer at –10 kilobases	Promoter Enhancer	HNF1, C/EBP, DBP HNF3, C/EBP	NFY, NF1 NF1	 GATA4 (development)	Bossard *et al.*, 1997; Bossard and Zaret, 1998; Herbomel *et al.*, 1989; Mueller *et al.*, 1990
α-Fetoprotein (AFP)	Serum protein	High-level expression in fetal liver, liver stem cells, and primitive endoderm; silenced in adult, but reactivated in liver cancer and liver regeneration; developmental regulation of promoter but not three strong upstream enhancers	Promoter Enhancer 3	HNF1, C/EBP, Lrh1 HNF3, C/EBP, HNF6	NF1	Nkx2.8 (development)	Apergis *et al.*, 1998; Galameau *et al.*, 1996; Groupp *et al.*, 1994; Samadani and Costa, 1996; Zhang *et al.*, 1991
Transthyretin	Serum protein	High-level expression in fetal and adult liver; low-level expression in fetal gut	Promoter Enhancer	HNF4, HNF3, HNF1, HNF6 HNF3, C/EBP, HNF4	AP1		Samadani and Costa, 1996
Apolipoprotein B	Serum protein	High-level expression in liver, intestine and placenta; regulated by promoter and downstream enhancers in introns 2 and 3	Promoter Second intron enhancer	HNF3, HNF4,C/EBP HNF-1, C/EBP, HNF3	COUP-TF, RXR		Kardassis *et al.*, 1992; Levy-Wilson, 1995
Hepatitis B (HBV)	Virus	Expressed only in liver; has four promoters and two enhancers	X promoter Enhancer I	C/EBP HNF3, HNF4, C/EBP, PPAR	NF1 RFX1, XPBP, NFI, AP1, RXR, COUP-TF, p53	CREB (cAMP)	Buckwold *et al.*, 1997; Fukai *et al.*, 1997; Ori and Shaul, 1995; Ori *et al.*, 1998; Trujillo *et al.*, 1991; Yen, 1993

L-pyruvate kinase	Metabolic enzyme	Expressed in liver; small intestine, kidney, endocrine pancreas, and fetal exocrine pancreas; expression parallels HNF4 levels in these tissues	Promoter	HNF1, HNF4	NFI, USF, COUP-TF	MLTF (insulin)	Kennedy *et al*., 1997; Miquerol *et al*., 1996; Miquerol *et al*., 1994
Phospho-enolpyruvate carboxykinase (PEPCK)	Metabolic enzyme	Expressed in liver; small intestine, kidney and adipocytes; key gluconeogenic enzyme induced at birth; strongly and rapidly induced by cAMP glucocorticoid and thyroid hormone; inhibited by insulin	Proximal promoter	DBP, C/EBP	NF1, AP1,	CREB (cAMP)	Croniger *et al*., 1998; Hanson and Reshef, 1997
			Upstream elements	C/EBP, HNF1, HNF3, HNF4, PPAR	TR, AP1, RXR	TR, RAR, GR (gluco-corticoid), CREB, (cAMP), AF2 (insulin)	
Tyrosine amino-transferase (TAT)	Metabolic enzyme	Glucocorticoid-regulated gene induced in neonatal liver	Promoter	HNF1	NFY, NF1		Sassi *et al*., 1995, 1998; Schweizer-Groyer *et al*., 1994
			Enhancers (GRUs)	C/EBP, HNF3	Ets	GR (glucocorticoid)	
C/EBPα	Trans-cription factor	Murine promoter directly autoregulated by binding of C/EBP factors: human promoter lacks a C/EBP site but is indirectly regulated by C/EBP through modulation of upstream stimulatory factor (USF)	Promoter	C/EBP	NFI, USF, Myc/Max, NF-κB, BTEB, Sp1, AP2		Antonson *et al*., 1995; Legraverend *et al*., 1993; Tang *et al*., 1999; Timchenko *et al*., 1995
HNF4α	Trans-cription factor	Regulatory loop with HNF1	Promoter	HNF1, HNF3, HNF6	NF1, AP1		Furuta *et al*., 1997
HNF1α	Trans-cription factor	Regulatory loop with HNF4	Promoter	HNF4			Gragnoli *et al*., 1997

Second, some enhancers also function as promoters. Such functional relationships are expected to become clearer as liver-specific promoters and enhancers become more fully defined. Nevertheless, it appears that the liver-enriched factors (*Figure 1*) provide a variety of essential functions that are complemented by housekeeping factors. Thus the liver-specific regulatory systems can only function properly in a transcriptional environment that contains the liver-enriched factors. This transcriptional environment is gradually established through the sequential appearance of transcription factors during development (*Figure 2*). Tissue-specific inducible responses, e.g., to glucocorticoid or cAMP, are mediated by *cis*-binding sites that are embedded in modules containing sites for liver-enriched factors. Finally, the set of liver-enriched transcription factors is linked to the distinctive control of growth and proliferation that characterizes a liver cell.

2.2 Hepatocyte nuclear factors HNF3 and HNF6

HNF3 binding sites, among the most common in liver genes, are associated with activating factors that are among the earliest to appear in liver development. Liver contains three closely related factors, HNF3α, β, and γ (*Figure 1a*), which have a

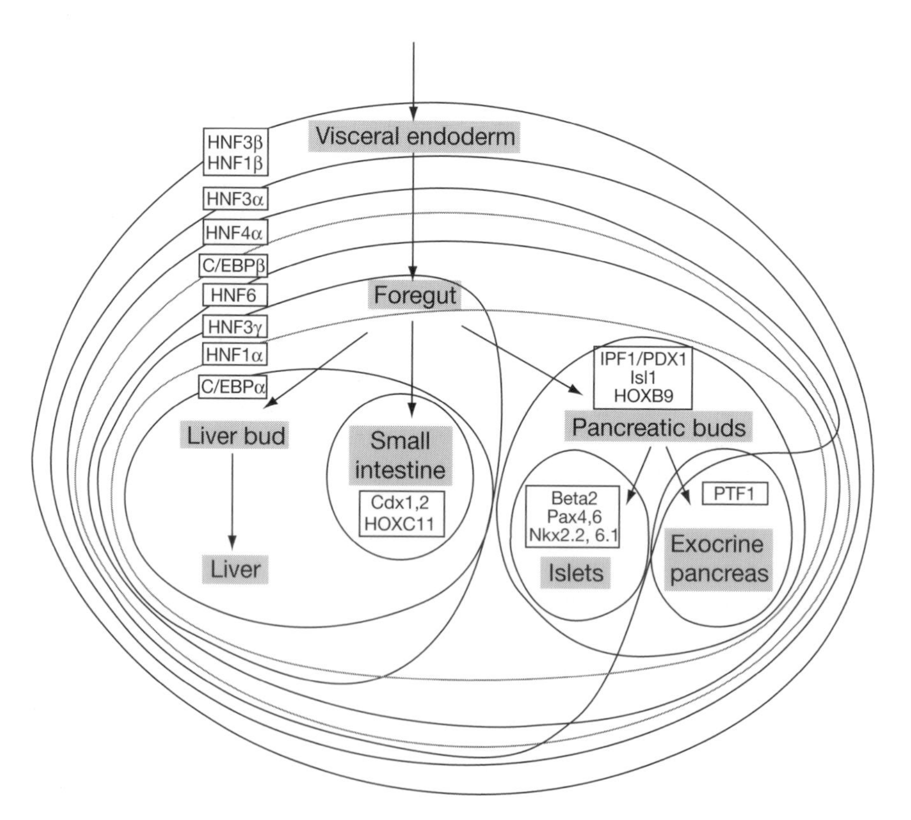

Figure 2. Transcription factor expression during development of liver and related tissues. The relationships are approximated from the references described in the text. Only limited information is available for C/EBPβ, but the presence in multiple tissues suggests a pattern of expression similar to hepatocyte nuclear factors HNF3γ and HNF6.

winged helix DNA-binding domain closely related to *Drosophila forkhead* (Costa *et al.*, 1988; Lai *et al.*, 1993; 1991). All three HNF3 factors have nearly identical DNA binding (Peterson *et al.*, 1997), so that the activity at most binding sites reflects the sum of contributions from all three, although weak sites have been described with a selective affinity for HNF3α (Samadani *et al.*, 1996). Expression of all three precedes formation of the liver. HNF3β is expressed first, in early gastrula, while α appears at a later gastrula stage and γ only in the differentiated foregut. In the adult, all three genes are expressed in liver and intestine; α and β are also expressed in pancreas, stomach, and lung (Cereghini, 1996). Deletion of the HNF3β gene results in death of the mouse embryo before formation of the foregut (Ang and Rossant, 1994; Weinstein *et al.*, 1994), so its contribution to the liver phenotype cannot be assessed. Deletions of α or γ produce much milder changes. Surprisingly, since it was first characterized in liver, HNF3α deletion has no effect on liver gene expression, and indeed has no effect on prenatal development. Postnatally, the animals show growth retardation and severe impairment of islet proglucagon expression from pancreatic islets (Kaestner *et al.*, 1999; Shih, 1999). Deletion of the γ gene has even milder effects. These animals have normal development but reduced expression of liver genes known to be HNF3 targets. However, the minimal effects are partly compensated by upregulation of HNF3α and β (Kaestner *et al.*, 1998).

Peptide deletion studies of HNF3α have demonstrated four regions that contribute to activator function, in addition to region I, which is the winged helix DNA-binding domain. Regions IV and V comprise most of the *N*-terminal, while regions II and III are small segments near the *C*-terminal (Pani *et al.*, 1992; Qian and Costa, 1995). Comparison of sequences demonstrates that regions II, III, and IV are highly conserved and region V partially conserved in HNF3β, suggesting that α and β activate transcription by the same mechanism. These regions are less conserved in γ. Studies to date have not identified HNF3 interactions with co-activators or the basal transcription factors. However, studies of the albumin gene enhancer suggest that HNF3 has a role in organizing chromatin architecture, acting to position the nucleosome cores and establishing the natural spacing observed in liver nuclei (Shim *et al.*, 1998). Since the winged helix domain is structurally related to linker histone protein (Clark *et al.*, 1993), this architectural effect on chromatin structure may be a common function for HNF3 members. In related studies, it was demonstrated that HNF3 members bind the albumin enhancer in the foregut, prior to gene activation (Bossard and Zaret, 1998). HNF3 activator function appeared only in combination with other factors that bound at a later developmental stage.

Though HNF6 is abundant, its discovery was delayed because its activity was obscured by HNF3 (*Figure 1a*). Both have similar behavior in gel-shift assays and bind many of the same DNA sites (Lemaigre *et al.*, 1996; Samadani and Costa, 1996). However, HNF6 is a distinctive factor that binds DNA through its single Cut and homeodomains (Rastegar *et al.*, 1998). During development, HNF6 appears about 2 days after HNF3β and shows a similar pattern of localization in the foregut and its derivatives (Lemaigre *et al.*, 1996; Rausa *et al.*, 1997; Vaisse *et al.*, 1997). In the liver, expression is dependent on growth hormone and is therefore sexually dimorphic, with low intermittent levels in males and continuous high levels in females (Lahuna *et al.*, 1997). This continuous high-level expression partially accounts for the female-specific expression of the *CYP2C12* gene, which is regulated through a site that has high HNF6 and lower HNF3 affinity. With only a single Cut domain, HNF6 is the prototype of a unique ONECUT class of transcription factors (Lannoy *et al.*, 1998). A second ONECUT factor, OC2 has also been recently identified in liver (Jacquemin *et al.*, 1999). Both HNF6 and OC2 are transcriptional activators, but except for their DNA binding domains, the activation domains, binding partners, and transcriptional activation mechanism have not yet been defined. In liver genes, function of most 'HNF3 sites' thus represents a composite activity of at least three HNF3 species and two ONECUT proteins. Their individual contributions are probably additive and somewhat redundant, but analysis might also demonstrate distinctive activation mechanisms acting through a single binding site.

2.3 C/EBP and DBP-related factors

C/EBP binding sites are as frequent as HNF3 sites and often found together with them in liver regulatory modules. Liver contains constitutive high levels of the bZIP factors C/EBPα and C/EBPβ (also liver-enriched activator protein [LAP] or nuclear factor IL-6 [NFIL6]) (*Figure 1b*), and lower levels of C/EBPδ (Ig/EBP) (Takiguchi, 1998). C/EBPα is also expressed in epidermis, myeloid cells, and adipose tissue. Whether it is expressed in the pancreas is controversial, but expression has been reported in some studies (Antonson and Xanthopoulos, 1995; Dabeva *et al.*, 1995). C/EBPβ expression is widespread, but prominent in liver. Expression is clear in fetal liver, and it is also expressed in pancreatic acini and islets (Dabeva *et al.*, 1995; Lu *et al.*, 1997).

These bZIP factors bind DNA as homo and heterodimers. Each subunit binds a DNA motif related to ATTGC, and the dimer binds two adjacent sites on opposite DNA strands (ATTGC GCAAT). Variation is tolerated at most of these bases, so that the range of potential binding sites remains poorly defined (Johnson, 1993; Ryden and Beemon, 1988). C/EBPα, β, and δ are all transcriptional activators, and, *in vitro*, the various dimers have similar activating effects on most target genes. *In vivo*, they have biologically distinct effects and are separately regulated. Several additional peptides that act as dominant negative regulators are also binding partners. One of these, liver-enriched inhibitory protein (LIP) (*Figure 1b*), is produced by translation of C/EBPβ mRNA from an internal ATG codon, and proteolytic processing produces similar isoforms (Descombes and Schibler, 1991; Welm *et al.*, 1999). LIP acts as a dominant negative because it binds DNA as homo and heterodimers but lacks the *N*-terminal activation domains. The ratio of C/EBPβ and LIP is dynamically regulated. For example, dietary protein restriction decreases C/EBPβ and increases LIP (Marten *et al.*, 1996). C/EBPα also has *N*-terminal-truncated dominant negative forms (Ossipow *et al.*, 1993). C/EBP homologous protein (CHOP) or Gadd153, another dominant negative molecule that lacks *N*-terminal activation domains (Park *et al.*, 1992), is produced from an inducible gene that is expressed in most cells, including hepatocytes (Wolfgang *et al.*, 1997). CHOP is induced by a variety of cell injuries, including DNA damage, glucose deprivation, amino acid deprivation, liver cell injury by carbon tetrachloride, and distention of the endoplasmic reticulum (Wang *et al.*, 1996; Wolfgang *et al.*, 1997). CHOP can also activate apoptosis, suggesting that positive activation of some C/EBP-responsive genes is necessary to maintain cell integrity (Zinszner *et al.*, 1998). Modulation of C/EBP isoforms appears to be an important mechanism for increasing or decreasing absolute levels of many liver proteins without significantly changing the overall phenotype.

Constitutive expression of liver genes is regulated by both C/EBPα and β, so that loss of either reduces gene expression without changing phenotype. The *C/EBPα* knockout mouse dies shortly after birth from postnatal reduction of critical liver mRNAs (e.g. glycogen synthase mRNA is reduced by 30% to 50%), causing failure of energy homeostasis and death (Flodby *et al.*, 1996; Wang *et al.*, 1995). Nevertheless, the reduction of gene expression is limited, and liver development is essentially normal although neonates have increased cell proliferation (Flodby *et al.*, 1996). In a knockout selectively targeted to adult liver, the phenotype is non-lethal, but expression of a number of genes is reduced (Lee *et al.*, 1997). The C/EBPβ knockout is completely viable and has normal liver development. Usually, a few liver genes, like phosphoenolpyruvate carboxykinase, show moderately reduced expression, though some animals fail to express phosphoenolpyruvate carboxykinase and die of hypoglycemia (Croniger *et al.*, 1997).

C/EBPα is most abundant in quiescent hepatocytes while β has a distinctive role in proliferation. This role has been confirmed by study of liver regeneration in the *C/EBPβ* knockout mouse, which has a greatly reduced proliferative response (Greenbaum *et al.*, 1998). In normal animals, the α and β levels are about equal in quiescent hepatocytes, but regeneration rapidly causes a decrease in α and a marked increase in β, though total C/EBP activity remains relatively constant (Diehl, 1998; Greenbaum *et al.*, 1995; Rana *et al.*, 1995). C/EBPβ is induced by interleukin 6 (IL6), which is an important mediator of the acute phase response. The elevated β of liver regeneration is also attributed to IL6 stimulation. Moreover, the increased expression of serum proteins in both the acute phase reaction and liver regeneration is generally attributed to increased C/EBPβ. Similar changes of gene expression are observed in liver regeneration, where the increased β is also attributed to IL6 (Burgess-Beusse and Darlington, 1998; Greenbaum *et al.*, 1998; 1995; Milland *et al.*, 1990).

Although there is limited structural conservation outside of the bZIP domains, studies suggest that C/EBP*a* and β have both common and distinct mechanisms of transcriptional activation. In C/EBPβ, three *N*-terminal regions have been mapped as activation domains; together these have been shown to bind p300/CBP at its E1A-binding domain (Mink *et al.*, 1997). A comparable *N*-terminal region of C/EBPα shows clear conservation of region 2 (Williams *et al.*, 1995). Nevertheless, in contrast to C/EBPβ, the *N*-terminal region of α has been shown to bind directly to the basal transcription factors TBP and TFIIB (Nerlov and Ziff, 1995). In addition, C/EBPα has an internal activation domain not found in β, which presumably mediates a different set of interactions (Nerlov and Ziff, 1994). The highly conserved leucine zippers are also associated with important binding interactions with transcriptional regulators. These include the coactivator, Tif1β (Chang *et al.*, 1998), NF-κB subunits, the glucocorticoid receptor (Takiguchi, 1998), and SRF (Hanlon and Sealy, 1999). Thus, although they are both transcriptional activators, it appears that C/EBPα and β activate by multiple mechanisms, some common and some distinct. Their activities appear complementary or even synergistic.

Beyond transcriptional activation, other molecular interactions confirm reciprocal roles for C/EBP factors in cell-cycle regulation. C/EBPα, β, and δ all have a domain near region 2 that binds hypophosphorylated Rb (Chen *et al.*, 1996). C/EBPα, but not β, also binds p107 and p130 complexes through this same domain. In neonatal liver, binding of p107 prevents the latter from binding to E2F. As a result, C/EBPα has a selective antiproliferative effect in hepatocytes (Timchenko *et al.*, 1999a; 1999b). C/EBPα also binds and stabilizes p21 through an unlocalized domain (Timchenko *et al.*, 1997).

Another family of bZIP factors, the proline and amino acid-rich (PAR)-domain factors, also regulates liver genes. The PAR-domain factors generally bind to a subset of C/EBP sites (Falvey *et al.*, 1996). Three factors, D-site binding protein (DBP), hepatic leukemia factor (HLF), and transcription enhancer factor/vitellogenin gene-binding protein (TEF/VBP), are expressed in liver. They can homo and heterodimerize with each other, but not with C/EBP factors (Falvey *et al.*, 1995; Inaba *et al.*, 1994; Mueller *et al.*, 1990). DBP, the most abundant PAR factor in liver, was initially characterized from its strong binding of the albumin promoter D site, which is also a strong C/EBP binding site. The PAR domain has activating function and binds p300/CBP at the E1A site (Lamprecht and Mueller, 1999). Although structurally distinct, the PAR family factors appear to activate genes by the same mechanism as C/EBPβ. The PAR-domain factors show circadian regulation and, at their peak levels, significantly augment the transcription of numerous liver-specific genes (Fonjallaz *et al.*, 1996; Lavery and Schibler, 1993; Wuarin *et al.*, 1992). These factors are expressed only very late in development. Hence the DBP knockout mouse has normal liver development and shows only moderate changes in liver gene expression, notably the loss of circadian expression of some genes, e.g., steroid 15 alpha-hydroxylase (CyP2a4) and coumarin 7-hydroxylase (CyP2a5) (Lavery *et al.*, 1999; Lopez-Molina *et al.*, 1997).

Gene regulation through C/EBP-binding sites accounts for a major fraction of total liver gene expression. Through these binding sites, expression levels are modulated by many stimuli that change the mix and levels of all of these activating and dominant forms. In addition, switching between α and β, required because of their different effects on proliferation, maintains the high overall level of activating forms that is necessary for high transcription.

2.4 HNF1

HNF1α and β (*Figure 1c*) are both abundant liver-enriched factors with a unique role in establishing the liver phenotype. HNF1-binding sites are commonly found in proximal promoters, and deletion of these proximal sites profoundly reduces or abolishes *in vitro* gene expression (Song *et al.*, 1998; Tronche *et al.*, 1990; Zhang *et al.*, 1991). This effect is generally greater than for other single binding sites, suggesting that HNF1 has a fundamental role in the regulation of liver genes that differs from the role of other liver-enriched factors. Two isoforms,

HNF1α and β are expressed from different genes, but have nearly identical DNA-binding and dimerization domains and thus can bind as homo or heterodimers. The DNA-binding domains are formed from a divergent homeodomain that has fused with a partial POU domain (Cereghini, 1996) and each dimeric subunit binds to a 7-base motif derived from GGTTAAT (Tronche *et al.*, 1997).

In development, the appearance of HNF1α roughly coincides with liver specification. Nevertheless, the *HNF1α* knockout mouse has essentially normal liver development, though with reduced expression of many liver-specific genes. The most important effects of the knockout are non-hepatic. The mice eventually die from renal failure and also have defective insulin secretion from β-cells (Pontoglio *et al.*, 1996; 1998). The contrast between the loss of function in promoter studies and mild phenotypic change in the HNF1α knockout is partly explained by the residual expression of β. Its levels are sufficient to maintain nearly normal function of most liver genes, with a few exceptions like phenylalanine hydroxylase (Pontoglio *et al.*, 1997). Unlike HNF1α, HNF1β appears before liver development and its expression is more widespread (Coffinier *et al.*, 1999a). Thus, the targeted ablation of the *HNF1β* gene produces an early embryonic lethal phenotype prior to liver formation. In these mouse embryos, visceral endoderm formation is distorted and development does not progress (Coffinier *et al.*, 1999b).

The mechanism of transcriptional activation by HNF1α has not been determined, but activation function has been localized to three *C*-terminal regions. Activation domain ADI is serine-rich, ADII proline-rich, and ADIII glutamine-rich (Cereghini, 1996; Toniatti *et al.*, 1993). HNF1β may not have the same activation function as α, as the HNF1β *C*-terminal region is shorter and divergent, with partial conservation (61%) only of ADIII (Cereghini, 1996). Indeed, a recent study of the vitamin-D-binding protein gene has demonstrated distinctive functions for HNF1α and β (Song *et al.*, 1998). In this study, HNF1α was required for promoter activation from a distant enhancer, while HNF1β was a *trans*-dominant inhibitor of enhancer activity. Like other factor combinations, it thus appears that HNF1α and β can contribute both common and distinct functions to a liver gene while activating through a single *cis*-binding site.

2.5 HNF4 and DR1-binding nuclear receptors

HNF4 (*Figure 1d*) has been shown to be a fundamental regulator of hepatocyte phenotype, both through it own activity and by regulation of the HNF1α gene promoter through a strong binding site (Kuo *et al.*, 1992; Tian and Schibler, 1991). There is also an HNF1 site in the promoter of the *HNF4α* gene (Furuta *et al.*, 1997), suggesting a full regulatory loop. Studies of de-differentiated hepatocytic cells have confirmed the importance of the HNF4/HNF1 relationship. Such cells lack both HNF1α and HNF4α, while C/EBP and HNF3 remain abundant and transfection of an HNF4-expressing plasmid can restore the phenotype. This treatment replaces the missing HNF4α and activates HNF1α synthesis. In combination, the two factors reactivate hepatocyte-specific genes (Spath and Weiss, 1998; 1997). Even so, HNF4α transfection does not reverse silencing of the endogenous *HNF4α* gene, indicating that the differentiation block is at another level of control.

HNF4α is also expressed in intestine, kidney and islets, but not exocrine pancreas (Miquerol *et al.*, 1994; Sladek *et al.*, 1990). Mammals also express a second gene, *HNF4γ*, but not in the liver (Drewes *et al.*, 1996). Since *HNF4α* is first expressed in early endoderm, gene knockout prevents normal gastrulation and is thus an early embryonic lethal (Chen *et al.*, 1994; Duncan *et al.*, 1994).

HNF4 is a member of the nuclear receptor family of zinc finger transcription factors, most closely related to the RXRs (Nuclear Receptors Nomenclature Committee, 1999). HNF4 is unusual, however, because it only binds DNA as a homodimer, not as a heterodimer with RXR species (Jiang and Sladek, 1997). The dimer binds DNA at two 6-base direct-repeat half-sites separated by one base (DR1), a binding site that is closely related to those of RXRs and other members of the nuclear receptor family (Duncan *et al.*, 1994; Sladek *et al.*, 1990). Like similar receptors, HNF4 has a typical zinc finger DNA-binding and dimerization domain, and also a putative ligand-binding domain (Sladek *et al.*, 1999; Wang *et al.*, 1998). Hertz *et al.* (1998) have recently reported that fatty acyl-CoA thioesters act as ligands as these compounds bind to HNF4 and increase gene activity in transfection experiments. Binding was observed with both saturated and unsaturated fatty acids, but shorter fatty acids activated transcription while longer fatty acids suppressed it. However, because of the variety of effects, the heterogeneity and unusual nature of these ligands, and the failure to demonstrate that ligand produced a conformational change or promoted coactivator binding (Sladek *et al.*, 1999), the putative ligands have remained controversial. The existence and nature of a true ligand is a question of fundamental importance, because the ligand would regulate hepatic phenotype through its activity on HNF4. One property is likely. The putative ligand must be endogenous in most hepatocytic cells, since HNF4 generally functions as a strong activator.

The structure of HNF4 and its interactions with coactivators are similar to those of other nuclear receptors. Two activation regions have been localized. AF1 lies in the first 24 *N*-terminal amino acids, while AF2 extends from aa 128 to 370 (Dell and Hadzopoulou-Cladaras, 1999). AF2 binds to p300/CBP and, as in other nuclear receptors, strongly binds to a domain (aa271 to 451) that contains two characteristic nuclear-receptor-binding LXXLL motifs (Heery *et al.*, 1997), and to another LXXLL motif in the glutamine-rich domain near the *C*-terminal (Dell and Hadzopoulou-Cladaras, 1999; Yoshida *et al.*, 1997). These binding sites on p300/CBP are different from the 'E1A-binding site' that binds C/EBPβ. This implies that HNF4 can synergize with C/EBPβ to recruit p300/CBP, which appears to integrate the effects of diverse signaling pathways (Mink *et al.*, 1997; Perissi *et al.*, 1999). The AF2 domain also binds to LXXLL motifs of Src1 and Grip1, additional coactivators that associate with nuclear receptors and p300/CBP (Wang *et al.*, 1998; McInerney *et al.*, 1998). Again, the question of a ligand is particularly relevant, because the coactivator-binding sites of most nuclear receptors bind corepressors in the absence of a ligand (Hu and Lazar, 1999). Thus HNF4 potentially has a dual effect on hepatic phenotype: repression in the absence, but activation in the presence, of the ligand.

Two other interactions are also important. AF2 has also been reported to bind TFIIB and thus to directly facilitate transcription complex assembly (Malik and Karathanasis, 1996). Also, HNF4 activity is modified by protein kinase A phosphorylation, which decreases DNA-binding activity and provides a linkage to cAMP activation that is known to downregulate hepatocyte gene expression (Viollet *et al.*, 1997).

The hepatocyte contains several other abundant DR1-binding nuclear receptor transcription factors (*Figure 3*). RXR and PPAR are transcriptional activators with coactivator interactions similar to HNF4. In contrast, chick ovalbumin upstream promoter transcription factor (COUP-TF) is generally a negative transcriptional regulator. Individual sites that bind HnF4 are frequently reported to bind these other factors (see *Table 10.1*) Galson *et al.*, 1995; Hall *et al.*, 1995; Jiang *et al.*, 1995; Mietus-Snyder *et al.*, 1992; Yu and Mertz, 1997). Thus, these factors may potentiate, antagonize, or even replace the function of HNF4 in establishment of phenotype. Hepatocytes contain RXRα, β, and γ. RXRα predominates, and, in fact, liver has the highest level of any tissue (Mangelsdorf *et al.*, 1992). RXRs bind in combination with numerous nuclear receptors, but bind as RXR dimers at DR1 sites in the presence of the ligand 9-*cis*-retinoic acid. This ligand, and other retinoids, has been shown to strongly regulate HNF1 and HNF4 in Hep3B cells (Magee *et al.*, 1998). This is an intriguing relationship because retinoids are differentiating agents in many tissues, where their transcriptional effects are comparable to those of HNF4 in liver.

PPAR (Desvergne and Wahli, 1999) are also important in the liver. Liver contains PPARα, δ (also called β), and γ, although α is considerably more abundant than the others. In the presence of natural (unsaturated fatty acids) or synthetic (drugs that induce hypolipidemia) ligands, PPAR form heterodimers with RXRs and bind at DR1 sites. PPAR typically regulate genes associated with fatty acid metabolism and transport, including genes of the peroxisomal β-oxidation pathway, but common binding with HNF4 is known at other genes, including those of hepatitis B virus (Yu and Mertz, 1997). PPAR ligands have been found to induce liver hyperplasia and eventually cancer, indicating a link between gene activation through DR1-binding sites and regulation of liver cell growth. 9-*cis*-retinoic acid has also been

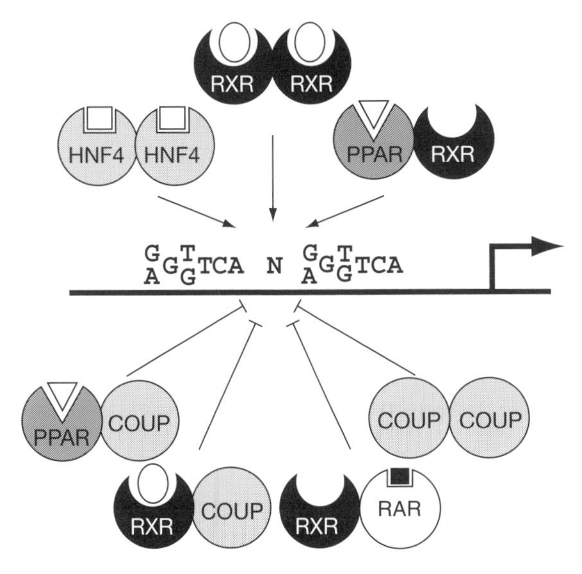

Figure 3. DR1-binding nuclear receptors of hepatocytes.

shown to be a hepatocyte mitogen (Ohmura *et al.*, 1996). The PPARα knockout mouse (Lee *et al.*, 1995) has normal liver development and minimal alteration of baseline gene expression, but lacks the growth stimulatory effects of hypolipidemic drugs. As in liver, fat cell gene expression, especially of metabolic regulatory genes that are also expressed in liver, is regulated by PPAR factors in combination with C/EBPα and β (Darlington *et al.*, 1998; Mandrup and Lane, 1997).

Liver also contains significant levels of the orphan receptors COUP-TFI and TFII (Tsai and Tsai, 1997). These bind DR1 sites as COUP-TF dimers and as RXR/COUP heterodimers. Although COUP-TFI and TFII have a highly conserved ligand-binding domain, no potential ligands have been found. Like other unliganded receptors, COUP-TF interacts with corepressors (Bailey *et al.*, 1998). COUP-TF factors generally act as negative regulators. This negative action presumably indicates that a natural COUP ligand is not usually present in liver, but there is still the possibility of a ligand-induced activating state. As a negative regulator, COUP-TF modulates the positive effects of HNF4, RXR and PPAR. COUP-TF expression is significantly increased by retinoids, which may represent a significant secondary mechanism for limiting the transcriptional stimulatory effects of these agents.

2.6 Regulators of additional inducible transcriptional responses in hepatocytes

Beyond the basic set of liver-enriched transcription factors just described, there are several more specialized regulators that control important aspects of liver gene regulation. These additional factors have several common features: they are receptors that interact with various endogenous ligands or with drugs and other agents; they mediate inducible responses that are not necessary for normal liver development or the basic hepatocytic phenotype; and they regulate inducible metabolic or detoxifying enzymes, many of which are forms of cytochrome p450 (CyP). In cytochrome p450 genes, these inducible regulators generally bind to sites within regulatory modules that also bind liver-enriched factors.

The aryl hydrocarbon receptor binds dioxin and other xenobiotic agents, and has been shown to regulate enzymes that metabolize xenobiotic agents, including CyP1A1, CyP1A2, and CyP40 (Meyer *et al.*, 1998). Aryl hydrocarbon receptor heterodimerizes with its close relative, the aryl hydrocarbon nuclear transporter, a widely expressed factor. Aryl hydrocarbon receptor is the only liganded member of the helix-loop-helix (HLH) family and is also the only liver-enriched HLH factor yet described. This is unusual because tissue-enriched HLH factors have a critical role in establishing the phenotype of a wide range of other tissues, from pancreas to skeletal muscle. AHR, perhaps in combination with an unidentified endogenous ligand, may have a role in normal liver growth and development, since the *AHR* knockout mouse has decreased levels of aldehyde dehydrogenase, decreased growth, and liver fibrosis (Fernandez-Salguero *et al.*, 1995; Gonzalez and Fernandez-Salguero, 1998).

CAR, LXR, FXR, and PXR are all nuclear receptors that heterodimerize with RXR, and presumably share common activation (and repressing?) mechanisms with HNF4 and similar nuclear receptors. CAR is an orphan receptor that heterodimerizes with RXR on a DR4 binding site. Activated by phenobarbital, CAR regulates the CyP2B10 gene. Other ligands include 1,4-bis[2-(3,5-dichloropyridyloxy)] benzene (TCPOBOP), polychlorinated biphenyls, chlorinated pesticides, and organic solvents (Honkakoski *et al.*, 1998). Phenobarbital and TCPOBOP stimulate liver-growth (Ledda-Columbano *et al.*, 1998), as does thyroid hormone (Oren *et al.*, 1999). The TR also binds to a DR4 element as an RXR heterodimer, suggesting that some critical liver-growth regulatory genes are activated by DR4-binding factors. LXR, when liganded by oxidized derivatives of cholesterol, also binds to DR4 elements as an RXR heterodimer. LXR is an important regulator of cholesterol and bile metabolism, through its activation of genes like *cholesterol-7α-hydroxylase* (CyP7A) (Peet *et al.*, 1998). FXR is a nuclear receptor for bile acids that heterodimerizes with RXR on an inverted repeat element, and acts as a negative regulator of CyP7A. LXR and FXR therefore reciprocally regulate CyP7A, which is the rate-limiting enzyme in bile acid metabolism (Makishima *et al.*, 1999). PXR, which heterodimerizes with RXR on a DR3 binding site, activates the important drug-metabolizing enzyme CyP3A4. PXR ligands include many pharmacologically important drugs and hormones (Kliewer *et al.*, 1998; Lehmann *et al.*, 1998). The last of these liver-enriched regulators is the orphan nuclear receptor, liver receptor homolog 1 (LRH1) (also fetoprotein transcription factor [FTF] and CyP7A promotor binding factor [CPF]. LRH1 differs from the other receptors discussed here because it binds as a monomer and has no known ligands. Lrh1 also regulates the *CyP7A* gene, where it synergizes with LXR (Mangelsdorf and Evans, 1995; Nitta *et al.*, 1999). A developmental role is also possible for LRH1, as it is also an important regulator of the α-fetoprotein gene (Galarneau *et al.*, 1996). These specialized liver-enriched transcription factors act on a limited set of genes, but they all function within liver-regulatory modules that are controlled by the common liver-enriched factors described earlier in section 10.2.

3. Developmental relationship to pancreas and intestine

Liver is an endodermal tissue that initially buds from the posterior foregut. Pancreas also arises from the posterior foregut, as ventral and dorsal buds adjacent to the liver. The posterior foregut itself persists and differentiates into the upper small intestine. Liver, pancreas and small intestine have a common sequence of transcription-factor expression during development (*Figure 2*), and some overlap of gene expression that is greater during development. Nevertheless, the epithelia (liver: hepatocyte plates, bile ducts; pancreas: acini, ducts, and islets of Langerhans; intestine: enteric mucosa) have distinct phenotypes that cannot simply be explained by HNF1, HNF3, HNF4, HNF6, and C/EBP family transcription factors. For these factors, there are few major differences among these epithelia. One difference is the lack of expression of C/EBPα in pancreas, but β is clearly expressed and even α is apparent in some studies, e.g., Dabeva *et al.*, 1995. A second difference is disappearance of HNF4α from mature exocrine pancreas, though HNF1α is somehow maintained. A third difference is the absence of HNF3γ in the pancreas, which nevertheless strongly expresses α and β.

The differences from liver are thus limited, and the liver-enriched factors are clearly important in pancreas and intestine. For example, the L-pyruvate kinase gene is expressed in adult liver, small intestine and islets, and also in fetal exocrine pancreas. In all three tissues the gene is regulated by HNF1α and β, and HNF4α. The level of L-pyruvate kinase expression is proportional to the level of HNF4α, which selectively disappears from adult exocrine pancreas (Miquerol *et al.*, 1994). HNF4 and HNF1 are also important regulators of the insulin gene in pancreatic islet β-cells. Thus MODY1 (mild onset diabetes of the young) is an autosomal dominant syndrome resulting from mutations in the HNF4α gene that result in haplo-insufficiency. The resultant HNF4α deficiency causes β-cell dysfunction and downregulation of genes that regulate glucose transport and metabolism (Navas *et al.*, 1999; Stoffel and Duncan, 1997; Yamagata *et al.*, 1996). Similarly, the HNF1α gene is the site of MODY3 mutations, which also reduce insulin production (Gragnoli *et al.*, 1997). Like HNF1 and HNF4, the HNF3 group is also important in pancreas. HNF3β and γ regulate pancreatic α-amylase in collaborate with pancreas transcription factor (PTF1) (Cockell *et al.*, 1995). Finally, C/EBP factors are also important pancreatic regulators. In islets, C/EBPβ is increased by chronic high glucose levels, and acts as an important negative regulator of insulin gene (Lu *et al.*, 1997).

The most important difference between liver and pancreatic gene expression appears to be the presence of several additional pancreatic factors, which have critical developmental roles and persist as phenotype regulators (Edlund, 1998; Sander and German, 1997). In exocrine pancreas, one critical factor is the HLH-dimer factor Ptfl, which consists of heterodimers of p48 with E2A factors. The latter are widely expressed HLH factors (Sharma *et al.*, 1997), but p48 is restricted to fetal and adult pancreas. PTF1 regulates acinar genes like elastase through E-Box (CANNTG) binding sites (Rose *et al.*, 1994). Notably, p48 knockout completely abolishes formation of exocrine pancreas (Krapp *et al.*, 1998; 1996). Pancreatic islets express β2/neuro D, another HLH factor that also forms E2A heterodimers and binds E boxes. β2 knockout mice have defective islet formation and reduced insulin production (Mutoh *et al.*, 1998; Naya *et al.*, 1997). The presence of this islet-specific factor illustrates a common problem in explaining phenotype. Since the functions of β2 and p48 are so similar, and their binding partners are common to both islets and acini, it is not clear why they do not activate common target genes in both cell types.

Pancreatic development and gene expression is also regulated by a sequence of specialized homeobox factors. First is IPF1 (also Pdx1, Idx1, Stf1), a regulator of the earliest stages of pancreatic development. IPF1 is first expressed in dorsal and ventral gut epithelium where pancreatic buds will form, and expression persists in the buds. Knockout causes complete pancreatic agenesis (Edlund, 1998; Stoffers *et al.*, 1997). Notably, HNF3β is required for transcription of the Ipfl gene (Wu *et al.*, 1997). Two more factors, Isl1 and hoxb9, also affect both exocrine and endocrine tissues, but act at slightly later stages. Knockout of Isl1, a LIM homeodomain factor, results in complete absence of the dorsal pancreas, and absence of islets in the ventral pancreas (Ahlgren *et al.*, 1997). *Hoxb9* knockouts also lack the dorsal pancreas, and have fewer islets in the ventral (Harrison *et al.*, 1999).

Several other homeobox factors selectively regulate islet cells. Pax4 and 6 are required for the appearance of differentiated cell types in islets, and bind to elements in the glucagon, insulin, and somatostatin promoters (Andersen *et al.*,

1999; Hill *et al.*, 1999; Sander *et al.*, 1997; Sosa-Pineda *et al.*, 1997; St-Onge *et al.*, 1997). Nkx2.2-knockout mice lack β-cells and have reduction of other islet cells (Sussel *et al.*, 1998). Finally, Nkx6.1-knockout mice have fewer β cells (Inoue *et al.*, 1997; Jensen *et al.*, 1996).

In small intestine, other homeobox factors, Cdxl and Hoxcl1, collaborate with 'liver-enriched' factors to produce the differentiated phenotype. Thus Cdx2, HNF1, and HNF4 collaborate in duodenum to regulate the calbindin *D9k* gene (Lorentz *et al.*, 1997; Subramanian *et al.*, 1998; Tamai *et al.*, 1999). The related gene Cdx2 is also expressed in small intestine (Colnot *et al.*, 1998). Similarly, Hoxcl1 coregulates the lactase-phlorizin hydrolase promoter with HNF1 (Mitchelmore *et al.*, 1998).

The importance of the developmental factors that are expressed in the pancreas and small intestine highlights the apparent absence of similar factors in the liver. Perhaps liver is different from the other tissues, but it is more likely that additional regulators will be found. Indeed, studies have turned up more liver developmental regulatory factors. One is the zinc-finger factor Gata4 (Bossard and Zaret, 1998; Zaret, 1999), which regulates the albumin gene enhancer. Gata4 binds and synergistically regulates this enhancer with HNF3. The activity persists in adult liver and is required for differentiated gene expression, but notably, both Gata4 and HNF3 sites are occupied in foregut before the albumin gene is activated, indicating that additional, more liver-specific regulators must also be involved. Moreover, Gata4 and related factors regulate the entire ventral foregut that gives rise to liver, small intestine and pancreas (Kuo *et al.*, 1997; Molkentin *et al.*, 1997). The closely related factor Gata6 has an even earlier role. In the early embryo, Gata6 regulates the HNF4α gene and has been shown to be essential for the initial differentiation of visceral endoderm (Morrisey *et al.*, 1998).

Other recent observations have demonstrated a specialized factor, Nkx2.8, which regulates the α-fetoprotein in fetal liver (Apergis *et al.*, 1998). Nkx2.8 disappears after birth and is closely related to other factors that regulate organo-genesis, i.e., Nkx2.2 (pancreatic islets), Nkx2.5 (heart), and thyroid transcription factor 1 (TTF1)/Nkx2.1 (lung and thyroid). However, beyond the α-fetoprotein gene, the function of Nkx2.8 in liver has yet to be established.

4. Future directions

Much is known about tissue-specific gene expression but detailed examination has led to a paradox. The unique expression of genes appears more tissue-specific than the transcription factors that regulate them. Current knowledge is insufficient to resolve this paradox and future research should clarify at least four overlapping areas.

First, virtually no tissue-specific promoters or enhancers have been totally characterized. Footprints and gel shifts generally show more sites than the ones that are well characterized, and even more sites may be embedded in known binding regions. Older studies (e.g. Cordle *et al.*, 1991; Nir *et al.*, 1986) have led to the concept that many regulatory modules contain *cis* acting regions that repress transcription in non-expressing cells. Whether these effects represent specific repressors or merely incompatible factor interactions remains to be resolved. In either case, subtle differences in the transcription factor milieu may prove critical.

Second, new factors will almost certainly be discovered. The known tissue-enriched transcription factors are usually abundant and regulate a relatively large set of target genes. Once these abundant factors appear, they generally persist throughout development to regulate their target genes continuously. Relatively few known factors silence like Nkx2.8. Thus the missing factors are likely to be less abundant, and to have more transient expression and a limited number of target genes. A more complete understanding of the functions necessary for tissue-specific transcription regulatory modules in general could be used to make predictions about missing components in individual modules.

There is also a hidden program of sequential interactions that might be built into each promoter or enhancer. Proper activation could require a particular order in which factors modify promoter or enhancer structure to get it ready for activation. Subtle differences in abundance or timing, absence of one critical factor, or addition of one new factor could redirect a regulatory module.

Finally, there may be a higher level of cell-specific chromosome regulation. Chromatin condensation into heterochromatin or decondensation into euchromatin probably regulates the accessibility of genes to the transcription machinery. For some genes, this process may be developmental and inflexible, while for others it may be dynamically regulated.

It is almost certain that all of these possibilities contribute to tissue-specific gene expression. The general picture has been established, but how transcription generates specific cells and tissues will not be understood until the details are filled in.

Acknowledgments

Work in my laboratory is supported by NIH grants CA68440 and CA76354.

References

Ahlgren, U., Pfaff, S.L., Jessell, T.M., Edlund, T. and Edlund, H. (1997) Independent requirement for ISL1 in formation of pancreatic mesenchyme and islet cells. *Nature* **385:** 257–260.

Andersen, F.G., Heller, R.S., Petersen, H.V., Jensen, J., Madsen, O.D. and Serup, P. (1999) Pax6 and Cdx2/3 form a functional complex on the rat glucagon gene promoter G1-element. *FEBS Lett.* **445:** 306–310.

Ang, S.L. and Rossant, J. (1994) HNF-3β is essential for node and notochord formation in mouse development. *Cell* **78:** 561–574.

Antonson P., Pray M.G., Jacobsson A. and Xanthopoulos K.G. (1995) Myc inhibits CCAAT/enhancer-binding protein α-gene expression in HIB-1B hibernoma cells through interactions with the core promoter region. *Eur. J. Biochem.* **232:** 397–403.

Antonson P. and Xanthopoulos K.G. (1995) Molecular cloning, sequence, and expression patterns of the human gene encoding CCAAT/enhancer binding protein α (C/EBPα). *Biochem. Biophys. Res. Commun.* **215:** 106–113.

Apergis, G.A, Crawford, N., Ghosh, D. *et al.* (1998) A novel nk-2-related transcription factor associated with human fetal liver and hepatocellular carcinoma. *J. Biol. Chem.* **273:** 2917–2925.

Bailey, P., Sartorelli, V., Hamamori, Y. and Muscat, G.E. (1998) The orphan nuclear receptor, COUP-TF II, inhibits myogenesis by post-transcriptional regulation of MyoD function: COUP-TF II directly interacts with p300 and MyoD. *Nucl. Acids Res.* **26:** 5501–5510.

Bossard, P. and Zaret, K.S. (1998) GATA transcription factors as potentiators of gut endoderm differentiation. *Development* **125:** 4909–4917.

Bossard, P., McPherson, C.E. and Zaret, K.S. (1997) *In vivo* footprinting with limiting amounts of embryo tissues: a role for C/EBPβ in early hepatic development. *Methods* **11:** 180–188.

Buckwold, V.E., Chen, M. and Ou, J.H. (1997) Interaction of transcription factors RFX1 and MIBP1 with the gamma motif of the negative regulatory element of the hepatitis B virus core promoter. *Virology* **227:** 515–518.

Burgess-Beusse, B.L. and Darlington, G.J. (1998) C/EBPα is critical for the neonatal acute-phase response to inflammation. *Mol. Cell. Biol.* **18:** 7269–7277.

Carey, M. (1998) The enhanceosome and transcriptional synergy. *Cell* **92:** 5–8.

Cereghini, S. (1996) Liver-enriched transcription factors and hepatocyte differentiation. *FASEB J.* **10:** 267–282.

Chang, C.J., Chen, Y.L., and Lee, S.C. (1998) Coactivator TIF1β interacts with transcription factor C/EBPβ and glucocorticoid receptor to induce *α1-acid glycoprotein* gene expression. *Mol. Cell. Biol.* **18:** 5880–5887.

Chen, P.L., Riley, D.J., Chen, Y. and Lee, W.H. (1996) Retinoblastoma protein positively regulates terminal adipocyte differentiation through direct interaction with C/EBPs. *Genes Dev.* **10:** 2794–2804.

Chen, W.S., Manova, K., Weinstein, D.C. *et al.* (1994) Disruption of the *HNF-4* gene, expressed in visceral endoderm, leads to cell death in embryonic ectoderm and impaired gastrulation of mouse embryos. *Genes Dev.* **8:** 2466–2477.

Clark, K.L., Halay, E.D., Lai, E. and Burley, S.K. (1993) Co-crystal structure of the HNF-3/fork head DNA-recognition motif resembles histone H5. *Nature* **364:** 412–420.

Cockell, M., Stolarczyk, D., Frutiger, S., Hughes, G.J., Hagenbuchle, O. and Wellauer, P.K. (1995) Binding sites for hepatocyte nuclear factor 3 β or 3 γ and pancreas transcription factor 1 are required for efficient expression of the gene encoding pancreatic α-amylase. *Mol. Cell. Biol.* **15:** 1933–1941.

Coffinier C., Barra J., Babinet, C. and Yaniv, M. (1999a) Expression of the vHNF1/HNF1β homeoprotein gene during mouse organogenesis. *Mech. Dev.* **89:** 211–213.

Coffinier, C., Thepot, D., Babinet, C., Yaniv, M. and Barra, J. (1999b) Essential role for the homeoprotein vHNF1/HNF1β in visceral endoderm differentiation. *Development* **126,** 4785–4794.

Colnot, S., Romagnolo, B., Lambert, M. (1998) Intestinal expression of the *calbindin-D9K* gene in transgenic mice. Requirement for a Cdx2-binding site in a distal activator region. *J. Biol. Chem.* **273:** 31939–31946.

Cordle, S.R., Whelan, J., Henderson, E., Masuoka, H., Weil, P.A. and Stein, R. (1991) Insulin gene expression in nonexpressing cells appears to be regulated by multiple distinct negative-acting control elements. *Mol. Cell. Biol.* **11:** 2881–2886.

Costa, R.H., Lai, E., Grayson, D.R. and Darnell, J.E. (1988) The cell-specific enhancer of the mouse *transythretin* (prealbumin) gene binds a common factor at one site and a liver-specific factor(s) at two others. *Mol. Cell. Biol.* **8:** 81–90.

Croniger, C., Trus, M., Lysek-Stupp, K. *et al.* (1997) Role of the isoforms of CCAAT/enhancer-binding protein in the initiation of phosphoenolpyruvate carboxykinase (*GTP*) gene transcription at birth. *J. Biol. Chem.* **272:** 26306–26312.

Croniger, C., Leahy, P., Reshef, L. and Hanson, R.W. (1998) C/EBP and the control of phosphoenolpyruvate carboxykinase gene transcription in the liver. *J. Biol. Chem.* **273:** 31629–31632.

Dabeva, M.D., Hurston, E. and Shafritz, D.A. (1995) Transcription factor and liver-specific mRNA expression in facultative epithelial progenitor cells of liver and pancreas. *Am. J. Pathol.* **147:** 1633–1648.

Darlington, G.J., Ross, S.E. and MacDougald, O.A. (1998) The role of C/EBP genes in adipocyte differentiation. *J. Biol. Chem.* **273:** 30057–30060.

Dell, H. and Hadzopoulou-Cladaras, M. (1999) CREB-binding protein is a transcriptional coactivator for hepatocyte nuclear factor-4 and enhances apolipoprotein gene expression. *J. Biol. Chem.* **274:** 9013–9021.

Descombes, P. and Schibler, U. (1991) A liver-enriched transcriptional activator protein, LAP, and a transcriptional inhibitory protein, LIP, are translated from the same mRNA. *Cell* **67**: 569–579.
Desvergne, B. and Wahli, W. (1999) Peroxisome proliferator-activated receptors: nuclear control of metabolism. *Endocr. Rev.* **20**: 649–688.
Diehl, AM. (1998) Roles of CCAAT/enhancer-binding proteins in regulation of liver regenerative growth. *J. Biol. Chem.* **273**: 30843–30846.
Drewes, T., Senkel, S., Holewa, B. and Ryffel, G.U. (1996) Human hepatocyte nuclear factor 4 isoforms are encoded by distinct and differentially expressed genes. *Mol. Cell. Biol.* **16**: 925–931.
Duncan, S.A., Manova, K., Chen, W. S. *et al.* (1994) Expression of transcription factor HNF-4 in the extraembryonic endoderm, gut, and nephrogenic tissue of the developing mouse embryo: HNF-4 is a marker for primary endoderm in the implanting blastocyst. *Proc. Natl Acad. Sci. USA* **91**: 7598–7602.
Edlund, H. (1998) Transcribing pancreas. *Diabetes* **47**: 1817–1823.
Falvey, E., Marcacci, L. and Schibler, U. (1996) DNA-binding specificity of PAR and C/EBP leucine zipper proteins: a single amino acid substitution in the C/EBP DNA-binding domain confers PAR-like specificity to C/EBP. *Biol. Chem.* **377**: 797–809.
Falvey, E., Fleury-Olela, F. and Schibler, U. (1995) The rat hepatic leukemia factor (*HLF*) gene encodes two transcriptional activators with distinct circadian rhythms, tissue distributions and target preferences. *EMBO J.* **14**: 4307–4317.
Fernandez-Salguero, P., Pineau, T., Hilbert., D.M. *et al.* (1995) Immune system impairment and hepatic fibrosis in mice lacking the dioxin-binding Ah receptor. *Science* **268**: 722–726.
Flodby, P., Barlow, C., Kylefjord, H., Ahrlund-Richter, L. and Xanthopoulos, K.G. (1996) Increased hepatic cell proliferation and lung abnormalities in mice deficient in CCAAT/enhancer binding protein α. *J. Biol. Chem.* **271**: 24753–24760.
Fonjallaz, P., Ossipow, V., Wanner, G. and Schibler, U. (1996) The two PAR leucine zipper proteins, TEF and DBP, display similar circadian and tissue-specific expression, but have different target promoter preferences. *EMBO J.* **15**: 351–362.
Fukai, K., Takada, S., Yokosuka, O., Saisho, H., Omata, M. and Koike, K. (1997) Characterization of a specific region in the hepatitis B virus enhancer I for the efficient expression of X gene in the hepatic cell. *Virology* **236**: 279–287.
Furuta, H., Iwasaki, N., Oda, N. *et al.* (1997) Organization and partial sequence of the hepatocyte nuclear factor-4 α/MODY1 gene and identification of a missense mutation, R127W, in a Japanese family with MODY. *Diabetes* **46**: 1652–1657.
Galarneau, L., Pare, J. F. and Allard, D. (1996) The α1-fetoprotein locus is activated by a nuclear receptor of the *Drosophila* FTZ-F1 family. *Mol. Cell. Biol.* **16**: 3853–3865.
Galson, D.L, Tsuchiya, T., Tendler, D.S. *et al.* (1995) The orphan receptor hepatic nuclear factor 4 functions as a transcriptional activator for tissue-specific and hypoxia-specific erythropoietin gene expression and is antagonized by EAR3/COUP-TF1. *Mol. Cell. Biol.* **15**: 2135–2144.
Gonzalez, F. J. and Fernandez-Salguero, P. (1998) The aryl hydrocarbon receptor: studies using the AHR-null mice. *Drug Metab. Dispos.* **26**: 1194–1198.
Gragnoli, C., Lindner, T., Cockburn, B.N. *et al.* (1997) Maturity-onset diabetes of the young due to a mutation in the hepatocyte nuclear factor-4 α binding site in the promoter of the *hepatocyte nuclear factor-1 α* gene. *Diabetes* **46**: 1648–1651.
Greenbaum, L.E., Li, W., Cressman, D.E. *et al.* (1998) CCAAT enhancer-binding protein β is required for normal hepatocyte proliferation in mice after partial hepatectomy. *J. Clin. Invest.* **102**: 996–1007.
Greenbaum, L.E, Cressman, D.E., Haber, B.A. and Taub, R. (1995) Coexistence of C/EBPα, β, growth-induced proteins and DNA synthesis in hepatocytes during liver regeneration. Implications for maintenance of the differentiated state during liver growth. *J. Clin. Invest.* **96**: 1351–1365.
Groupp, E.R., Crawford, N. and Locker, J. (1994) Characterization of the distal α-fetoprotein enhancer, a strong, long distance, liver-specific activator. *J. Biol. Chem.* **269**: 22178–22187.
Hall, R.K., Sladek, F.M. and Granner, D.K. (1995) The orphan receptors COUP-TF and HNF-4 serve as accessory factors required for induction of phosphoenolpyruvate carboxykinase gene transcription by glucocorticoids. *Proc. Natl Acad. Sci. USA* **92**: 412–416.
Hanlon, M. and Sealy, L. (1999) Ras regulates the association of serum response factor and CCAAT/enhancer-binding protein beta. *J. Biol. Chem.* **274**: 14224–14228.

Hanson, R.W. and Reshef, L. (1997) Regulation of phosphoenolpyruvate carboxykinase (*GTP*) gene expression. *Annu. Rev. Biochem.* **66:** 581–611.

Harrison, K.A., Thaler, J., Pfaff, S.L., Gu, H. and Kehrl, J. H. (1999) Pancreas dorsal lobe agenesis and abnormal islets of Langerhans in Hlxb9-deficient mice. *Nat. Genet.* **23:** 71–75.

Heery, D.M., Kalkhoven, E., Hoare, S. and Parker, M.G. (1997) A signature motif in transcriptional co-activators mediates binding to nuclear receptors. *Nature* **387:** 733–736.

Herbomel, P., Rollier, A., Tronche, F., Ott, M.-O., Yaniv, M. and Weiss, M.C. (1989) The rat albumin promoter is composed of six distinct positive elements within 130 nucleotides. *Mol. Cell. Biol.* **9**: 4750–4758.

Hertz, R., Magenheim, J., Berman, I. and Bar-Tana, J. (1998) Fatty acyl-CoA thioesters are ligands of hepatic nuclear factor-4α. *Nature* **392:** 512–516.

Hill, M.E., Asa, S.L. and Drucker, D.J. (1999) Essential requirement for Pax6 in control of enteroendocrine proglucagon gene transcription. *Mol. Endocrinol.* **13:** 1474–1486.

Honkakoski, P., Zelko, I., Sueyoshi, T. and Negishi, M. (1998) The nuclear orphan receptor CAR-retinoid X receptor heterodimer activates the phenobarbital-responsive enhancer module of the *CYP2B* gene. *Mol. Cell. Biol.* **18:** 5652–5658.

Hu, X. and Lazar, M.A. (1999) The CoRNR motif controls the recruitment of corepressors by nuclear hormone receptors. *Nature* **402:** 93–96.

Inaba, T., Shapiro, L.H., Funabiki, T. *et al.* (1994) DNA-binding specificity and *trans*-activating potential of the leukemia-associated E2A-hepatic leukemia factor fusion protein. *Mol. Cell. Biol.* **14**: 3403–3414.

Inoue, H., Rudnick, A., German, M.S., Veile, R., Donis-Keller, H. and Permutt, M.A. (1997) Isolation, characterization, and chromosomal mapping of the human Nkx6.1 gene (*NKX6A*), a new pancreatic islet homeobox gene. *Genomics* **40:** 367–370.

Jacquemin, P., Lannoy, V.J., Rousseau, G.G. and Lemaigre, F.P. (1999) OC-2, a novel mammalian member of the ONECUT class of homeodomain transcription factors whose function in liver partially overlaps with that of hepatocyte nuclear factor-6. *J. Biol. Chem.* **274:** 2665–2671.

Jensen, J., Serup, P., Karlsen, C., Nielsen, T.F. and Madsen, O.D. (1996) mRNA profiling of rat islet tumors reveals nkx 6.1 as a β-cell-specific homeodomain transcription factor. *J. Biol. Chem.* **271:** 18749–18758.

Jiang, G., Nepomuceno, L., Hopkins, K. and Sladek, F.M. (1995) Exclusive homodimerization of the orphan receptor hepatocyte nuclear factor 4 defines a new subclass of nuclear receptors. *Mol. Cell. Biol.* **15:** 5131–5143.

Jiang, G. and Sladek, F.M. (1997) The DNA binding domain of hepatocyte nuclear factor 4 mediates cooperative, specific binding to DNA and heterodimerization with the retinoid X receptor alpha. *J. Biol. Chem.* **272:** 1218–1225.

Johnson, P. (1993) Identification of C/EBP basic region residues involved in DNA sequence recognition and half-site spacing preference. *Mol. Cell. Biol.* **13**: 6919–6930.

Kaestner, K.H., Katz, J., Liu, Y., Drucker, D.J. and Schutz, G. (1999) Inactivation of the winged helix transcription factor HNF3α affects glucose homeostasis and islet glucagon gene expression *in vivo*. *Genes Dev.* **13:** 495–504.

Kaestner, K.H., Hiemisch, H. and Schutz, G. (1998) Targeted disruption of the gene encoding hepatocyte nuclear factor 3γ results in reduced transcription of hepatocyte-specific genes. *Mol. Cell. Biol.* **18:** 4245–4251.

Kardassis, D., Zannis, V.I. and Cladaras, C. (1992) Organization of the regulatory elements and nuclear activities participating in the transcription of the human apolipoprotein B gene. *J. Biol. Chem.* **267**: 2622–2632.

Kennedy, H.J., Viollet, B., Rafiq, I., Kahn, A. and Rutter, G.A. (1997) Upstream stimulatory factor-2 (USF2) activity is required for glucose stimulation of L-pyruvate kinase promoter activity in single living islet beta-cells. *J. Biol. Chem.* **272:** 20636–20640.

Kim, T.K. and Maniatis, T. (1997) The mechanism of transcriptional synergy of an *in vitro* assembled interferon-beta enhanceosome. *Mol. Cell* **1:** 119–129.

Kliewer, S.A., Moore, J.T., Wade, L. *et al.* (1998) An orphan nuclear receptor activated by pregnanes defines a novel steroid signaling pathway. *Cell* **92:** 73–82.

Krapp, A., Knofler, M., Ledermann, B. *et al.* (1998) The bHLH protein PTF1-p48 is essential for the formation of the exocrine and the correct spatial organization of the endocrine pancreas. *Genes Dev.* **12:** 3752–3763.

Krapp, A., Knofler, M., Frutiger, S., Hughes, G.J., Hagenbuchle, O. and Wellauer, P.K. (1996) The p48 DNA-binding subunit of transcription factor PTF1 is a new exocrine pancreas-specific basic helix-loop-helix protein. *EMBO J.* **15:** 4317–4329.

Kuo, C.J., Conley, P.B., Chen, L., Sladek, F.M., Darnell, J.E. Jr. and Crabtree, G.R. (1992) A transcriptional hierarchy involved in mammalian cell-type specification. *Nature* **355:** 457–461.

Kuo, C.T., Morrisey, E.E., Anandappa, R. *et al.* (1997) GATA4 transcription factor is required for ventral morphogenesis and heart tube formation. *Genes Dev.* **11:** 1048–1060.

Lahuna, O., Fernandez, L., Karlsson, H. *et al.* (1997) Expression of hepatocyte nuclear factor 6 in rat liver is sex-dependent and regulated by growth hormone. *Proc. Natl Acad. Sci. USA* **94:** 12309–12313.

Lai, E., Clark, K.L., Burley, S.K. and Darnell, J.E. Jr. (1993) Hepatocyte nuclear factor 3/fork head or 'winged helix' proteins: a family of transcription factors of diverse biologic function. *Proc. Natl Acad. Sci. USA* **90:** 10421–10423.

Lai, E., Prezioso, V.R., Tao, W., Chen, W.S. and Darnell, J.E. (1991) Hepatocyte nuclear factor 3α belongs to a gene family in mammals that is homologous to the *Drosophila* homeotic gene fork head. *Genes Dev.* 5: 416–427.

Lamprecht, C. and Mueller, C.R. (1999) D-site binding protein transactivation requires the proline- and acid-rich domain and involves the coactivator p300. *J. Biol. Chem.* **274:** 17643–17648.

Lannoy, V.J., Burglin, T.R., Rousseau, G.G. and Lemaigre, F.P. (1998) Isoforms of hepatocyte nuclear factor-6 differ in DNA-binding properties, contain a bifunctional homeodomain, and define the new ONECUT class of homeodomain proteins. *J. Biol. Chem.* **273:** 13552–13562.

Lavery, D.J., Lopez-Molina, L., Margueron, R. *et al.* (1999) Circadian expression of the steroid 15 α-hydroxylase (Cyp2a4) and coumarin 7-hydroxylase (Cyp2a5) genes in mouse liver is regulated by the PAR leucine zipper transcription factor DBP. *Mol. Cell. Biol.* **19:** 6488–6499.

Lavery, D. and Schibler, U. (1993) Circadian transcription of the cholesterol 7α hydroxylase gene may involve the liver-enriched bZIP protein DBP. *Genes Dev.* **7:** 1871–1884.

Ledda-Columbano, G.M, Curto, M., Piga, R. *et al.* (1998) *In vivo* hepatocyte proliferation is inducible through a TNF and IL-6-independent pathway. *Oncogene* **17:** 1039–1044.

Lee, S.S., Pineau, T., Drago, J. *et al.* (1995) Targeted disruption of the α isoform of the peroxisome proliferator-activated receptor gene in mice results in abolishment of the pleiotropic effects of peroxisome proliferators. *Mol. Cell. Biol.* **15:** 3012–3022.

Lee, Y.H., Sauer, B., Johnson, P.F. and Gonzalez, F.J. (1997) Disruption of the c/ebp α gene in adult mouse liver. *Mol. Cell. Biol.* **17:** 6014–6022.

Legraverend, C., Antonson, P., Flodby, P. and Xanthopoulos, K.G. (1993) High level activity of the mouse CCAAT/enhancer binding protein (C/EBPα) gene promoter involves autoregulation and several ubiquitous transcription factors. *Nucl. Acids Res.* **21:** 1735–1742.

Lehmann, J.M., McKee, D.D., Watson, M.A., Willson, T.M., Moore, J.T. and Kliewer, S.A. (1998) The human orphan nuclear receptor PXR is activated by compounds that regulate CYP3A4 gene expression and cause drug interactions. *J. Clin. Invest.* **102:** 1016–1023.

Lemaigre, F.P., Durviaux, S.M., Truong, O., Hsuan, J.J. and Rousseau, G.G. (1996) Hepatocyte nuclear factor 6, a transcription factor that contains a novel type of homeodomain and a single cut domain. *Proc. Natl Acad. Sci. USA* **93:** 9460–9464.

Lemaigre, F.P. and Rousseau, G.G. (1994) Transcriptional control of genes that regulate glycolysis and gluconeogenesis in adult liver. *Biochem. J.* **303:** 1–14.

Levy-Wilson, B. (1995) Transcriptional control of the human apolipoprotein B gene in cell culture and in transgenic animals. *Prog. Nucl. Acid Res. Mol. Biol.* **50:** 161–190.

Lopez-Molina, L., Conquet, F., Dubois-Dauphin, M. and Schibler, U. (1997) The DBP gene is expressed according to a circadian rhythm in the suprachiasmatic nucleus and influences circadian behavior. *EMBO J.* **16:** 6762–6771.

Lorentz, O., Duluc, I., Arcangelis, A.D., Simon-Assmann, P., Kedinger, M. and Freund, J.N. (1997) Key role of the Cdx2 homeobox gene in extracellular matrix-mediated intestinal cell differentiation. *J. Cell. Biol.* **139:** 1553–1565.

Lu, M., Seufert, J. and Habener, J.F. (1997) Pancreatic β-cell-specific repression of insulin gene transcription by CCAAT/enhancer-binding protein β. Inhibitory interactions with basic helix-loop-helix transcription factor E47. *J. Biol. Chem.* **272:** 28349–28359.

Magee, T.R., Cai, Y., El-Houseini, M.E., Locker, J. and Wan, Y.J. (1998) Retinoic acid mediates down-regulation of the α-fetoprotein gene through decreased expression of hepatocyte nuclear factors. *J. Biol. Chem.* **273:** 30024–30032.

Makishima, M., Okamoto, A.Y., Repa, J. *et al.* (1999) Identification of a nuclear receptor for bile acids. *Science* **284:** 1362–1365.

Malik, S. and Karathanasis, S.K. (1996) TFIIB-directed transcriptional activation by the orphan nuclear receptor hepatocyte nuclear factor 4. *Mol. Cell. Biol.* **16:** 1824–1831.

Mandrup, S. and Lane, M.D. (1997) Regulating adipogenesis. *J. Biol. Chem.* **272:** 5367–5370.

Mangelsdorf, D.J. and Evans, R. M. (1995) The RXR heterodimers and orphan receptors. *Cell* **83:** 841–850.

Mangelsdorf, D.J., Borgmeyer, U., Heyman, R.A. *et al.* (1992) Characterization of three RXR genes that mediate the action of 9-*cis* retinoic acid. *Genes Dev.* **6:** 329–344.

Marten, N.W., Sladek, F.M. and Straus, D.S. (1996) Effect of dietary protein restriction on liver transcription factors. *Biochem. J.* **317:** 361–370.

McInerney, E.M., Rose, D.W., Flynn, S.E. *et al.* (1998) Determinants of coactivator LXXLL motif specificity in nuclear receptor transcriptional activation. *Genes Dev.* **12:** 3357–3368.

Meyer, B.K., Pray-Grant, M.G., Vanden Heuvel, J.P. and Perdew, G.H. (1998) Hepatitis B virus X-associated protein 2 is a subunit of the unliganded aryl hydrocarbon receptor core complex and exhibits transcriptional enhancer activity. *Mol. Cell. Biol.* **18:** 978–988.

Mietus-Snyder, M., Sladek, F.M., Ginsburg, G.S. *et al.* (1992) Antagonism between apolipoprotein AI regulatory protein 1, Ear3/COUP-TF, and hepatocyte nuclear factor 4 modulates apolipoprotein CIII gene expression in liver and intestinal cells. *Mol. Cell. Biol.* **12:** 1708–1718.

Milland, J., Tsykin, A., Thomas, T., Aldred, A.R., Cole, T. and Schreiber, G. (1990) Gene expression in regenerating and acute-phase rat liver. *Am. J. Physiol.* **259:** G340-G347.

Mink, S., Haenig, B. and Klempnauer, K.H. (1997) Interaction and functional collaboration of p300 and C/EBPβ. *Mol. Cell. Biol.* **17:** 6609–6617.

Miquerol, L., Cluzeaud, F., Porteu, A., Alexandre, Y., Vandewalle, A. and Kahn, A. (1996) Tissue specificity of L-pyruvate kinase transgenes results from the combinatorial effect of proximal promoter and distal activator regions. *Gene Expr.* **5:** 315–330.

Miquerol, L., Lopez, S., Cartier, N., Tulliez, M., Raymondjean, M. and Kahn, A. (1994) Expression of the L-type pyruvate kinase gene and the hepatocyte nuclear factor 4 transcription factor in exocrine and endocrine pancreas. *J. Biol. Chem.* **269:** 8944–8951.

Mitchelmore, C., Troelsen, J.T., Sjostrom, H. and Noren, O. (1998) The HOXC11 homeodomain protein interacts with the lactase-phlorizin hydrolase promoter and stimulates HNF1α-dependent transcription. *J. Biol. Chem.* **273:** 13297–13306.

Molkentin, J.D., Lin, Q., Duncan, S.A. and Olson, E.N. (1997) Requirement of the transcription factor GATA4 for heart tube formation and ventral morphogenesis. *Genes Dev.* **11:** 1061–1072.

Morrisey, E.E., Tang, Z., Sigrist, K. *et al.* (1998) GATA6 regulates HNF4 and is required for differentiation of visceral endoderm in the mouse embryo. *Genes Dev.* **12:** 3579–3590.

Mueller, C., Maire, P. and Schibler, U. (1990) DBP, a liver-enriched transcriptional activator, is expressed late in ontogeny and its tissue specificity is determined posttranscriptionally. *Cell* **61:** 279–291.

Mutoh, H., Naya, F.J., Tsai, M.J. and Leiter, A.B. (1998) The basic helix-loop-helix protein BETA2 interacts with p300 to coordinate differentiation of secretin-expressing enteroendocrine cells. *Genes Dev.* **12:** 820–830.

Navas, M.A., Munoz-Elias, E.J., Kim, J., Shih, D. and Stoffel, M. (1999) Functional characterization of the MODY1 gene mutations HNF4(R127W), HNF4(V255M), and HNF4(E276Q) *Diabetes* **48:** 1459–1465.

Naya, F.J., Huang, H.P., Qiu, Y. *et al.* (1997) Diabetes, defective pancreatic morphogenesis, and abnormal enteroendocrine differentiation in BETA2/neuroD-deficient mice. *Genes Dev.* **11:** 2323–2334.

Nerlov, C. and Ziff, E.B. (1994) Three levels of functional interaction determine the activity of CCAAT/enhancer binding protein-α on the serum albumin promoter. *Genes Dev.* **8:** 350–362.

Nerlov, C. and Ziff, E.B. (1995) CCAAT/enhancer binding protein-alpha amino acid motifs with dual TBP and TFIIB binding ability co-operate to activate transcription in both yeast and mammalian cells. *EMBO J.* **14:** 4318–4328.

Nir, U., Walker, M.D. and Rutter, W.J. (1986) Regulation of rat insulin 1 gene expression: evidence for negative regulation in nonpancreatic cells. *Proc. Natl Acad. Sci. USA* **83:** 3180–3184.

Nitta, M., Ku, S., Brown, C., Okamoto, A.Y. and Shan, B. (1999) CPF: an orphan nuclear receptor that regulates liver-specific expression of the human cholesterol 7α-hydroxylase gene. *Proc. Natl Acad. Sci. USA* **96:** 6660–6665.

Nuclear Receptors Nomenclature Committee (1999) A unified nomenclature system for the nuclear receptor superfamily. *Cell* **97:** 161–163.

Ohmura, T., Columbano, G.L., Columbano, A., Katyal, S.L., Locker, J. and Shinozuka, H. (1996) 9-*cis* retinoic acid is a direct hepatocyte mitogen in rats. *Life Sci.* **58:** L211-L216.

Oren, R., Dabeva, M.D., Karnezis, A.N. *et al.* (1999) Role of thyroid hormone in stimulating liver repopulation in the rat by transplanted hepatocytes. *Hepatology* **30:** 903–913.

Ori, A. and Shaul, Y. (1995) Hepatitis B virus enhancer binds and is activated by the Hepatocyte nuclear factor 3. *Virology* **207:** 98–106.

Ori, A., Zauberman, A., Doitsh, G., Paran, N., Oren, M. and Shaul, Y. (1998) p53 binds and represses the HBV enhancer: an adjacent enhancer element can reverse the transcription effect of p53. *EMBO J.* **17:** 544–553.

Ossipow, V., Descombes, P. and Schibler, U. (1993) CCAAT/enhancer-binding protein mRNA is translated into multiple proteins with different transcription activation potentials. *Proc. Natl Acad. Sci. USA* **90:** 8219–8223.

Pani, L., Quian, X.B., Clevidence, D. and Costa, R.H. (1992) The restricted promoter activity of the liver transcription factor hepatocyte nuclear factor 3 β involves a cell-specific factor and positive autoactivation. *Mol. Cell. Biol.* **12:** 552–562.

Park, J.S., Luethy, J.D., Wang, M.G. *et al.* (1992) Isolation, characterization and chromosomal localization of the human *GADD153* gene. *Gene* **116:** 259–267.

Peet, D.J., Janowski, B.A. and Mangelsdorf, D.J. (1998) The LXRs: a new class of oxysterol receptors. *Curr. Opin. Genet. Dev.* **8:** 571–575.

Perissi, V., Dasen, J.S., Kurokawa, R. *et al.* (1999) Factor-specific modulation of CREB-binding protein acetyltransferase activity. *Proc. Natl. Acad. Sci. USA* **96:** 3652–3657.

Peterson, R.S., Clevidence, D.E., Ye, H. and Costa, R.H. (1997) Hepatocyte nuclear factor-3 α promoter regulation involves recognition by cell-specific factors, thyroid transcription factor-1, and autoactivation. *Cell Growth Differ.* **8:** 69–82.

Pontoglio, M., Barra, J., Hadchouel, M. *et al.* (1996) Hepatocyte nuclear factor 1 inactivation results in hepatic dysfunction, phenylketonuria, and renal Fanconi syndrome. *Cell* **84:** 575–585.

Pontoglio, M., Sreenan, S., Roe, M. *et al.* (1998) Defective insulin secretion in hepatocyte nuclear factor 1α-deficient mice. *J. Clin. Invest.* **101:** 2215–2222.

Pontoglio, M., Faust, D.M., Doyen, A., Yaniv, M. and Weiss, M.C. (1997) Hepatocyte nuclear factor 1α gene inactivation impairs chromatin remodeling and demethylation of the phenylalanine hydroxylase gene. *Mol. Cell. Biol.* **17:** 4948–4956.

Qian, X. and Costa, R.H. (1995) Analysis of hepatocyte nuclear factor-3 β protein domains required for transcriptional activation and nuclear targeting. *Nucl. Acids Res.* **23:** 1184–1191.

Rana, B., Xie, Y., Mischoulon, D., Bucher, N.L. and Farmer, S.R. (1995) The DNA binding activity of C/EBP transcription factor is regulated in the G1 phase of the hepatocyte cell cycle. *J. Biol. Chem.* **270:** 18123–18132.

Rastegar, M., Szpirer, C., Rousseau, G.G. and Lemaigre, F.P. (1998) Hepatocyte nuclear factor 6: organization and chromosomal assignment of the rat gene and characterization of its promoter. *Biochem. J.* **334:** 565–569.

Rausa, F., Samadani, U., Ye, H. *et al.* (1997) The cut-homeodomain transcriptional activator HNF-6 is coexpressed with its target gene HNF-3β in the developing murine liver and pancreas. *Dev. Biol.* **192:** 228–246.

Rose, S.D., Kruse, F., Swift, G.H., MacDonald, R.J. and Hammer, R.E. (1994) A single element of the elastase I enhancer is sufficient to direct transcription selectively to the pancreas and gut. *Mol. Cell. Biol.* **14:** 2048–2057.

Ryden, T. and Beemon, K. (1988) Avian retroviral long terminal repeats bind CCAAT/enhancer-binding protein. *Mol. Cell. Biol.* **9:** 1155–1164.

Samadani, U. and Costa, R.H. (1996) The transcriptional activator hepatocyte nuclear factor 6 regulates liver gene expression. *Mol. Cell. Biol.* **16:** 6273–6284.

Samadani, U., Qian, X. and Costa, R.H. (1996) Identification of a transthyretin enhancer site that selectively binds the hepatocyte nuclear factor-3β isoform. *Gene Expr.* **6:** 23–33.

Sander, M. and German, M.S. (1997) The β cell transcription factors and development of the pancreas. *J. Mol. Med.* **75:** 327–340.

Sander, M., Neubuser, A., Kalamaras, J., Ee, H.C., Martin, G.R. and German, M.S. (1997) Genetic analysis reveals that PAX6 is required for normal transcription of pancreatic hormone genes and islet development. *Genes Dev.* **11:** 1662–1673.

Sassi, H., Fromont-Racine, M., Grange, T. and Pictet, R. (1995) Tissue specificity of a glucocorticoid-dependent enhancer in transgenic mice. *Proc. Natl Acad. Sci. USA* **92:** 7197–7201.

Sassi, H., Pictet, R. and Grange, T. (1998) Glucocorticoids are insufficient for neonatal gene induction in the liver. *Proc. Natl Acad. Sci. USA* **95:** 5621–5625.

Schweizer-Groyer, G., Groyer, A., Cadepond, F., Grange, T., Baulieu, E.E. and Pictet, R. (1994) Expression from the tyrosine aminotransferase promoter (nt –350 to +1) is liver-specific and dependent on the binding of both liver-enriched and ubiquitous *trans*-acting factors. *Nucl. Acids Res.* **22:** 1583–1592.

Sharma, A., Henderson, E., Gamer, L., Zhuang, Y. and Stein, R. (1997) Analysis of the role of E2A-encoded proteins in insulin gene transcription. *Mol. Endocrinol.* **11:** 1608–1617.

Shih, D.Q., Navas, M.A., Kuwajima, S., Duncan, S.A. and Stoffal, M. (1999) Impaired glucose homeostasis and neonatal mortality in hepatocyte nuclear factor 3a-deficient mice. *Proc. Natl Acad. Sci. USA* **96:** 10152–10157.

Shim, E.Y., Woodcock, C. and Zaret, K.S. (1998) Nucleosome positioning by the winged helix transcription factor HNF3. *Genes Dev.* **12:** 5–10.

Sladek, F.M., Ruse, M.D. Jr., Nepomuceno, L., Huang, S.M. and Stallcup, M.R. (1999) Modulation of transcriptional activation and coactivator interaction by a splicing variation in the F domain of nuclear receptor hepatocyte nuclear factor 4α1. *Mol. Cell. Biol.* **19:** 6509–6522.

Sladek, F., Zhong, W., Lai, E. and Darnell, J. (1990) Liver-enriched transcription factor HNF-4 is a novel member of the steroid hormone receptor superfamily. *Genes Devel.* **4:** 2353–2365.

Song, Y.H., Ray, K., Liebhaber, S.A. and Cooke, N.E. (1998) Vitamin D-binding protein gene transcription is regulated by the relative abundance of hepatocyte nuclear factors 1α and 1β. *J. Biol. Chem.* **273:** 28408–28418.

Sosa-Pineda, B., Chowdhury, K., Torres, M., Oliver, G. and Gruss, P. (1997) The *Pax4* gene is essential for differentiation of insulin-producing beta cells in the mammalian pancreas. *Nature* **386:** 399–402.

Spath, G.F. and Weiss, M.C. (1998) Hepatocyte nuclear factor 4 provokes expression of epithelial marker genes, acting as a morphogen in dedifferentiated hepatoma cells. *J. Cell. Biol.* **140:** 935–946.

Spath, G.F. and Weiss, M.C. (1997) Hepatocyte nuclear factor 4 expression overcomes repression of the hepatic phenotype in dedifferentiated hepatoma cells. *Mol. Cell. Biol.* **17:** 1913–1922.

Stoffel, M. and Duncan, S.A. (1997) The maturity-onset diabetes of the young (MODY1) transcription factor HNF4α regulates expression of genes required for glucose transport and metabolism. *Proc. Natl Acad. Sci. USA* **94:** 13209–13214.

Stoffers, D.A., Zinkin, N.T., Stanojevic, V., Clarke, W.L. and Habener, J.F. (1997) Pancreatic agenesis attributable to a single nucleotide deletion in the human IPF1 gene coding sequence. *Nat. Genet.* **15:** 106–110.

St-Onge, L., Sosa-Pineda, B., Chowdhury, K., Mansouri, A. and Gruss, P. (1997) Pax6 is required for differentiation of glucagon-producing α-cells in mouse pancreas. *Nature* **387:** 406–409.

Subramanian, V., Meyer, B. and Evans, G.S. (1998) The murine Cdx1 gene product localises to the proliferative compartment in the developing and regenerating intestinal epithelium. *Differentiation* **64:** 11–18.

Sussel, L., Kalamaras, J., Hartigan-O'Connor, D.J. *et al.* (1998).Mice lacking the homeodomain transcription factor Nkx2.2 have diabetes due to arrested differentiation of pancreatic beta cells. *Development* **125:** 2213–2221.

Takiguchi, M. (1998) The C/EBP family of transcription factors in the liver and other organs. *Int. J. Exp. Pathol.* **79:** 369–391.

Tamai, Y., Nakajima, R., Ishikawa, T., Takaku, K., Seldin, M. F. and Taketo, M.M. (1999) Colonic hamartoma development by anomalous duplication in Cdx2 knockout mice. *Cancer Res.* **59:** 2965–2970.

Tang, Q.Q., Jiang, M.S. and Lane, M.D. (1999) Repressive effect of Sp1 on the C/EBPα gene promoter: role in adipocyte differentiation. *Mol. Cell. Biol.* **19:** 4855–4865.

Tian, J. and Schibler, U. (1991) Tissue-specific expression of the gene encoding hepatocyte nuclear factor 1 may involve hepatocyte nuclear factor 4. *Genes Dev.* **5:** 2225–2234.

Timchenko, N.A., Wilde, M. and Darlington, G.J. (1999a) C/EBPα regulates formation of S-phase-specific E2F-p107 complexes in livers of newborn mice. *Mol. Cell. Biol.* **19:** 2936–2945.

Timchenko, N.A., Wilde, M., Iakova, P., Albrecht, J.H. and Darlington, G.J. (1999b) E2F/p107 and E2F/p130 complexes are regulated by C/EBPα in 3T3-L1 adipocytes. *Nucl. Acids Res.* **27:** 3621–3630.

Timchenko, N.A., Harris, T.E., Wilde, M. *et al.* (1997) CCAAT/enhancer binding protein α regulates p21 protein and hepatocyte proliferation in newborn mice. *Mol. Cell. Biol.* **17:** 7353–7361.

Timchenko, N., Wilson, D.R., Taylor, L.R. *et al.* (1995) Autoregulation of the human C/EBP α gene by stimulation of upstream stimulatory factor binding. *Mol. Cell. Biol.* **15:** 1192–1202.

Toniatti, C., Monaci, P., Nicosia, A., Cortese, R. and Ciliberto, G. (1993) A bipartite activation domain is responsible for the activity of transcription factor HNF1/LFB1 in cells of hepatic and nonhepatic origin. *DNA Cell Biol.* **12:** 199–208.

Tronche, F., Ringeisen, F., Blumenfeld, M., Yaniv, M. and Pontoglio, M. (1997) Analysis of the distribution of binding sites for a tissue-specific transcription factor in the vertebrate genome. *J. Mol. Biol.* **266:** 231–245.

Tronche, F., Rollier, A., Herbomel, P. (1990) Anatomy of the rat albumin promoter. *Mol. Biol. Med.* 7: 173–185.

Trujillo, M., Letovsky, J., MaGuire, H., Lopez-Cabrera, M. and Siddiqui, A. (1991) Functional analysis of a liver-specific enhancer of the hepatitis B virus. *Proc. Natl Acad. Sci. USA* **88:** 3797–3801.

Tsai, S.Y. and Tsai, M.J. (1997) Chick ovalbumin upstream promoter-transcription factors (COUP-TFs): coming of age. *Endocr. Rev.* **18:** 229–240.

Vaisse, C., Kim, J., Espinosa, R. 3rd, Le Beau, M.M. and Stoffel, M. (1997) Pancreatic islet expression studies and polymorphic DNA markers in the genes encoding hepatocyte nuclear factor-3α, -3β, -3γ, - 4γ, and -6. *Diabetes* **46:** 1364–1367.

Viollet, B., Kahn, A. and Raymondjean, M. (1997) Protein kinase A-dependent phosphorylation modulates DNA-binding activity of hepatocyte nuclear factor 4. *Mol. Cell. Biol.* **17:** 4208–4219.

Wang, J.C., Stafford, J.M. and Granner, D.K. (1998) SRC-1 and GRIP1 coactivate transcription with hepatocyte nuclear factor 4. *J. Biol. Chem.* **273:** 30847–30850.

Wang, N.D., Finegold, M.J., Bradley, A. (1995) Impaired energy homeostasis in C/EBPα knockout mice. *Science* **269:** 1108–1112.

Wang, W. and Gralla, J.D. (1991) Differential ability of proximal and remote element pairs to cooperate in activating RNA polymerase II transcription. *Mol. Cell. Biol.* **11:** 4561–4571.

Wang, X.Z., Lawson, B., Brewer, J.W. *et al.* (1996) Signals from the stressed endoplasmic reticulum induce C/EBP-homologous protein (CHOP/GADD153) *Mol. Cell. Biol.* **16:** 4273–4280.

Weinstein, D.C., Ruiz, I., Altaba, A. *et al.* (1994) The winged-helix transcription factor HNF-3β is required for notochord development in the mouse embryo. *Cell* **78:** 575–588.

Welm, A.L., Timchenko, N.A. and Darlington, G.J. (1999) C/EBPα regulates generation of C/EBPβ isoforms through activation of specific proteolytic cleavage. *Mol. Cell. Biol.* **19:** 1695–1704.

Williams, S.C., Baer, M., Dillner, A.J. and Johnson, P.F. (1995) CRP2 (C/EBPβ) contains a bipartite regulatory domain that controls transcriptional activation, DNA binding and cell specificity. *EMBO J.* **14:** 3170–3183.

Wolfgang, C.D., Chen, B.P., Martindale, J.L., Holbrook, N.J. and Hai, T. (1997) *gadd153/Chop10*, a potential target gene of the transcriptional repressor ATF3. *Mol. Cell. Biol.* **17:** 6700–6707.

Wu, K.L., Gannon, M., Peshavaria, M. *et al.* (1997) Hepatocyte nuclear factor 3β is involved in pancreatic β-cell-specific transcription of the *pdx-1* gene. *Mol. Cell. Biol.* **17:** 6002–6013.

Wuarin, J., Falvey, E., Lavery, D. *et al.* (1992) The role of the transcriptional activator protein DBP in circadian liver gene expression. *J. Cell Sci. Suppl.* **16:** 123–127.

Yamagata, K., Furuta, H., Oda, N. *et al.* (1996) Mutations in the hepatocyte nuclear factor-4α gene in maturity-onset diabetes of the young (MODY1) *Nature* **384:** 458–460.

Yen, T. S. B. (1993) Regulation of hepatitis B gene expression. *Sem. Virol.* **4**, 33–42.

Yoshida, E., Aratani, S., Itou, H. *et al.* (1997) Functional association between CBP and HNF4 in *trans*-activation. *Biochem. Biophys. Res. Commun.* **241:** 664–669.

Yu, X. and Mertz, J. E. (1997) Differential regulation of the pre-C and pregenomic promoters of human hepatitis B virus by members of the nuclear receptor superfamily. *J. Virol.* **71:** 9366–9374.

Zaret, K. (1999) Developmental competence of the gut endoderm: genetic potentiation by GATA and HNF3/fork head proteins. *Dev. Biol.* **209:** 1–10.

Zhang, D.-E., Ge, X., Rabek, J. P. and Papaconstantinou, J. (1991) Functional analysis of the *trans*-acting factor binding sites of the mouse α-fetoprotein proximal promoter by site-directed mutagenesis. *J. Biol. Chem.* **266**: 21179–21185.

Zinszner, H., Kuroda, M., Wang, X. *et al.* (1998) CHOP is implicated in programmed cell death in response to impaired function of the endoplasmic reticulum. *Genes Dev.* **12:** 982–995.

11

Neural transcription factors

Robert Bowser and A. Paula Monaghan

1. Introduction

The vertebrate nervous system represents a highly compartmentalized but interconnected network that regulates most other organ systems. Development and maintenance of this complex system requires an interplay of numerous intrinsic and extrinsic signals that culminates in a transcriptional response. All neuron and glial cells in the mature nervous system are derived from a pseudostratified neural epithelium whose formation is initiated on embryonic day 6.5 (E6.5) during gastrulation with formation of the neural plate in the mouse embryo. The neural plate then buckles at its midline to form the neural fold and a floor plate forms at the midline that remains in contact with the underlying notochord. The neural tube forms when the tips of the neural folds fuse. Cells in the region of the fusion form the roof plate.

In the developing central nervous system (CNS), regional specification occurs along the rostral caudal (anterior/posterior AP), dorso/ventral (DV) and medio/lateral (ML) axes. The AP axis contains the four major subdivisions: prosencephalon (forebrain), mesencephalon (mid brain), rhombencephalon (hindbrain), and spinal cord, each specialized to perform a unique function and to coordinate interactions with other regions of the CNS. The dorsoventral axis is defined by the roof plate dorsally and the floor plate ventrally. Developmentally the ML axis consists of: 1) early mitotically active progenitor cells in the ventricular zone (VZ) adjacent to the lumen; 2) late mitotically active progenitor cells in the subventricular zone (SVZ), adjacent to the VZ; 3) migrating precursors adjacent to the SVZ in the intermediate zone (IZ); and 4) differentiated cells in the superficial marginal zone (MZ). Regional specification in these domains is mediated by regional differences in the patterns of cell proliferation, survival, migration and differentiation, which ultimately lead to cellular specification.

There are an increasing number of transcription factors (TFs) expressed in the brain but fortunately TFs which fall into the same structural class are often involved in similar developmental processes. Development of the nervous system parallels that of many other tissues in the body requiring local and long distance

Transcription Factors, edited by J. Locker.

inductive interactions, therefore many TFs important for brain development are also required for the development of non-neuronal tissue. Mutational analyses in mice and humans have highlighted the critical concepts of TF action; 1) a TF hierarchy is required for cell birth, survival and death; 2) TFs provide a combinatorial code for regional specification; 3) TFs can directly and indirectly, positively or negatively, influence a cell's behavior. We will limit our discussion to the roles which individual TFs play in specific functions in the development of the nervous system and in neurological disease in mice and humans.

2. TFs in neural induction

2.1 TFs in early neural induction

Neural induction was first demonstrated by Spemann when it was shown that the dorsal lip of the blastopore in amphibians (the organizer), which normally gives rise to mesoderm has the potential to induce neural tissue in adjacent ectoderm (Spemann and Mangold, 1924). The organizer of higher vertebrates, such as Hensen's node in avian embryos and the node in mammals, functions much like the organizer of lower vertebrates. Since the identification of the first organizer gene, *goosecoid*, more than 25 organizer-specific genes have been characterized and many of these encode transcription factors and are conserved between amphibians and mammals. Many of these genes operate by inhibiting antineuralizing activities of so-called bone morphogenetic proteins (BMPs) in the ectoderm.

Traditional experiments in *Xenopus* embryos have indicated that both planar and vertical signals between germ layers assign anteroposterior identity on the central nervous system. The identification of a number of genes and their characterization has confirmed Spemann's original proposition that the vertebrate organizer is subdivided into separate domains: the head, trunk and tail organizers. The importance of ectoderm and mesoderm germ layers for neural induction was realized early in the twentieth century but more recently adjacent endoderm tissue has come to the forefront in patterning of the nervous system. In mouse anterior visceral endoderm, an extraembryonic tissue, expresses several genes essential for normal development of the head and has been termed the head organizer. Several transcription factor gene families have been implicated in this process including members of the homeotic gene superfamily (*Otx-1* and *Otx-2*) and cascades of neuronal basic helix-loop helix (bHLH) genes have been shown to promote differentiation whereas anti-neuronal bHLH genes repress them. This family of transcription factors act antagonistically and are essential not only for CNS patterning but also for the generation of the proper number of neurons and for morphogenesis of the nervous system. This review will focus on recent findings on the role that transcription factor families play in CNS induction in mice and the insights these studies provide into CNS patterning.

The anterior–posterior axis of the mouse embryo is established on E6 with the formation of the primitive streak at the future caudal pole of the embryo. Further patterning of the neural ectoderm (NE) along the anteroposterior axis has been attributed to inductive signals arising from tissue produced by the primitive streak. Both planar and vertical signals are involved in neural induction and axial patterning (Reviewed in Nieuwkoop, 1997). Influences from adjacent tissues mediated in part by key families of TFs initiate NE formation. A number of key TFs have been identified in this process.

One of the earliest TFs implicated in anterior-posterior NE patterning is a member of the homeodomain containing superfamily of proteins. Otx-2, a mammalian homologue of the *Drosophila* orthodenticle gene, is transcribed throughout the epiblast and visceral endoderm prior to primitive streak formation E3.5 to E6 (Cohen and Jurgens, 1990; Finkelstein and Perrimon, 1990). Once gastrulation is initiated, its expression becomes progressively restricted to the anterior end of the embryo in progenitor cells of all three germ layers, the endoderm, the endomesoderm and the ectoderm, thus defining the future anterior end of the embryo. Between E7 and E7.5 there is a progressive confinement of the protein to the anterior ectoderm corresponding to the forming headfold (*Figure 1*). At late gastrula

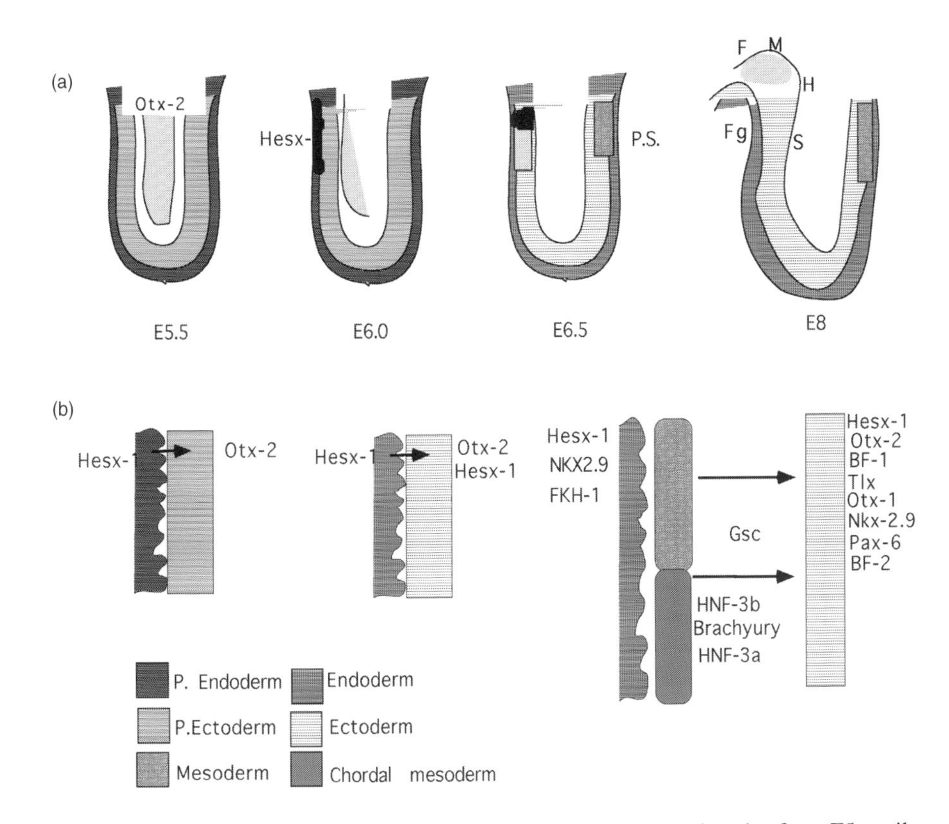

Figure 1. (a) Schematic representation of egg cylinder development in mice from E5 until E8. Otx-2 expression (stippled gray areas) and Hesx1 (black box) are indicated. Otx-2 is initially expressed throughout the embryo but gradually becomes restricted to the anterior end of the embryo in all three germ layers. Hesx1 expression is initiated in the primitive endoderm and later is activated in the neural ectoderm. Primitive streak (P.S.) formation on E6.5 is indicated in the rostral end of the embryo (black spotted box). F, forebrain; M, midbrain; H, hindbrain; S, spinal cord; Fg, foregut. (b) Schematic representation of tissue inductive interactions and the transcription factors implicated in individual tissues.

(from E8) stages the transcriptional domain of *Otx-2* is confined to the anterior NE where its caudal limit demarks the first boundary formed in the brain, the midbrain–hindbrain boundary (mesencephalon/metencephalon) (Simeone *et al.*, 1992). Later transcription of this gene becomes restricted to distinct cell groups in the adult, the choroid plexus, lateral septum, thalamic and pretectal nuclei, four colliculi and cerebellum (Frantz *et al.*, 1994).

Otx-2 heterozygous newborns suffer variable craniofacial and brain malformations in defined genetic backgrounds (Matsuo *et al.*, 1995). *Otx-2* homozygous mutant animals suffer from anterior truncations (Acampora *et al.*, 1995; Matsuo *et al.*, 1995, Ang, 1996), several animals exhibiting complete head loss anterior to rhombomere 3, indicating an essential requirement for the *Otx-2* gene in primary head formation. Analyses of mutant embryos have indicted: 1) that *Otx-2* is required for its own transcription in the epiblast but not in the visceral endoderm, 2) that endomesodermal cells fail to locate anteriorly, and 3) that ectoderm expression is extinguished by late streak stage (Acampora *et al.*, 1995).

The related gene, *Otx-1*, is expressed from E8, also in the presumptive forebrain and midbrain progenitor cells. In older embryos, transcripts of *Otx-1* are localized to the cortical plate and postnatally are found in layers 5 and 6 of the cerebral cortex (Simeone *et al.*, 1992, Frantz *et al.*, 1994, see below). Outside the neural epithelium, the olfactory placode, optic and otic vesicle highly express *Otx-1*. *Otx-1* homozygous mutant mice exhibit a milder phenotype, manifested in reduced temporal and perirhinal areas correlated to a reduced rate of cellular proliferation on E9.5. In older mutant animals, the acoustic and visual sense organs are reduced and the pituitary is also impaired. As a consequence of these developmentally induced structural changes, mutant animals survive but suffer from focal and generalized seizures (Acampora *et al.*, 1996, 1998).

Human *Otx-1* 'knocked-in' to the *Otx-2* locus can rescue anterior neural plate formation, but from E8 onward animals fail to maintain fore–midbrain identities and ultimately display a headless phenotype. Interestingly, the human *Otx-1* is only found in the visceral endoderm (VE) which indicates that *Otx-1* and *Otx-2* are equivalent in this tissue and that differences are due to a specific requirement for the *Otx-2* protein in the mesendoderm and fore/midbrain in late gestation (Acampora *et al.*, 1998). Normal anatomy and function is restored to *Otx1–/–* brains when replaced with the human (or *Drosophila*) *Otx-2* gene (Acampora *et al.*, 1997). A unique role for these genes in NE patterning at the midbrain–hindbrain boundary is implicated through the analysis of compound mutant animals. Vertebrate midbrain development depends on an organizing center located at the isthmus, a constriction in the embryonic mid/hindbrain region. Fibroblast growth factor 8 (FGF-8) is expressed at this boundary in a manner consistent with it providing a polarizing signal for the developing midbrain. In *Otx-1–/–*, *Otx-2+/–* mice the expression domain of FGF-8 is expanded indicating that there is a threshold requirement for these genes to correctly position the isthmic organizer, thereby influencing midbrain development (Acampora *et al.*, 1997). This family of transcription factors is therefore not only important in neural induction but also in patterning and regional identity in the CNS.

2.2 Role of endoderm in neural induction

Elegant experiments by Rhinn *et al.* (1998) have dissected the roles of these genes in individual tissues. Chimeric analyses of *Otx–/–* animals have provided insights into the tissues required to induce NE. Anterior neural induction can be rescued when the visceral endoderm is derived from wildtype cells in *Otx–/–* chimeric animals, indicating that a prepattern for head induction exists prior to the formation of mesendoderm and that anterior visceral endoderm patterns anterior NE (Rhinn *et al.*, 1998). These and other studies have shown that the primitive endoderm is responsible for the induction of anterior structures and for the definition of the future prosencephalon. Another homeobox containing gene, *Hesx-1*, is induced in the primitive endoderm opposite to the site of primitive streak formation (Thomas and Beddington, 1996). One day later Hesx-1 expression is detected in the anterior ectoderm of the future forebrain. Expression in the NE is dependent upon inductive signals from the primitive endoderm since removing this primitive endoderm region prevents expression of Hesx-1 in the NE, leading to a reduced anterior NE which is abnormally patterned. Once the forebrain is formed, Hesx-1 expression is maintained in the NE. Eventually Hesx-1 becomes localized to Rathke's pouch, the primordium of the anterior pituitary gland (Thomas and Beddington, 1996, Dattani *et al.*, 1998).

There is now a growing list of TFs which are expressed in these early endoderm cells and subsequently in the NE, including the homeodomain containing protein Nkx2,9, the winged helix containing gene *fkh-2* and members of the *Sox* gene family (sry-like HMG-box) indicating that a transcriptional hierarchy in endoderm is required for anterior head formation (Kaestner *et al.*, 1995; Pabst *et al.*, 1998).

Mice lacking Hesx-1 exhibit variable anterior CNS defects and pituitary dysplasia. Mutants have a reduced prosencephalon, anopthalmia or micropthalmia, defective olfactory development and bifurcations in Rathke's pouch. Neonates exhibit abnormalities in the corpus callosum, the anterior and hippocampal commissures, and the septum pellucidum (Ishibashi *et al.*, 1995, 1994). Mutations in the human *Hesx-1* gene lead to similar abnormalities characterized by septo-optic dysplasia (Dattani *et al.*, 1998). In null mice, Hesx-1 up-regulates other TFs such as Mash-1, a member of the bHLH gene superfamily, and accelerates neuronal differentiation, leading to the defects observed in mice and humans. Interestingly, Hesx-1 is negatively autoregulated, repressing its own transcription by binding to its own promoter and downregulating expression during development. Forced expression of Hesx-1 in precursor cells blocks neuronal differentiation in the brain and retina, further supporting the proposal that Hesx-1 is a negative regulator of neuronal differentiation. This leads to a scenario where Hesx-1 regulates brain morphogenesis by preventing precocious neuronal differentiation during early developmental stages and its downregulation must be precisely temporally controlled for normal development of the nervous system. Recent studies indicate that Hesx-1 expression is controlled by RBP-J, a mammalian homologue of Suppressor of Hairless [Su(H)], and Notch, a key membrane protein that may regulate lateral specification through RBP-J during neural development (Kageyama *et al.*, 1997; Issack and Ziff 1998, Jarriault *et al.*, 1998). Thus, the Notch–RBP-J–Hesx-1–Mash1 pathway may play a critical role the proper timing of neurogenesis and regulate neural tube morphogenesis.

2.3 Role of mesoderm in neural induction

Classic neural induction studies in *Urodeles* have implicated the mesoderm in primary NE induction. Distinct populations of mesoderm underlie the developing NE, pre-chordal mesoderm in rostral regions and the notochord in caudal regions. A unique population of cells in the primitive ectoderm, termed the organizer, delaminates to form both the embryonic mesoderm and endoderm. Early organizer cells produce anterior precordal mesoderm and are able to induce anterior properties in NE. Late organizer cell populations produce the node and mesoderm that pattern posterior NE such as the spinal cord (Lemaire and Kodjabachian 1996). Several TFs are expressed in the organizer, node, and in derived mesoderm populations, including the homeodomain containing proteins *Otx-2*, Lim-1 and Goosecoid (gsc); brachyury; and the winged helix factors HNF-3α and HNF-3β. All are expressed in prechordal mesoderm from early gastrulation. Lim-1 and HNF-3α and β are also found in the notochord (Beddington *et al.*, 1992; Gaunt *et al.*, 1993; Monaghan *et al.*, 1993; Shawlot and Behringer, 1995).

A precise role for the Lim-1 homeoprotein in neurogenesis was obtained through the analysis of null mutants. These animals do not form a recognizable node, prechordal mesoderm or head process at early gastrula stages. Like the *Otx-2* gene disruption, older embryos exhibit complete lack of head structures anterior to rhombomere 3. In contrast, posterior structures are relatively normal and a node can be identified from E7 suggesting a clear distinction between young and old node cells and anterior and posterior axis formation. This is supported by the observation that in *Otx-2–/–* and *Lim-1–/–* animals, expression of HNF-3β, brachyury, Lim-1 and Otx-2 are absent in the prechordal mesoderm but are present in posterior regions (Ang and Rossant, 1994; Matsuo *et al.*, 1995; Shawlot and Behringer, 1995; Acampora *et al.*, 1996). This suggests that other molecules can functionally substitute for these genes in the formation of posterior neural and cordal tissue.

HNF3β–/– embryos which do not have axial mesendoderm have normal anterior axial patterning (Ang and Rossant, 1994) but also have shortened axes primarily affecting the rostral head. They lack a notocord leading to dorso/ventral patterning defects in the spinal cord and also lack a foregut. In *Lim-1*, *Otx-2* and *HNF-3β* null animals scattered gsc-positive cells are obvious suggesting that an early organizer is present but one that cannot organize anterior structures. Otx-2, Lim-1 and HNF-3β may therefore be part of a genetic hierarchy which are required to organize early node activity. Surprisingly goosecoid, which was the first organizer transcription factor identified and which has been shown to induce and dorsalize mesoderm and initiate a crawling behavior in *Xenopus* cells has only a mild phenotype when mutated in mice (Niehrs *et al.*, 1993). gsc-null animals have normal node activity and exhibit only craniofacial and rib defects reminiscent of later gsc expression (Rivera-Perez *et al.*, 1995; Yamada *et al.*, 1995). This indicates that there are most likely redundant activities in organizer activity and that other as yet unidentified factors are required for normal pattern formation.

3. TFs in brain patterning

3.1 Forebrain patterning

Once the anterior end of the embryo is specified through inductive interactions from the primitive endoderm and mesoderm, the NE undergoes serial subdivision into three distinct rostro/caudal regions: forebrain, midbrain and hindbrain. These distinct anterior–posterior domains have different responses to the inductive signals from the prechordal plate, bone morphogenic proteins (BMP 4, BMP7), Sonic Hedgehog (SHH), the anterior neural ridge and FGF8 (Ruiz I Altaba and Jessell, 1993; Barth and Wilson, 1995; Furuta *et al.*, 1997; Shimamura and Rubenstein 1997). Many of these mechanisms are conserved throughout the vertebrate lineage indicating that relatively few factors are required to pattern the forebrain. Morphological changes are facilitated by differential patterns of proliferation, differentiation and cellular migration. Segmentation is achieved by the sequential formation of boundary/zones in the developing brain. Between E7 and E8.5, the prosencephalic/ mesencephalic boundary is established by the bulging of the mesencephalic hemisphere. The prosencephalon of the early embryo then subdivides into the secondary prosencephalon rostrally, from which the telencephalon differentiates, and the diencephalon caudally. Subsequently, the forebrain and hindbrain undergo patterns of refined segmentation. Patterning results from the appearance of transient segment-like territories, defined as prosomeres in the forebrain and rhombomeres in the hindbrain. Boundaries between neuromeres are distinguished by low rates of cellular proliferation, can be defined morphologically by the patterns of axons, ventricular ridges and sulci; in cell lineage studies and at the molecular level by the spatially restricted patterns of gene expression (Lumsden and Keynes, 1989; Fraser *et al.*, 1990; Lumsden, 1990; Guthrie and Lumsden, 1991; 1993; Bulfone *et al.*, 1993; Figdor and Stern, 1993). Sulci are often used as boundary markers but more recently the expression limits of a number of TFs have been shown to be more accurate predictors of coincident neuronal domains. These boundaries pose restrictions to cell or molecule mixing, cell migration, encompass fields of different cell proliferation capabilities and often define possible organizing centers in the brain (midbrain/hindbrain or isthmus, discussed below). This process of serial subdivision is therefore an important mechanism for regionalization, patterning and cell specification.

The forebrain gives rise to the telencephalon, diencephalon, hypothalamus, and retina. Each in turn is subdivided into distinct morphological and functional domains. Six transverse forebrain prosomeres numbered from caudal to rostral (P1 to P6; Bulfone *et al.*, 1993; Puelles and Rubinstein, 1993) have been identified through analyses of transcriptional domains (*Figure* 2). The longitudinal axis is divided into four domains: the roof, alar, basal and floor plates. Prosomeres with distinct gene expression appear early (E8–E11) and later (E12–E17) and their numerical order roughly reflects their timing during development (reviewed in Bulfone *et al.*, 1993).

Once the anterior neural plate has been specified by early TFs such as *Otx-2* and Hesx-1, other TFs can be found to be regionally localized on E8. BF-1, a member of the winged helix gene superfamily, is one of the earliest telencephalon markers, first detected on E8 (Xuan *et al.*, 1995). The BF-1 gene is required for normal proliferation in the telencephalon and the timing of neurogenic events as *BF-1–/–* animals have hypoplastic cerebral hemispheres, which is more severe in structures derived from the ventral telencephalon (Xuan *et al.*, 1995). The mouse homologue of the *Drosophila* tailless gene, *Tlx*, which is a member of the nuclear receptor gene super family, delimits the forebrain on E8 (Monaghan *et al.*, 1995). Targeted disruption of this gene does not lead to early patterning defects but to a reduction of a set of cell populations in the limbic system. *Tlx* mutant animals survive but have neuroanatomical defects in the limbic system and exhibit severe aggressive behavioral abnormalities (Monaghan *et al.*, 1997). These studies indicate the profound consequences that developmental defects in TFs can have on behavioral abnormalities.

3.2 Role of TFs in dorsal/ventral telencephalon formation

The neocortex derives from the dorsolateral wall, basal ganglia from the ventral and the hippocampus and choroid plexus from the medial wall of the telencephalon.

(a)

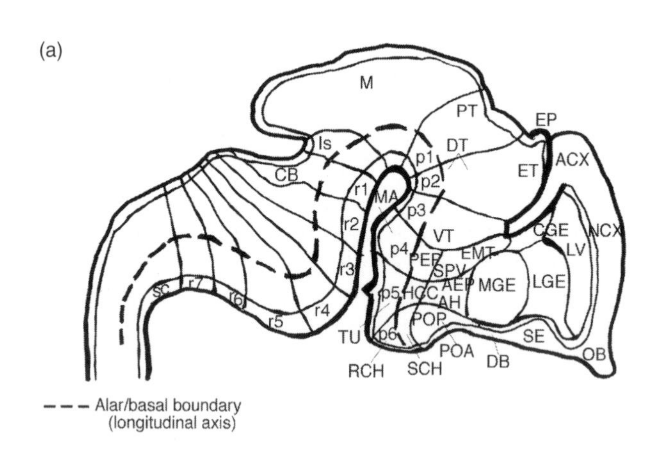

(b)

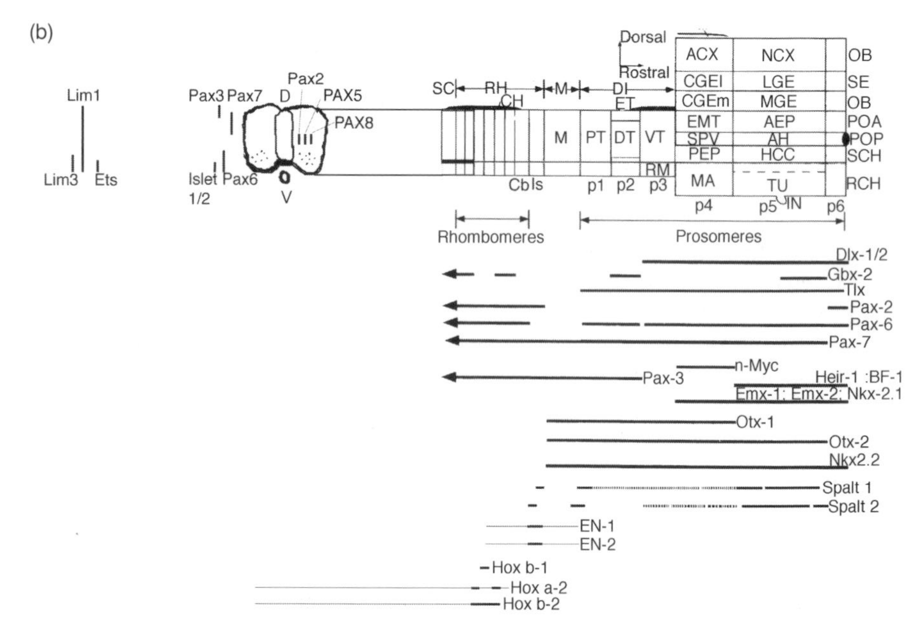

Two homeoproteins related to *Drosophila* empty spiracles, Emx-1 and Emx-2, define subdomains in the dorsal/medial telencephalon (Simeone *et al.*, 1992). Emx-1 is exclusively restricted to the dorsal telencephalon whereas Emx-2 is found in a subpopulation of dorsal telencephalic cells but is also detected in the diencephalon. *Emx-2* mutant newborn animals exhibit defects in structures derived from the dorsal/medial telencephalon, such as the dentate gyrus, hippocampus and medial limbic cortex. In contrast, *Emx-1* mutant animals suffer only mild defects in the adult brain suggesting that Emx-2 compensates for Emx-1 in regions of overlap (Yoshida *et al.*, 1997).

The boundary between the dorsal and ventral telencephalon is identified by the expression patterns of several TFs. As mentioned above, Emx-2 is confined to the primordia of the cerebral cortex in dorsal structures, the striatum primordium (LGE) is marked by the dorsal limit of the homeoproteins Dlx-1 and Dlx-2, which are widely expressed in strial progenitor cells (Price *et al.*, 1991). The junction between these two expression patterns defines the corticostrial boundary. Subsequently, distinct proliferative zones appear within the ganglionic eminence (GE), the dorsal (striatum, LGE) and the ventral (pallidum, MGE). The boundary between the lateral (LGE) and medial ganglionic eminences (MGE) is defined by the expression of the homeoprotein Gbx2 which is expressed in four domains in the brain, two of which coincide with forebrain boundaries. The homeoprotein Nkx-2.1 defined the domain of the MGE (Porteus *et al.*, 1991; Price *et al.*, 1991; Bulfone *et al.*, 1993). Another dorsally expressed gene is *pax-6* whereas *Vax-1* and member of the distal-less gene family (Dlx1, 2, 4, 5) are ventrally expressed (Bulfone *et al.*, 1993; Puelles and Rubenstein, 1993; and Pax6, Hallonet *et al.*, 1998; Casarosa *et al.*, 1999, reviewed in Rubenstein *et al.*, 1998). The fact that these TFs

---

Figure 2. (a) Schematic representation of a mid sagittal section through the brain of an E12.5 mouse. The transverse neuromeric divisions proposed by Bulfone *et al.* (1993) are indicated. Forebrain prosomeres (p1 to p6) and hindbrain rhombomeres (r1 to r7) are labeled. The Alar/basal plate division is indicated by a dotted line. The roof and floor plates are located superficially. (b) Deconvoluted representation of the brain in (a) from the rostral telencephalon to the spinal cord showing the expression patterns and boundaries of transcription factors in the A/P axis (Modified from Bulfone, A. *et al.* (1993) Spatially restricted expression of Dlx-1, Dlx-2 (Tes-1), Gbx-2 and Wnt-3 in the embryonic day 12.5 mouse forebrain defines potential transverse and longitudinal segmental boundaries. *J. Neurosci* 13: 3155–3172, with permission from the Society for Neuroscience). On the left hand side a transverse section through the spinal cord is shown. The dorsal (D) and ventral (V) expression patterns of key transcription factors in the spinal cord are shown. *Heir-1* and *n-myc* gene expression were described in Ellmeier *et al.*, (1992), additional references are in the text. Abbreviations: ACX, archecortex; AEP, Anterior entopeduncular area; AH, anterior hypothalamus; CB, cerebellum; CGE, caudal ganglionic eminence (L, lateral; M, Medial); CH choroid plexus; DB diagonal band; DI, diencephalon; DT, dorsal thalamus; EMT, eminentia thalami; EP, epiphysis; ET, epithalamus; HCC hypothalamic cell cord; IN, infundibulum; IS, isthmus; LGE, lateral ganglionic eminence; LV, lateral ventricle; MA, mammillary area; M, mesencephalon; MGE, medial ganglionic eminence; NCX, neocortex; OB, olfactory bulb; PEP, posterior entopeduncular area; POA, anterior preoptic area; POP, posterior preoptic area; PT pretectum; RCH, retrochiasmatic area; RH, rhombencephalon; SC, spinal cord; SE, septum; SCH, suprachiasmatic area; SPV, supraoptic area; TU, tuberal hypothalamus; VT, ventral thalamus.

respect morphological boundaries provides strong evidence for a functional role in prosomeric formation. In *Pax-6* mutant animals the segregation of cortical and strial cells observed in an *in vitro* aggregate assay is lost. In the absence of pax-6 the corticostrial boundary does not form and the normal expression of the cell surface molecules such as, R-cadherin whose expression ceases abruptly at the boundary between lateral cortex and LGE, and tenascin-C is lost (Stoykova *et al.*, 1997). Targeted disruption of Nkx2.1 has also indicated that it essential for ventral forebrain structures since the MGE does not form in mutant animals but is replaced with an LGE-like structure (Sussel *et al.*, 1999). Null animals with mutations in *Pax-6* and *Nkx2.2* show patterning defects which indicates that these genes are functionally required for compartmentalization and boundary definition.

Cascades of bHLH proteins also define dorsal and ventral boundaries in the forebrain. NeuroD, Math2, Nscl1 and Nscl2, Neurogenin 1 and Neurogenin 2 (Ngn1,2) are located dorsally whereas Mash-1 defines a complementary ventral domain (Lee, 1997) More recently this family of TFs have been shown to play a role in the specification of dorsal ventral neuronal identity. In *Ngn1/2* double mutants ventral markers are ectopically expressed in dorsal domains, furthermore ectopic expression of the ventral gene *Mash-1* in dorsal domains confers ventral characteristics to dorsal neurons (Fode *et al.*, 2000).

The ganglionic eminence gives rise to populations of interneurons which migrate into the dorsal telencephalon to form mature interneurons of the adult cortex and olfactory bulb. In addition to the corticostrial boundary, the dlx-gene family define different mediolateral cell populations in GE differentiation, dlx-1 and *Dlx-2* are localized in VZ and SVZ populations, *Dlx-5* in SVZ and mantle with *Dlx-6* in the mantle zone. A critical role for the *Dlx* genes in ventral ganglionic development has been identified through the analysis of mutant animals. Analyses of the *Dlx-1* and *Dlx-2* compound homozygous mutant animals demonstrate that these genes are required for differentiation of late born striatal neurons of the SVZ and for the development of a component of the GABAergic interneuron populations in the neocortex and olfactory bulb (Anderson and Jan, 1997; Bulfone *et al.*, 1998). *Dlx-1/dlx-2* compound mutants have a time-dependent block in strial differentiation. Neurons are born from early progenitor cells normally but late-born neurons accumulate within the proliferating region, the VZ and SVZ (Anderson and Jan, 1997). A number of genes are not expressed in these mutants including, *Dlx-5*, *Oct-6*, *Brn-4* while other are misexpressed e.g. *Lhx-2* (Anderson and Jan, 1997). The striatum in *Dlx-1* and *Dlx-2* individual mutants appear normal implying that the function of these genes is redundant in the forebrain. However, Dlx-2 affects the development of tyrosine hydroxylase periglomular cells interneurons in the olfactory bulb that they are not fully redundant (Qiu *et al.*, 1995). These studies highlight the importance of TFs in the timing of proliferation and differentiation in the neural epithelium.

3.3 Role of TFs in diencephalon development

Three distinct morphological stages of diencephalic differentiation have been identified in the developing mouse brain (Niimi *et al.*, 1962). In the first stage (day 10 p.c. until day 12 p.c.), the fields of the epithalamus, dorsal thalamus, ventral thalamus and hypothalamus are defined. The second stage (E13 until E15) is the period of layer differentiation in which ventricular (or germinal), mantle and marginal layers are defined. Finally, in the third stage, from E16 until birth, the anlage of the diencephalic nuclei is formed (Bhide, 1996; Sheth and Bhide, 1997). One of the first prosomeric boundaries to be formed is the diencephalic p2/p3 boundary, from which the zona limitans intrathalamica (ZLI) forms (D1/D2 defined by Figdor and Stern, 1993) and is proposed to be the boundary between the ventral and dorsal thalamus (Niimi *et al.*, 1962; Bulfone *et al.*, 1993; Figdor and Stern, 1993; Puelles and Rubenstein, 1993). Several TF expression zones abut at this boundary. The rostral and caudal boundaries of expression of BF-2, Tlx and BF-1 respectively is at the zona limitans intrathalamica (Hatini *et al.*, 1994). The caudal limit of *Dlx1* transcription is at the ZLI also. The boundary between dorsal and epithalamus is defined by the dorsal limit of expression of Gbx2. These patterns of expression suggest that TFs are required to establish boundaries in the rostro/caudal and dorsa/ventral axes, essentially defining different functional domains in the developing brain.

3.4 Role of TFs in hindbrain development

The midbrain/hindbrain boundary contributes to the colliculi, isthmic nuclei, cerebellum and pons. Developmental restriction occurs late in development in this region and is controlled transcriptionally. The homeoproteins engrailed 1 and engrailed 2; the paired box proteins Pax-2, Pax-5, Pax-8 and the zinc finger containing proteins Msal-1 and Msal-2 are expressed in a band of cells that will give rise to the midbrain–hindbrain boundary. Their temporal expression pattern Pax-2>En-1, Msal-1, Msal-2>Pax-5 suggests that a transcriptional hierarchy regulates formation of this morphogenic region. Direct evidence for the roles of Pax-5, En-1 and En-2 in midbrain–hindbrain specification has come from the analyses of targeted mutation in mice (Davis and Joyner, 1988; Joyner *et al.*, 1991; Wurst *et al.*, 1994, Rowitch and McMahon, 1995; Monaghan and Ott, unpublished observations).

The hindbrain is the only part of the nervous system that is obviously metamerized into eight segmental units, the rhombomeres. In contrast to the forebrain, this segmentation is obvious from E8.5 and similar to the forebrain, is reflected at the molecular level by the segmental restricted patterns of gene expression. The cell constituents of each rhombomere do not mix with neighboring rhombomeres but cells from even or odd rhombomeres mix more easily with each other than with adjacent segments (Guthrie *et al.*, 1993). Individual rhombomeres define distinct functional units exemplified by the observation that non-overlapping rhombomere (r) pairs contribute to cranial nerves, each pair innervating one adjacent branchial arch (Lumsden and Keynes, 1989). The 5th cranial nerve, the trigeminal, originates from r2 and r3 and innervates first branchial arch muscles (BAMs); the 7th cranial nerve, the facial, originates from r4/5 and innervates second BAMs while the glossopharyngeal originates from r6,7 and innervates the third BAMs.

During development, members of the *Hox* gene superfamily (Chapter 9) are spatially restricted, have different responses to retinoic acid and have sharp anterior boundaries in rhombomeres in a two-segment periodicity, suggesting that they are involved in establishing this anterior-posterior periodicity (Keyes and

Krumlauf, 1994; Krumlauf, 1994). In vertebrates, genes in individual *Hox* clusters are oriented in the same 5′ to 3′ direction with respect to transcription. There is a collinear relationship between the position of genes in parologous groups and their distinct A–P boundaries of expression in the CNS. Genes at the 3′ end of the cluster are expressed in the NE and neural crest of the hindbrain. Their anterior boundaries coincide with the segment boundaries of individual rhombomeres. For example, 3′ members of parologous groups 2, 3 and 4 terminate at r2/3, r4/5 and r6/7 respectively.

Mutation analysis of several TFs with rhombomere specific or restricted expression patterns has demonstrated that they are required for establishing segmentation pattern and functional identity in individual rhombomeres. Genes at the 3′ end of the cluster such as *Hoxa-1*, *Hoxa-2* and *Hoxa-3*, which are the earliest and most anterior, have defects in various craniofacial and upper thoracic regions (reviewed in Maconochie *et al.*, 1996). In *Hoxa-1* mutant animals for example, r5 is lost or missing and adjacent rhombomeres display abnormal axonal projections (Chisaka *et al.*, 1992; Carpenter *et al.*, 1993). Mutations in *Hoxa-5* and *Hoxa-6* indicate that these genes also contribute to the development of the hindbrain and neck (Chisaka and Capecchi, 1991; Lufkin *et al.*, 1991; Chisaka *et al.*, 1992). *Hoxb-1* is precisely expressed in r4 only (Murphy *et al.*, 1989) and *Hoxb-1* mutants exhibit altered development of r4 branchiomotor neurons and have neural crest migratory defects (Goddard *et al.*, 1996, Studer *et al.*, 1996). Another TF, GATA-2, which is a putative downstream target of *Hoxb-1*, is expressed as early as E9 in the hindbrain, in ventral r4 and transiently in ventral (r2) (Bell *et al.*, 1999; Nardelli *et al.*, 1999). *GATA-2–/–* mice exhibit cranial nerve defects and decreases Hox-b1 expression in r4. Mutants have fewer axonal projections exiting r2 and r4 with lack of or reduced facial motor neurons from r5. Other TFs such as Nkx2.2 and Nkx6.1 have expanded expression patterns in r4 (Qiu *et al.*, 1998). Global overexpression of Hoxb1 in chick embryos results in the transformation of r2 to r4, on the basis of molecular markers and motor neuronal migration (Bell *et al.*, 1999). Also, ectopic expression of Hoxa-1 using the beta-actin promoter leads to embryonic lethality but interestingly leads to ectopic expression of Hoxb-1 in r2 (and to a lesser extent in r1, r3, r6) with alterations in a subpopulation of neural crest cells from r2 (Zhang *et al.*, 1994). Thus combinatorial codes of *Hox* genes confers positional identity on individual rhombomeres.

Krox-20 is a zinc finger containing transcription factor that is expressed in r3 and r5. Mice homozygous for a mutation in this allele die within two weeks after birth. Disruption of *Krox-20* leads to progressive elimination of r3 and r5 leading to fusion's in the trigeminal, facial, glossopharyngeal and vagus nerves (Swiatek and Gridley, 1993, Schneider-Maunoury *et al.*, 1997). Krox-20 regulates at least five different regulatory genes in the hindbrain in a rhombomere specific pattern. Binding sites for Krox-20 have been identified in the *Hoxb-2* and the paralogous gene *Hoxa-2* promoter that is responsible for r3/r5 expression (Nonchev *et al.*, 1996). Hoxb-3 expression in r5 and the receptor tyrosine kinase gene *Sek-1* in r3 and r5 is disrupted in *Krox-20* mutant animals (Seitanidou *et al.*, 1997). Therefore Krox-20 is part of the upstream cascade that regulates rhombomere specific gene expression. These studies indicate that transcription factor cascades provide a starting point for understanding the molecular basis of segmental patterning in the hindbrain.

4. TF functions during spinal cord development

4.1 Overview

The formation of the vertebrate spinal cord is initiated by signaling events that first occur during neural plate formation (McConnell, 1995). Epithelial cells receiving these signals from the underlying mesoderm form the neural plate. While the exact signaling pathways remain uncharacterized, they include four classes of mesodermal derived proteins, follistatin, noggin, cordin, and gremlin (Hemmati-Brivanlou *et al.*, 1994; Knecht *et al.*, 1995; Sasai *et al.*, 1995; Hsu *et al.*, 1998). After formation of the neural tube, signals for cell proliferation and differentiation originate from the roof plate, floor plate, and notochord to activate specific sets of TFs and give rise to the fully differentiated spinal cord (Bang and Goulding, 1996).

4.2 Ventral signals, A/P axis and TFs

The notochord and floor plate express a protein called sonic hedgehog (Shh) that induces the formation and differentiation of motor neurons and interneurons in the ventral region of the neural tube via interactions with cell surface receptors (Chapter 6). These receptors include Patched (PTCH and PTCH2) and a co-factor called smoothened (Smo) that function via serine-threonine kinases to induce downstream signals (Carpenter *et al.*, 1998; Motoyama *et al.*, 1998). One early response to Shh is the expression and altered processing of members of the Gli (Gli-1, 2, 3) family of zinc-finger containing TFs. Shh binding to Ptch/Smo receptors blocks proteolytic processing of Gli proteins, resulting in downstream activation of Shh target genes. In the absence of Shh, Gli undergoes proteolysis releasing an N-terminal fragment that functions as a transcriptional repressor. In mice, Gli-2 is required for induction of floor plate and adjacent cells, but not for motor neuron or interneuron formation, whereas Gli-1 and Gli-3 are required for dorsoventral patterning (Matise *et al.*, 1998). Motor neurons and interneurons continue to differentiate in *Gli-2–/–* mice, implying the existence of a Gli-independent pathway that is downstream of Shh required for the formation of ventral spinal cord neurons (Tanabe *et al.*, 1995; Matise *et al.*, 1998).

In addition to its effects on Gli expression and function, high levels of Shh induce expression of the transcription factor HNF-3β via transcriptional activation by Gli proteins (Sasaki *et al.*, 1997). HNF-3β induction leads to formation of notochord, induction of floor plate differentiation, and the initiation of Shh expression in the floor plate. Mice lacking HNF-3β fail to form a notochord (Ang and Rossant, 1994; Weinstein *et al.*, 1994), while targeted expression of HNF-3β in the dorsal neural tube induces generation of floor plate cells (Sasaki and Hogan, 1994; Ruiz i Altaba *et al.*, 1995). Therefore, early induction of HNF-3β via Gli proteins is required for formation of both the notochord and floor plate.

Lower concentrations of Shh lead to one more round of cell division and the expression of Isl-1 and motor neuron differentiation. Isl-1 is a homeodomain containing transcription factor that possesses an amino-terminal pair of LIM domains (Karlsson *et al.*, 1990). Elimination of Isl-1 expression via expression of mutant protein or treating chick neural tube explants with antisense oligonucleotides inhibits formation of motor neurons (Tsuchida *et al.*, 1994). Therefore, Isl-1 plays a critical role in regulating motor neuron differentiation. In the absence of Isl-1, prospective motor neurons undergo apoptosis (Pfaff *et al.*, 1996).

The anteroposterior (A/P) axis also develops within the developing hindbrain and spinal cord. Motor neurons represent a prominent cell type that exhibits subclass specification along the A/P axis of the hindbrain and spinal cord. Columnar subclasses of motor neurons can be identified by the coordinated expression of Isl-1/ Isl-2 and specific members of the Ets and Lim homeodomain proteins (Tsuchida *et al.*, 1994; Lin *et al.*, 1998). These TFs confer motor neuron specification and selection of axon guidance pathways that insure proper sensory and motor innervation.

Interneurons are generated via expression of specific homeobox genes such as *En-1*, *Lim-3*, *Lim-2*, and *Gsh-4*. Shh does not appear to the primary signal for induction of interneurons, but instead the presence of motor neurons appear required for the formation of interneurons. In addition to loss of motor neurons, *Isl-1* mutant mice also lack interneurons. This implies a complex signaling pathway that is yet to be fully defined but likely involves specific TFs to regulate cell fate decisions.

4.3 Dorsal signals and TFs

The primary signals for the induction of dorsal cell fate are members of the bone morphogenetic protein (BMP) family. Four members of the BMP family, BMP-4, BMP-5, BMP-7, and Dsl1 are induced in the dorsal midline cells as the neural tube forms. Members of this family of TGF-β-like proteins that act via cell surface receptors to induce their downstream signal to regulate dorsal cell differentiation and formation of neural crest cells, roof plate cells, and dorsal interneurons (reviewed in Lee and Jessell, 1999).

The intrinsic factors that determine dorsal cell fates include several TFs that exhibit elevated expression upon BMP binding. These include the paired homeobox factors *Pax-3*, *Pax-6* and *Pax-7*, and *Msx1–3* (Liem *et al.*, 1997). All *Pax* genes are expressed during embryonic development and within the developing spinal cord individual family members are expressed in dorsoventrally restricted regions (Gruss and Walther, 1992; Goulding *et al.*, 1993). Pax-3/Pax-7 expression induces the expression of a zinc finger transcription factor called SLUG that functions during neural crest cell specification (Dickinson *et al.*, 1995). Neural crest cells are a migratory cell population that arises during neural tube formation and generates neurons and glia of the peripheral nervous system, melanocytes, and other non-neuronal cell types. While expression of *Slug* is insufficient for neural crest cell formation, additional TFs including the Zic family, Zic1–3, and AP2 also function during neural crest cell differentiation (Zhang *et al.*, 1996; Nakata *et al.*, 1997). Neural crest cells also express the transcription factors MASH-1 and Phox2a and Phox2b that are required for proliferation and survival of these cells (Anderson and Jan, 1997; Morin *et al.*, 1997; Pattyn *et al.*, 1999). BMP signaling induces a rapid proteosome-mediated degradation of MASH-1 that results in terminal differentiation (Fishell, 1999). Therefore, neural crest induction involves a multistep process including numerous TFs that progressively restrict the cell fate of these neural cells.

The dorsal spinal cord contains interneurons that in the chick are defined as two classes termed D1 and D2 (Liem *et al.*, 1997). To generate dorsal interneurons, BMP signals from the ectoderm epithelial cells first signal the induction of dorsal midline cells to form the roof plate cells, identified by expression of the transcription factor MAFB (Liem *et al.*, 1997). The roof plate then provides a secondary signal that generates distinct classes of interneurons and involves both temporal and qualitative differences in the response of the progenitor cells. D1 type interneurons form near the roof plate and express the LIM homeodomain proteins LH2A and LH2B, whereas D2 type interneurons are generated further from the roof plate and express Isl-1 (Liem *et al.*, 1997; Lee and Jessell, 1999). Genetic ablation of the roof plate indicates that it is required for specifying multiple classes of interneurons within the dorsal spinal cord (Lee *et al.*, 2000). The loss of the roof plate did not effect proper dorsoventral expression of Pax genes but did prevent expression of the Math1 and Neurogenin1 transcription factors in neural progenitor cells in the dorsal spinal cord. Both Math1 and Neurogenin1 function in cell fate determination of interneurons (Ma *et al.*, 1996; Helms and Johnson, 1998). Signaling molecules derived from the roof plate (including BMPs) are therefore crucial for the expression of transcription factors required for the generation of interneurons in the dorsal spinal cord.

Therefore, the SHH pathway represents the main signal transduction pathway utilized for development of the ventral spinal cord, and members of the BMP family function as signals for formation of the dorsal spinal cord. Whereas Shh appears to control ventral patterning by induction of specific cell types at different Shh concentrations, dorsal patterning involves the interplay of functional differences in multiple members of the BMP family along with temporal differences in the responses of dorsal cells to BMPs. These signals induce intrinsic expression and activation of TFs that determine cell fate and differentiation. While specific TFs have been identified that demarcate different cell types of the spinal cord, the downstream genes regulated by these factors that result in cell fate determination remain unclear.

5. TFs in neurologic disorders and neurodegeneration

5.1 Mutations in TFs that lead to nervous system defects

As discussed above, TFs play a critical role in the development of the nervous system. Therefore, the inactivation of TFs as a result of an inheritable disease is usually incompatible with survival (Latchman, 1996). However, mutations in several TFs permit survival and result in nervous system defects (usually the visual system) or abnormalities in derivatives of neural crest cells, resulting in either a developmental disorder or oncogenesis. The *retinoblastoma (RB)* gene and members of the *PAX* and *SOX* gene families are also associated with eye development and are discussed further in a comprehensive review of transcriptions factors in the eye (Freund *et al.*, 1996).

The *Rb* gene product was the first identified tumor suppressor protein and is a member of a gene family that includes the pRb-related proteins p107 and p130 (see Chapter 5 and reviewed in Mulligan and Jacks, 1998). Transcriptional regulation via pRb proteins is complex and includes interactions with numerous proteins, including members of the E2F gene family and histone deacetylase proteins that dictate transcriptional activation or repression (Weintraub *et al.*, 1992; Brehm *et al.*, 1998; Ferreira *et al.*, 1998; Magnaghi-Jaulin *et al.*, 1998).

Various RB mutations, including deletions, duplications and point mutations can inactivate pRb (Bookstein and Lee, 1991). pRb is expressed in most tissues but mutations induce specific CNS pathology, particularly in the retina. Somatic mutations are commonly found in both *RB* alleles within patients with retinoblastoma, leading to loss of protein expression or abnormal function (Fung *et al.*, 1987; Lee *et al.*, 1987; Bookstein and Lee, 1991). Complete loss of pRb function in transgenic animals results in embryonic lethality and developmental defects in both the hematopoietic and nervous systems, with ectopic mitoses and massive cell loss in the hindbrain, spinal cord, and dorsal root ganglia (Lee *et al.*, 1992). As demonstrated by these transgenic animals, pRb is not essential for basic pattern formation, cellular proliferation or early differentiation in most organ systems. However, within the nervous system pRb appears to function during the migration and/or differentiation of postmitotic neurons, as massive cell death occurs within the intermediate zone where postmitotic neuronal precursors migrate and continue to differentiate.

Mutations in members of the *Pax* gene family also result in nervous system abnormalities and have been identified in human disease. *Pax* genes contain a paired box and are organized into different classes depending on whether they also encode a complete or partial homeodomain, an octapeptide or class-specific amino acids and whether or not their genes share similar exon/intron boundaries. Class I (Pax-1, -9) have an octapeptide only; Class II (Pax-3, -7) have an octapeptide and a homeodomain; Class III (Pax-2, -5, -8) contain only a partial homeodomain and Class IV (Pax-4, -6) contain a complete homeodomain. Pax-3, -6, and -7 expression is restricted to mitotically active cells along the A/P axis of the neural tube, before the onset of differentiation. Pax2, 5, and 8 expression begins during neural differentiation.

Several mutation phenotypes in mice have been associated with alterations in *Pax* genes. *Pax-1* mutations produce undulated axial skeleton structures. The *Pax-2* gene is found in the brain, optic stalk, auditory vesicle, hindbrain, spinal cord, and kidney (Dressler *et al.*, 1990). Mutations in the *Pax-2* gene are associated with kidney and retinal abnormalities in the embryonic lethal krd mice (Otteson *et al.*, 1998). Mutations in the *Pax-2* gene have been identified in a family with eye and kidney malformations (Sanyanusin *et al.*, 1995). The *Pax-6* gene is a master regulator of eye development in both vertebrates and invertebrates. Eye development is exquisitely sensitive to the dose and type of Pax-6 protein, as indicated by the variety of clinical phenotypes that result from different levels and forms of the Pax-6 protein. During murine development, Pax-6 is expressed in the entire anterior posterior axis and expression extends from rostral telencephalon to the myelencephalon, sparing the roof of the mesencephalon from E8. Pax-6 is found in optic cup, lens and surface ectoderm nasal placode from E9 (Hanson and Van Heyningen, 1995). From E15.5 onwards Pax-6 is detected in the hindbrain with a boundary between the diencephalon and mesencephalon.

Mutations in the murine *Pax-6* gene lead to the semidominant small eye (*Sey*) phenotype in heterozygotes and a complete lack of eyes and nasal cavities in homozygous mutant animals, resulting in death soon after birth. Four *Sey* alleles have been identified in mice, Small-eye (*Pax-6 Sey*) that contains a base-pair mutation and results in a truncated protein before the homeodomain; Dickie small eye (*Pax-6 Sey^{Dey}*) that is a deletion of the *Pax-6* locus; Small-eye Harwell (*Pax-6 Sey^{H}*) that also is a deletion of the *Pax-6* locus; and Small-eye Neyherberg (*Pax-6 Sey^{Neu}*) with a splice donor mutation in exon 10 (Hogan *et al.*, 1986; Hill *et al.*, 1991; reviewed in Stuart *et al.*, 1993 and Freud *et al.*, 1996). In vertebrates, a 14 amino acid insertion in the paired domain of Pax-6 results in a protein with different DNA binding specificities and different expression levels in the eye and brain of mice (Walther and Gruss, 1991). Interestingly, increased expression of this extended form of Pax-6 in humans results in an unusual eye phenotype with abnormal development of the anterior aspect of the iris (Epstein *et al.*, 1994). *Pax-6* mutations in humans lead to developmental eye defects such as aniridia and Peters' anomaly and are inherited as autosomal dominant traits (Ton *et al.*, 1991; Glaser *et al.*, 1992). To date all *Pax-6* mutations identified in aniridia or Peter's anomaly are similar to those described for the murine small eye mutations, either point mutations that result in truncated protein or large deletions of the gene locus. An interesting aspect of *Pax-6* mutations is that they result in the inactivation of one allele and therefore the mutant phenotype results from haploid insufficiency, in which the amount of protein produced by one functional allele is not sufficient to supply functional transcriptional regulation.

Waardenburg syndrome is a human autosomal dominant disorder that results in pigmentary abnormalities, deafness and retinal abnormalities due to the absence of melanocytes and neural crest developmental abnormalities caused by mutations in the *Pax-3* gene (Tassabehji *et al.*, 1992). Over 50 mutations have been found in the *Pax-3* gene that result in Waardenburg syndrome, including deletions and frameshift mutations that result in null alleles, amino acid substitutions in the paired box that effect DNA binding, and substitutions in the homeobox that are known to alter DNA binding (Read and Newton, 1997). Since all melanocytes, except those in the retina, are derived from neural crest cells (see above) the *Pax-3* gene product is required either for differentiation in the neural crest, for migration of melanoblasts, or for terminal differentiation and survive in their final location. These results again suggest that the mutant phenotype is a result of haploid insufficiency and indicates the importance of the *Pax* gene family in neural crest cell differentiation and development of the visual system.

Additional mutations in TFs that function in neural crest cell differentiation also induce human disease. These include members of the *Sox* gene family that function as modulators of other TFs. Sox proteins contain a HMG box DNA binding domain. Mutations in *Sox-10* have been found in patients with a subtype of Waardenburg syndrome called Type IV Shah–Waardenburg syndrome with Hirschsprung disease (Pingault *et al.*, 1998). Therefore mutations in TFs that function during neural crest cell differentiation result in a number of disorders involving melanocytes and other neural crest cell derivatives.

The POU (Pit-Oct-Unc) family of TFs also functions in eye development. This gene family was originally defined on the basis of a common DNA binding domain in the mammalian TFs Pit-1, Oct-1, the universal octamer binding protein Oct-2, and the nematode protein Unc-86 (Herr *et al.*, 1988). Many members of the POU family have been shown to play critical roles in the development and function of the nervous system. POU containing proteins fall into five classes based on sequence similarity: POU-I (Pit-1), POU-II (Oct-1, Oct-2), POU-III (Brn-1, Brn-2, Tst-1, CF1), POU-IV (Brn-3, Unc-86, I-POU) and POU-V (Oct3/4) (Treacy and Rosenfeld, 1992). *POU-III* gene products are detected in early regionalization of the developing brain and also in the adult brain. One of these genes, *Pit-1*, is involved in the development of the pituitary and controls production of growth hormone in somatotrophs, prolactin in lactotrophs, and TSH in thyrotrophs. Pit-1 can enhance the activation of genes involved in cell cycle initiation as it has been shown to bind to the c-fos serum responsive element and enhance c-fos expression. This activity is dependent upon phosphorylation of the Pit-1 protein, suggesting a mechanism for pit-1 activation of somato-lactotroph cell proliferation (Gaiddon *et al.*, 1999). Synergistic interaction between pit-1 and estrogen receptor has also been demonstrated, allowing cell and stage specific effects on target genes in the developing pituitary (Simmons *et al.*, 1990).

Several distinct *Brn-3* genes are associated with eye abnormalities, and over-expression is observed in specific tumor cell lines (Budhram-Mahadeo *et al.*, 1999). Expression of *Brn-3* class genes occurs mainly in the retinal ganglion cell layer, somatosensory neurons in the dorsal root ganglia, spinal cord and brain stem (Turner *et al.*, 1994; Xiang *et al.*, 1997). Differentiation into a neuron-type cell *in vivo* and *in vitro* is associated with an increase in Brn-3a expression and a decrease in Brn-3b while the levels of Brn-3c remains unchanged (Lillycrop *et al.*, 1992; Budhram-Mahadeo *et al.*, 1995a; Smith *et al.*, 1997). These observations are reflected in their opposing activities on the SNAP-25, α-internexin and neuro-filament promoters, where Brn-3a activates transcription and Brn-3b represses transcription (Budhram-Mahadeo *et al.*, 1995b; Lakin *et al.*, 1995; Smith *et al.*, 1997). *Brn-3b* mutant animals are viable but have a 70% reduction in the number of ganglion cells in the retina probably due to inadequate proliferation of ganglion progenitor cells (Gan *et al.*, 1996). Mice carrying a targeted deletion of the *Brn-3c* gene are deaf and have impaired balance. *Brn-3a* null animals exhibit widespread loss of sensory and motor neurons, effecting neurotrophin and neurotrophin-receptor gene expression resulting in death shortly after birth (McEvilly *et al.*, 1996; Xiang *et al.*, 1997). Mutations in the human *Brn-3c* gene have also been found in a family with progressive deafness. These observations indicate the critical roles that the *Brn-3* genes play in sensory nerve development (reviewed in Latchman 1999).

5.2 TFs regulate neuronal survival during injury and disease

Neuronal cell death is a feature common to neurodevelopment, neurological diseases and brain injury (Oppenheim, 1991). Apoptosis, one form of cell death, is generally believed to play an important and major role in the neuronal cell death mechanisms (Linnik, 1996). Transcriptional regulation also plays an important role in regulating neuronal apoptosis observed during neurodevelopment, brain injury and disease (Hughes *et al.*, 1999). Neurological diseases such as Alzheimer's disease (AD) exhibit alterations to the extracellular environment (including amyloid plaques and immunological responses) sensed by neurons and glia. Signal transduction pathways that initiate at the cell surface are activated in response to changes of extracellular stimuli and dictate cellular responses. Downstream signals are received in the nucleus, resulting in altered activity of transcription factors. Cellular responses via regulated gene expression play a pivotal role in determining cell survival during times of neurologic stress, injury or disease (Ambron and Walters, 1996; Herdegen and Leah, 1998).

The role of TFs in regulating neuronal cell death has been demonstrated by studies using either cultured rat superior cervical ganglion (SCG) neurons that are nerve growth factor (NGF) dependent sympathetic neurons, PC12 cells, a rat pheochromocytoma cell line that in response to NGF differentiates into sympathetic neuron-like cells, or cerebellar granule neurons that are grown in the presence of high extracellular K^+ to support survival (Greene and Tischler, 1976; Batistatou *et al.*, 1992; DöMello *et al.*, 1993). When these cells are deprived of either NGF or placed in low K^+ they undergo apoptotic cell death (Kerr and Harmon, 1991). Inhibitors of RNA or protein synthesis (actinomycin D and cycloheximide) block cell death, indicating that transcriptional processes are required for apoptosis in these model systems. Staurosporine, a protein kinase inhibitor, induces apoptosis independent of transcription since it acts directly on intracellular kinases that function in apoptosis (Philpott *et al.*, 1996). These data indicate that cultured neurons normally possess the proteins required for inducing apoptosis, but signals that support cell survival inhibit activation of the apoptotic pathway. Under conditions that reduce neurotrophic factor support or alter extracellular signals required for cell survival, additional transcriptional events may be necessary for apoptosis to occur.

Induction of gene expression and transcriptional activity has been shown to occur in cultured neurons in response to numerous insults, including oxidative stress, addition of glutamate or cytokines, trophic factor removal or serum starvation (reviewed in Tong *et al.*, 1998). One example is the hydrogen peroxide-induced expression, phosphorylation and enhanced activity of the c-Jun and Fos family members with resulting transient increases in AP-1 transcriptional activity and induction of neuronal apoptosis (Estus *et al.*, 1994; Tong and Perez-Polo, 1996). The transcription factor NF-κB has been shown to function in antiapoptotic pathways in PC12 cells. Inhibition of NF-κB nuclear translocation by the addition of excess oligonucleotides containing the NF-κB binding site to PC12 decreases NF-κB activity and increases apoptosis (Taglialatela *et al.*, 1997).

Recent *in vitro* studies have shown that the addition of amyloid β (Aβ) peptide, a component of AD lesions, to neuronal cultures induces the hyperphosphorylation of pRb and activation of the E2F transcription factor (Giovanni *et al.*, 1999). Inhibition of E2F activity via expression of a dominant negative form of DP-1 (its binding partner required for transcriptional activity) reduces Aβ induced cell death. Within this system E2F-1 functions via a p53-independent, and bax- and caspase 3-dependent, manner in the death of cortical cultures (Giovanni *et al.*, 2000). pRb and E2F have also been shown to play an important role in neuronal death evoked by DNA damage (Park *et al.*, 2000).

Overexpression of E2F-1 has also been shown to induce aberrant entry into S-phase and apoptosis of postmitotic neurons (Azuma-Hara *et al.*, 1999). E2F-1 induces neuronal cell death via blocking antiapoptotic signaling pathways regulated by cell surface death receptors such as Fas, TNF receptor or p75 (Nagata, 1997; Phillips *et al.*, 1999). Recent data indicates that E2F-1 binds to the TRAF2 protein to inhibit activation of antiapoptotic signals including NF-κB (Phillips *et al.*, 1999). The ability to block signaling by cell surface death receptors and bind members of the TRAF family of cytokine signaling proteins indicates that E2F-1 localizes to the cytoplasm under appropriate conditions to induce neuronal cell death.

Transcriptional regulators, including hyperphosphorylation of pRb and subsequent E2F activation may also contribute to neuronal cell death mechanisms during human neurodegenerative diseases or after brain injury. AD is the most common neurodegenerative disease and results in neuronal loss in specific brain regions and dementia (Braak and Braak, 1991; Mirra *et al.*, 1991). While a neurological disease is a much more complex system than the cell culture systems described above, including multiple cell types and several types of neuropathological lesions, a central concept is the crucial role of neuronal apoptosis in the pathogenesis of the disease (Su *et al.*, 1994; Smale *et al.*, 1995). Numerous cell cycle and apoptotic proteins are alternatively expressed in AD, suggesting that transcription factors exhibit altered functional activity during AD (Cohen *et al.*, 1988; Lukiw *et al.*, 1996; Vincent *et al.*, 1996; MacGibbon *et al.*, 1997a,b; Busser *et al.*, 1998; Desjardins and Ledoux, 1998; Kitamura *et al.*, 1998; Callahan *et al.*, 1999; Raina *et al.*, 1999). *Figure* 3 shows extracellular stimuli that occur during AD and the downstream pathways leading to altered activity of transcription factors and gene expression.

Recent studies have demonstrated alterations in pRb phosphorylation and E2F-1 subcellular distribution in AD brain (Ranganathan *et al.*, 2000). Hyperphosphorylated pRb was observed in neuronal nuclei in AD brain and E2F-1 immunoreactivity was observed in the cytoplasm of both neurons and glial cells. Simian immunodeficiency virus (SIV) induced neurodegeneration also induces re-distribution of E2F-1 into the cytoplasm of neurons (Jordan-Sciutto *et al.*, 2000b). Therefore, both *in vivo* and *in vitro* data suggest that the activation or altered subcellular distribution of cell cycle regulated transcription factors contribute to neuronal cell death during neurodegenerative diseases.

NF-κB represents a centrally important transcription factor for brain signaling in both neurons and glia during development, inflammation, injury, neurological diseases, viral infection, synaptic transmission and neuronal plasticity (O'Neill and Kaltschmidt, 1997). In AD brain NF-κB is overexpressed in the nuclei and cytoplasm of neurons and glia surrounding and within Aβ-containing diffuse plaques (Kaltschmidt *et al.*, 1997). NF-κB can be activated by a variety of extracellular stimuli that occur during AD, including increased levels of cytokines (IL-1, TNF), neurotransmitters (glutamate), oxidative stress or neurotoxic peptides (Aβ) (Hunot *et al.*, 1997). NF-κB functions in both antiapoptotic and proapoptotic pathways and the outcome of its activation likely depends on the cellular context.

In addition to NF-κB, other transcription factors exhibit altered expression and functional activity during AD, including Krox-24 and FAC1 (Schoonover *et al.*, 1996; MacGibbon *et al.*, 1997; Jordan-Sciutto *et al.*, 1999). Krox-24 and FAC1 function during brain development and each is re-expressed during AD. Krox-24, also called Egr-1 or zif268, contains three zinc-finger motifs and functions as an immediate-early transcription factor induced by mitogenic stimuli that in the brain also may function to sustain differentiated cell types (Lemaire *et al.*, 1988; Gashler *et al.*, 1993). In AD brain Krox-24 exhibits increased immunoreactivity in pyramidal neurons of the CA1 region of the hippocampus, with corresponding increases in mRNA levels as shown by *in situ* hybridization (MacGibbon *et al.*, 1997). FAC1 was identified as a gene product that is expressed at high levels during brain development, at lower levels in adult and is re-expressed during AD (Bowser *et al.*, 1995; Schoonover *et al.*, 1996). FAC1 protein is redistributed into the cytoplasm of neurons and activated microglia during AD (Bowser and Reilly, 1998; Schoonover *et al.*,

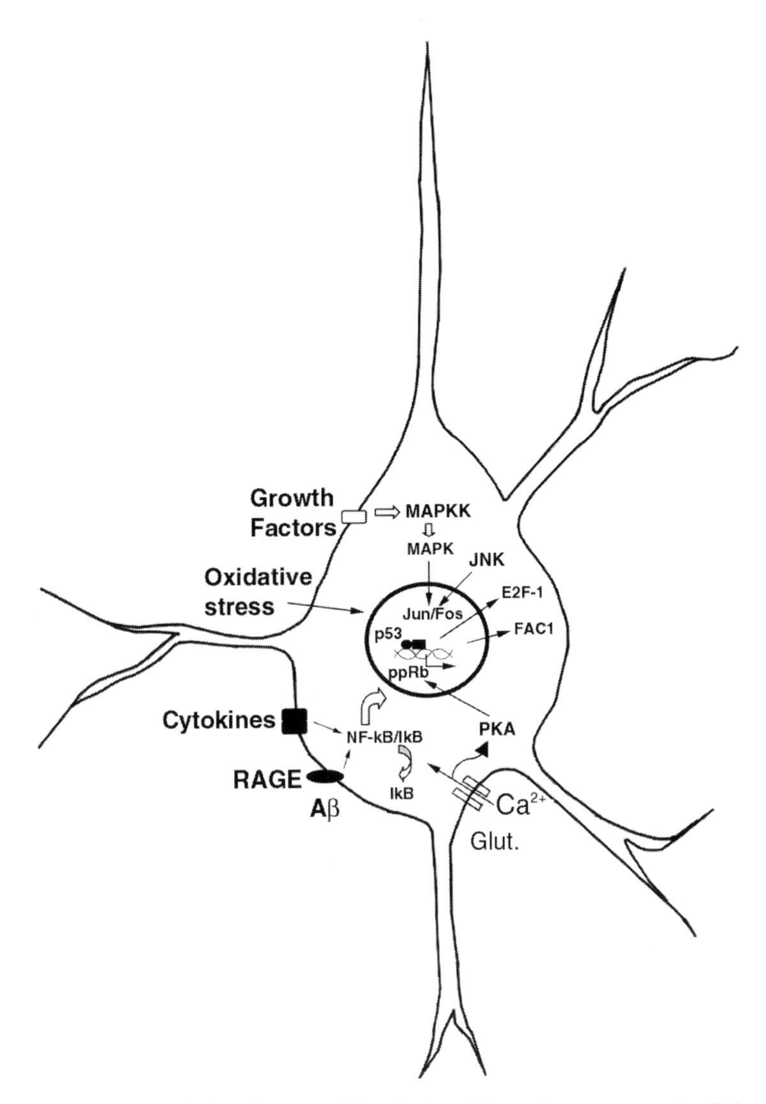

Figure 3. Altered transcription factor activity during AD regulates neuronal cell death. Numerous signal transduction pathways are activated during neurologic diseases that induce altered expression or activity of downstream transcription factors. The presence or absence of various growth factors and/or receptors induces alterations in intracellular kinase activities (MAPK, JNK, PKC, PKA) or altered intracellular levels of Ca^{2+} or cAMP. Binding of Aβ peptide to cell surface receptors (RAGE, p75) induces activation and nuclear translocation of NF-κB and hyperphosphorylation of pRb. Cytokine release from glial cells also induces activation of NF-κB and downstream changes in gene expression. Stimulation of glutamate receptors induces a rise in intracellular Ca^{2+} that also leads to activation of protein kinases, NF-κB activation and oxidative injury leads to DNA damage/p53 induction and mitochondrial damage with subsequent energy deficits. Additional signals culminate in the nucleus to induce altered transcriptional responses via Jun/Fos family members, p53, pRb/E2F-1, and FAC1. Abbreviations: MAPK, mitogen activated protein kinase; JNK, c-Jun N-terminal kinases; PKC, protein kinase C; PKA, protein kinase A; RAGE, receptor for advanced glycation end products.

1996). FAC1 binds a specific DNA sequence and function as a transcriptional repressor (Jordan-Sciutto *et al.*, 1999). FAC1 can also interact with other transcription factors (Jordan-Sciutto and Bowser, 1998; Jordan-Sciutto *et al.*, 2000), potentially regulating the overall function of numerous transcription factor protein complexes. The DNA binding sites for FAC1, Krox-24, and NF-κB have been localized in the promoter elements of genes relevant to AD, including APP and presenilin-1.

These results indicate that multiple transcription factors regulated by extracellular stimuli induce many of the changes associated with neurodegenerative diseases. Overall, the activity of numerous signal transduction pathways and the interplay of multiple transcription factors determine the transcriptional responses that regulate degeneration, regeneration and survival during a neurological disease.

6. Conclusions

We have provided an overview of TFs that function during development and disease of the central nervous system. While incomplete, this overview highlights many of the TFs that are critical to the correct development and proper maintenance of a functional nervous system. Further understanding of transcription factor function during development and disease will lead to novel therapies for developmental abnormalities and neurologic disease. It will be important to explore the possibility that inhibitors or activators of specific transcriptional complexes have therapeutic value in the treatment of neurodegenerative disorders or brain injury.

References

Acampora, D., Mazan, S., Lallemand, Y., Avantaggiato, V., Maury, M., Simeone, A. and Brulet, P. (1995) Forebrain and midbrain regions are deleted in Otx-2–/– mutants due to a defective anterior NE specification during gastrulation. *Development* **121**: 3279–3290.

Acampora, D., Mazan, S., Avantaggiato, V., Barone, P., Tuorto, F., Lallemand, Y., Brulet, P. and Simeone, A. (1996) Epilepsy and brain abnormalities in mice lacking the Otx1 gene. *Nat Genet* **14**: 218–222.

Acampora, D., Avantaggiato, V., Tuorto, F. and Simeone, A. (1997) Genetic control of brain morphogenesis through Otx gene dosage requirement. *Development* **124**: 3639–3650.

Acampora, D., Avantaggiato, V., Tuorto, F., Briata, P., Corte, G. and Simeone, A. (1998) Visceral endoderm-restricted translation of Otx1 mediates recovery of Otx-2. *Development* **125**: 5091–5104.

Akins, P.T., Liu, P.K. and Hsu, C.Y. (1996) Immediate early gene expression in response to cerebral ischemia: friend or foe? *Stroke* **27**: 1682–1687.

Alvarez-Bolado, G., Rosenfeld, M.G. and Swanson, L.W. (1995) Model of forebrain regionalization based on spatiotemporal patterns of POU-III homeobox gene expression, birthdates, and morphological features. *J. Comp. Neurol.* **355**: 237–295.

Ambron, R.T. and Walters, E.T. (1996) Priming events and retrograde injury signals. *Mol. Neurobiol.* **13**: 61–79.

Anderson, D.J. and Jan, Y.N. (1997) The determination of the neuronal phenotype. In *Molecular and Cellular Approaches to Neural Development*, W. M. Cowan, T. M. Jessell and S. L. Zipursky, eds. (Oxford, UK: Oxford Univ. Press), pp. 26–63.

Anderson, A.J., Cummings, B.J. and Cotman, C.W. (1994) Increased immunoreactivity for Jun- and Fos-related proteins in Alzheimer's disease: association with pathology. *Exp. Neurol.* **125**: 286–295.

Anderson, A.J., Su, J.H. and Cotman, C.W. (1996) DNA damage and apoptosis in Alzheimer's disease: colocalization with c-Jun immunoreactivity, relationship to brain area, and effect of postmortem delay. *J. Neurosci.* **16**: 1710–1719.

Ang S.L.(1996) The brain organization. *Nature* **380(6569)**: 25–27.

Ang, S.L. and Rossant, J. (1994) HNF-3β is essential for node and notochord formation in mouse development. *Cell* **78**: 561–574.

Azuma-Hara, M., Taniura, H., Uetsuki, T., Niinobe, M. and Yoshikawa, K. (1999) Regulation and deregulation of E2F1 in postmitotic neurons differentiated from embryonal carcinoma P19 cells. *Exp. Cell Res.* **251**: 442–451.

Bang, A.G. and Goulding, M.D. (1996) Regulation of vertebrate neural cell fate by transcription factors. *Curr. Opin. Neurobiol.* **6**: 25–32.

Barth, K.A. and Wilson, S.W. (1995) Expression of zebrafish nkx2.2 is influenced by sonic hedgehog/vertebrate hedgehog-1 and demarcates a zone of neuronal differentiation in the embryonic forebrain. *Development* **121**: 1755–1768.

Batistatou, A., Volonte, C. and Greene, L.A. (1992) Nerve growth factor employs multiple pathways to induce primary response genes in PC12 cells. *Mol. Biol. Cell* **3**: 363–371.

Beddington, R.S., Rashbass, P. and Wilson, V. (1992) Brachyury – a gene affecting mouse gastrulation and early organogenesis. *Dev. Suppl.* **1992**: 157–165.

Bell, E., Wingate, R.J. and Lumsden, A. (1999) Homeotic transformation of rhombomere identity after localized Hoxb1 misexpression. *Science* **284**: 2168–2171.

Bhide, P.G. (1996) Cell cycle kinetics in the embryonic mouse corpus striatum. *J. Comp. Neurol.* **374**: 506–522.

Bohr, P.J. and Anson, R.M. (1995) DNA damage, mutation and fine structure DNA repair in aging. *Mutat. Res.* **338**: 25–34.

Bookstein, R. and Lee, W.H. (1991) Molecular genetics of the retinoblastoma suppressor gene. *CRC Crit. Rev. Oncogenesis* **2**: 211–227.

Bowser, R. and Reilly, S. (1998) Expression of FAC1 in activated microglia during Alzheimer's disease. *Neurosci. Lett.* **253**: 163–166.

Bowser, R., Giambrone, A. and Davies, P. (1995) FAC1, a novel gene identified with the monoclonal antibody Alz50, is developmentally regulated in human brain. *Develop. Neurosci.* **17**: 20–37.

Braak, H. and Braak, E. (1991) Neuropathological stageing of Alzheimer-related changes. *Acta Neuropathol.* **82**: 239–259.

Brehm, A., Miska, E.A., McCance, D.J., Reid, J.L., Bannister, A.J. and Kouzarides, T. (1998) Retinoblastoma protein recruits histone deacetylase to repress transcription. *Nature* **391**: 597–601.

Budhram-Mahadeo, V., Morris, P.J., Lakin, N.D., Theil, T., Ching, G.Y., Lillycrop, K.A., Moroy, T., Liem, R.K. and Latchman, D.S. (1995a) Activation of the α-internexin promoter by the Brn-3a transcription factor is dependent on the N-terminal region of the protein. *J. Biol. Chem.* **270**: 2853–2858.

Budhram-Mahadeo, V., Lillycrop, K.A. and Latchman, D.S. (1995b) The levels of the antagonistic POU family transcription factors Brn-3a and Brn-3b in neuronal cells are regulated in opposite directions by serum growth factors. *Neurosci. Lett.* **185**: 48–51.

Budhram-Mahadeo, V., Morris, P.J., Smith, M.D., Midgley, C.A., Boxer, L.M. and Latchman, D.S. (1999) p53 suppresses the activation of the bcl-2 promoter by the brn-3a POU family transcription factor. *J. Biol. Chem.* **274**: 15237–15244.

Bulfone, A., Puelles, L., Porteus, M.H., Frohman, M.A., Martin, G.R. and Rubenstein, J.L. (1993) Spatially restricted expression of Dlx-1, Dlx-2 (Tes-1), Gbx-2, and Wnt-3 in the embryonic day 12.5 mouse forebrain defines potential transverse and longitudinal segmental boundaries. *J. Neurosci.* **13**: 3155–3172.

Bulfone, A., Wang, F., Hevner, R., Anderson, S., Cutforth, T., Chen, S., Meneses, J., Pedersen, R., Axel, R. and Rubenstein, J.L. (1998) An olfactory sensory map develops in the absence of normal projection neurons or GABAergic interneurons. *Neuron.* **21**: 1273–1282.

Busser, J., Geldmacher, D.S. and Herrup, K. (1998) Ectopic cell cycle proteins predict the sites of neuronal cell death in Alzheimer's disease brain. *J. Neurosci.* **18**: 2801–2807.

Callahan, L.M., Vaules, W.A. and Coleman, P.D. (1999) Quantitative decrease in synaptophysin message expression and increase in cathepsin D message expression in Alzheimer disease neurons containing neurofibrillary tangles. *J. Neuropath. Exp. Neurol.* **58**: 275–287.

Carpenter, D., Stone, D.M., Brush, J., Ryan, A., Armanini, M., Frantz, G., Rosenthal, A. and de Sauvage, F.J. (1998) Characterization of two patched receptors for the vertebrate hedgehog protein family. *Proc. Natl Acad. Sci. USA* **95**: 13630–13634.

Carpenter, E.M., Goddard, J.M., Chisaka, O., Manley, N.R. and Capecchi, M.R. (1993) Loss of Hox-A1 (Hox-1.6) function results in the reorganization of the murine hindbrain. *Development* **118**: 1063–1075.

Casarosa, S., Fode, C. and Guillemot, F. (1999) Mash1 regulates neurogenesis in the ventral telencephalon. *Development* **126**: 525–534.

Chisaka, O. and Capecchi, M.R. (1991) Regionally restricted developmental defects resulting from targeted disruption of the mouse homeobox gene hox-1.5. *Nature* **350**: 473–479.

Chisaka, O., Musci, T.S. and Capecchi, M.R. (1992) Developmental defects of the ear, cranial nerves and hindbrain resulting from targeted disruption of the mouse homeobox gene Hox-1.6. *Nature* **355**: 516–520.

Cohen, S.M. and Jurgens, G. (1990) Mediation of Drosophila head development by gap-like segmentation genes. *Nature* **346**: 482–485.

Cohen, M.L., Golde, T.E., Usiak, M.F., Younkin, L.H. and Younkin, S.G. (1988) In situ hybridization of nucleus basalis neurons shows increased beta-amyloid mRNA in Alzheimer disease. *Proc. Natl Acad. Sci. USA* **85**: 1227–1231.

Dattani, M.T., Martinez-Barbera, J.P., Thomas, P.Q., Brickman, J.M., Gupta, R., Martensson, I.L., Toresson, H., Fox, M., Wales, J.K., Hindmarsh, P.C., Krauss, S., Beddington, R.S. and Robinson, I.C. (1998) Mutations in the homeobox gene HESX1/Hesx1 associated with septo-optic dysplasia in human and mouse. *Nat. Genet.* **19**: 125–133.

Davis, C.A. and Joyner, A.L. (1988) Expression patterns of the homeo box-containing genes En-1 and En-2 and the proto-oncogene int-1 diverge during mouse development. *Genes Dev.* **2**: 1736–1744.

de la Monte, S.M., Sohn, Y.K. and Wands, J.R. (1997) Correlates of p53- and Fas (CD95)-mediated apoptosis in Alzheimer's disease. *J. Neurol. Sci.* **152**: 73–83.

Desjardins, P. and Ledoux, S. (1998) Expression of ced-3 and ced-9 homologs in Alzheimer's disease cerebral cortex. *Neurosci. Lett.* **244**: 69–72.

Dickinson, M.E., Selleck, M.A., McMahon, A.P. and Bronner-Fraser, M. (1995) Dorsalization of the neural tube by the non-neural ectoderm. *Development* **121**: 2099–2106.

D'Mello, S.R., Galli, C., Ciotti, T. and Callissano, P. (1993) Induction of apoptosis in cerebellar granule neurons by low potassium: inhibition of death by insulin-like growth factor 1 and cAMP. *Proc. Natl Acad. Sci. USA* **90**: 10989–10993.

Dressler, G.R., Deutsch, U., Chowdhury, K., Nornes, H.O. and Gruss, P. (1990) Pax2, a new murine paired-box-containing gene and its expression in the developing excretory system. *Development* **109**: 787–795.

el-Deiry, W.S., Tokino, T., Velculescu, V.E., Levy, D.B., Parsons, R., Trent, J.M., Lin, D., Mercer, W.E., Kinzler, K.W. and Vogelstein, B. (1993) WAF1, a potential mediator of p53 tumor suppression. *Cell* **75**: 817–825.

Ellmeier, W., Aguzzi, A., Kleiner, E., Kurzbauer, R. and Weith, A. (1992) Mutually exclusive expression of a helix-loop-helix gene and N-myc in human neuroblastomas and in normal development. *EMBO J.* **11**: 2563–2571.

Epstein, J.A., Glaser, T., Cai, J., Jepeal, L., Walton, D.S. and Maas, R.L. (1994) Two independent and interactive DNA-binding subdomains of the Pax6 paired domain are regulated by alternative splicing. *Genes Dev* **8**: 2022–2034.

Estus, S., Zaks, W., Freeman, R., Gruda, M., Bravo, R. and EM Johnson, J. (1994) Altered gene expression in neurons during programmed cell death: Identification of c-jun as necessary for neuronal apoptosis. *J. Cell Biol.* **127**: 1717–1727.

Fode, C., Ma, Q., Casarosa, S., Ang, S.L., Anderson, D.J. and Guillemot, F. (2000) A role for neural determination genes in specifying the dorsoventral identity of telencephalic neurons. *Genes Dev.* **14**: 67–80.

Ferreira, R., Magnaghi-Jaulin, L., Robin, P., Harel-Bellan, A. and Trouche, D. (1998) The three members of the pocket proteins family share the ability to repress E2F activity through recruitment of a histone deacetylase. *Proc. Natl Acad. Sci. USA* **95**: 10493–10498.

Figdor, M.C. and Stern, C.D. (1993) Segmental organisation of embryonic diencephalon. *Nature* **363**: 630–634.

Finkelstein, R. and Perrimon, N. (1990) The orthodenticle gene is regulated by bicoid and torso and specifies Drosophila head development. *Nature* **346**: 485–488.

Fishell, G. (1999) BMPs: time to murder and create? *Nature Neurosci.* **2**: 301–303.

Frantz, G.D., Weimann, J.M., Levin, M.E. and McConnell, S.K. (1994) Otx1 and Otx2 define layers and regions in developing cerebral cortex and cerebellum. *J. Neurosci.* **14**: 5725–5740.

Fraser, S., Keynes, R. and Lumsden, A. (1990) Segmentation in the chick embryo hindbrain is defined by cell lineage restrictions. *Nature* **344**: 431–435.

Freeman, R.S., Estus, S. and Johnson Jr., E.M. (1994) Analysis of cell cycle-related gene expression in postmitotic neurons: selective induction of cyclin D1 during programmed cell death. *Neuron.* 12: 343–355.

Freund, C., Horsford, D.J. and McInnes, R.R. (1996) Transcription factor genes and the developing eye: a genetic perspective *Hum. Mol. Genet.* 5 Spec No: 1471–1488.

Fung, Y.K.T., Murphree, A.L., T'Ang, A., Qian, J., Hinrichs, S.H. and Benedict, W.F. (1987) Structural evidence for the authenticity of the human retinoblastoma gene. *Science* **236**: 1657–1661.

Furuta, Y., Piston, D.W. and Hogan, B.L. (1997) Bone morphogenetic proteins (BMPs) as regulators of dorsal forebrain development. *Development* **124**: 2203–2212.

Gaiddon, C., de Tapia, M. and Loeffler, J.P. (1999) The tissue-specific transcription factor Pit-1/GHF-1 binds to the c-fos serum response element and activates c-fos transcription. *Mol. Endocrinol.* **13**: 742–751.

Gan, L., Xiang, M., Zhou, L., Wagner, D.S., Klein, W.H. and Nathans, J. (1996) POU domain factor Brn-3b is required for the development of a large set of retinal ganglion cells. *Proc. Natl Acad. Sci. USA* **93**: 3920–3925.

Gashler, A.L., Swaminathan, S. and Sukhatme, V.P. (1993) A novel repression module, an extensive activation domain, and a bipartite nuclear localization signal defined in the immediate-early transcription factor Egr-1. *Mol. Cell Biol.* **13**: 4556–4571.

Gaunt, S.J., Blum, M. and De Robertis, E.M. (1993) Expression of the mouse goosecoid gene during mid-embryogenesis may mark mesenchymal cell lineages in the developing head, limbs and body wall. *Development* **117**: 769–778.

Giovanni, A., Wirtz-Brugger, F., Keramaris, E., Slack, R. and Park, D.S. (1999) Involvement of cell cycle elements, cyclin-dependent kinases, pRb, and E2F-DP, in B-amyloid-induced neuronal death. *J. Biol. Chem.* **274**: 19011–19016.

Giovanni, A., Keramaris, E., Morris, E.J., Hou, S.T., O'Hare, M., Dyson, N., Robertson, G.S., Slack, R.S. and Park, D.S. (2000) E2F1 mediates death of B-amyloid-treated cortical neurons in a manner independent of p53 and dependent on Bax and caspase 3. *J. Biol. Chem.* **275**: 11553–11560.

Glaser, T., Walton, D.S. and Maas, D.L. (1992) Genomic structure, evolutionary conservation and aniridia mutations in the human PAX6 gene. *Nat. Genet.* **2**: 232–239.

Goddard, J.M., Rossel, M., Manley, N.R. and Capecchi, M.R. (1996) Mice with targeted disruption of Hoxb-1 fail to form the motor nucleus of the VIIth nerve *Development* **122**: 3217–3228.

Goulding, M.D., Lumsden, A. and Gruss, P. (1993) Signals from the notochord and floor plate regulate the region-specific expression of two Pax genes in the developing spinal cord. *Development* **117**: 1001–1016.

Greene, L.A. and Tischler, A.S. (1976) Establishment of a noradrenergic clonal line of rat adrenal pheochromocytoma cells which respond to nerve growth factor. *J. Biol. Chem.* **260**: 14101–14110.

Gruss, P. and Walther, C. (1992) Pax in development. *Cell* **69**: 719–722.

Guthrie, S. and Lumsden, A. (1991) Formation and regeneration of rhombomere boundaries in the developing chick hindbrain. *Development* **112**: 221–229.

Guthrie, S., Prince, V. and Lumsden, A. (1993) Selective dispersal of avian rhombomere cells in orthotopic and heterotopic grafts. *Development* **118**: 527–538.

Hallonet, M., Hollemann, T., Wehr, R., Jenkins, N.A., Copeland, N.G., Pieler, T. and Gruss, P. (1998) Vax1 is a novel homeobox-containing gene expressed in the developing anterior ventral forebrain. *Development* **125**: 2599–2610.

Ham, J., Babij, C., Whitfield, J., Pfarr, C.M., Lallemand, D., Yani, M. and Rubin, L.L. (1995) A c-Jun dominant negative mutant protects sympathetic neurons against programmed cell death. *Neuron.* **14**: 927–939.

Hanson, I. and Van Heyningen, V. (1995) Pax6: more than meets the eye. *Trends Genet.* **11**: 268–272.

Hatini, V., Tao, W. and Lai, E. (1994) Expression of winged helix genes, BF-1 and BF-2, define adjacent domains within the developing forebrain and retina. *J. Neurobiol.* **25**: 1293–1309.

Helms, A.W. and Johnson, J.E. (1998) Progenitors of dorsal commissural interneurons are defined by Math1 expression. *Development* **125**: 919–928.

Hemmati-Brivanlou, A., Kelly, O.G. and Melton, D.A. (1994) Follistatin, an antagonist of activin, is expressed in the Spemann organizer and displays direct neuralizing activity. *Cell* **77**: 283–295.

Herdegen, T. and Leah, J.D. (1998) Inducible and constitutive transcription factors in the mammalian nervous system: control of gene expression by Jun, Fos, and Krox, and CREB/ATF proteins. *Brain Res. Rev.* **28**: 370–490.

Herr, W., Sturm, R.A., Clerc, R.G., Corcoran, L.M., Baltimore, D., Sharp, P.A., Ingraham, H.A., Rosenfeld, M.G. and Finney, M. (1988) The POU domain: a large conserved region in the mammalian pit-1, oct-1, oct-2, and *Caenorhabditis elegans* unc-86 gene products. *Genes Dev.* **2**: 1513–1516.

Hill, R.E., Favor, J., Hogan, B.L., Ton, C.C., Saunders, G.F., Hanson, I.M., Prosser, J., Jordan, T., Hastie, N.D. and van Heyningen, V. (1991) Mouse small eye results from mutations in a paired-like homeobox-containing gene. *Nature* 19–26; **354**: 522–525.

Hogan, B.L., Horsburgh, G., Cohen, J., Hetherington, C.M., Fisher, G. and Lyon, M.F. (1986) Small eyes (Sey): a homozygous lethal mutation on chromosome 2 which affects the differentiation of both lens and nasal placodes in the mouse. *J. Embryol. Exp. Morphol.* **97**: 95–110.

Hsu, D.R., Economides, A.N., Wang, X., Eimon, P.M. and Harland, R.M. (1998) The Xenopus dorsalizing factor Gremlin identifies a novel family of secreted proteins that antagonize BMP activities. *Mol. Cell* **1**: 673–683.

Hughes, P.E., Alexi, T., Walton, M., Williams, C.E., Dragunow, M., Clark, R.G. and Gluckman, P.D. (1999) Activity and injury-dependent expression of inducible transcription factors, growth factors and apoptosis-related genes within the central nervous system. *Prog. Neurobiol.* **57**: 421–450.

Hunot, S., Brugg, B., Ricard, D., Michel, P.P., Muriel, M. P., Ruberg, M., Faucheux, B.A., Agid, Y. and Hirsch, E.C. (1997) Nuclear translocation of NF-kB is increased in dopaminergic neurons of patients with Parkinson disease. *Proc. Natl Acad. Sci. USA* **94**: 7531–7536.

Ishibashi, M., Moriyoshi, K., Sasai, Y., Shiota, K., Nakanishi, S. and Kageyama, R. (1994) Persistent expression of helix-loop-helix factor HES-1 prevents mammalian neural differentiation in the central nervous system. *EMBO J.* **13**: 1799–1805.

Ishibashi, M., Ang, S.L., Shiota, K., Nakanishi, S., Kageyama, R. and Guillemot, F. (1995) Targeted disruption of mammalian hairy and Enhancer of split homolog-1 (HES-1) leads to up-regulation of neural helix-loop-helix factors, premature neurogenesis, and severe neural tube defects. *Genes Dev.* **9**: 3136–3148.

Issack, P.S. and Ziff, E.B. (1998) Genetic elements regulating HES-1 induction in Wnt-1-transformed PC12 cells. *Cell Growth Differ.* **9**: 827–836.

Jarriault, S., Le Bail, O., Hirsinger, E., Pourquie, O., Logeat, F., Strong, C.F., Brou, C., Seidah, N.G. and Isra, I.A. (1998) Delta-1 activation of notch-1 signaling results in HES-1 transactivation. *Mol. Cell. Biol.* **18**: 7423–7431.

Jordan, J., Galindo, M.F., Prehn, J.H.M., Weichselbaum, R.R., Beckett, M., Ghadge, G.D., Roos, R.P., Leiden, J.M. and Miller, R.J. (1997) p53 expression induces apoptosis in hippocampal pyramidal neuron cultures. *J. Neurosci.* **17**: 1397–1405.

Jordan-Sciutto, K.L. and Bowser, R. (1998) Alzheimer's disease and brain development: Common molecular pathways. *Frontiers Bioscience* **3**: 100–112.

Jordan-Sciutto, K.L., Dragich, J.M. and Bowser, R. (1999) DNA binding activity of the fetal Alz-50 clone 1 (FAC1) protein is enhanced by phosphorylation. *Biochem. Biophys. Res. Comm.* **260**: 785–789.

Jordan-Sciutto, K.L., Dragich, J.M., Caltagarone, J.M., Hall, D.J. and Bowser, R. (2000a) Fetal Alz-50 clone1 (FAC1) protein interacts with the zinc finger protein (ZF87/MAZ) and alters its transcriptional activity. *Biochemistry* **39**: 3206–3215.

Jordan-Sciutto, K.L., Wang, G., Murphy-Corb, M. and Wiley, C.A. (2000b) Induction of cell-cycle regulators in simian immunodeficiency virus encephalitis. *Am. J. Pathol.* **157**: 497–507.

Joyner, A.L., Herrup, K., Auerbach, B.A., Davis, C.A. and Rossant, J. (1991) Subtle cerebeller phenotype in mice homozygous for a targeted deletion of the En-2 homeobox. *Science* **251**: 1239–1243.

Kaestner, K.H., Monaghan, A.P., Kern, H., Ang, S.L., Weitz, S., Lichter, P. and Schutz, G. (1995) The mouse fkh-2 gene. Implications for notochord, foregut, and midbrain regionalization. *J. Biol. Chem.* **270**: 30029–30035.

Kageyama, R., Ishibashi, M., Takebayashi, K. and Tomita, K. (1997) bHLH transcription factors and mammalian neuronal differentiation. *Int. J. Biochem. Cell. Biol.* **29**: 1389–1399.

Kaltschmidt, B., Uherek, M., Volk, B., Baeuerle, P. and Kaltschmidt, C. (1997) Transcription factor NF-kB is activated in primary neurons by amyloid B peptides and in neurons surrounding early plaques from patients with Alzheimer's disease. *Proc. Natl Acad. Sci. USA* **94**: 2642–2647.

Karlsson, O., Thor, S., Norbert, T., Ohlsson, H. and Edlund, T. (1990) Insulin gene enhancer binding protein Isl-1 is a member of a novel class of proteins containing both a homeo and a Cys-His domain. *Nature* **344**: 879–882.

Kastan, M.B., Onyekwere, O., Sidransky, D., Volgelstein, B. and Craig, R.W. (1991) Participation of p53 protein in the cellular response to DNA damage. *Cancer Res.* **51**: 6304–6311.

Kerr, J.F.R. and Harmon, B.V. (1991) Definition and incidence of apoptosis: An historical perspective. In *Apoptosis: the molecular basis of cell death*, L.D. Tomei and F.O. Cope, eds. (New York: Cold Spring Harbor Laboratory Press), pp. 5–30.

Keynes R. and Krumlauf, R. (1994) Hox genes and regionalization of the nervous system. *Annu. Rev. Neurosci.* **17**: 109–132.

Kitamura, Y., Shimohama, S., Ota, T., Matsuoka, Y., Nomura, Y. and Taniguchi, T. (1997) Alteration of transcription factors NF-kB and STAT1 in Alzheimer's disease brains. *Neurosci. Lett.* **237**: 17–20.

Kitamura, Y., Shimohama, S., Kamoshima, W., Ota, T., Matsuoka, Y., Nomura, Y., Smith, M.A., Perry, G., Whitehouse, P.J. and Taniguchi, T. (1998) Alteration of proteins regulating apoptosis, Bcl-2, Bcl-x, Bax, Bad, ICH-1 and CPP32, in Alzheimer's disease. *Brain Res.* **780**: 260–269.

Knecht, A.K., Good, P.J., Dawid, I.B. and Harland, R.M. (1995) Dorsal-ventral patterning and differentiation of noggin-induced neural tissue in the absence of mesoderm. *Development* **121**: 1927–1935.

Krumlauf, R. (1994) Hox genes in vertebrate development. *Cell* **78(2)**: 191–201.

Lakin, N.D., Morris, P.J., Theil, T., Sato, T.N., Moroy, T., Wilson, M.C. and Latchman, D.S. (1995) Regulation of neurite outgrowth and SNAP-25 gene expression by the Brn-3a transcription factor. *J. Biol. Chem.* **270**: 15858–15863.

Latchman, D.S. (1996) Transcription-factor mutations and disease. *New Engl. J. Med.* **334**: 28–33.

Latchman, D.S. (1999) POU family transcription factors in the nervous system. *J. Cell. Physiol.* **179**: 126–133.

Lee, E.H., Chang, C.Y., Hu, N., Wang, Y.C.J., Lai, C.C., Herrup, K., Lee, W.H. and Bradley, A. (1992) Mice deficient for Rb are nonviable and show defects in neurogenesis and haematopoiesis. *Nature* **359**: 288–294.

Lee, J.E. (1997) Basic helix-loop-helix genes in neural development. *Curr. Opin. Neurobiol.* **7**: 13–20.

Lee, K.J. and Jessell, T.M. (1999) The specification of dorsal cell fates in the vertebrate central nervous system. In *Ann. Rev. Neurosci.*, W.M. Cowan, E.M. Shooter, C.F. Stevens and R.F. Thompson, eds. (Palo Alto: Annual Reviews), pp. 261–294.

Lee, K.J., Dietrich, P. and Jessell, T.M. (2000) Genetic ablation reveals that the roof plate is essential for dorsal interneuron specification. *Nature* **403**: 734–740.

Lee, W.H., Bookstein, R., Hong, F., Young, L.J., Shew, J.Y. and Lee, E.H. (1987) Human retinoblastoma susceptibility gene: cloning, identification, and sequence. *Science* **235**: 1394–1399.

Lemaire, P. and Kodjabachian, L. (1996) The vertebrate organizer: structure and molecules. *Trends Genet.* **12**: 525–531.

Lemaire, P., Revelant, O., Bravo, R. and Charnay, P. (1988) Two mouse genes encoding potential transcription factors with identical DNA-binding domains are activated by growth factors in cultured cells. *Proc. Natl Acad. Sci. USA* **85**: 4691–4695.

Liem, K.F., Jr., Tremml, G. and Jessell, T.M. (1997) A role for the roof plate and its resident TGFβ-related proteins in neuronal patterning in the dorsal spinal cord. *Cell* **91**: 127–138.

Linnik, M.D. (1996) Role of apoptosis in acute neurodegenerative disorders. *Restor. Neurol. Neurosci.* **9**: 219–225.

Lillycrop, K.A., Budrahan, V.S., Lakin, N.D., Terrenghi, G., Wood, J.N., Polak, J.M. and Latchman, D.S. (1992) A novel POU family transcription factor is closely related to Brn-3 but has a distinct expression pattern in neuronal cells. *Nucleic Acids Res.* **20**: 5093–5096.

Lin, J.H., Saito, T., Anderson, D.J., Lance-Jones, C., Jessell, T.M. and Arber, S. (1998) Functionally related motor neuron pool and muscle sensory afferent subtypes defined by coordinate ETS gene expression. *Cell* **95**: 393–407.

Lufkin, T., Dierich, A., LeMeur, M., Mark, M. and Chambon, P. (1991) Disruption of the Hox-1.6 homeobox gene results in defects in a region corresponding to its rostral domain of expression. *Cell* **66**: 1105–1119.

Lukiw, W.J., Rogaev, E.I. and Bazan, N.G. (1996) Synaptic and cytoskeletal RNA message levels in sporadic Alzheimer neocortex. *Alzheimer's Res.* **2**: 221–228.

Lumsden, A. (1990) The cellular basis of segmentation in the developing hindbrain. *Trends Neurosci.* **13**: 329–335.

Lumsden, A. and Keynes, R. (1989) Segmental patterns of neuronal development in the chick hindbrain. *Nature* **337**: 424–428.

Ma, Q., Kintner, C. and Anderson, D.J. (1996) Identification of neurogenin, a vertebrate neuronal determination gene. *Cell* **87**: 43–52.

MacGibbon, G.A., Lawlor, P.A., Sirimanne, E.S., Walton, M.R., Connor, B., Young, D., Williams, C., Gluckman, P., Faull, R.L.M., Hughes, P. and Dragunow, M. (1997a) Bax expression in mammalian neurons undergoing apoptosis, and in Alzheimer's disease hippocampus. *Brain Res.* **750**: 223–234.

MacGibbon, G.A., Lawlor, P.A., Walton, M., Sirimanne, E., Faull, R.M.L., Synek, B., Mee, E., Conner, B. and Dragunow, M. (1997b) Expression of Fos, Jun, and Krox family proteins in Alzheimer's disease. *Exp. Neurol.* **147**: 316–332.

Maconochie, M., Nonchev, S., Morrison, A. and Krumlauf, R. (1996) Paralogous Hox genes: function and regulation. *Annu. Rev. Genet.* **30**: 529–556.

Magnaghi-Jaulin, L., Groisman, R., Naguibneva, I., Robin, P., Lorain, S., LeVillain, J.P., Troalen, F., Trouche, D. and Harel-Bellan, A. (1998) Retinoblastoma protein represses transcription by recruiting a histone deacetylase. *Nature* **391**: 601–605.

Matise, M.P., Epstein, D.J., Park, H.L., Platt, K.A. and Joyner, A.L. (1998) *Gli-2* is required for induction of floor plate and adjacent cells, but not most ventral neurons in the mouse central nervous system. *Development* **125**: 2759–2770.

Matsuo, I., Kuratani, S., Kimura, C., Takeda, N. and Aizawa, S. (1995) Mouse Otx2 functions in the formation and patterning of rostral head. *Genes Dev.* **9**: 2646–2658.

McConnell, S.K. (1995) Strategies for the generation of neuronal diversity in the developing central nervous system. *J. Neurosci.* **15**: 6987–6998.

McEvilly, R.J., Erkman, L., Luo, L., Sawchenko, P.E., Ryan, A.F. and Rosenfeld, M.G. (1996) Requirement for Brn-3.0 in differentiation and survival of sensory and motor neurons. *Nature* **384**: 574–577.

Mirra, S.S., Heyman, A. and McKeel, D. (1991) The Consortium to Establish Registry for Alzheimer's Disease (CERAD) II. Standardization of the neuropathological assessment of Alzheimer's disease. *Neurology* **41**: 479–486.

Monaghan A.P., Kaestner, K.H., Grau, E. and Schutz, G. (1993) Postimplantation expression patterns indicate a role for the mouse winged helix/HNF-3 α, β and γ genes in determination of the definitive endoderm, chordamesoderm and NE. *Development* **119**: 567–578.

Monaghan, A.P., Grau, E., Bock, D. and Schutz, G. (1995) The mouse homolog of the orphan nuclear receptor tailless is expressed in the developing forebrain. *Development* **121**: 839–853.

Monaghan, A.P., Bock, D., Gass, P., Schwager, A., Wolfer, D.P., Lipp, H.P. and Schutz, G. (1997) Defective limbic system in mice lacking the tailless gene. *Nature* **390**: 515–517.

Morin, X., Cremer, H., Hirsch, M.R, Kapur, R.P, Goridis, C. and Brunet J.F. (1997) Defects in sensory and autonomic ganglia and absence of locus coeruleus in mice deficient for the homeobox gene Phox2a. *Neuron* **18**: 411–423.

Motoyama, J., Heng, H., Crackower, M.A., Takabatake, T., Takeshima, K., Tsui, L.C. and Hui, C. (1998) Overlapping and non-overlapping Ptch2 expression with Shh during mouse embryogenesis. *Mech. Dev.* **78**: 81–84.

Mulligan, G. and Jacks, T. (1998) The retinoblastoma gene family: cousins with overlapping interests. *Trends Genet.* **14**: 223–229.

Murphy, P., Davidson, D.R. and Hill, R.E. (1989) Segment-specific expression of a homoeobox-containing gene in the mouse hindbrain. *Nature* **341**: 156–159.

Nagata, S. (1997) Apoptosis by death factor. *Cell* **88**: 355–365.

Nakata, K., Nagai, T., Aruga, J. and Mikoshiba, K. (1997) Xenopus Zic3, a primary regulator both in neural and neural crest development. *Proc. Natl Acad. Sci. USA* **94**: 11980–11985.

Nardelli, J., Thiesson, D., Fujiwara, Y., Tsai, F.Y. and Orkin, S.H. (1999) Expression and genetic interaction of transcription factors GATA-2 and GATA-3 during development of the mouse central nervous system. *Dev. Biol.* **210**: 305–321.

Niehrs, C., Keller, R., Cho, K.W., De Robertis, E.M. (1993) The homeobox gene goosecoid controls cell migration in Xenopus embryos. *Cell* **72**: 491–503.

Nieuwkoop, P.D. (1997) Short historical survey of pattern formation in the endo-mesoderm and the neural anlage in the vertebrates: the role of vertical and planar inductive actions. *Cell Mol. Life Sci.* **53**: 305–318.

Niimi, K., Harada, I., Kusaka, Y. and Kishi, S. (1962) The ontogenetic development of the diencephalon of the mouse. *Tokushima J. Exp. Med.* **8**: 203–238.

Nonchev, S., Vesque, C., Maconochie, M., Seitanidou, T., Ariza-McNaughton, L., Frain, M., Marshall, H., Sham, M.H., Krumlauf, R. and Charnay, P. (1996) Segmental expression of Hoxa-2 in the hindbrain is directly regulated by Krox-20. *Development* **122**: 543–554.

O'Neill, L.A.J. and Kaltschmidt, C. (1997) NF-kB: a crucial transcription factor for glial and neuronal cell function. *Trends Neurosci.* **20**: 252–258.

Oppenheim, R.W. (1991) Cell death during development of the nervous system. *Annu. Rev. Neurosci.* **14**: 453–501.

Otteson, D.C., Shelden, E., Jones, J.M., Kameoka, J. and Hitchcock, P.F. (1998) Pax2 expression and retinal morphogenesis in the normal and Krd mouse. *Dev. Biol.* **193**: 209–224.

Pabst, O., Herbrand, H. and Arnold, H.H. (1998) Nkx2–9 is a novel homeobox transcription factor which demarcates ventral domains in the developing mouse CNS. 1998 Apr **73**: 85–93.

Park, D. S., Morris, E. J., Bremner, R., Keramaris, E. *et al.* (2000) Involvement of retinoblastoma family members and E2F/DP complexes in the death of neurons evoked by DNA damage. *J. Neurosci.* **20**: 3104–3114.

Park, D.S., Morris, E.J., Stefanis, L., Troy, C.M., Shelanski, M.L., Geller, H.M. and Greene, L.A. (1998) Multiple pathways of neuronal death induced by DNA-damaging agents, NGF deprivation, and oxidative stress. *J. Neurosci.* **18**: 830–840.

Pattyn A., Morin X., Cremer H., Goridis C. and Brunet J.F. (1999) The homeobox gene Phox2b is essential for the development of autonomic neural crest derivatives. *Nature* **399**: 366–370.

Pfaff, S.L., Mendelsohn, M., Stewart, C.L., Edlund, T. and Jessell, T.M. (1996) Requirement for LIM homeobox gene Isl1 in motor neuron generation reveals a motor neuron-dependent step in interneuron differentiation. *Cell* **84**: 1–20.

Phillips, A.C., Ernst, M.K., Bates, S., Rice, N.R. and Vousden, K.H. (1999) E2F-1 potentiates cell death by blocking antiapoptotic signaling pathways. *Mol. Cell* **4**: 771–781.

Philpott, K.L., McCarthy, M.J., Becker, D., Gatchalian, C. and Rubin, L.L. (1996) Morphological and biochemical changes in neurons: apoptosis versus mitosis. *Eur. J. Neurosci.* **8**: 1906–1915.

Pingault, V., Bondurand, N., Kuhlbrodt, K., Goerich, D. E., Prehu, M.-O., Puliti, A., Herbarth, B., Hermans-Borgmeyer, I., Legius, E., Matthijs, G., Amiel, J., Lyonnet, S., Ceccherini, I., Romeo, G., Smith, J.C., Read, A.P., Wegner, M. and Goossens, M. (1998) SOX10 mutations in patients with Waardenburg-Hirschsprung disease. *Nat. Genet.* **18**: 171–173.

Porteus, M.H., Bulfone, A., Ciaranello, R.D. and Rubenstein, J.L. (1991) Isolation and characterization of a novel cDNA clone encoding a homeodomain that is developmentally regulated in the ventral forebrain. *Neuron.* **7**: 221–229.

Price, M., Lemaistre, M., Pischetola, M., Di Lauro, R. and Duboule, D. (1991) A mouse gene related to Distal-less shows a restricted expression in the developing forebrain. *Nature* **351**: 748–751.

Puelles, L. and Rubenstein, J.L. (1993) Expression patterns of homeobox and other putative regulatory genes in the embryonic mouse forebrain suggest a neuromeric organization. *Trends Neurosci.* **16**: 472–479.

Qiu, M., Bulfone, A., Martinez, S., Meneses, J.J., Shimamura, K., Pedersen, R.A. and Rubenstein, J.L. (1995) Null mutation of Dlx-2 results in abnormal morphogenesis of proximal first and second branchial arch derivatives and abnormal differentiation in the forebrain. *Genes Dev.* **9**: 2523–2538.

Qiu, M., Shimamura, K., Sussel, L., Chen, S. and Rubenstein, J.L. (1998) Control of anteroposterior and dorsoventral domains of Nkx-6.1 gene expression relative to other Nkx genes during vertebrate CNS development. *Mech. Dev.* **72**: 77–88.

Raganathan, S., Scudiere, S. and Bowser, R. (2000) Hyperphosphorylation of the retinoblastoma gene product and altered subcellular distribution of E2F-1 in Alzheimer's disease and amyotrophic lateral sclerosis. *J. Alzheimer's Dis.* (in press).

Raina, A.K., Monteiro, M.J., McShea, A. and Smith, M.A. (1999) The role of cell cycle-mediated events in Alzheimer's disease. *Int. J. Exp. Path.* **80**: 71–76.

Read, A.P. and Newton, V.E. (1997) Waardenburg syndrome. *J. Med. Genet.* **34**: 656–665.

Rhinn, M., Dierich, A., Shawlot, W., Behringer, R.R., Le Meur, M. and Ang, S.L. (1998) Sequential roles for Otx-2 in visceral endoderm and NE for forebrain and midbrain induction and specification. *Development* **125**: 845–856.

Rivera-Perez, J.A., Mallo, M., Gendron-Maguire, M., Gridley, T. and Behringer, R.R. (1995) Goosecoid is not an essential component of the mouse gastrula organizer but is required for craniofacial and rib development. *Development* **121**: 3005–3012.

Rowitch, D.H. and McMahon, A.P. (1995) Pax-2 expression in the murine neural plate precedes and encompasses the expression domains of Wnt-1 and En-1. *Mech. Dev.* **52**: 3–8.

Rubenstein, J.L., Shimamura, K., Martinez, S. and Puelles, L. (1998) Regionalization of the prosencephalic neural plate. *Annu. Rev. Neurosci.* **21**: 445–477. Review.

Ruiz i Altaba, A. and Jessell, T.M. (1993) Midline cells and the organization of the vertebrate neuraxis. *Curr. Opin. Genet. Dev.* **3**: 633–640.

Ruiz i Altaba, A., Placzek, M., Baldassare, M., Dodd, J. and Jessell, T.M. (1995) Early stages of notochord and floor plate development in the chick embryo defined by normal and induced expression of HNF3b. *Develop. Biol.* **170**: 299–313.

Ruvkun, G. ***et al.*** (1988) The POU domain: a large conserved region in the mammalian pit-1, oct-1, oct-2, and *Caenorhabditis elegans* unc-86 gene products. *Genes Dev.* **2**: 1513–1516.

Sanyanusin, P., Schimmenti, L.A., McNoe, L.A., Ward, T.A., Pierpont, M.E.M., Sullivan, M.J., Dobyns, W.B. and Eccles, M.R. (1995) Mutation of the PAX2 gene in a family with optic nerve colobomas, renal anomalies and vesicoureteral reflux. *Nat. Genet.* **9**: 358–363.

Sasai, Y., Lu, B., Steinbeisser, H. and DeRobertis, E. (1995) Regulation of neural induction by the Chd and Bmp-4 antagonistic patterning signals in Xenopus. *Nature* **376**: 333–336.

Sasaki, H. and Hogan, B.L. (1994) HNF-3 β as a regulator of floor plate development. *Cell* **76**: 103–115.

Sasaki, H., Hui, C.-C., Nakafuku, M. and Kondoh, H. (1997) A binding site for Gli proteins is essential for HNF-3b floor plate enhancer activity in transgenics and can respond to Shh *in vitro*. *Development* **124**: 1313–1322.

Schneider-Maunoury, S., Seitanidou, T., Charnay, P. and Lumsden, A. (1997) Segmental and neuronal architecture of the hindbrain of Krox-20 mouse mutants. *Development* **124**: 1215–1226.

Schoonover, S., Davies, P. and Bowser, R. (1996) Immunolocalization and redistribution of the FAC1 protein in Alzheimer's disease. *J. Neuropathol. Exp. Neurol.* 55: 444–455.

Seitanidou, T., Schneider-Maunoury, S., Desmarquet, C., Wilkinson, D.G. and Charnay, P. (1997) Krox-20 is a key regulator of rhombomere-specific gene expression in the developing hindbrain. *Mech. Dev.* **65**: 31–42.

Shawlot, W. and Behringer, R.R. (1995) Requirement for Lim1 in head-organizer function. *Nature* **374**: 425–430.

Sheth, A.N. and Bhide, P.G. (1997) Concurrent cellular output from two proliferative populations in the early embryonic mouse corpus striatum. *J. Comp. Neurol.* **383**: 220–230.

Shimamura, K. and Rubenstein, J.L. (1997) Inductive interactions direct early regionalization of the mouse forebrain. *Development* **124**: 2709–2718.

Shimamura, K., Hartigan, D.J., Martinez, S., Puelles, L. and Rubenstein, J.L. (1995) Longitudinal organization of the anterior neural plate and neural tube. *Development* **121**: 3923–3933.

Simeone, A., Acampora, D., Gulisano, M., Stornaiuolo, A. and Boncinelli, E. (1992) Nested expression domains of four homeobox genes in developing rostral brain. *Nature* **358**: 687–690.

Simmons, D.M., Voss, J.W., Ingraham, H.A., Holloway, J.M., Broide, R.S., Rosenfeld, M.G. and Swanson, L.W. (1990) Pituitary cell phenotypes involve cell-specific Pit-1 mRNA translation and synergistic interactions with other classes of transcription factors. *Genes. Dev.* **4**: 695–711.

Smale, G., Nichols, N.R., Brady, D.R., Finch, C.E. and Horton, W.E. (1995) Evidence for apoptotic cell death in Alzheimer's disease. *Exp. Neurol.* **133**: 225–230.

Smith, M.D., Dawson, S.J. and Latchman, D.S. (1997) Inhibition of neuronal process outgrowth and neuronal specific gene activation by the Brn-3b transcription factor. *J. Biol. Chem.* **272**: 1382–1388.

Spemann, H. and Mangold, H. (1924) Über Induktion von Embryonanlagen durch Implantation artfremder Organisatoren. *Arch. Mikr. Anat. Entwicklngsmech.* **100**: 599–638.

Stoykova, A., Gotz, M., Gruss, P. and Price, J. (1997) Pax6-dependent regulation of adhesive patterning, R-cadherin expression and boundary formation in developing forebrain. *Development* **124**: 3765–3777.

Stuart, E.T., Kioussi, C. and Gruss, P. (1993) Mammalian PAX genes. *Annu. Rev. Genet.* **27**: 219–236.

Studer, M., Lumsden, A., Ariza-McNaughton, L., Bradley. A. and Krumlauf, R. (1996) Altered segmental identity and abnormal migration of motor neurons in mice lacking Hoxb-1. *Nature* **384**: 630–634.

Su, J.H., Anderson, A.J., Cummings, B.J. and Cotman, C.W. (1994) Immunohistochemical evidence for apoptosis in Alzheimer's disease. *NeuroReport* 5: 2529–2533.

Sussel L., Marin O., Kimura S. and Rubenstein J.L. (1999) Loss of Nkx2.1 homeobox gene function results in a ventral to dorsal molecular respecification within the basal telencephalon: evidence for a transformation of the pallidum into the striatum. *Development* **126**: 3359–3370.

Swiatek P.J. and Gridley, T. (1993) Perinatal lethality and defects in hindbrain development in mice homozygous for a targeted mutation of the zinc finger gene Krox20. *Genes Dev.* 7: 2071–2084.

Taglialatela, G., Robinson, R. and Perez-Polo, J.R. (1997) Inhibition of nuclear factor kappa B (NFkB) activity induces NGF-resistant apoptosis in PC12 cells. *J. Neurosci. Res.* **47**: 155–162.

Tanabe, Y., Roelink, H. and Jessell, T.M. (1995) Induction of motor neurons by Sonci hedgehog is independent of floor plate differentiation. *Current Biol.* 5: 651–658.

Tao, W. and Lai, E. (1992) Telencephalon restricted expression of BF-1, a new member of the HNF-3/*forkhead* gene family, in the developing rat brain. *Neuron.* **8**: 957–966.

Tassabehji, M., Read, A.P., Newton, V.E., Harris, R., Balling, R., Gruss, P. and Strachan, T. (1992) Waardenburg's syndrome patients have mutations in the human homologue of the Pax-3 paired box gene. *Nature* **355**: 635–636.

Thomas, P. and Beddington, R. (1996) Anterior primitive endoderm may be responsible for patterning the anterior neural plate in the mouse embryo. *Curr. Biol.* **6(11)**: 1487–1496.

Ton, C.C., Hirvonen, H., Miwa, H., Weil, M.M., Monaghan, P., Jordan, T., van Heyningen, V., Hastie, N.D., Meijers-Heijboer, H., Drechsler, M. *et al.* (1991) Positional cloning and characterization of a paired box- and homeobox-containing gene from the aniridia region. *Cell* **67**: 1059–1074.

Tong, L. and Perez-Polo, J.R. (1996) Effect of nerve growth factor on AP-1, NF-kB, and Oct DNA binding activity in apoptotic PC12 cells: extrinsic and intrinsic elements. *J. Neurosci. Res.* **45**: 1–12.

Tong, L., Toliver-Kinsky, T., Taglialatela, G., Werrbach-Perez, K., Wood, T. and Perez-Polo, J.R. (1998) Signal transduction in neuronal death. *J. Neurochem.* **71**: 447–459.

Torp, R., Su, J.H., Deng, G., and Cotman, C.W. (1998) *GADD45* is induced in Alzheimer's disease, and protects against apoptosis *in vitro*. *Neurobiol. Disease* 5: 245–252.

Treacy, M.N. and Rosenfeld, M.G. (1992) Expression of a family of POU-domain protein regulatory genes during development of the central nervous system. *Annu. Rev. Neurosci.* **15**: 139–165.

Tsuchida, T., Ensini, M., Morton, S.B., Baldassare, M., Edlund, T., Jessell, T.M. and Pfaff, S.L. (1994) Topographic organization of embryonic motor neurons defined by expression of LIM homeobox genes. *Cell* **79**: 957–970.

Turner, E.E., Jenne, K.J. and Rosenfeld, M.G. (1994) Brn-3.2: a Brn-3-related transcription factor with distinctive central nervous system ex.pression and regulation by retinoic acid. *Neuron.* **12(1)**: 205–218.

Vincent, I., Rosado, M. and Davies, P. (1996) Mitotic mechanisms in Alzheimer's Disease? *J. Cell. Biol.* **132**: 413–425.

Walther, C. and Gruss, P. (1991) Pax-6, a murine paired box gene, is expressed in the developing CNS. *Development* **113**: 1435–1449.

Weinstein, D.C., Ruiz i Albaba, A., Chen, W.S., Hoodless, P., Prezioso, V.R., Jessell, T.M. and Darnell, J.J. (1994) The winged-helix transcription factor HNF-3 beta is required for notochord development in the mouse embryo. *Cell* **78**: 575–588.

Weintraub, S.J., Prater, C.A. and Dean, D.C. (1992) Retinoblastoma protein switches the E2F site from positive to negative element. *Nature* **358**: 259–261.

Wood, K.A. and Youle, R.J. (1995) The role of free radicals and p53 in neuron apoptosis *in vivo*. *J. Neurosci.* **15**: 5851–5857.

Wurst, W., Auerbach, A.B. and Joyner, A.L. (1994) Multiple developmental defects in Engrailed-1 mutant mice: an early mid-hindbrain deletion and patterning defects in forelimbs and sternum. *Development* **120**: 2065–2075.

Xiang, H., Hochman, D.W., Saya, H., Fujiwara, T., Schwartzkroin, P.A. and Morrison, R.S. (1997) Evidence for p53-mediated modulation of neuronal viability. *J. Neurosci.* **16**: 6753–6765.

Xiang, M., Zhou, L., Macke, J.P., Yoshioka, T., Hendry, S.H., Eddy, R.L., Shows, T.B. and Nathans, J. (1995) The Brn-3 family of POU-domain factors: primary structure, binding specificity, and expression in subsets of retinal ganglion cells and somatosensory neurons. *J. Neurosci.* **15**: 4762–4785.

Xiang, M., Gan, L., Li, D., Chen, Z.Y., Zhou, L., O'Malley, B.W. Jr, Klein, W. and Nathans, J. (1997) Essential role of POU-domain factor Brn-3c in auditory and vestibular hair cell development. *Proc. Natl Acad. Sci. USA* **94(17)**: 9445–9450.

Xuan, S., Baptista, C.A., Balas, G., Tao, W., Soares, V.C. and Lai, E. (1995) Winged helix transcription factor BF-1 is essential for the development of the cerebral hemispheres. *Neuron.* **14**: 1141–1152.

Yamada, G., Mansouri, A., Torres, M., Stuart, E.T., Blum, M., Schultz, M., De Robertis, E.M. and Gruss, P. (1995) Targeted mutation of the murine goosecoid gene results in craniofacial defects and neonatal death. *Development* **121**: 2917–2922.

Yoshida, M., Suda, Y., Matsuo, I., Miyamoto, N., Takeda, N., Kuratani, S. and Aizawa, S. (1997) Emx1 and Emx2 functions in development of dorsal telencephalon. *Development* **124**: 101–111.

Zhang, P., Hirsch, E.C., Damier, P., Duyckaerts, C. and Javoy-Agid, F. (1992) c-fos protein-like immunoreactivity: Distribution in the human brain and over-expression in the hippocampus of patients with Alzheimer's disease. *Neuroscience* **46**: 9–21.

Zhang, M., Kim, H.J., Marshall, H., Gendron-Maguire, M., Lucas, D.A., Baron, A., Gudas, L.J., Gridley, T., Krumlauf, R. and Grippo, J.F. (1994) Ectopic Hoxa-1 induces rhombomere transformation in mouse hindbrain. *Development* **120**: 2431–2442.

Zhang, J., Hagopian-Donaldson, S., Serbedzija, G., Elsemore, J. and Plehn-Dujowich, D. (1996) Neural tube, skeletal and body wall defects in mice lacking transcription factor AP-2. *Nature* **381**: 238–241.

12

Abnormal transcription factors produced by chromosome translocations in human cancer

Stephen P. Hunger

1. Introduction

Human cancer is the phenotypic manifestation of an accumulation of mutations, most somatic but some hereditary, that lead to alterations in control of cell proliferation, differentiation, and survival. These alterations result from changes in complex patterns of gene expression. Over the past 15 years, a wide variety of oncogenes and tumor suppressor genes that are mutated in human cancer have been identified. The protein products of these genes can be grouped into a number of broad classes, the largest of which is transcription factors. For example, *TP53*, the most commonly mutated gene in human cancer, encodes a nuclear transcription factor believed to play a critical role in cell-cycle checkpoint control, and mutations observed in cancer cells alter its transcriptional regulatory properties (Kern *et al.*, 1992).

Oncogenes are frequently activated in leukemias, lymphomas and certain mesenchymal-derived sarcomas by exchanges of genetic material between chromosomes that are known as chromosomal translocations (Cleary, 1991; Rabbitts, 1994). Circumstantial evidence suggests that some translocations found in lymphoid malignancies arise via mistakes in somatic recombination of immunoglobulin (*Ig*) or T-cell receptor (*tcr*) genes (Finger *et al.*, 1986). The etiology of other translocations is less certain, but may involve recombination between repetitive genomic elements. Translocations convert cellular proto-oncogenes to oncogenes by one of two general mechanisms. One class of translocations brings a proto-oncogene into the vicinity of an *Ig* or *tcr* gene, leading to dysregulated expression

Transcription Factors, edited by J. Locker.

of a structurally intact protein whose cellular presence is normally under tight homeostatic control. With a few exceptions, the proteins encoded by this class of translocations are structurally normal and their oncogenic activity appears to be a direct consequence of altered patterns of expression and not due to fundamental alterations in function. The second class of translocations includes those that create fusion genes that produce fusion mRNAs coding for novel chimeric proteins. Most of these fusion proteins possess structural features and functional properties of chimeric transcription factors. Transcription factors are typically modular, containing discrete motifs responsible for protein dimerization and DNA binding, and effector domains that mediate protein–protein interaction with components of the basal transcription apparatus. Chimeric transcription factors often contain a DNA-binding domain of one protein joined to an effector domain, such as a transcriptional activation domain, from another, and therefore possess functional properties that are distinct from those of the parental wild-type proteins.

In this chapter, I will focus on chimeric oncoproteins that have been created by chromosomal translocations in acute leukemias and certain sarcomas. Many of these chimeras have been demonstrated to behave as oncogenes in experimental model systems *in vitro* and *in vivo*. Recent advances are beginning to shed light on how these chimeric transcription factors alter the expression of downstream genes and genetic programs, providing important insights into the molecular details of oncogenesis and identifying new targets for therapeutic intervention.

2. Chimeric transcription factors in human leukemias

Translocations are detected by standard cytogenetic analyses and/or targeted molecular studies in 50–75% of human leukemias. Many of the recurring translocations are tightly associated with specific subtypes of leukemia (Rowley, 1973). In the early 1980s, this association between genotype and phenotype provided a compelling impetus for investigators to use emerging molecular biology techniques to analyze non-random chromosomal abnormalities in human leukemias. Using this approach, several groups found that *c-myc*, a mammalian cellular homologue of a known avian retrovirus-transforming gene, was translocated into an *Ig* locus in human leukemias and mouse plasmacytomas (Dalla-Favera *et al.*, 1983; Taub *et al.*, 1982). The molecular details of approximately 100 chromosome translocations have been defined in the ensuing 15 years (Rabbitts, 1994).

Translocations that occur in human leukemias or lymphomas and create proteins with structural features and/or functional properties of chimeric transcription factors are listed in *Table 1*. This list will certainly be outdated by the time this book is published, a testament to the rapid expansion of knowledge in molecular oncology. The proteins targeted by these translocations include major members of well-recognized transcription factor families, including bHLH proteins, nuclear hormone receptors, homeobox proteins, zinc-finger proteins, bZIP proteins, and many others. It is notable that many of these proteins are mammalian homologues of 'master regulatory proteins' of embryonic development in primitive organisms that are major model systems used in developmental biology, such as *Drosophila* and the nematode *Caenorhabditis elegans* (Look, 1997; Rabbitts, 1994).

The structural organization of many chimeric transcription factors suggests a relatively straightforward model for how they could contribute to leukemogenesis. This model predicts that the DNA-binding domain specifies the target sites recognized by the chimera, while the effector domain specifies the net result on gene transcription. Fusion to a transcriptional activation domain should activate transcription, while fusion to a repression domain should repress transcription. Many elements of this 'simplistic' model are likely correct. However, it is remarkable that we still know very little about the downstream target genes and genetic programs altered by oncogenic transcription factors. It is anticipated that new and rapidly evolving techniques in gene expression analysis should provide a wealth of information in this area during the next 5–10 years. This should help answer two critical questions that have important therapeutic implications. First, is there a single, or a small number of downstream target genes that mediate the oncogenic effects of a given transcription factor? Second, does the wide variety of different chimeric transcription factors converge to regulate a few specific pathways critical to oncogenesis, or does each contribute to transformation in its own unique manner?

In recent years, a more complex, and arguably more elegant, view of the function of chimeric transcription factors has begun to emerge as the complexity of transcriptional regulation has been characterized better. Particularly exciting is the recognition that many chimeric oncoproteins interact with proteins responsible for remodeling chromatin, or in some cases are themselves directly responsible for remodeling chromatin, through histone acetylation (Redner *et al.*, 1999). This has raised the hope that new therapies can be devised to treat leukemia by modifying chromatin structure, or interfering with the effects of leukemogenic oncoproteins on chromatin structure.

I will now consider chimeric transcription factors produced by several of the major leukemia-associated translocations. Each provides important insights into how alterations in transcriptional regulation contribute to malignant transformation.

2.1 ALL and E2A protein chimeras: All may not be as it seems

The *E2A* gene (see also Chapter 7) encodes two highly similar transcription factors, termed E12 and E47 (which I will refer to collectively as E2A), of the bHLH family via alternative splicing of a single exon that codes for the bHLH domain (Murre *et al.*, 1989a, 1989b). The bHLH domain is located in the carboxy terminus of E2A, while the amino terminal region includes two discrete transcriptional activation domains (*Figure 1*) (Aronheim *et al.*, 1993; Henthorn *et al.*, 1990; Murre, *et al.*, 1989a; Quong, *et al.*, 1993). E2A proteins play an important general role in transcriptional regulation by serving as obligate heterodimerization partners for various tissue-specific bHLH proteins such as MyoD. However, their most critical function is in regulation of B-cell development, where all evidence indicates that they do not heterodimerize with other proteins and that E2A homodimers are the critical entity. *E2A*-null mutant mice do not develop any mature B-cells and most die shortly after birth (Bain *et al.*, 1994; Zhuang *et al.*, 1994). B-cell maturation is blocked early in differentiation in the *E2A*-null mice, as they do not contain any cells with DJ rearrangements of the *Ig* genes.

Table 1. Chimeric transcription factors in human leukemias and lymphomas

Disease	Translocation	Fusion protein	Reference
MOZ chimeras			
AML	t(8;16)(p11;p13)	MOZ/CBP	Borrow *et al.*, 1996
AML	inv(8)(p11q13)	MOZ/Tif2	Carapeti *et al.*, 1998; Liang *et al.*, 1998
MLL chimeras			
ALL	t(1;11)(p32;q23)	MLL/AF1p	Bernard *et al.*, 1994
AML/ALL	t(1;11) (q21;q23)	MLL/AF1q	Tse *et al.*, 1995
ALL/AML	t(4;11)(q21;q23)	MLL/AF4	Gu *et al.*, 1992
AML/ALL	t(6;11)(q27;q23)	MLL/AF6	Prasad *et al.*, 1993
AML	t(6;11)(q21;q23)	MLL/AF6q21	Hillion *et al.*, 1997
AML/ALL	t(9;11)(p21;q23)	MLL/AF9	Nakamura *et al.*, 1993
AML/ALL	t(10;11)(p12;q23)	MLL/AF10	Chaplin *et al.*, 1995
AML	t(10;11)(p11.2;q23)	MLL/Abl1	Taki *et al.*, 1998
AML	t(11;16)(q23;p13.3)	MLL/CBP	Sobulo *et al.*, 1997
AML	t(11;17)(q23;q15)	MLL/MSF	Osaka *et al.*, 1999
AML	t(11;17)(q23;q21)	MLL/AF17	Prasad *et al.*, 1994
AML/ALL	t(11;19)(q23;p13.1)	MLL/ELL	Thirman *et al.*, 1994
ALL/AML	t(11;19)(q23;p13.3)	MLL/ENL	Tkachuk *et al.*, 1992
AML	t(11;19)(q23;p13)	MLL/EEN	So *et al.*, 1997
AML	t(11;22)(q23;q11.2)	MLL/hCDCrel	Megonigal *et al.*, 1998
AML	t(11;22)(q23;q13)	MLL/p300	Ida *et al.*, 1997
ALL	t(X;11)(q13;q23)	MLL/AFX1	Corral *et al.*, 1993
NUP98 chimeras			
AML	t(1;11)(q23;p15)	NUP98/Pmx1	Nakamura *et al.*, 1999
AML	t(2;11)(q31;p15)	NUP98/Hoxd13	Raza-Egilmez *et al.*, 1998
AML	t(7;11)(p15;p15)	NUP98/Hoxa9	Borrow *et al.*, 1996; Nakamura *et al.*, 1996
AML	inv(11)(p15q22)	NUP98/Ddx10	Arai *et al.*, 1997
ETV6 chimeras			
MDS	t(3;12)(q26;p13)	Etv6/EVI1	Raynaud *et al.*, 1996
CMML	t(5;12)(q33;p13)	Etv6/PDGFRβ	Golub *et al.*, 1994
ALL	t(6;12)(q23;p13)	Etv6/STL	Suto *et al.*, 1997
ALL	t(9;12)(q34;p13)	Etv6/ABL	Papadopoulos *et al.*, 1995
ALL/AML	t(9;12)(p24;p13)	Etv6/Jak2	Lacronique *et al.*, 1997; Peeters *et al.*, 1997
AML	t(12;13)(p13;q12)	Etv6/Cdx2	Chase *et al.*, 1999
AML; congenital fibrosarcoma	t(12;15)(p13;q25)	Etv6/Trkc	Eguchi *et al.*, 1999; Knezevich *et al.*, 1998
ALL	t(12;21)(p13;q22)	Etv6/CBFα2	Golub *et al.*, 1995; Romana *et al.*, 1995
MDS	t(12;22) (p13;q11)	Etv6/MN1	Buijs *et al.*, 1995
RARα chimeras			
APML	t(5;17)(q32;q12)	NPM/RARα	Redner *et al.*, 1996
APML	t(11;17)(q23;q21)	PLZF/RARα	Chen *et al.*, 1993
APML	t(11;17)(q13;q21)	NuMA/RARα	Wells *et al.*, 1997
APML	t(15;17(q22;q21)	PML/RARα	de Thé *et al.*, 1990; Borrow *et al.*, 1990; Longo *et al.*, 1990)
E2A chimeras			
ALL	t(1;19)(q23;p13)	E2A/Pbx1	Kamps *et al.*, 1990; Nourse *et al.*, 1990

ALL	t(17;19)(q21–22;p13)	E2A/HLF	Hunger *et al.*, 1992; Inaba *et al.*, 1992
ALL	inv(19)	E2A/FB1	Brambillasca *et al.*, 1999
CBF chimeras			
AML	t(8;21)(q22;q22)	CBFα2/ETO	Erickson *et al.*, 1992
MDS	t(3;21)(q26;q22)	CBFα2/EAP CBFα2/EVI1	Mitani *et al.*, 1994; Nucifora *et al.*, 1993
AML	t(16;21)(q24;q22)	CBFα2/Mtg16	Gamou *et al.*, 1998
AML	inv(16)(p13q22) t(16;16)(p13;q22)	CBFβ/Myh11	Liu *et al.*, 1993
Other chimeras			
NHL	ins(2;2)(p13;p11.2–14)	REL/NRG	Lu *et al.*, 1991
AML	t(16;21)(p11;q22)	FUS/ERG	Ichikawa *et al.*, 1994

NHL, non-Hodgkin's lymphoma; ALL, acute lymphoblastic leukemia; AML, acute myeloid leukemia; MDS, myelodysplastic syndrome; APML, acute promyelocytic leukemia; CMML, chronic myelomonocytic leukemia. Full and alternative names for specific genes are given in the text or the primary references.

Shortly after *E2A* was cloned, it was shown to be interrupted by the t(1;19)(q23;p13), which occurs in about 5% of acute lymphoblastic leukemias (ALLs), and fused to a gene termed *Pbx1* that encodes a novel homeobox domain-containing protein (Kamps *et al.*, 1990; Mellentin *et al.*, 1989; Nourse *et al.*, 1990). Subsequently, Pbx1 and closely-related Pbx2 and Pbx3 proteins have been recognized to be mammalian homologues of fly *exd*, and to serve as critical cofactors that modify DNA-binding specificities of mammalian HOX proteins (Chan *et al.*, 1994; Monica *et al.*, 1991; Rauskolb *et al.*, 1993; Rauskolb & Wieschaus, 1994; van Dijk & Murre, 1994).

In E2A/Pbx1, the first chimeric transcription factor linked to human cancer, a large portion of Pbx1, including the homeodomain, replaces the carboxy-terminal portion of E2A, including its bHLH domains (*Figure 1*). This immediately suggested the parsimonious explanation that E2A/Pbx1 likely contributes to leukemogenesis by binding to target genes recognized by the Pbx1 homeodomain and altering their transcription through critical E2A effector domains located within its amino terminus (Kamps *et al.*, 1990; Nourse *et al.*, 1990). Several subsequent developments provided compelling support for this hypothesis. A high affinity Pbx1-binding site was identified, and E2A/Pbx1 was demonstrated to activate *in vitro* transcription of reporter constructs located downstream of this site (LeBrun & Cleary, 1994; Lu *et al.*, 1994; van Dijk *et al.*, 1993). E2A/Pbx1 was also demonstrated to have oncogenic properties in several model systems. It induces transformation and loss of contact inhibition in NIH 3T3 murine fibroblasts, causes acute myeloid leukemia (AML) when lethally irradiated mice are reconstituted with marrow progenitors infected with retroviruses that express E2A/Pbx1, and causes T-cell lymphomas following expression in transgenic mice (Dedera *et al.*, 1993; Kamps and Baltimore, 1993; Kamps *et al.*, 1991). However, detailed structure–function analyses gave the surprising result that while one or both of the E2A transcriptional activation domains were required for transformation, constructs from which the Pbx1 homeodomain were deleted retained many of the

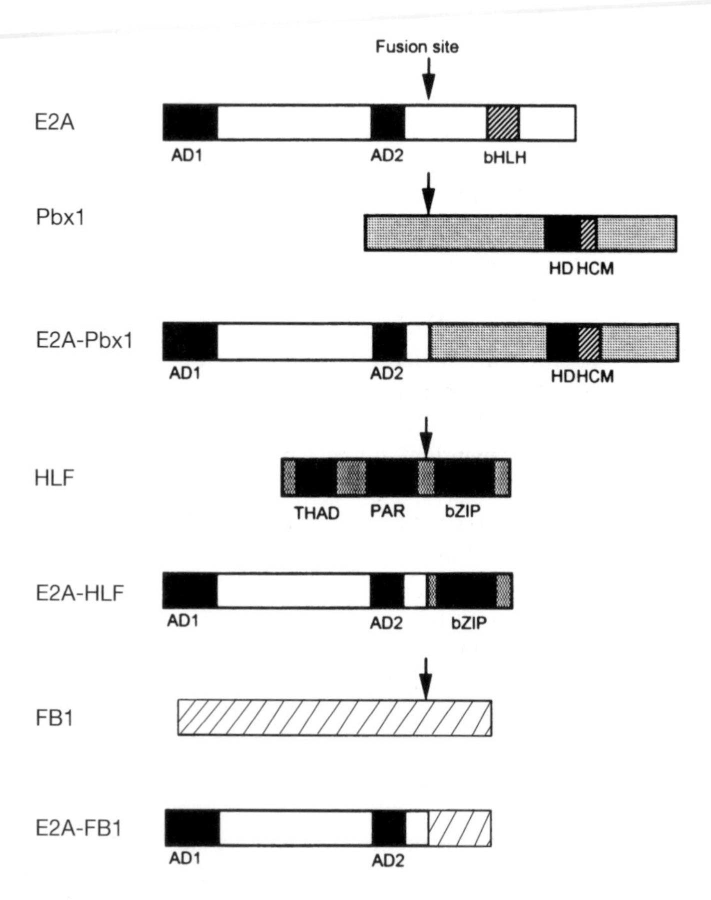

Figure 1. Schematic depiction of wild type and chimeric proteins involved in E2A translocations. Functional domains are depicted schematically. The sites of protein fusion are indicated by arrowheads. Abbreviations: AD1/AD2, transcriptional activation domains 1 and 2; bHLH, basic-helix-loop-helix; HD, homeodomain; HCM, HOX cooperativity motif; THAD, TEF/HLF activation domain; PAR, proline and acidic amino acid-rich domain; bZIP, basic-leucine-zipper domain.

oncogenic properties of E2A/Pbx1 and were still able to transform NIH 3T3 cells and cause lymphomas in transgenic mice (Kamps *et al.*, 1996; Monica *et al.*, 1994).

How does E2A/Pbx1 contribute to leukemogenesis if its DNA-binding domain is non-essential? Chang *et al.* found that the HOX cooperativity motif, a small portion of Pbx1 immediately *C*-terminal to the homeodomain, is absolutely required for transformation, and suggested that this may allow the homeodomain-deleted constructs to be tethered to DNA via interactions with other HOX proteins (Chang *et al.*, 1997). This still allows for the possibility that altered transcriptional regulation of specific, currently unidentified target genes may be the critical transforming event. However, several other facts raise additional possibilities.

Another translocation that occurs in 0.5–1.0% of ALLs, the t(17;19)(q22;p13), creates a structurally analogous chimera consisting of the same amino terminal portion of E2A fused to the carboxy terminus of HLF that contains a bZIP DNA-binding and protein-dimerization domain (*Figure 1*) (Hunger *et al.*, 1992; Inaba *et al.*, 1992). The normal function of HLF is unknown. *HLF* expression is restricted to certain cell types, including hepatocytes and cells of certain regions of the central nervous system (CNS), but not normal lymphoid or hematopoietic cells (Hunger *et al.*, 1992; Inaba *et al.*, 1992). E2A/HLF binds to target sites recognized by HLF and related bZIP proteins (such as DBP, TEF and C/EBP proteins) and activates target gene transcription, suggesting that it contributes to leukemogenesis by activating transcription of target genes normally regulated by HLF or other bZIP proteins (Hunger *et al.*, 1994; Inaba *et al.*, 1994). Like E2A/Pbx1, E2A/HLF, but not wild-type HLF, which can activate transcription via its TEF/HLF activation domain (THAD) in some, but not all experimental assays (Hunger *et al.*, 1996), displays properties of an oncogene in 3T3 and transgenic mouse systems (Honda *et al.*, 1999; Smith *et al.*, 1999; Yoshihara *et al.*, 1995). When expressed in IL3-dependent murine pro-B cell lines, E2A/HLF blocks apoptosis that is normally induced by growth factor withdrawal, but does not substitute for the IL3 proliferative signal (Inaba *et al.*, 1996). An anti-apoptotic role for E2A/HLF is further supported by the observation that the DNA-binding domain of HLF is highly homologous to that of the *C. elegans* cell-death specification-2 (Ces2) protein, and both bind optimally to the identical canonical DNA sequence (Metzstein *et al.*, 1996). Taken together, these observations support the hypothesis that E2A/HLF contributes to oncogenesis by directly activating transcription of critical downstream target genes, whose identity remains unknown, and suggests that at least some of the critical target genes may encode proteins involved in an evolutionarily conserved cell death pathway that is active in pro-B cells.

However, it is difficult to reconcile the anti-apoptotic effects of E2A/HLF observed *in vitro* with the fact that *E2A/HLF* transgenic mice have increased cell death (Honda *et al.*, 1999; Smith *et al.*, 1999). Furthermore, as E2A/HLF and E2A/Pbx1 bind to completely different target sequences, it is perplexing that transgenic mice expressing either chimera have a very similar and distinctive phenotype (Dedera *et al.*, 1993; Honda *et al.*, 1999; Smith *et al.*, 1999). Leukemias in *E2A/Pbx1* and *E2A/HLF* transgenic mice are T-lineage, not B-lineage as seen in humans with a t(1;19) or t(17;19). In the pre-malignant phase, transgenic mice display a decreased number of lymphocytes due to increased apoptosis, and there are very similar blocks in lymphoid differentiation. Results from detailed structure–function studies of E2A/HLF further cloud this issue. While early studies found that both of the E2A transcriptional activation domains and the HLF leucine zipper were absolutely required for transformation of 3T3 cells, more recent studies showed that mutant constructs containing basic-region point mutations which crippled DNA-binding of E2A/HLF were still able to prolong survival of factor-dependent pro-B cell lines following IL3 withdrawal as long as at least one E2A transcriptional activation domain remained intact (Inukai *et al.*, 1998; Yoshihara *et al.*, 1995).

Taken together, these data suggest a different potential role for E2A chimeras in leukemia. Both the t(1;19) and t(17;19) lead to expression of the amino terminal region of E2A, stripped of its bHLH domain, in the nucleus. The integrity of at least one of the E2A transcriptional activation domains, which it is assumed interact with multiprotein transcriptional complexes, is required for transformation in all systems. Thus, E2A chimeras could potentially interfere, in a dominant negative manner, with normal E2A function. This effect could be compounded by loss of one *E2A* allele through translocation, potentially lowering the concentration of E2A within the nucleus.

This second model predicts that a, and perhaps the, critical leukemogenic function of E2A fusion proteins in ALL is to interfere with the function of wild-type E2A proteins. Several observations from human leukemias and experimental models support this alternative hypothesis. While almost all *E2A*-null mice die shortly after birth, a small percentage survive, most of which later develop T-cell malignancies (Bain *et al.*, 1997; Yan *et al.*, 1997). *E2A* has some functions of a tumor suppressor gene, because expression of either E12 or E47 in cell lines derived from lymphomas that arose spontaneously in E2A-deficient mice caused apoptosis (Engel and Murre, 1999). Intriguingly, *Tal1/SCL* (or one of its two close homologues *Lyl1* or *Tal2*) is overexpressed by chromosomal rearrangements in over 30% of childhood T-ALLs (Baer, 1993). Tal1, Tal2, and Lyl1 are bHLH proteins that form heterodimers with E2A (Hsu *et al.*, 1994). TAL/E2A heterodimers bind DNA as part of a large complex that also includes Rbtn2, another oncoprotein that is overexpressed due to chromosomal translocations in T-ALL (Valge-Archer *et al.*, 1994). Mice transgenic for both *Tal1* and *Rbtn1* have a T-cell phenotype strikingly similar to that seen in *E2A*-null mice with a pre-malignant phase that is characterized by reduced cell numbers which exhibit increases in both proliferation and apoptosis. This is followed by clonal expansion and development of T-cell malignancies (Chervinsky *et al.*, 1999). Taken together, these data suggest that overexpression of TAL or RBTN proteins leads to sequestration of E2A (or other critical components of this complex), thereby interfering with critical functions and contributing to cellular transformation. Indeed, E2A is complexed with Tal1, and E2A transcriptional activity is very low, in the human T-ALL cell line Jurkat (Park *et al.*, 1999). Restoration of E2A activity in Jurkat cells by expression of an E2A/Tal1 fusion protein resulted in growth arrest and apoptosis.

Lending further support to the idea that interference with normal E2A function is critical for leukemogenesis, a third E2A chimera has recently been identified in B-precursor ALLs. E2A/FB1, which is produced by a cryptic inversion of chromosome 19, contains the same portion of E2A present in E2A/Pbx1 and E2A/HLF fused to variable portions of FB1, a protein of unknown function (*Figure 1*) (Brambillasca *et al.*, 1999). In several cases, gene fusion events alter the *FB1* reading frame, which results in expression of proteins that are basically truncated E2A proteins containing only a few additional carboxy-terminal amino acid residues. Functional studies of E2A/FB1 have not been reported yet, but lack of an identifiable DNA-binding domain clearly suggests that E2A/FB1 cannot contribute to leukemogenesis by binding to DNA and altering target gene transcription.

Thus, the mechanism by which E2A chimeras contribute to leukemogenesis remains uncertain. The weight of the evidence now seems to favor the hypothesis that they interfere in some manner with normal functions of E2A, or protein complexes that interact with E2A. This model does not preclude the possibility that aberrant transcription of downstream genes induced by E2A/Pbx1 or E2A/HLF may have an important secondary role and might also influence the clinical phenotype of human leukemias with a t(1;19) or t(17;19).

2.2 *Histone deacetylases and chimeric oncoproteins: a newly recognized recurring theme*

Acute promyelocytic leukemia and RARα-fusion proteins. Acute promyelocytic leukemia (APML) is a clinically distinctive subtype of AML characterized by a clonal expansion of myeloid cells blocked at the promyelocyte stage of granulocyte differentiation (Fenaux *et al.*, 1997). By the early 1980s, cytogenetic studies demonstrated that a t(15;17)(q22;q11.2) was present in the malignant cells of almost all patients with APML, and that this translocation was not seen in any other subtypes of leukemia, with the exception of rare cases of chronic myeloid leukemia (CML) in promyelocytic blast crisis (Larson *et al.*, 1984). This is the tightest known linkage between genotype and phenotype in leukemia biology.

In the mid 1980s, reports emerged from China that patients with APML who had relapsed following treatment with chemotherapy could achieve remission, and perhaps even be cured, when treated with all-*trans*-retinoic acid (Huang *et al.*, 1988). These reports were initially greeted with guarded skepticism, until it was found that the t(15;17) fused a novel gene, named *PML*, to the RARα gene (*Rarα*) (Borrow *et al.*, 1990; de Thé *et al.*, 1990; Longo *et al.*, 1990). Subsequent molecular studies have demonstrated that *Pml/Rarα* fusion is present in almost all patients with APML. The rare cases that do not have a t(15;17) or *Pml-Rarα* fusion contain variant translocations that fuse *Rarα* to one of three other genes: *Plzf* (promyeloctyic leukemia zinc-finger protein), *Npm* (nucleophosmin), or *NuMA* (nuclear mitotic apparatus protein) (*Table 1*) (Chen *et al.*, 1993; Redner *et al.*, 1996; Wells *et al.*, 1997). RARα chimeras contain the same *C*-terminal portion of RARα joined to *N*-terminal portions of the fusion partner (*Figure 2a*). Initially it was recognized that this portion of RARα included domains responsible for DNA binding, protein dimerization, and ligand binding. Other critical functional domains have been recognized subsequently.

The structure of PML/RARα, and the *in vivo* and *in vitro* response of APML cells to all-*trans*-retinoic acid suggested that this chimeric oncoprotein interfered with transcriptional regulation by native RARα in a dominant negative manner to cause a block in myeloid differentiation at the promyelocyte stage (de Thé *et al.*, 1991). Today, we have a much better understanding of how PML/RARα contributes to leukemogenesis, and all-*trans*-retinoic acid has become the first non-cytotoxic agent to become an integral part of antileukemic therapy (Melnick and Licht, 1999; Tallman *et al.*, 1997). These advances have been aided enormously by analyses of the naturally occurring variants of PML/RARα, particularly PLZF/RARα produced by the t(11;17)(q23;q21), and by a better understanding of how transcription is regulated by nuclear hormone receptors. We are now beginning to understand the molecular details of how all-*trans*-retinoic acid works in most cases of APML, why it does not work in some cases, and have been provided with clues to other potential therapies for this subtype of AML.

A brief summary of our current understanding of how RARα and other nuclear hormone receptors regulate transcription is depicted schematically in *Figure 2b*

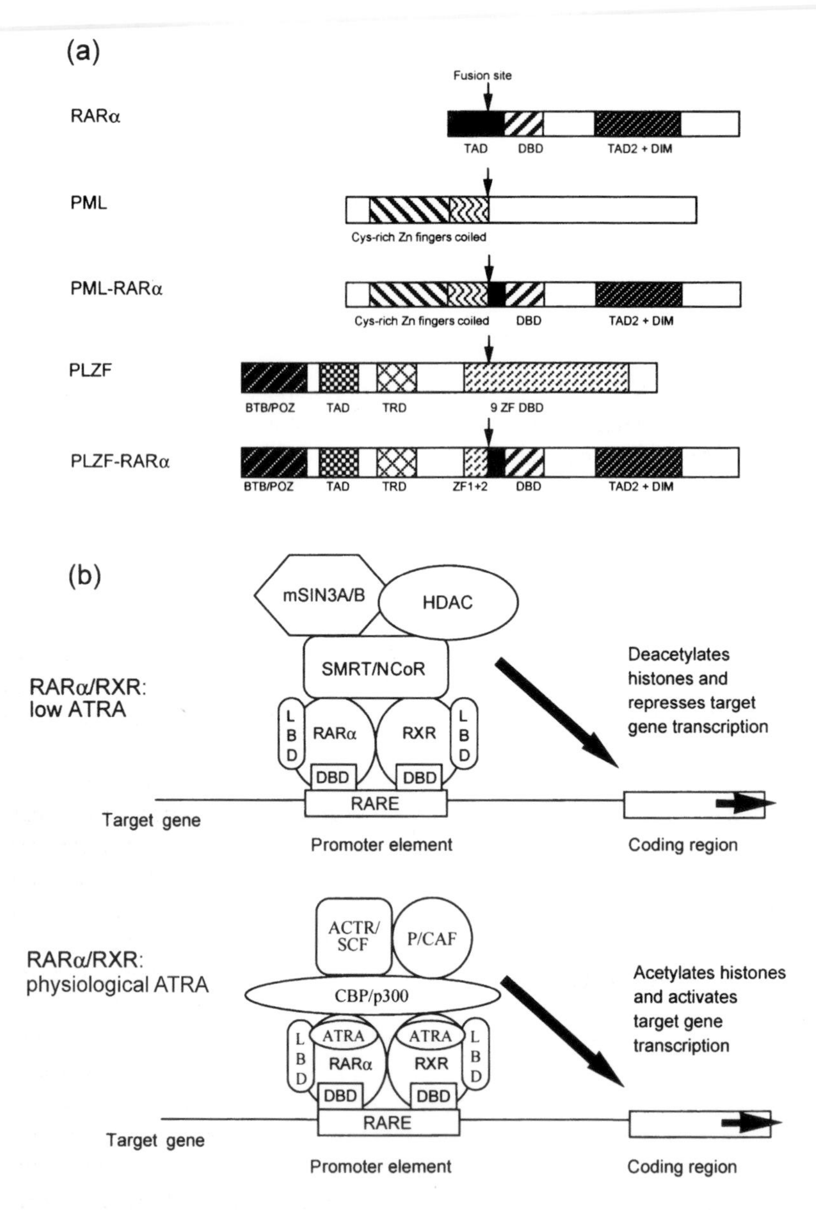

Figure 2. Structure of RARa proteins and proposed mechanism of action. (a) Functional domains are depicted schematically. The sites of protein fusion are indicated by arrowheads. The TAD2 + DIM domain of RARα is also responsible for ligand binding. Abbreviations: TAD, transcriptional activation domain; DBD, DNA-binding domain; DIM, dimerization domain; Cys-rich Zn Fingers, cysteine-rich zinc fingers; coiled, coil region; BTB/POZ, *Broad complex*, *tramtrack*, *Bric a Brac*/pox virus and zinc finger domain; TRD, transcriptional repression domain; ZF: zinc fingers. (b) Model showing repression of target gene transcription by RARα/RXR heterodimers under low concentrations of all-*trans*-retinoic acid (ATRA) and transcriptional activation under physiological concentrations. RAR, retinoic acid receptor; RXR, retinoid X receptor. See text for other abbreviations.

(see also Chapter 8). RARα binds to target DNA sequences (RARE) as a heterodimer with the related RXR protein (Umesono *et al.*, 1991). Like other nuclear hormone receptors, the transcriptional activity of the RARα:RXR heterodimer is regulated by ligands, the most important of which is all-*trans*-retinoic acid. In the absence of physiological levels of all-*trans*-retinoic acid, DNA-bound RARα:RXR heterodimers repress transcription by recruiting a corepressor complex that includes NCoR or SMRT, mSin3A or 3B, and histone deacetylase (Hdacl or 2) (Heinzel *et al.*, 1997; Nagy *et al.*, 1997). In the presence of ligand the conformation of RARα:RXR changes, which causes the corepressor complex to dissociate and be replaced by a coactivator complex that is composed of CBP or p300, activator of thyroid and retinoic acid receptors (ACTR) or steroid receptor activator 1 (SRC1) and P300/CBP associated factor (P/CAF), that has histone acetyltransferase activity and activates target gene transcription (Chen *et al.*, 1997).

To understand how RARα chimeras alter transcriptional regulation in APML, it is critical to recall that, in contrast to patients with APML and *PML/RARA* fusion, patients with *PLZF/RARα* fusion do not respond to all-*trans*-retinoic acid (Guidez *et al.*, 1994; Licht *et al.*, 1995). Similarly, while both *PML/RARα* and *PLZF/RARα* transgenic mice develop leukemia, cells from the former, but not the latter, respond to all-*trans*-retinoic acid (Brown *et al.*, 1997; Grisolano *et al.*, 1997; He *et al.*, 1998, 1997). Since PML/RARα and PLZF/RARα contain identical portions of RARα, these differences must be attributable to biological differences of the RARα chimera that are imposed on it by the fusion partner. Wild-type PML is an intriguing protein that itself has growth-suppressive properties and localizes to an intranuclear structure termed PML-oncogenic domains or nuclear bodies (Dyck *et al.*, 1994; Koken *et al.*, 1994; Weis *et al.*, 1994). While controversy continues over the potential contribution of PML haplo-insufficiency and the reciprocal RARα/PML fusion protein to the pathogenesis of APML (Melnick and Licht, 1999), this discussion will focus on the effect of RARα chimeras. The PLZF protein contains nine Krüppel-like zinc fingers at its *C*-terminus and an *N*-terminal poxvirus and zinc finger (POZ) or *Broad complex*, *tramtrack*, *Bric a brac* (BTB) domain. In the past several years, it has been discovered that the POZ is a protein–protein interaction domain that mediates interaction between wild-type PLZF and the corepressor complexes previously discussed that contain NCoR or SMRT (David *et al.*, 1998; Hong *et al.*, 1997; Wong & Privalsky, 1998).

Studies from several laboratories have demonstrated that PML/RARα:RXR heterodimers are more potent transcriptional repressors than RARα:RXR in the absence of ligand, and are substantially less sensitive to ligand than native RARα:RXR (de Thé *et al.*, 1991; Kakizuka *et al.*, 1991). Thus, physiological levels of ligand ($\sim 10^{-8}$ molar for all-*trans*-retinoic acid) in the bone marrow of patients with APML and a t(15;17) fail to dissociate the repressor complex and target genes (expression of which is apparently essential for myeloid differentiation to proceed past the promyelocyte stage) remain repressed. However, this differentiation block can be overcome by pharmacologic doses of all-*trans*-retinoic acid ($\sim 10^{-6}$ molar), which cause PML/RARα:RXR heterodimers to dissociate from the corepressor complex and be replaced by the coactivator complex. This leads to transcription of the critical target genes (Grignani *et al.*, 1998; Guidez *et al.*, 1998; Lin *et al.*, 1998). Indeed, when patients with APML are treated with all-*trans*-retinoic acid, the malignant promyelocytes are replaced by maturing granulocytes (which contain a t(15;17) indicating that they derived from the malignant clone) that subsequently die via apoptosis (Vyas *et al.*, 1996).

In contrast to PML/RARα, PLZF/RARα has a dual interaction with corepressor complexes, mediated by the all-*trans*-retinoic acid-sensitive repressor/activator binding domain of RARα and the all-*trans*-retinoic acid-insensitive POZ domain of PLZF (Grignani *et al.*, 1998; Guidez *et al.*, 1998; Lin *et al.*, 1998). Thus, a subfraction of PLZF/RARα remains associated with the NCoR:SMRT complex when exposed to pharmacological doses of all-*trans*-retinoic acid, and differentiation remains blocked. Two important experimental results underscore the critical role of histone deactylase recruitment in the pathobiology of APML. First, mutation of the RARα NCoR binding site in PML/RARα abrogates its ability to block myeloid differentiation (Grignani *et al.*, 1998). Second, histone deactylase inhibitors interfere with the oncogenic and transcriptional repression properties of PML/RARα and PLZF/RARα (Grignani *et al.*, 1998; Guidez *et al.*, 1998; Lin *et al.*, 1998).

By necessity, this summary neglects many of the other exciting new developments in understanding the pathobiology of APML. However, it highlights how RARα chimeras contribute to APML and why all-*trans*-retinoic acid is an effective therapy for patients with a t(15;17) but not for those with a t(11;17). These observations also suggest that pharmacological manipulation of histone actylase activity may be a rational approach for treatment of APML. The next vignette expands upon this theme.

Core binding factor chimeras in acute leukemia. Core-binding factor (CBF) is a heterodimeric transcription factor composed of α and β subunits, both of which are affected by translocations in human leukemia that create chimeric oncoproteins (*Table 1*). DNA-binding by the CBF complex is mediated by a domain of the α subunit, which shares significant homology to the *Drosophila* segmentation gene runt and hence is termed the runt-homology domain (RHD). This binds to the target sequence TGT/cGGT (Meyers *et al.*, 1993). By itself, the α subunit has a low DNA-binding affinity that is substantially increased by dimerization with the non-DNA-binding β subunit, also mediated by the RHD (Meyers *et al.*, 1993; Ogawa *et al.*, 1993).

The α subunit of CBF is encoded by at least three different genes in humans. Relevant to this discussion is *CbfA2* (also termed *AML1*). *CbfA2* was initially identified in 1991 as the target of the t(8;21)(q22;q22) that occurs in 10–15% of AMLs (Gao *et al.*, 1991; Miyoshi *et al.*, 1991). The t(8;21) fuses *CbfA2* to *Eto* (*Mtg8*), leading to production of a Cbfα2/ETO chimera that includes the RHD (Erickson *et al.*, 1992; Miyoshi *et al.*, 1993). Several additional Cbfα2 fusion-proteins that were created by translocations in leukemia have been identified subsequently. For our purposes, the most relevant of these other translocations is the t(12;21), which fuses *CbfA2* to the ETS gene *Etv6* (*Tel*) (Golub *et al.*, 1995; Romana *et al.*, 1995). Despite the fact that the t(12;21) is rarely detected by standard cytogenetics, molecular studies show that *Etv6*/*CbfA2* fusion is present in 20–30% of younger children with B-precursor ALL, but occurs rarely in infants, older children or adults with ALL (Aguiar *et al.*, 1996; Maloney *et al.*, 1997;

Rubnitz *et al.*, 1999). The inv(16)(p13;q22) or its molecular equivalent, the t(16;16)(p13;q22), occurs in 10–15% of AMLs, most of whom have a characteristic clinical phenotype with large dysplastic eosinophils (AML, FAB subtype M4Eo) (Bitter *et al.*, 1984). These rearrangements fuse the gene encoding the core binding factor β subunit (*Cbf*β) to the smooth muscle myosin heavy-chain gene (*SMMHC*) *Myh11* (Liu *et al.*, 1993). The Cbfβ/Myh11 chimera includes the portions of Cbfβ necessary for interaction with Cbfα2 and an oligomerization domain from Myh11. Taken together, translocations producing CBF-fusion proteins are the most common genetic abnormalities found in human leukemia.

CBF-binding sites are present in many genes encoding proteins involved in hematopoietic differentiation, including granulocyte-macrophage colony-stimulating factor (GM-CSF), IL3, CSF1, the T cell receptor and myeloperoxidase, suggesting that the CBF complex might play a critical role in hematopoiesis (Speck *et al.*, 1999). Current understanding of how CBF-fusion proteins contribute to leukemogenesis has been aided enormously by observations from studies of *CbfA2* knockout mice. Homozygous inactivation of *CbfA2* is embryonic-lethal in mice, with all *CbfA2*$^{-/-}$ mice dying around embryonic day (E) 11.5–12.5 with a unique pattern of CNS necrosis and hemorrhage (Okuda *et al.*, 1996; Wang *et al.*, 1996a). Prior to E11.5, *CbfA2*$^{-/-}$ embryos appear morphologically identical to normal and heterozygous litter-mates, except for a pale liver and anemia. E10–10.5, when the *CbfA2*-null mice start to die, is approximately the time at which the location of murine hematopoiesis switches from the yolk sac (embryonic) to the liver (fetal). Histological examination of *CbfA2*$^{-/-}$ embryos showed that their livers lacked hematopoietic cells, and their peripheral blood contained only primitive erythrocytes, which originate in the yolk sac, and not more mature erythrocytes or platelets (Okuda *et al.*, 1996; Wang *et al.*, 1996a). *CbfA2*$^{-/-}$ embryonic stem (ES) cells were morphologically identical to *CbfA2*$^{+/+}$ ES cells and had similar growth properties *in vitro*. When *in vitro* differentiation assays were performed, *CbfA2*$^{-/-}$, heterozygote and wild-type ES cells showed a similar ability to differentiate into primitive erythroid cells, but the *CbfA2*$^{-/-}$ ES cells were unable to generate more mature, definitive erythroid and myeloid cells (Okuda *et al.*, 1996; Wang *et al.*, 1996a). Furthermore, when chimeric mice were generated by injecting blastocysts with *CbfA2*$^{-/-}$ ES cells, the *CbfA2*$^{-/-}$ cells contributed to all non-hematopoietic organs but did not contribute to bone marrow, spleen, thymus or peripheral blood (Okuda *et al.*, 1996). An identical phenotype was seen with a homozygous null allele of *Cbf*β, and a homozygous *Cbf*β, mutation which interfered with its ability to heterodimerize with Cbfα2 (Niki *et al.*, 1997; Wang *et al.*, 1996b).

These studies demonstrated that CBF activity was essential for murine-definitive hematopoiesis, but was not needed for primitive hematopoiesis. Two groups have subsequently used a 'knock-in' strategy to create murine ES cells with one normal *CbfA2* allele and one that encodes a Cbfα2/ETO fusion protein analogous to the chimera observed in AML. The other group used an analogous strategy to create *Cbf*β/*Myh11* knock-in ES cells. Similar to *CbfA2*$^{-/-}$ and *Cbf*β$^{-/-}$ mice, the *CbfA2*/*Eto* and *Cbf*β/*Myh11* knock-in mice died during gestation at ~E13.5 (slightly later than the –/– embryos) with CNS necrosis and hemorrhage (Castilla *et al.*, 1996; Okuda *et al.*, 1998; Yergeau *et al.*, 1997). However, *CbfA2*/*Eto*

ES cells retained some ability to differentiate into definitive hematopoietic cells *in vitro,* and had an abnormally high self-renewal capacity (Okuda *et al.*, 1998; Yergeau *et al.*, 1997).

Taken together, these results indicate that a major effect of the CBF chimeras in leukemogenesis is mediated by dominant negative interference with normal CBF function. Through alternative splicing, *CbfA2* encodes at least three protein isoforms. Cbfα2b and Cbfα2c contain a *C*-terminal transcriptional activation domain and activate reporter gene transcription, while Cbfα2a lacks this domain and does not activate transcription (Miyoshi *et al.*, 1995). Both Cbfα2/ETO and Etv6-Cbfα2 block activation of the target gene transcription that is mediated by Cbfα2b (Hiebert *et al.*, 1996; Meyers *et al.*, 1995). Activation by the Cbfα2 transcriptional activation domain is mediated through interactions with the CBP/p300 coactivators, both of which are themselves targets of translocations in AML (see next section) (Kitabayashi *et al.*, 1998). Therefore, a rather straightforward dominant negative or competition model can be envisioned whereby Cbfα2b-Cbfβ heterodimers normally activate transcription, by recruiting CBP/p300, of critical target genes that are necessary for myeloid cells to differentiate and adopt a non-proliferative phenotype (*Figure 3a*). Following t(8;21), Cbfα2/ETO, complexed with Cbfβ, can bind these target gene promoters, but not activate transcription. This blocks differentiation and maintains the cells in a 'proliferative' state. As a normal *Cbfα2* allele is still present, one must hypothesize either that there is a critical 'dose' of Cbfα2 that is necessary for myeloid differentiation, that Cbfα2/ETO 'outcompetes' Cbfα2 for the CBFβ dimerization partner, that the Cbfα2/ETO/Cbfβ complex is more stable, or more stably bound, to promoters than wild-type Cbfα2b-Cbfβ.

Recently, the ETO component of the fusion protein has come under closer scrutiny. The fact that ETO itself can transform NIH 3T3 cells suggests that it is not simply playing a 'passive' role (Wang *et al.*, 1997). Several groups independently reported that ETO interacts directly with the same corepressors that have been implicated in APML, including NCoR, SMRT and mSin3A, and recruits histone deactylase activity (Gelmetti *et al.*, 1998; Lutterbach *et al.*, 1998; Wang *et al.*, 1998). This suggests that Cbfα2/ETO is not a passive dominant negative competitor, but rather binds to DNA and directly represses target gene transcription by recruiting a corepressor complex (*Figure 3b*). This mechanism bears strong similarities to that proposed for RARα chimeras in APML, and suggests that targeted use of histone deacetylase inhibitors could be an important therapeutic strategy in AML.

Although much less progress has been made in deciphering the mechanisms by which ETS variant gene-6 (Etv6)/Cbfα2 contributes to leukemogenesis, the available data suggest that it also represses transcription of Cbfα2 target genes. Etv6 contains an amino-terminal HLH (or pointed) domain, which mediates protein dimerization and protein–protein interactions, and a *C*-terminal DNA-binding domain that shares significant homology with other ETS-family proteins. As a consequence of the t(12;21), the amino-terminal-333 amino acids of Etv6 are fused to almost all of Cbfα2b, including its transcriptional activation domain. Based upon this organization, one might assume that Etv6/Cbfα2 would be a transcriptional activator, but this is not the case. Etv6/Cbfα2

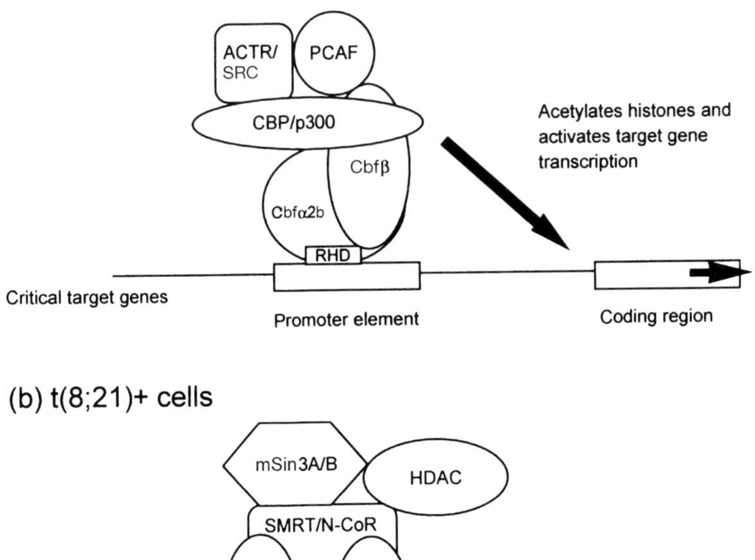

Figure 3. Proposed mechanism of action of core-binding factor (CBF) complexes during normal myeloid differentiation and in leukemia cells with a t(8;21). (a) During normal myeloid differentiation the core-binding factor, CBF complex composed of Cbfα2b and Cbfβ binds to promoter elements in critical target genes via the Cbfα runt homology domain (RHD). This complex then recruits a coactivator complex via interactions with CBP/p300 and activates target gene transcription. (b) In cells containing a t(8;21), Cbfα2/ETO dimerizes with Cbfβ and binds to target gene promoters via the intact Cbfα2 RHD. The ETO portion of this complex is able to bind and recruit a corepressor complex thereby repressing transcription of the target genes necessary for normal myeloid differentiation. PCAF, p300/CBP-associated factor; SMRT, silencing mediator for retinoid and thyroid hormone receptors; NCOR, nuclear corepressor; HDAC, histone deacetylase; ACTR, activator of thyroid and retinoic acid receptor; SRC, steroid receptor coactivator.

represses transcriptional activation that is mediated by Cbfα2b (Hiebert *et al.*, 1996; Uchida *et al.*, 1999). Repression requires both the Etv6/HLH domain and the Cbfα2/RHD. Like Cbfα2/ETO, Etv6/Cbfα2 repression occurs sub-stoichiometrically, suggesting that it is not mediated by competition for DNA-binding sites (Hiebert *et al.*, 1996). In the future, it will be important to determine whether Etv6/Cbfα2 also recruits a corepressor complex, as this could have important therapeutic implications.

Coactivator and histone acetyltransferase chimeras in leukemia. The leukemogenic properties of RARα and CBF chimeras appear to be mediated by altered interactions of corepressor complexes with DNA. Cloning of several other translocation breakpoints has directly implicated the opposite side of this transcriptional equation in human leukemia; that is, coactivators and proteins with intrinsic histone acetyltransferase activity (*Table 1*). The 11q23 gene myeloid/lymphoid leukemia, *MLL* (*HRX*, *All1*) is involved in translocations with more than 30 different partner genes in ALL and AML, and at least 17 different translocation partners are known at this time (*Table 1*) (Gu *et al.*, 1992; Hunger *et al.*, 1993; Thirman *et al.*, 1993; Tkachuk *et al.*, 1992). *MLL* encodes a large (>400 kiloDalton) protein that shares regions of high homology with the *Drosophila* transcriptional regulator, *trithorax* (*trx*). In the fly, *trx* is required for maintenance, but not initiation, of *Hox* gene expression (Mazo *et al.*, 1990). Gene-targeting studies demonstrate that MLL plays a similar role in mammalian development. It is required for proper segment identity and for maintaining expression of genes such as *Hoxa7* (Yu *et al.*, 1998, 1995). It seems likely that this role of MLL involves chromatin remodeling, because a *C*-terminal SET domain, conserved with *trx*, interacts with components of the SWI/SNF complex (Rozenblatt-Rosen *et al.*, 1998).

The MLL–partner chimera encoded by the der(11) fusion gene is critical for leukemogenesis, because the der(11) is always retained while the der(partner) chromosome is lost in about a quarter of cases (Rowley, 1992). These der(11)-encoded chimeras include MLL 'AT hooks' that are involved in minor-groove DNA-binding, a methyltransferase domain, and a putative transcriptional repression domain. Several experimental roadblocks and conundrums have impeded progress in understanding how MLL chimeras contribute to leukemogenesis. First, it is extremely difficult to exogenously express these proteins in cells at significant levels. Indeed, in one of the few systems that MLL chimeras have been demonstrated to have transforming properties, MLL/eleven-nineteen leukemia (ENL) expression was barely detectable (Lavau *et al.*, 1997). Another problem has been the lack of common features present in the portions of partner-gene-encoded proteins included in the leukemia-associated chimeras. While several partner proteins have regions classically associated with transcription factors, others do not, and some do not localize to the nucleus. However, several of the translocation partners link MLL chimeras to the emerging theme of altered interactions with proteins involved in chromatin remodeling, The portions of AF9 and ENL fused to MLL are highly homologous to one another, and to a component of the yeast SWI/SNF complex, suggesting that these chimeras may be able to remodel chromatin (Cairns *et al.*, 1996). In addition, the t(11;16)(q23;p13) and t(11;22)(q23;q13) translocations that occur in AML create MLL/CBP and MLL/p300 chimeras, respectively, which retain the CBP/p300 protein interaction and histone acetyltransferase domains (Ida *et al.*, 1997; Sobulo *et al.*, 1997).

Further extending the theme of abnormal histone acetyltransferases in leukemia, *CBP* is a target of the t(8;16)(p11;p13) in AML, which creates a MOZ/CBP fusion protein (Borrow *et al.*, 1996). MOZ/CBP contains zinc fingers and a putative histone acetyltransferase domain from MOZ fused to almost all of CBP, including its own histone acetyltransferase domain. In phenotypically similar AMLs with an inv(8)(p11;q13), MOZ is fused to Tif2, which itself interacts with p300/CBP and is predicted to have intrinsic histone acetyltransferase activity (Carapeti *et al.*, 1998; Liang *et al.*, 1998).

Potential therapeutic implications of alterations in histone acetylation in leukemia. In this section, I have summarized the remarkable, recently recognized connection between chimeric transcription factors in leukemia, particularly AMLs that display a prominent block in cellular differentiation, and alterations in regulation of histone acetylation. Potent inhibitors of histone deacetylase activity are available, and can alter transcriptional repression and differentiation blocks that are induced by PML/RARα and Cbfα2-ETO *in vitro* (Redner *et al.*, 1999). One case report has been published that describes the outstanding clinical response of an all-*trans*-retinoic acid-resistant patient with APML to all-*trans*-retinoic acid combined with phenylbutyrate, a relatively weak histone deacetylase inhibitor (Warrell *et al.*, 1998). During the next few years, it is anticipated that there will be a great expansion of activity in this area that should define whether pharmacological manipulation of histone deacetylase and/or histone acetyltransferase activity is a viable therapeutic strategy in patients with specific, genetically defined subtypes of leukemia.

3. Chimeric transcription factors in human sarcomas

The cancers commonly observed in adults (lung, colon, breast, prostate) are derived from epithelial cells and are caused by an accumulation of gain-of-function mutations in oncogenes and loss-of-function mutations in tumor-suppressor genes. Translocations are uncommon in these tumors, and chimeric oncoproteins are unknown. In contrast, mesenchymal-derived sarcomas, which typically occur in children and younger adults, have genetic alterations that are much more similar to those found in leukemias than to those found in epithelial-derived solid tumors. A variety of recurrent translocations exist and many of the involved genes have been identified (*Table 2*). These tumors display a genotype–phenotype correlation that is highly reminiscent of the link between the t(15;17) and APML. Almost all cases of these particular tumors possess a common translocation, or several different variations on a common theme, and these translocations are almost never seen in tumors of other histologies. Molecular analyses have demonstrated that almost all of the translocations seen in mesenchymal tumors involve transcription factors. Two major examples are the fusions between one of several different *Pax* (paired box) genes and a gene that encodes a member of the forkhead family of transcription factors (*Fkhr*) that are present in almost all alveolar rhabdomyosarcomas, and fusions that involve the Ewing's sarcoma gene, *EWS*, that are present in essentially all Ewing's sarcomas and peripheral neuroepitheliomas (PNET) as well as a variety of other sarcomas (Delattre *et al.*, 1992; Galili *et al.*, 1993; Shapiro *et al.*, 1993). The next section focuses on the nature and role of EWS fusion proteins in the pathogenesis of a variety of human sarcomas.

3.1 A diverse variety of sarcomas contain translocations that target EWS

For years, considerable debate took place regarding the cell of origin of Ewing's sarcoma and PNET and whether they represented two distinct entities, or a spectrum

Table 2. Chimeric transcription factors in human solid tumors

Disease	Translocation	Fusion protein	Reference
FKHR chimeras			
Alveolar RMS	t(1;13)(p36;q14)	Pax7/FKHR	Davis *et al.*, 1994
Alveolar RMS	t(2;13)(q35;q14)	Pax3/FKHR	Barr *et al.*, 1993; Shapiro *et al.*, 1993
EWS/TLS chimeras			
ESFT	t(7;22)(p22;q12)	EWS/Etv1	Jeon *et al.*, 1995
ESFT	t(11;22)(q24;q12)	EWS/Fli1	Delattre *et al.*, 1992
ESFT	t(17;22)(q12;q12)	EWS/Etv4	Ishida *et al.*, 1998
ESFT	t(21;22)(q22;q12)	EWS/ERG	Sorensen *et al.*, 1994
Extraskeletal myxoid chondrosarcoma	t(9;22) (q22;q12)	EWS/TEC	Labelle *et al.*, 1995
DSRCT	t(11;22)(p13;q12)	EWS/WT1	Ladanyi and Gerald, 1994
Melanoma of soft parts	t(12;22)(q13;q12)	EWS/Atf1	Zucman *et al.*, 1993
Myxoid liposarcoma	t(12;22)(q13;q12)	EWS/CHOP	Panagopoulos *et al.*, 1996
Myxoid liposarcoma	t(12;16)(q13;p11)	TLS/CHOP	Crozat *et al.*, 1993; Rabbitts *et al.*, 1993
HMGI-C chimeras			
Lipoma	t(3;12)(q29;q15)	HMGI-C/LPP	Ashar *et al.*, 1995
Pleomorphic adenomas of the salivary glands	ins(9;12)(p23;q12q15)	HMGI-C/NFIB	Geurts *et al.*, 1998
Lipoma	t(12;15)(q15;q24)	HMGI-C/Acidic protein	Ashar *et al.*, 1995
Other chimeras			
Congenital fibrosarcoma and AML	t(12;15)(p13;q25)	Etv6/TRKC	Eguchi *et al.*, 1999; Knezevich *et al.*, 1998
Synovial sarcoma	t(X;18)(p11.2;q11.2)	SYT/Ssx1 SYT/Ssx2	Clark *et al.*, 1994; Kawai *et al.*, 1998)

RMS, rhabdomyosarcoma; ESFT, Ewing's sarcoma family tumor; DSRCT, desmoplastic small round cell tumor. Full and alternative names for specific genes are given in the text or the primary references.

of a single tumor type. Several developments led to the currently accepted view that they are a continuum of a single disease (subsequently referred to as the Ewing's sarcoma family of tumors (ESFT)) that, despite their common involvement of bone and adjacent soft tissues, are derived from a pleuripotent cell of neural crest origin. Critical to this new view was the observation that both Ewing's sarcomas and PNETs commonly contained an identical t(11;22)(q24;q12) (Whang-Peng *et al.*, 1986). In 1992, the t(11;22) was shown to fuse a novel gene termed *Ews* to the *Fli1* gene (Delattre *et al.*, 1992). Subsequent molecular studies showed that ~85–90% of ESFT contained *Ews/Fli1* fusion (Delattre *et al.*, 1994). As these tumors may fall into the category of difficult-to-classify 'small round blue cell tumors' of childhood, demonstration of *Ews/Fli1* fusion is now regarded as diagnostic of an ESFT.

Ews encodes a previously unknown protein that contains a region rich in glutamine, serine and tyrosine residues, three glycine-rich segments, and a domain homologous to RNA-binding proteins. *Fli1* encodes a member of the ETS family of transcription factors, and had been implicated previously in oncogenesis because it was the insertion site for Friend leukemia virus in virally-induced murine erythroleukemias (Ben-David *et al.*, 1991). Like other ETS proteins, Fli1 has a conserved *C*-terminal DNA-binding domain and an amino-terminal transcriptional activation domain that is less conserved among family members. Ews/Fli1 contains the Fli1 ETS DNA-binding domain with the amino-terminal transcriptional activation domain replaced by the glutamine, serine and tyrosine-rich region of Ews. Early molecular studies showed that Ews/Fli1 transformed NIH 3T3 cells *in vitro*, was a more potent transcriptional activator than wild-type Fli1, and that transformation was dependent on the integrity of both the Fli1 DNA-binding domain and the EWS glutamine, serine and tyrosine-rich region that was shown to be a transcriptional activation domain (May *et al.*, 1993a, 1993b). Based on these results, the view emerged that Ews/Fli1 was a chimeric transcription factor and contributed to transformation by aberrantly regulating target gene transcription. Because there are a large number of ETS proteins with similar DNA-binding properties, it was unclear whether the critical target genes were normally regulated by Fli1, or by other ETS proteins. This is underscored by results of subsequent studies which showed that the ~10–15% of ESFTs that lacked *Ews/Fli1* fusion contained analogous fusions between *Ews* and genes that encoded other ETS family members (*Table 2*) (Ishida *et al.*, 1998; Jeon *et al.*, 1995; Sorensen *et al.*, 1994). The most common of these is the t(21;22)(q22;q12), which is present in 5–10% of cases and creates an EWS/ERG chimeric oncoprotein (Sorensen *et al.*, 1994).

Remarkably, a variety of other, relatively uncommon sarcomas also contain characteristic translocations that fuse *Ews* to portions of other genes that encode the DNA-binding domains of various transcription factors (*Table 2*). In two of these cases, the *Ews* translocation partners encode well-known transcription factors: *Atf1* in malignant melanoma of soft parts associated with t(12;22)(q13;q12), and *WT1* in intra-abdominal desmoplastic small round cell tumor (DSRCT) associated with t(11;22)(p13;q11–q12) (Ladanyi & Gerald, 1994; Zucman *et al.*, 1993). The role of *WT1* in oncogenesis is particularly intriguing. In a subset of familial and sporadic Wilm's tumors (a childhood kidney tumor), *WT1* is a classic tumor-suppressor gene, and tumorigenesis involves functional inactivation of both alleles (Call *et al.*, 1990; Gessler *et al.*, 1990). In contrast, the EWS/WT1 fusion protein in DSRCT appears to function as a dominant oncogene. Extending the important role of EWS proteins in the pathogenesis of sarcomas, it is notable that the t(12;16) observed in malignant liposarcomas fuses *Tls* (*Fus*), which encodes a protein closely related to EWS, to *Chop* a gene that codes for a dominant negative member of the C/EBP family of transcription factors (Crozat *et al.*, 1993; Rabbitts *et al.*, 1993). This theme extends further to leukemias, as the t(16;21), an uncommon translocation present in some AMLs, fuses *Tls* to the *Ews* fusion partner, *Erg* (Ichikawa *et al.*, 1994).

How do EWS and TLS-fusion proteins contribute to oncogenesis? The essential contribution of EWS to the pathogenesis of Ewing's and other sarcomas appears to be its capacity to activate target gene transcription. In this respect it is critical that transcriptional activation is not considered as a generic function. Many transcriptional activation domains are protein–protein interaction domains that aid in recruitment and/or stabilization of an active TFIID complex (see Chapter 2). As discussed in Chapter 1, TFIID includes TBP and TAFs. Coactivators and corepressors are also critical to the activity of TFIID and Pol II. Structure–function studies have demonstrated that the portion of EWS present in oncogenic chimeras can be divided into two regions; amino terminal domain A that is transformation-competent but has limited transactivation potential, and the more carboxy domain B that is a potent transcriptional activation domain but has limited transformation potential (Lessnick *et al.*, 1995). Further studies showed the ability of EWS/FLI1 to transform NIH 3T3 cells was preserved if the EWS portion was replaced by some heterologous transcriptional activation domains, but not by others (Lessnick, *et al.*, 1995). These observations suggest that the portion of EWS present in the chimeric transcription factors involved in the pathogenesis of various sarcomas mediates interactions with a critical protein complex that is transactivation-competent, and it is recruitment of this complex, not simply any transactivation-competent complex, that is crucial for oncogenesis.

Results of other studies initially seemed to clarify this picture, but more detailed experiments have raised additional questions. $hTAF_{II}68$ was cloned and shown to share extensive homology with EWS and FUS. Homologous regions include the carboxy-terminal consensus RNA-binding domain that binds both RNA and single-stranded DNA, and the amino-terminal transcriptional activation domain (Bertolotti *et al.*, 1996). Several subpopulations of TFIID complexes were found. Each contained, on a mutually exclusive basis, a single member of the $hTAF_{II}68$/EWS/TLS family, suggesting that the different complexes might be important for transcription of different genes (Bertolotti *et al.*, 1996). Subsequent studies from this group showed that EWS can associate stably with both the TFIID and pol II complexes, but that EWS/Fli1 is not stably associated with either of these complexes in nuclear extracts of Ewing's sarcoma cells (Bertolotti *et al.*, 1998). These results do not preclude a transient interaction between EWS/Fli1 and TFIID or Pol II. More work needs to be done to define exactly how wild-type EWS/TLS proteins and their oncogenic variants interact with components of the general transcription apparatus.

The current challenge is to integrate information on the role of wild-type EWS in transcriptional regulation with that of mutant EWS proteins in transformation. If it is true that direct effects on transcription of specific target genes mediate the oncogenic properties of EWS fusion proteins, then it should be possible to identify these genes and potentially to experimentally bypass a need for EWS/Fli1. Previous generation tools, such as representational difference analysis, have identified a handful of target genes whose transcription is altered by EWS/Fli1 (Braun *et al.*, 1995). Among these, *manic fringe* is particularly interesting, as its overexpression can transform NIH 3T3 cells (May *et al.*, 1997). It is hoped that new DNA chip and array technologies will be able to refine this view of target genes altered by EWS, and define how their altered transcription contributes to malignant transformation.

4. Summary

Genes that encode DNA-binding transcription factors are the largest class of genes involved in human oncogenesis. In leukemias and mesenchymal-derived sarcomas, oncogenic transcription factors are frequently created by chromosomal translocations that produce fusion proteins with unique structural features and functional properties. Many of these fusion proteins contain DNA-binding domains that are derived from one protein joined to effector domains, such as transcriptional activation domains, that are derived from another. While this suggests a simple model of direct target gene activation, the critical oncogenic properties of these chimeras often appear to be more complicated. Altered interactions with transcriptional corepressors are a fundamental part of the transforming properties of chimeras such as PML/RARα and Cbfα2-ETO, raising the possibility of novel therapeutic interventions. While candidate target genes are beginning to be identified for many oncogenic transcription factors, their immediate biochemical effects remain largely unknown. As new technologies facilitate the identification of candidate target genes, it will be critical to define which target genes and pathways specific chimeras alter, and whether these are unique to each oncoprotein or whether there are common targets that could be manipulated pharmacologically.

References

Aguiar, R.C., Sohal, J., van Rhee, F., *et al.* (1996) TEL-AML1 fusion in acute lymphoblastic leukaemia of adults. M.R.C. Adult Leukaemia Working Party. *Br. J. Haematol.* **95**: 673–677.

Arai, Y., Hosoda, F., Kobayashi, H., *et al.* (1997) The inv(11)(p15q22) chromosome translocation of *de novo* and therapy-related myeloid malignancies results in fusion of the nucleoporin gene, NUP98, with the putative RNA helicase gene, DDX10. *Blood* **89**: 3936–3944.

Aronheim, A., Shiran, R., Rosen, A. and Walker, M.D. (1993) The E2A gene product contains two separable and functionally distinct transcription activation domains. *Proc. Natl Acad. Sci. USA* **90**: 8063–8067.

Ashar, H.R., Fejzo, M.S., Tkachenko, A. *et al.* (1995) Disruption of the architectural factor HMGI-C: DNA-binding AT hook motifs fused in lipomas to distinct transcriptional regulatory domains. *Cell* **82**: 57–65.

Baer, R. (1993) TAL1, TAL2 and LYL1: a family of basic helix-loop-helix proteins implicated in T cell acute leukaemia. *Semin. Cancer Biol.* **4**: 341–347.

Bain, G., Engel, I., Robanus Maandag, E.C. *et al.* (1997) E2A deficiency leads to abnormalities in alphabeta T-cell development and to rapid development of T-cell lymphomas. *Mol. Cell. Biol.* **17**: 4782–4791.

Bain, G., Maandag, E.C., Izon, D.J. *et al.* (1994) E2A proteins are required for proper B cell development and initiation of immunoglobulin gene rearrangements. *Cell* **79**: 885–892.

Barr, F.G., Galili, N., Holick, J., Biegel, J.A., Rovera, G. and Emanuel, B.S. (1993) Rearrangement of the PAX3 paired box gene in the paediatric solid tumour alveolar rhabdomyosarcoma. *Nature Genet.* **3**: 113–117.

Ben-David, Y., Giddens, E.B., Letwin, K. and Bernstein, A. (1991) Erythroleukemia induction by Friend murine leukemia virus: insertional activation of a new member of the ets gene family, Fli-1, closely linked to c-ets-1. *Genes and Dev.* **5**: 908–918.

Bernard, O.A., Mauchauffe, M., Mecucci, C., Van den Berghe, H. and Berger, R. (1994) A novel gene, AF-1p, fused to HRX in t(1;11)(p32;q23), is not related to AF-4, AF-9 nor ENL. *Oncogene* **9**: 1039–1045.

Bertolotti, A., Melot, T., Acker, J., Vigneron, M., Delattre, O. and Tora, L. (1998) EWS, but not EWS-FLI-1, is associated with both TFIID and RNA polymerase II: interactions between two members of the TET family, EWS and hTAFII68, and subunits of TFIID and RNA polymerase II complexes. *Mol. Cell. Biol.* **18**: 1489–1497.

Bertolotti, A., Lutz, Y., Heard, D.J., Chambon, P. and Tora, L. (1996) hTAF(II)68, a novel RNA/ssDNA-binding protein with homology to the pro-oncoproteins TLS/FUS and EWS is associated with both TFIID and RNA polymerase II. *EMBO J.* **15**: 5022–5031.

Bitter, M.A., Le Beau, M.M., Larson, R.A. *et al.* (1984) A morphologic and cytochemical study of acute myelomonocytic leukemia with abnormal marrow eosinophils associated with inv(16)(p13q22). *Am. J. Clin. Path.* **81**: 733–741.

Borrow, J., Shearman, A.M., Stanton, V.P. *et al.* (1996) The t(7;11)(p15;p15) translocation in acute myeloid leukaemia fuses the genes for nucleoporin NUP98 and class I homeoprotein HOXA9. *Nature Genet.* **12**: 159–167.

Borrow, J., Stanton, V.P. Jr., Andresen, J.M. *et al.* (1996) The translocation t(8;16)(p11;p13) of acute myeloid leukaemia fuses a putative acetyltransferase to the CREB-binding protein. *Nature Genet.* **14**: 33–41.

Borrow, J., Goddard, A.D., Sheer, D. and Solomon, E. (1990) Molecular analysis of acute promyelocytic leukemia breakpoint cluster region on chromosome 17. *Science* **249**: 1577–1580.

Brambillasca, F., Mosna, G., Colombo, M. *et al.* (1999) Identification of a novel molecular partner of the E2A gene in childhood leukemia. *Leukemia* **13**: 369–375.

Braun, B.S., Frieden, R., Lessnick, S.L., May, W.A. and Denny, C.T. (1995) Identification of target genes for the Ewing's sarcoma EWS/FLI fusion protein by representational difference analysis. *Mol. Cell. Biol.* **15**: 4623–4630.

Brown, D., Kogan, S., Lagasse, E. *et al.* (1997) A *PMLRAR*α transgene initiates murine acute promyelocytic leukemia. *Proc. Natl Acad. Sci. USA* **94**: 2551–2556.

Buijs, A., Sherr, S., van Baal, S. *et al.* (1995) Translocation (12;22) (p13;q11) in myeloproliferative disorders results in fusion of the ETS-like TEL gene on 12p13 to the MN1 gene on 22q11. *Oncogene* **10**: 1511–1519.

Cairns, B.R., Henry, N.L. and Kornberg, R.D. (1996) TFG/TAF30/ANC1, a component of the yeast SWI/SNF complex that is similar to the leukemogenic proteins ENL and AF-9. *Mol. Cell. Biol.* **16**: 3308–3316.

Call, K.M., Glaser, T., Ito, C.Y. *et al.* (1990) Isolation and characterization of a zinc finger polypeptide gene at the human chromosome 11 Wilms' tumor locus. *Cell* **60**: 509–520.

Carapeti, M., Aguiar, R.C., Goldman, J.M. and Cross, N.C. (1998) A novel fusion between MOZ and the nuclear receptor coactivator TIF2 in acute myeloid leukemia. *Blood* **91**: 3127–3133.

Castilla, L.H., Wijmenga, C., Wang, Q. *et al.* (1996) Failure of embryonic hematopoiesis and lethal hemorrhages in mouse embryos heterozygous for a knocked-in leukemia gene CBFB-MYH11. *Cell* **87**: 687–696.

Chan, S.K., Jaffe, L., Capovilla, M., Botas, J. and Mann, R.S. (1994) The DNA binding specificity of Ultrabithorax is modulated by cooperative interactions with extradenticle, another homeoprotein. *Cell* **78**: 603–615.

Chang, C.P., de Vivo, I. and Cleary, M.L. (1997) The Hox cooperativity motif of the chimeric oncoprotein E2a-Pbx1 is necessary and sufficient for oncogenesis. *Mol. Cell. Biol.* **17**: 81–88.

Chaplin, T., Ayton, P., Bernard, O.A. *et al.* (1995) A novel class of zinc finger/leucine zipper genes identified from the molecular cloning of the t(10;11) translocation in acute leukemia. *Blood* **85**: 1435–1441.

Chase, A., Reiter, A., Burci, L. *et al.* (1999) Fusion of ETV6 to the caudal-related homeobox gene CDX2 in acute myeloid leukemia with the t(12;13)(p13;q12). *Blood* **93**: 1025–1031.

Chen, H., Lin, R.J., Schiltz, R.L. *et al.* (1997) Nuclear receptor coactivator ACTR is a novel histone acetyltransferase and forms a multimeric activation complex with P/CAF and CBP/p300. *Cell* **90**: 569–580.

Chen, Z., Brand, N.J., Chen, A. *et al.* (1993) Fusion between a novel Krüppel-like zinc finger gene and the retinoic acid receptor-alpha locus due to a variant t(11;17) translocation associated with acute promyelocytic leukaemia. *EMBO J.* **12**: 1161–1167.

Chervinsky, D.S., Zhao, X.F., Lam, D.H., Ellsworth, M., Gross, K.W. and Aplan, P.D. (1999) Disordered T-cell development and T-cell malignancies in SCL LMO1 double-transgenic mice: parallels with E2A-deficient mice. *Mol. Cell. Biol.* **19**: 5025–5035.

Clark, J., Rocques, P.J., Crew, A.J. *et al.* (1994) Identification of novel genes, SYT and SSX, involved in the t(X;18)(p11.2;q11.2) translocation found in human synovial sarcoma. *Nature Genet.* **7**: 502–508.

Cleary, M.L. (1991) Oncogenic conversion of transcription factors by chromosomal translocations. *Cell* **66**: 619–622.

Corral, J., Forster, A., Thompson, S. *et al.* (1993) Acute leukemias of different lineages have similar MLL gene fusions encoding related chimeric proteins resulting from chromosomal translocation. *Proc. Natl Acad. Sci. USA* **90**: 8538–8542.

Crozat, A., Aman, P., Mandahl, N. and Ron, D. (1993) Fusion of CHOP to a novel RNA-binding protein in human myxoid liposarcoma. Nature **363**: 640–644.

Dalla-Favera, R., Martinotti, S., Gallo, R.C., Erikson, J. and Croce, C.M. (1983) Translocation and rearrangements of the *c-myc* oncogene locus in human undifferentiated B-cell lymphomas. *Science* **219**: 963–967.

David, G., Alland, L., Hong, S.H., Wong, C.W., DePinho, R.A. and Dejean, A. (1998) Histone deacetylase associated with mSin3A mediates repression by the acute promyelocytic leukemia-associated PLZF protein. *Oncogene* **16**: 2549–2556.

Davis, R.J., D'Cruz, C.M., Lovell, M.A., Biegel, J.A. and Barr, F.G. (1994) Fusion of PAX7 to FKHR by the variant t(1;13)(p36;q14) translocation in alveolar rhabdomyosarcoma. *Cancer Res.* **54**: 2869–2872.

de Thé, H., Lavau, C., Marchio, A., Chomienne, C., Degos, L. and Dejean, A. (1991) The PML–RAR alpha fusion mRNA generated by the t(15;17) translocation in acute promyelocytic leukemia encodes a functionally altered RAR. *Cell* **66**: 675–684.

de Thé, H., Chomienne, C., Lanotte, M., Degos, L. and Dejean, A. (1990) The t(15;17) translocation of acute promyelocytic leukaemia fuses the retinoic acid receptor alpha gene to a novel transcribed locus. *Nature* **347**: 558–561.

Dedera, D.A., Waller, E.K., LeBrun, D.P. *et al.* (1993) Chimeric homeobox gene E2A-PBX1 induces proliferation, apoptosis, and malignant lymphomas in transgenic mice. *Cell* **74**: 833–843.

Delattre, O., Zucman, J., Melot, T. *et al.* (1994) The Ewing family of tumors – a subgroup of small-round-cell tumors defined by specific chimeric transcripts. *New Engl. J. Med.* **331**: 294–299.

Delattre, O., Zucman, J., Plougastel, B. *et al.* (1992) Gene fusion with an ETS DNA-binding domain caused by chromosome translocation in human tumours. *Nature* **359**: 162–165.

Dyck, J.A., Maul, G.G., Miller, W.H., Jr., Chen, J.D., Kakizuka, A. and Evans, R.M. (1994) A novel macromolecular structure is a target of the promyelocyte-retinoic acid receptor oncoprotein. *Cell* **76**: 333–343.

Eguchi, M., Eguchi-Ishimae, M., Tojo, A. *et al.* (1999) Fusion of ETV6 to neurotrophin-3 receptor TRKC in acute myeloid leukemia with t(12;15)(p13;q25). *Blood* **93**: 1355–1363.

Engel, I. and Murre, C. (1999) Ectopic expression of E47 or E12 promotes the death of E2A-deficient lymphomas. *Proc. Natl Acad. Sci. USA* **96**: 996–1001.

Erickson, P., Gao, J., Chang, K.S. *et al.* (1992) Identification of breakpoints in t(8;21) acute myelogenous leukemia and isolation of a fusion transcript, AML1/ETO, with similarity to *Drosophila* segmentation gene, *runt*. *Blood* **80**: 1825–1831.

Fenaux, P., Chomienne, C. and Degos, L. (1997) Acute promyelocytic leukemia: biology and treatment. *Semin. Oncol.* **24**: 92–102.

Finger, L.R., Harvey, R.C., Moore, R.C., Showe, L.C. and Croce, C.M. (1986) A common mechanism of chromosomal translocation in T- and B-cell neoplasia. *Science* **234**: 982–985.

Galili, N., Davis, R.J., Fredericks, W.J. *et al.* (1993) Fusion of a *fork head* domain gene to PAX3 in the solid tumour alveolar rhabdomyosarcoma. *Nature Genet.* **5**: 230–235.

Gamou, T., Kitamura, E., Hosoda, F. *et al.* (1998) The partner gene of AML1 in t(16;21) myeloid malignancies is a novel member of the MTG8(ETO) family. *Blood* **91**: 4028–4037.

Gao, J., Erickson, P., Gardiner, K. *et al.* (1991) Isolation of a yeast artificial chromosome spanning the 8;21 translocation breakpoint t(8;21)(q22;q22.3) in acute myelogenous leukemia. *Proc. Natl Acad. Sci USA* **88**: 4882–4886.

Gelmetti, V., Zhang, J., Fanelli, M., Minucci, S., Pelicci, P.G. and Lazar, M.A. (1998) Aberrant recruitment of the nuclear receptor corepressor-histone deacetylase complex by the acute myeloid leukemia fusion partner ETO. *Mol. Cell. Biol.* **18**: 7185–7191.

Gessler, M., Poustka, A., Cavenee, W., Neve, R.L., Orkin, S.H. and Bruns, G.A. (1990) Homozygous deletion in Wilms tumours of a zinc-finger gene identified by chromosome jumping. *Nature* **343**: 774–778.

Geurts, J.M., Schoenmakers, E.F., Roijer, E., Astrom, A.K., Stenman, G. and van de Ven, W.J. (1998) Identification of NFIB as recurrent translocation partner gene of HMGIC in pleomorphic adenomas. *Oncogene* **16**: 865–872.

Golub, T.R., Barker, G.F., Bohlander, S.K. *et al.* (1995) Fusion of the TEL gene on 12p13 to the AML1 gene on 21q22 in acute lymphoblastic leukemia. *Proc. Natl Acad. Sci. USA* **92**: 4917–4921.

Golub, T.R., Barker, G.F., Lovett, M. and Gilliland, D.G. (1994) Fusion of PDGF receptor beta to a novel ets-like gene, *tel*, in chronic myelomonocytic leukemia with t(5;12) chromosomal translocation. *Cell* **77**: 307–316.

Grignani, F., De Matteis, S., Nervi, C. *et al.* (1998) Fusion proteins of the retinoic acid receptor-alpha recruit histone deacetylase in promyelocytic leukaemia. *Nature* **391**: 815–818.

Grisolano, J.L., Wesselschmidt, R.L., Pilicci, P.G. and Ley, T.J. (1997) Altered myeloid development and acute leukemia in transgenic mice expressing PML-RARα under control of cathepsin G regulatory sequences. *Blood* **89**: 376–387.

Gu, Y., Nakamura, T., Alder, H. *et al.* (1992) The t(4;11) chromosome translocation of human acute leukemias fuses the ALL-1 gene, related to *Drosophila trithorax*, to the AF-4 gene. *Cell* **71**: 701–708.

Guidez, F., Ivins, S., Zhu, J., Soderstrom, M., Waxman, S. and Zelent, A. (1998) Reduced retinoic acid-sensitivities of nuclear receptor corepressor binding to PML- and PLZF-RARα underlie molecular pathogenesis and treatment of acute promyelocytic leukemia. *Blood* **91**: 2634–2642.

Guidez, F., Huang, W., Tong, J.H. *et al.* (1994) Poor response to all-*trans* retinoic acid therapy in a t(11;17) PLZF/RAR alpha patient. *Leukemia* **8**: 312–317.

He, L.Z., Guidez, F., Tribioli, C. *et al.* (1998) Distinct interactions of PML-RARα and PLZF-RARα with co-repressors determine differential responses to RA in APL. *Nature Genet.* **18**: 126–135.

He, L.Z., Tribioli, C., Rivi, R. *et al.* (1997) Acute leukemia with promyelocytic features in PML/RARα transgenic mice. *Proc. Natl Acad. Sci. USA* **94**: 5302–5307.

Heinzel, T., Lavinsky, R.M., Mullen, T.M. *et al.* (1997) A complex containing N-CoR, mSin3 and histone deacetylase mediates transcriptional repression. *Nature* **387**: 43–48.

Henthorn, P., Kiledjian, M. and Kadesch, T. (1990) Two distinct transcription factors that bind the immunoglobulin enhancer μE5/κ 2 motif. *Science* **247**: 467–470.

Hiebert, S.W., Sun, W., Davis, J.N. *et al.* (1996) The t(12;21) translocation converts AML-1B from an activator to a repressor of transcription. *Mol. Cell. Biol.* **16**: 1349–1355.

Hillion, J., Le Coniat, M., Jonveaux, P., Berger, R. and Bernard, O. A. (1997) AF6q21, a novel partner of the MLL gene in t(6;11)(q21;q23), defines a forkhead transcriptional factor subfamily. *Blood* **90**: 3714–3719.

Honda, H., Inaba, T., Suzuki, T. *et al.* (1999) Expression of E2A-HLF chimeric protein induced T-cell apoptosis, B-cell maturation arrest, and development of acute lymphoblastic leukemia. *Blood* **93**: 2780–2790.

Hong, S.H., David, G., Wong, C.W., Dejean, A. and Privalsky, M.L. (1997) SMRT corepressor interacts with PLZF and with the PML-retinoic acid receptor alpha (RARα) and PLZF-RARα oncoproteins associated with acute promyelocytic leukemia. *Proc. Natl Acad. Sci. USA* **94**: 9028–9033.

Hsu, H.L., Wadman, I. and Baer, R. (1994) Formation of *in vivo* complexes between the TAL1 and E2A polypeptides of leukemic T cells. *Proc. Natl Acad. Sci. USA* **91**: 3181–3185.

Huang, M.E., Yu-chen, Y., Shu-rong, C. *et al.* (1988) Use of all-*trans* retinoic acid in the treatment of acute promyelocytic leukemia. *Blood* **72**: 567–572.

Hunger, S.P., Tkachuk, D.C., Amylon, M.D. *et al.* (1993) HRX involvement in *de novo* and secondary leukemias with diverse chromosome 11q23 abnormalities. *Blood* **81**: 3197–3203.

Hunger, S.P., Li, S., Fall, M.Z., Naumovski, L. and Cleary, M.L. (1996) The proto-oncogene HLF and the related basic leucine zipper protein TEF display highly similar DNA-binding and transcriptional regulatory properties. *Blood* **87**: 4607–4617.

Hunger, S.P., Brown, R. and Cleary, M.L. (1994) DNA-binding and transcriptional regulatory properties of hepatic leukemia factor (HLF) and the t(17;19) acute lymphoblastic leukemia chimera E2A-HLF. *Mol. Cell. Biol.* **14**: 5986–5996.

Hunger, S.P., Ohyashiki, K., Toyama, K. and Cleary, M.L. (1992) Hlf, a novel hepatic bZIP protein, shows altered DNA-binding properties following fusion to E2A in t(17;19) acute lymphoblastic leukemia. *Genes Dev.* **6**: 1608–1620.

Ichikawa, H., Shimizu, K., Hayashi, Y. and Ohki, M. (1994) An RNA-binding protein gene, TLS/FUS, is fused to ERG in human myeloid leukemia with t(16;21) chromosomal translocation. *Cancer Res.* **54**: 2865–2868.

Ida, K., Kitabayashi, I., Taki, T. *et al.* (1997) Adenoviral E1A-associated protein p300 is involved in acute myeloid leukemia with t(11;22)(q23;q13). *Blood* **90**: 4699–4704.

Inaba, T., Inukai, T., Yoshihara, T. *et al.* (1996) Reversal of apoptosis by the leukaemia-associated E2A-HLF chimaeric transcription factor. *Nature* **382**: 541–544.

Inaba, T., Shapiro, L.H., Funabiki, T. *et al.* (1994) DNA-binding specificity and trans-activating potential of the leukemia-associated E2A-hepatic leukemia factor fusion protein. *Mol. Cell. Biol.* **14**: 3403–3413.

Inaba, T., Roberts, W.M., Shapiro, L.H. *et al.* (1992) Fusion of the leucine zipper gene HLF to the *E2A* gene in human acute B-lineage leukemia. *Science* **257**: 531–534.

Inukai, T., Inaba, T., Ikushima, S. and Look, A.T. (1998) The AD1 and AD2 transactivation domains of E2A are essential for the antiapoptotic activity of the chimeric oncoprotein E2A-HLF. *Mol. Cell. Biol.* **18**: 6035–6043.

Ishida, S., Yoshida, K., Kaneko, Y. *et al.* (1998) The genomic breakpoint and chimeric transcripts in the EWSR1-ETV4/E1AF gene fusion in Ewing sarcoma. *Cytogenet. Cell Genet.* **82**: 278–283.

Jeon, I.S., Davis, J.N., Braun, B.S. *et al.* (1995) A variant Ewing's sarcoma translocation (7;22) fuses the EWS gene to the ETS gene ETV1. *Oncogene* **10**: 1229–1234.

Kakizuka, A., Miller, W.H., Jr., Umesono, K. *et al.* (1991) Chromosomal translocation t(15;17) in human acute promyelocytic leukemia fuses RARα with a novel putative transcription factor, PML. *Cell* **66**: 663–674.

Kamps, M.P., Wright, D.D. and Lu, Q. (1996) DNA-binding by oncoprotein E2a–Pbx1 is important for blocking differentiation but dispensable for fibroblast transformation. *Oncogene* **12**: 19–30.

Kamps, M.P. and Baltimore, D. (1993) E2A–Pbx1, the t(1;19) translocation protein of human pre-B-cell acute lymphocytic leukemia, causes acute myeloid leukemia in mice. *Mol. Cell. Biol.* **13**: 351–357.

Kamps, M.P., Look, A.T. and Baltimore, D. (1991) The human t(1;19) translocation in pre-B ALL produces multiple nuclear E2A–Pbx1 fusion proteins with differing transforming potentials. *Genes Dev.* 5: 358–368.

Kamps, M.P., Murre, C., Sun, X.H. and Baltimore, D. (1990) A new homeobox gene contributes the DNA binding domain of the t(1;19) translocation protein in pre-B ALL. *Cell* **60**: 547–555.

Kawai, A., Woodruff, J., Healey, J.H., Brennan, M.F., Antonescu, C.R. and Ladanyi, M. (1998) *SYT–SSX* gene fusion as a determinant of morphology and prognosis in synovial sarcoma. *New Engl. Med.* **338**: 153–160.

Kern, S.E., Pietenpol, J.A., Thiagalingam, S., Seymour, A., Kinzler, K.W. and Vogelstein, B. (1992) Oncogenic forms of p53 inhibit p53-regulated gene expression. *Science* **256**: 827–830.

Kitabayashi, I., Yokoyama, A., Shimizu, K. and Ohki, M. (1998) Interaction and functional cooperation of the leukemia-associated factors AML1 and p300 in myeloid cell differentiation. *EMBO J.* **17**: 2994–3004.

Knezevich, S.R., McFadden, D.E., Tao, W., Lim, J.F. and Sorensen, P.H. (1998) A novel *ETV6–NTRK3* gene fusion in congenital fibrosarcoma. *Nature Genet.* **18**: 184–187.

Koken, M.H., Puvion-Dutilleul, F., Guillemin, M.C. *et al.* (1994) The t(15;17) translocation alters a nuclear body in a retinoic acid-reversible fashion. *EMBO J.* **13**: 1073–1083.

Labelle, Y., Zucman, J., Stenman, G. *et al.* (1995) Oncogenic conversion of a novel orphan nuclear receptor by chromosome translocation. *Human Mol. Genet.* **4**: 2219–2226.

Lacronique, V., Boureux, A., Valle, V.D. *et al.* (1997) A TEL–JAK2 fusion protein with constitutive kinase activity in human leukemia. *Science* **278**: 1309–1312.

Ladanyi, M. and Gerald, W. (1994) Fusion of the EWS and WT1 genes in the desmoplastic small round cell tumor. *Cancer Res.* **54**: 2837–2840.

Larson, R.A., Kondo, K., Vardiman, J.W., Butler, A.E., Golomb, H.M. and Rowley, J.D. (1984) Evidence for a 15;17 translocation in every patient with acute promyelocytic leukemia. *Am. J. Med.* **76**: 827–841.

Lavau, C., Szilvassy, S.J., Slany, R. and Cleary, M.L. (1997) Immortalization and leukemic transformation of a myelomonocytic precursor by retrovirally transduced HRX–ENL. *EMBO J.* **16**: 4226–4237.

LeBrun, D.P. and Cleary, M.L. (1994) Fusion with E2A alters the transcriptional properties of the homeodomain protein PBX1 in t(1;19) leukemias. *Oncogene* **9**: 1641–1647.

Lessnick, S.L., Braun, B.S., Denny, C.T. and May, W.A. (1995) Multiple domains mediate transformation by the Ewing's sarcoma EWS/FLI-1 fusion gene. *Oncogene* **10**: 423–431.

Liang, J., Prouty, L., Williams, B.J., Dayton, M.A. and Blanchard, K.L. (1998) Acute mixed lineage leukemia with an inv(8)(p11q13) resulting in fusion of the genes for MOZ and TIF2. *Blood* **92**: 2118–2122.

Licht, J.D., Chomienne, C., Goy, A. *et al.* (1995) Clinical and molecular characterization of a rare syndrome of acute promyelocytic leukemia associated with translocation (11;17). *Blood* **85**: 1083–1094.

Lin, R.J., Nagy, L., Inoue, S., Shao, W., Miller, W.H., Jr. and Evans, R.M. (1998) Role of the histone deacetylase complex in acute promyelocytic leukaemia. *Nature* **391**: 811–814.

Liu, P., Tarle, S.A., Hajra, A. *et al.* (1993) Fusion between transcription factor CBF β/PEBP2 β and a myosin heavy chain in acute myeloid leukemia. *Science* **261**: 1041–1044.

Longo, L., Pandolfi, P.P., Biondi, A. ***et al.*** (1990) Rearrangements and aberrant expression of the retinoic acid receptor α gene in acute promyelocytic leukemias. *J. Exper. Med.* **172**: 1571–1575.
Look, A.T. (1997) Oncogenic transcription factors in the human acute leukemias. *Science* **278**: 1059–1064.
Lu, D., Thompson, J.D., Gorski, G.K., Rice, N.R., Mayer, M.G. and Yunis, J.J. (1991) Alterations at the *rel* locus in human lymphoma. Oncogene **6**: 1235–1241.
Lu, Q., Wright, D.D. and Kamps, M.P. (1994) Fusion with E2A converts the Pbx1 homeodomain protein into a constitutive transcriptional activator in human leukemias carrying the t(1;19) translocation. *Mol. Cell. Biol.* **14**: 3938–3948.
Lutterbach, B., Westendorf, J.J., Linggi, B. ***et al.*** (1998) ETO, a target of t(8;21) in acute leukemia, interacts with the N-CoR and mSin3 corepressors. *Mol. Cell. Biol.* **18**: 7176–7184.
Maloney, K.W., Rubnitz, J.E., Cleary, M.L. ***et al.*** (1997) Lack of ETV6 (TEL) gene rearrangements or p16INK4A/p15INK4B homozygous gene deletions in infant acute lymphoblastic leukemia. *Leukemia* **11**: 979–983.
May, W.A., Arvand, A., Thompson, A.D., Braun, B.S., Wright, M. and Denny, C.T. (1997) EWS/FLI1-induced manic fringe renders NIH 3T3 cells tumorigenic. *Nature Genet.* **17**: 495–497.
May, W.A., Gishizky, M.L., Lessnick, S.L. ***et al.*** (1993a) Ewing sarcoma 11;22 translocation produces a chimeric transcription factor that requires the DNA-binding domain encoded by FLI1 for transformation. *Proc. Natl Acad. Sci. USA* **90**: 5752–5756.
May, W.A., Lessnick, S.L., Braun, B.S. ***et al.*** (1993b) The Ewing's sarcoma EWS/FLI-1 fusion gene encodes a more potent transcriptional activator and is a more powerful transforming gene than FLI-1. *Mol. Cell. Biol.* **13**: 7393–7398.
Mazo, A.M., Huang, D.H., Mozer, B.A. and Dawid, I.B. (1990) The *trithorax* gene, a *trans*-acting regulator of the bithorax complex in *Drosophila*, encodes a protein with zinc-binding domains. *Proc. Natl Acad. Sci. USA* **87**: 2112–2116.
Megonigal, M.D., Rappaport, E.F., Jones, D.H. ***et al.*** (1998) t(11;22)(q23;q11.2) In acute myeloid leukemia of infant twins fuses MLL with hCDCrel, a cell division cycle gene in the genomic region of deletion in DiGeorge and velocardiofacial syndromes. *Proc. Natl Acad. Sci. USA* **95**: 6413–6418.
Mellentin, J.D., Murre, C., Donlon, T.A. ***et al.*** (1989) The gene for enhancer binding proteins E12/E47 lies at the t(1;19) breakpoint in acute leukemias. *Science* **246**: 379–382.
Melnick, A. and Licht, J.D. (1999) Deconstructing a disease: RARα, its fusion partners, and their roles in the pathogenesis of acute promyelocytic leukemia. *Blood* **93**: 3167–3215.
Metzstein, M.M., Hengartner, M.O., Tsung, N., Ellis, R.E. and Horvitz, H.R. (1996) Transcriptional regulator of programmed cell death encoded by *Caenorhabditis elegans* gene *ces-2*. *Nature* **382**: 545–547.
Meyers, S., Lenny, N. and Hiebert, S.W. (1995) The t(8;21) fusion protein interferes with AML-1B-dependent transcriptional activation. *Mol. Cell. Biol.* **15**: 1974–1982.
Meyers, S., Downing, J.R. and Hiebert, S.W. (1993) Identification of AML-1 and the (8;21) translocation protein (AML-1/ETO) as sequence-specific DNA-binding proteins: the runt homology domain is required for DNA binding and protein–protein interactions. *Mol. Cell. Biol.* **13**: 6336–6345.
Mitani, K., Ogawa, S., Tanaka, T. ***et al.*** (1994) Generation of the AML1-EVI-1 fusion gene in the t(3;21)(q26;q22) causes blastic crisis in chronic myelocytic leukemia. *EMBO J.* **13**: 504–510.
Miyoshi, H., Ohira, M., Shimizu, K. ***et al.*** (1995) Alternative splicing and genomic structure of the *AML1* gene involved in acute myeloid leukemia. *Nucl. Acids Res.* **23**: 2762–2769.
Miyoshi, H., Kozu, T., Shimizu, K. ***et al.*** (1993) The t(8;21) translocation in acute myeloid leukemia results in production of an AML1–MTG8 fusion transcript. *EMBO J.* **12**: 2715–2721.
Miyoshi, H., Shimizu, K., Kozu, T., Maseki, N., Kaneko, Y. and Ohki, M. (1991) t(8;21) breakpoints on chromosome 21 in acute myeloid leukemia are clustered within a limited region of a single gene, *AML1*. *Proc. Natl Acad. Sci. USA* **88**: 10431–10434.
Monica, K., LeBrun, D.P., Dedera, D.A., Brown, R. and Cleary, M.L. (1994) Transformation properties of the E2a–Pbx1 chimeric oncoprotein: fusion with E2a is essential, but the Pbx1 homeodomain is dispensable. *Mol. Cell. Biol.* **14**: 8304–8314.
Monica, K., Galili, N., Nourse, J., Saltman, D. and Cleary, M.L. (1991) *PBX2* and *PBX3*, new homeobox genes with extensive homology to the human proto-oncogene *PBX1*. *Mol. Cell. Biol.* **11**: 6149–6157.
Murre, C., McCaw, P.S. and Baltimore, D. (1989a) A new DNA binding and dimerization motif in immunoglobulin enhancer binding, daughterless, MyoD, and myc proteins. *Cell* **56**: 777–783.

Murre, C., McCaw, P.S., Vaessin, H. *et al.* (1989b) Interactions between heterologous helix-loop-helix proteins generate complexes that bind specifically to a common DNA sequence. *Cell* **58**: 537–544.

Nagy, L., Kao, H.Y., Chakravarti, D. *et al.* (1997) Nuclear receptor repression mediated by a complex containing SMRT, mSin3A, and histone deacetylase. *Cell* **89**: 373–380.

Nakamura, T., Yamazaki, Y., Hatano, Y. and Miura, I. (1999) NUP98 is fused to PMX1 homeobox gene in human acute myelogenous leukemia with chromosome translocation t(1;11)(q23;p15). *Blood* **94**: 741–747.

Nakamura, T., Largaespada, D.A., Lee, M.P. *et al.* (1996) Fusion of the nucleoporin gene NUP98 to HOXA9 by the chromosome translocation t(7;11)(p15;p15) in human myeloid leukaemia. *Nature Genet.* **12**: 154–158.

Nakamura, T., Alder, H., Gu, Y. *et al.* (1993) Genes on chromosomes 4, 9, and 19 involved in 11q23 abnormalities in acute leukemia share sequence homology and/or common motifs. *Proc. Natl Acad. Sci. USA* **90**: 4631–4635.

Niki, M., Okada, H., Takano, H. *et al.* (1997) Hematopoiesis in the fetal liver is impaired by targeted mutagenesis of a gene encoding a non-DNA binding subunit of the transcription factor, polyomavirus enhancer binding protein 2/core binding factor. *Proc. Natl Acad. Sci. USA* **94**: 5697–5702.

Nourse, J., Mellentin, J.D., Galili, N. *et al.* (1990) Chromosomal translocation t(1;19) results in synthesis of a homeobox fusion mRNA that codes for a potential chimeric transcription factor. *Cell* **60**: 535–545.

Nucifora, G., Begy, C.R., Erickson, P., Drabkin, H.A. and Rowley, J.D. (1993) The 3;21 translocation in myelodysplasia results in a fusion transcript between the *AML1* gene and the gene for *EAP*, a highly conserved protein associated with the Epstein-Barr virus small RNA EBER 1. *Proc. Natl Acad. Sci. USA* **90**: 7784–7788.

Ogawa, E., Inuzuka, M., Maruyama, M. *et al.* (1993) Molecular cloning and characterization of PEBP2 beta, the heterodimeric partner of a novel *Drosophila* runt-related DNA binding protein PEBP2 alpha. *Virology* **194**: 314–331.

Okuda, T., Cai, Z., Yang, S. *et al.* (1998) Expression of a knocked-in *AML1–ETO* leukemia gene inhibits the establishment of normal definitive hematopoiesis and directly generates dysplastic hematopoietic progenitors. *Blood* **91**: 3134–3143.

Okuda, T., van Deursen, J., Hiebert, S.W., Grosveld, G. and Downing, J.R. (1996) *AML1*, the target of multiple chromosomal translocations in human leukemia, is essential for normal fetal liver hematopoiesis. *Cell* **84**: 321–330.

Osaka, M., Rowley, J.D. and Zeleznik-Le, N.J. (1999) *MSF* (MLL septin-like fusion), a fusion partner gene of *MLL*, in a therapy-related acute myeloid leukemia with a t(11;17)(q23;q25). *Proc. Natl Acad. Sci. USA* **96**: 6428–6433.

Panagopoulos, I., Hoglund, M., Mertens, F., Mandahl, N., Mitelman, F. and Aman, P. (1996) Fusion of the EWS and CHOP genes in myxoid liposarcoma. *Oncogene* **12**: 489–494.

Papadopoulos, P., Ridge, S.A., Boucher, C.A., Stocking, C. and Wiedemann, L.M. (1995) The novel activation of ABL by fusion to an ets-related gene, *TEL*. *Cancer Res.* **55**: 34–38.

Park, S.T., Nolan, G.P. and Sun, X.H. (1999) Growth inhibition and apoptosis due to restoration of E2A activity in T cell acute lymphoblastic leukemia cells. *J. Exper. Med.* **189**: 501–508.

Peeters, P., Raynaud, S.D., Cools, J. *et al.* (1997) Fusion of *TEL*, the ETS-variant gene 6 (*ETV6*), to the receptor-associated kinase JAK2 as a result of t(9;12) in a lymphoid and t(9;15;12) in a myeloid leukemia. *Blood* **90**: 2535–2540.

Prasad, R., Leshkowitz, D., Gu, Y. *et al.* (1994) Leucine-zipper dimerization motif encoded by the *AF17* gene fused to ALL-1 (MLL) in acute leukemia. *Proc. Natl Acad. Sci. USA* **91**: 8107–8111.

Prasad, R., Gu, Y., Alder, H., Nakamura, T. *et al.* (1993) Cloning of the ALL-1 fusion partner, the *AF-6* gene, involved in acute myeloid leukemias with the t(6;11) chromosome translocation. *Cancer Res.* **53**: 5624–5628.

Quong, M.W., Massari, M.E., Zwart, R. and Murre, C. (1993) A new transcriptional-activation motif restricted to a class of helix-loop-helix proteins is functionally conserved in both yeast and mammalian cells. *Mol. Cell. Biol.* **13**: 792–800.

Rabbitts, T.H. (1994) Chromosomal translocations in human cancer. *Nature* **372**: 143–149.

Rabbitts, T.H., Forster, A., Larson, R. and Nathan, P. (1993) Fusion of the dominant negative transcription regulator CHOP with a novel gene *FUS* by translocation t(12;16) in malignant liposarcoma. *Nature Genet.* **4**: 175–180.

Rauskolb, C. and Wieschaus, E. (1994) Coordinate regulation of downstream genes by extradenticle and the homeotic selector proteins. *EMBO J.* **13**: 3561–3569.

Rauskolb, C., Peifer, M. and Wieschaus, E. (1993) *extradenticle*, a regulator of homeotic gene activity, is a homolog of the homeobox-containing human proto-oncogene *pbx1*. *Cell* **74**: 1101–1112.

Raynaud, S.D., Baens, M., Grosgeorge, J. *et al.* (1996) Fluorescence *in situ* hybridization analysis of t(3; 12)(q26; p13): a recurring chromosomal abnormality involving the *TEL* gene (*ETV6*) in myelodysplastic syndromes. *Blood* **88**: 682–689.

Raza-Egilmez, S.Z., Jani-Sait, S.N., Grossi, M., Higgins, M.J., Shows, T.B. and Aplan, P.D. (1998) *NUP98–HOXD13* gene fusion in therapy-related acute myelogenous leukemia. *Cancer Res.* **58**: 4269–4273.

Redner, R.A., Wang, J. and Liu, J.M. (1999) Chromatin remodeling and leukemia: new therapeutic paradigms. *Blood* **94**: 417–428.

Redner, R.L., Rush, E.A., Faas, S., Rudert, W.A. and Corey, S.J. (1996) The t(5;17) variant of acute promyelocytic leukemia expresses a nucleophosmin–retinoic acid receptor fusion. *Blood* **87**: 882–886.

Romana, S.P., Mauchauffe, M., Le Coniat, M. *et al.* (1995) The t(12;21) of acute lymphoblastic leukemia results in a *tel-AML1* gene fusion. *Blood* **85**: 3662–3670.

Rowley, J.D. (1973) Letter: a new consistent chromosomal abnormality in chronic myelogenous leukaemia identified by quinacrine fluorescence and Giemsa staining. *Nature* **243**: 290–293.

Rowley, J.D. (1992) The der(11) chromosome contains the critical breakpoint junction in the 4;11, 9;11, and 11;19 translocations in acute leukemia. *Genes Chromos. Cancer* **5**: 264–266.

Rozenblatt-Rosen, O., Rozovskaia, T., Burakov, D. *et al.* (1998) The *C*-terminal SET domains of ALL-1 and TRITHORAX interact with the INI1 and SNR1 proteins, components of the SWI/SNF complex. *Proc. Natl Acad. Sci. USA* **95**: 4152–4157.

Rubnitz, J.E., Pui, C.H. and Downing, J.R. (1999) The role of *TEL* fusion genes in pediatric leukemias. *Leukemia* **13**: 6–13.

Shapiro, D.N., Sublett, J.E., Li, B., Downing, J.R. and Naeve, C.W. (1993) Fusion of PAX3 to a member of the forkhead family of transcription factors in human alveolar rhabdomyosarcoma. *Cancer Res.* **53**: 5108–5112.

Smith, K.S., Rhee, J.W., Naumovski, L. and Cleary, M.L. (1999) Disrupted differentiation and oncogenic transformation of lymphoid progenitors in E2A–HLF transgenic mice. *Mol. Cell. Biol.* **19**: 4443–4451.

So, C.W., Caldas, C., Liu, M.M. *et al.* (1997) *EEN* encodes for a member of a new family of proteins containing an Src homology 3 domain and is the third gene located on chromosome 19p13 that fuses to MLL in human leukemia. *Proc. Natl Acad Sci. USA* **94**: 2563–2568.

Sobulo, O.M., Borrow, J., Tomek, R., Reshmi, S., Harden, A., Schlegelberger, B., Housman, D., Doggett, N.A., Rowley, J.D. and Zeleznik-Le, N.J. (1997) MLL is fused to CBP, a histone acetyltransferase, in therapy-related acute myeloid leukemia with a t(11;16)(q23;p13.3). *Proc. Natl Acad. Sci. USA* **94**: 8732–8737.

Sorensen, P.H., Lessnick, S.L., Lopez-Terrada, D., Liu, X.F., Triche, T.J. and Denny, C.T. (1994) A second Ewing's sarcoma translocation, t(21;22), fuses the EWS gene to another ETS-family transcription factor, ERG. *Nature Genet.* **6**: 146–151.

Speck, N.A., Stacy, T., Wang, Q. *et al.* (1999) Core-binding factor: a central player in hematopoiesis and leukemia. *Cancer Res.* **59**: 1789s–1793s.

Suto, Y., Sato, Y., Smith, S.D., Rowley, J.D. and Bohlander, S.K. (1997) A t(6;12)(q23;p13) results in the fusion of *ETV6* to a novel gene, *STL*, in a B-cell *ALL* cell line. *Genes Chromos. Cancer* **18**: 254–268.

Taki, T., Shibuya, N., Taniwaki, M. *et al.* (1998) ABI-1, a human homolog to mouse Abl-interactor 1, fuses the *MLL* gene in acute myeloid leukemia with t(10;11)(p11.2;q23). *Blood* **92**: 1125–1130.

Tallman, M.S., Andersen, J.W., Schiffer, C.A. *et al.* (1997) All-*trans*-retinoic acid in acute promyelocytic leukemia. *New Engl. J. Med.* **337**: 1021–1028.

Taub, R., Kirsch, I., Morton, C. *et al.* (1982) Translocation of the *c-myc* gene into the immunoglobulin heavy chain locus in human Burkitt lymphoma and murine plasmacytoma cells. *Proc. Natl Acad. Sci. USA* **79**:, 7837–7841.

Thirman, M.J., Gill, H.J., Burnett, R.C. *et al.* (1993) Rearrangement of the *MLL* gene in acute lymphoblastic and acute myeloid leukemias with 11q23 chromosomal translocations. *N. Engl. J. Med.* **329**: 909–914.

Thirman, M.J., Levitan, D.A., Kobayashi, H., Simon, M.C. and Rowley, J.D. (1994) Cloning of *ELL*, a gene that fuses to MLL in a t(11;19)(q23;p13.1) in acute myeloid leukemia. *Proc. Natl Acad. Sci. USA* **91**: 12110–12114.

Tkachuk, D.C., Kohler, S. and Cleary, M.L. (1992) Involvement of a homolog of *Drosophila trithorax* by 11q23 chromosomal translocations in acute leukemias. *Cell* **71**: 691–700.

Tse, W., Zhu, W., Chen, H.S. and Cohen, A. (1995) A novel gene, *AF1q*, fused to MLL in t(1;11) (q21;q23), is specifically expressed in leukemic and immature hematopoietic cells. *Blood* **85**: 650–656.

Uchida, H., Downing, J.R., Miyazaki, Y., Frank, R., Zhang, J. and Nimer, S.D. (1999) Three distinct domains in TEL–AML1 are required for transcriptional repression of the IL-3 promoter. *Oncogene* **18**: 1015–1022.

Umesono, K., Murakami, K.K., Thompson, C.C. and Evans, R.M. (1991) Direct repeats as selective response elements for the thyroid hormone, retinoic acid, and vitamin D3 receptors. *Cell* **65**: 1255–1266.

Valge-Archer, V.E., Osada, H., Warren, A.J. *et al.* (1994) The LIM protein RBTN2 and the basic helix-loop-helix protein TAL1 are present in a complex in erythroid cells. *Proc. Natl Acad. Sci. USA* **91**: 8617–8621.

van Dijk, M.A. and Murre, C. (1994) *extradenticle* raises the DNA binding specificity of homeotic selector gene products. *Cell* **78**: 617–624.

van Dijk, M.A., Voorhoeve, P.M. and Murre, C. (1993) Pbx1 is converted into a transcriptional activator upon acquiring the *N*-terminal region of E2A in pre-B-cell acute lymphoblastoid leukemia. *Proc. Natl Acad. Sci. USA* **90**: 6061–6065.

Vyas, R.C., Frankel, S.R., Agbor, P., Miller, W.H., Jr., Warrell, R.P., Jr. and Hittelman, W.N. (1996) Probing the pathobiology of response to all-*trans* retinoic acid in acute promyelocytic leukemia: premature chromosome condensation/fluorescence *in situ* hybridization analysis. *Blood* **87**: 218–226.

Wang, J., Hoshino, T., Redner, R.L., Kajigaya, S. and Liu, J.M. (1998) ETO, fusion partner in t(8;21) acute myeloid leukemia, represses transcription by interaction with the human *N*-CoR/mSin3/HDAC1 complex. *Proc. Natl Acad. Sci. USA* **95**: 10860–10865.

Wang, J., Wang, M. and Liu, J.M. (1997) Transformation properties of the *ETO* gene, fusion partner in t(8: 21) leukemias. *Cancer Res.* **57**: 2951–2955.

Wang, Q., Stacy, T., Binder, M., Marin-Padilla, M., Sharpe, A.H. and Speck, N.A. (1996a) Disruption of the *Cbfa2* gene causes necrosis and hemorrhaging in the central nervous system and blocks definitive hematopoiesis. *Proc. Natl Acad. Sci. USA* **93**: 3444–3449.

Wang, Q., Stacy, T., Miller, J.D. *et al.* (1996b) The CBFβ subunit is essential for CBFα2 (*AML1*) function *in vivo*. *Cell* **87**: 697–708.

Warrell, R.P., Jr., He, L.Z., Richon, V., Calleja, E. and Pandolfi, P.P. (1998) Therapeutic targeting of transcription in acute promyelocytic leukemia by use of an inhibitor of histone deacetylase. *J. Natl Cancer Instit.* **90**: 1621–1625.

Weis, K., Rambaud, S., Lavau, C. *et al.* (1994) Retinoic acid regulates aberrant nuclear localization of PML–RARα in acute promyelocytic leukemia cells. *Cell* **76**: 345–356.

Wells, R.A., Catzavelos, C. and Kamel-Reid, S. (1997) Fusion of retinoic acid receptor alpha to NuMA, the nuclear mitotic apparatus protein, by a variant translocation in acute promyelocytic leukaemia. *Nature Genet.* **17**: 109–113.

Whang-Peng, J., Triche, T.J., Knutsen, T. *et al.* (1986) Cytogenetic characterization of selected small round cell tumors of childhood. *Cancer Genet. Cytogenet.* **21**: 185–208.

Wong, C.W. and Privalsky, M.L. (1998) Components of the SMRT corepressor complex exhibit distinctive interactions with the POZ domain oncoproteins PLZF, PLZF-RARα, and BCL-6. *J. Biol. Chem.* **273**: 27695–27702.

Yan, W., Young, A.Z., Soares, V.C., Kelley, R., Benezra, R. and Zhuang, Y. (1997) High incidence of T-cell tumors in E2A-null mice and E2A/Id1 double-knockout mice. *Mol. Cell. Biol.* **17**: 7317–7327.

Yergeau, D.A., Hetherington, C.J., Wang, Q. *et al.* (1997) Embryonic lethality and impairment of haematopoiesis in mice heterozygous for an *AML1–ETO* fusion gene. *Nature Genet.* **15**: 303–306.

Yoshihara, T., Inaba, T., Shapiro, L.H., Kato, J.Y. and Look, A.T. (1995) E2A–HLF-mediated cell transformation requires both the *trans*-activation domains of E2A and the leucine zipper dimerization domain of HLF. *Mol. Cell. Biol.* **15**: 3247–3255.

Yu, B.D., Hanson, R.D., Hess, J.L., Horning, S.E. and Korsmeyer, S.J. (1998) *MLL*, a mammalian trithorax-group gene, functions as a transcriptional maintenance factor in morphogenesis. *Proc. Natl Acad. Sci. USA* **95**: 10632–10636.

Yu, B.D., Hess, J.L., Horning, S.E., Brown, G.A. and Korsmeyer, S.J. (1995) Altered Hox expression and segmental identity in Mll-mutant mice. *Nature* **378**: 505–508.
Zhuang, Y., Soriano, P. and Weintraub, H. (1994) The helix-loop-helix gene E2A is required for B cell formation. *Cell* **79**: 875–884.
Zucman, J., Delattre, O., Desmaze, C. *et al.* (1993) *EWS* and *ATF-1* gene fusion induced by t(12;22) translocation in malignant melanoma of soft parts. *Nature Genet.* **4**: 341–345.

Index